MW00835470

Noncommutative Geometry, Arithmetic, and Related Topics

Noncommutative Geometry, Arithmetic, and Related Topics

Proceedings of the Twenty-First Meeting
of the Japan-U.S. Mathematics Institute

Edited by Caterina Consani and Alain Connes

The Johns Hopkins University Press

Baltimore

© 2011 The Johns Hopkins University Press
All rights reserved. Published 2011
Printed in the United States of America on acid-free paper
2 4 6 8 9 7 5 3 1

The Johns Hopkins University Press
2715 North Charles Street
Baltimore, Maryland 21218-4363
www.press.jhu.edu

Library of Congress Control Number: 2011938337

ISBN 13: 978-1-4214-0352-6
ISBN 10: 1-4214-0352-8

A catalog record for this book is available from the British Library.

*Special discounts are available for bulk purchases of this book. For more information,
please contact Special Sales at 410-516-6936 or specialsales@press.jhu.edu.*

The Johns Hopkins University Press uses environmentally friendly book materials,
including recycled text paper that is composed of at least 30 percent post-consumer waste,
whenever possible.

CONTENTS

PREFACE

The interaction between the fields of noncommutative geometry, arithmetic and mathematical physics is a quite new area of mathematics which has developed rapidly in the past few years producing very exciting results. This volume collects together several original papers written by leading experts in these fields. Most of the contributors to this volume spoke at the 21st JAMI Conference held at Johns Hopkins University during the week of March 23–26, 2009.

This book aims to fill a big gap among frontiers of research in noncommutative geometry and seeks to fulfill the demand for a guiding text, suitable to researchers willing to gain a thorough understanding of the subject.

The broad field of Noncommutative Geometry originated in the early 80's with the goal to extend the classical correspondence between geometric spaces and commutative algebras in the framework of analysis and with the aim to study certain geometric objects which arise naturally both in mathematics and in quantum physics and whose associated coordinates do not commute (e.g., spaces of leaves of foliations, phase spaces in quantum mechanics, the space of irreducible unitary representations of a discrete group, quantum groups, etc.).

While some aspects of this general theory are concerned with a reinterpretation and a generalization of some classical geometrical constructions to noncommutative spaces, a branch of this field focuses on the study of totally new phenomena with no counterpart in the classical "commutative world." The typical example arises as a natural consequence of the well-known result in functional analysis stating that an arbitrary noncommutative von Neumann algebra admits a unique one parameter group of automorphisms, up to inner automorphisms, which measures the obstruction to the existence of a trace in the algebra. By applying techniques from the theory of KMS-states in quantum mechanics, it is then possible to perform certain thermodynamical operations on a noncommutative space, and in particular that of changing its temperature, with the goal of studying the corresponding structural variation of the space.

An example to which this thermodynamical operation is applied is the Bost-Connes system. The number-theoretic interest for this system arises from the fact that the BC-system has the Riemann zeta function as its partition function, which appears in the description of the KMS_β (extremal) states when $\beta > 1$. The ubiquitous nature of the noncommutative Hecke algebra associated to this system and its importance for the understanding of the so-called absolute arithmetic has started to emerge only very recently. Important interactions between noncommutative geometry and number theory started in the late 90's, with the paper of A. Connes *"Trace formula in noncommutative geometry and the zeros of the Riemann zeta function"* which develops the framework of a conjectural Selberg-type trace formula over a noncommutative space: the adèle class space. The validity of such a formula would imply the positivity of the Weil distribution and hence the validity of the Riemann hypothesis. By suitably relaxing the function space one obtains in particular a spectral realization of the zeros of L-functions and a trace-formula interpretation of the classical Riemann-Weil explicit formulae, as shown e.g. by R. Meyer. A few years later, in the paper of A. Connes, C. Consani and M. Marcolli *"Noncommutative geometry and motives: the thermodynamics of endomotives,"* these results were given a cohomological interpretation by performing several operations on the BC-system, inclusive of a cooling procedure combined with the use of cyclic (co)homological techniques and the application of a newly developed theory of noncommutative motives (the theory of endomotives).

In most recent years these constructions have started to merge with the so-called theory of absolute arithmetic, or arithmetic over $\mathbb{F}_1$. The existence of such a basic algebraic structure was first proposed in the seminal work of J. Tits, *"Sur les analogues algébriques des groupes semi-simples complexes."* Very slowly, the original idea of Tits that certain geometric structures over finite fields $\mathbb{F}_q$ should retain a meaning also in the limit case "$q = 1$" has been taking more and more substance and has finally given rise to a number of different and interesting approaches, starting with the Japanese school of absolute arithmetic over $\mathbb{F}_1$. We refer to the two papers by S. Koyama and N. Kurokawa (*"Absolute Modular Forms"* and *"Absolute Zeta Functions and Absolute Tensor Products"*) contained in this volume. The survey paper by J. López-Peña and O. Lorscheid, *"Mapping $\mathbb{F}_1$-land: An Overview of Geometries over the Field with One Element,"* also contained in this volume, provides an overview of the various approaches towards $\mathbb{F}_1$-geometry due to various mathematicians, and links between them.

The need for a field of characteristic one has also emerged in Arakelov geometry (see *e.g.* the paper by Y. Manin *"Lectures on zeta functions and motives (according to Deninger and Kurokawa)"*) in the context of a geometric interpretation of the zeros of zeta or L-functions. The idea is to try to imitate the Weil proof of the Riemann hypothesis for function fields in the number field case. In Weil's proof the key geometric object is the self-fiber product $C \times_{\mathbb{F}_q} C$ of the curve C over the finite field $\mathbb{F}_q$, whose field of functions is the given global field. For number fields, the idea is to postulate the existence of the "absolute point," namely, $\operatorname{Spec} \mathbb{F}_1$ over which any scheme would map. In particular this would be the case for $\operatorname{Spec} \mathbb{Z}$, and then one would be able to use the algebraic spectrum of the tensor product $\mathbb{Z} \times_{\mathbb{F}_1} \mathbb{Z}$ as a substitute for the above product $C \times_{\mathbb{F}_q} C$. The relation between these constructions and the Riemann zeta function has remained so far unsettled, despite the original guess, formulated in the abovementioned lectures, that the tensor product $\mathbb{Z} \otimes_{\mathbb{F}_1} \mathbb{Z}$ should represent a non-trivial analog of the product of the curve by itself.

A precise relation between the tower of algebraic extensions of $\mathbb{F}_1$ and the Riemann zeta function was first shown in the paper of A. Connes, C. Consani and M. Marcolli, *"Fun with $\mathbb{F}_1$,"* where the BC-system and the associated algebraic endomotive appear from first principles, by studying the algebraic extensions of $\mathbb{F}_1$ and implementing some of the techniques developed in the earlier work of C. Soulé on a theory of algebraic varieties over $\mathbb{F}_1$, *"Les variétés sur le corps à un élément."* The paper of C. Soulé, *"Lectures on Algebraic Varieties over $\mathbb{F}_1$"* in this volume, presents his approach to the study of varieties over the field with one element and their zeta functions. The idea is that the category of commutative rings (*i.e.* affine schemes) should be enhanced in such a way that the ring of integers ceases to be an initial object of this category. In fact its mysterious arithmetic should be recast by an algebra over $\mathbb{F}_1$. In line with A. Grothendieck's ideas in algebraic geometry, this development poses the following question: how to define schemes, and even more exotic types of spaces, over $\mathbb{F}_1$? In the recent paper by A. Connes and C. Consani, *"On the notion of geometry over $\mathbb{F}_1$,"* a solution to this problem is proposed by introducing a suitable refinement of the definition of algebraic varieties over $\mathbb{F}_1$ which combines C. Soulé's viewpoint with the geometry of monoids of K. Kato and A. Deitmar. This new approach is then tested with Chevalley group schemes by showing that these spaces can be defined over $\mathbb{F}_1$. The paper by A. Connes and C. Consani, *"Characteristic 1, Entropy and the Absolute Point"* contained in this volume, supplies detailed proofs of this proposed enhancement of the notion of scheme over $\mathbb{F}_1$ and its zeta function. The same paper also explains how the relation between quantum mechanics and its semi-classical limit (in particular the semi-classical limit of real numbers defined as the tropical number semi-field) suggests that idempotent analysis and tropical algebraic geometry provide an ideal set-up for developing mathematics in characteristic one. In the same paper it is shown in particular how to adapt the construction of the Witt ring in characteristic $p > 1$ to the limit case of characteristic one, and along the way one also discovers an interesting connection with entropy and thermodynamics.

This proceedings volume contains two number-theoretic papers. The paper by D. Goss, *"Zeta Phenomenology,"* collects evidence on the nature of symmetry group (functional equation, that is) of zeta functions in characteristic p. The paper by P. Tretkoff, *"Transcendence of Values of Transcendental Functions at Algebraic Points,"* surveys and explains various transcendence results, starting with Liouville's celebrated proof that such transcendence numbers exist at all.

Two papers in this volume, W. van Suijlekom's *"The Hopf Algebraic Structure of Perturbative Quantum Gauge Theories"* and L. Guo, S. Paycha and B. Zhang's *"Renormalization by Birkhoff-Hopf Factorization and by Generalized Evaluators: A Case Study"* treat questions of renormalization in quantum gauge field theory and number theory (multiple zeta values) via Hopf algebraic techniques. They are both inspired by the Connes-Kreimer theory of renormalization in quantum field theory via Hopf algebras.

Finally, there are three papers in this volume that are concerned with various questions in noncommutative geometry inspired either by arithmetics or by spectral geometry. The paper by A. Banerjee, *"Nearby Cycles and Periodicity in Cyclic Homology,"* relates the periodicity operator of cyclic homology (of a certain sheaf of differential operators) to the (local) monodromy operator acting on the complex of nearby cycles. The contribution by A. Carey, M. Marcolli, and A. Rennie, *"Modular Index Invariants of Mumford Curves,"* continues an investigation initiated by C. Consani and M. Marcolli on the relation between the algebraic geometry of p-adic Mumford curves and the noncommutative geometry of graph C^*-algebras associated to the action of the uniformizing p-adic Schottky group on the Bruhat-Tits tree. In *"The Gauss-Bonnet Theorem for the Noncommutative Two Torus,"* A. Connes and P. Tretkoff extend the classical Gauss-Bonnet theorem from Riemann surface theory to the noncommutative two torus, equipped with arbitrary Weyl factor at conformal class $\tau = i$.

The 21st JAMI Conference (and related year-long program) was supported by the National Science Foundation and by the JAMI Foundation. We are grateful for their generous financial support. We thank all the speakers and participants for their contributions, which made the conference most successful. We also thank all the referees for their contributions to this volume.

We hope that this proceedings volume will be of some use to workers in the fields of noncommutative geometry, arithmetic and mathematical physics and to mathematicians who want to gain a quick but deep and thorough impression about the issues that were addressed at this meeting. In particular, we hope that this collection of papers will stimulate the research of many young mathematicians by guiding them all the way through the investigation of the new areas of research outlined in this book.

A. Connes and C. Consani

Noncommutative Geometry, Arithmetic, and Related Topics

Nearby Cycles and Periodicity in Cyclic Homology

Abhishek Banerjee

ABSTRACT. In this paper, we consider an algebraic degeneration of a variety over a disc and its archimedean counterpart, which involves the fiber at infinity of an arithmetic variety. Furthermore, the monodromy operator in the case of the disc has an archimedean counterpart and we show that the monodromy operator (resp. its archimedean counterpart) may be seen as corresponding to the periodicity operator in cyclic homology (resp. cyclic cohomology). In the former case, we work with the periodicity on the cyclic (hyper)homology of a sheaf of differential operators. In the latter case, we work with the periodicity in the cyclic cohomology of the ring of differential operators. Finally, we use our understanding to construct an archimedean counterpart of the complex of nearby cycles.

1. Introduction

Cyclic cohomology, introduced by Connes [**Co1**] and Tsygan [**Tsy**], is the non-commutative analogue of de Rham homology. For commutative algebras, the connection between the cyclic (co)homology and the de Rham cohomology (and its Hodge decomposition) can be made explicit in a number of well-known cases (see for instance, [**Lod**, §**2.3**, §**4.6**]). It is therefore natural to consider whether the operators on cyclic (co)homology can be interpreted in the classical commutative framework.

In this paper, we show that the periodicity operator in cyclic (co)homology may be understood in terms of a monodromy operator. We demonstrate this in two separate situations.

The first situation is considered in Section 2. In Section 2, we consider the complex of nearby cycles for an algebraic degeneration of a variety X over the unit disc S (due to Deligne [**Del1**]). Further, we consider two resolutions of the nearby cycles complex, the former defined by Steenbrink [**St2**] which we shall refer to as the bicomplex A^{**} and the latter defined by Guillén and Navarro Aznar [**GN**] which we shall refer to as the bicomplex $\psi^{**}(\mathbb{C})$. Both bicomplexes A^{**} and $\psi^{**}(\mathbb{C})$ carry monodromy operators N_A and N_ψ respectively. Thereafter, Guillén and Navarro Aznar [**GN**] construct a quasi-isomorphism $\mu : \psi^{**}(\mathbb{C}) \longrightarrow A^{**}$ that commutes with the monodromy up to homotopy, i.e., the difference $N_A \circ \mu - \mu \circ N_\psi$ is homotopic to zero.

In this case, we start by considering the sheaf of holomorphic differential operators D_{X^*} on the locally Stein manifold X^*, $X^* := X \backslash Y$ being the complement in

X of the fiber Y over the point $\{0\}$ of the unit disc S. In Theorem 2.6, we adapt the proof of Wodzicki [**W**] to hypercohomology to demonstrate that the cyclic homology of D_{X^*} decomposes canonically as a direct sum of de Rham cohomologies of X^* and that the periodicity operator acts by dropping the top summand. Then we show that the filtration by kernels $Ker(N_\psi^j)$, $j \geq 0$ on $\psi^{**}(\mathbb{C})$ coincides with the filtration by columns on $\psi^{**}(\mathbb{C})$ and that the cohomologies of the graded pieces of the filtration are canonically isomorphic to the Hochschild homologies of the sheaf D_{X^*} (see Corollary 2.9) up to a shift in degree. Following this, we construct a triple complex $\mathcal{BC}^{***}$ by vertically stacking copies of $\psi^{**}(\mathbb{C})$ and again consider the "columns" of this triple complex $\mathcal{BC}^{***}$, which are bicomplexes of the form $\mathcal{BC}^{p**}$ for each fixed $p \in \mathbb{Z}$. We show that the cohomologies of these "columns" of the triple complex $\mathcal{BC}^{***}$ are canonically isomorphic to the cyclic homologies of D_{X^*} up to a shift in degree (see Theorem 2.10(a)). Then Theorem 2.10(b) shows that the periodicity operator on the cyclic homology of D_{X^*} coincides with the monodromy N induced on each $\mathcal{BC}^{p**}$ by the maps N_ψ on each of the copies of $\psi^{**}(\mathbb{C})$ stacked to form the triple complex $\mathcal{BC}^{***}$.

We now turn to the second situation. This is treated in Section 3. The algebraic degeneration over a disc has a counterpart that plays an important role in Arakelov geometry. This consists of the "fiber at archimedean infinity" of an arithmetic variety $\chi \longrightarrow Spec(\mathbb{Z})$, the complex points of which form a smooth projective manifold X over $\mathbb{C}$ or $\mathbb{R}$. For such a variety, Consani [**Con2**] has introduced a bicomplex (K^{**}, d', d'') with $K^{i,j} := \oplus_{k\geq 0} K^{i,j,k}$, $i, j, k \in \mathbb{Z}$, with a monodromy-like operator $N : K^{i,j,k} \longrightarrow K^{i+2,j,k+1}$ and differentials $d' : K^{i,j,k} \longrightarrow K^{i+1,j+1,k+1}$ and $d'' : K^{i,j,k} \longrightarrow K^{i+1,j+1,k}$. The terms $K^{i,j,k}$ are modules of real differential forms twisted by powers of $(2\pi i)$. The $K^{i,j,k}$'s at archimedean infinity play the same role as direct summands of the E_1-term of the spectral sequence associated to the "monodromy filtration" (see (2.10)) on Steenbrink's complex A^{**} (see [**BGS**] or [**GN**]).

For each fixed value of the twist $t \in \mathbb{Z}$, we choose, from among all the modules $K^{i,j,k}$, those terms in which the real differentials are twisted by $(2\pi i)^t$. It turns out that all the terms $K^{i,j,k}$ with a fixed value t of the twist form a bicomplex, which we call (K_t^{**}, d', d''). We now consider the ring $D(X)$ of C^∞-differential operators on X. Following Wodzicki [**W**], the cyclic homology of $D(X)$ decomposes canonically as a direct sum of de Rham cohomologies and the periodicity acts by dropping the top summand. In Proposition 3.8, we show that there are natural surjections from the Hochschild cohomology of $D(X)$ to the the cohomology of the graded pieces of the filtration on $Tot(K_t^{**})$ by $Im(N^j)$, $j \geq 0$. Further, the filtration on K_t^{**} by $Im(N^j)$, $j \geq 0$, is identical to the filtration by rows on K_t^{**}. Thereafter, we vertically stack the K_t^{**} to form a triple complex $\mathcal{K}^{***}$. Then the "rows" of $\mathcal{K}^{***}$ are bicomplexes of the form $\mathcal{K}^{*q*}$ for each $q \in \mathbb{Z}$. Then Proposition 3.10 shows that there are natural surjections from the cyclic cohomology of the ring $D(X)$ to the cohomology of the rows $\mathcal{K}^{*q*}$. Proposition 3.10 also shows that the periodicity operator on the cyclic cohomology of $D(X)$ corresponds to the induced monodromy on the row $\mathcal{K}^{*q*}$.

Finally, in Section 3.4, we construct a bicomplex φ^{**} that replaces the bicomplex $\psi^{**}(\mathbb{C})$ at archimedean infinity. We also construct a bicomplex B^{**} that acts as counterpart for Steenbrink's complex A^{**} mentioned above. The bicomplexes φ^{**} and B^{**} are equipped with monodromy-like operators N_φ and N_B respectively.

Then we define a natural morphism $\mu : \varphi^{**} \longrightarrow B^{**}$ and we show in Proposition 3.11 that $N_B \circ \mu - \mu \circ N_\varphi$ is homotopic to zero.

Acknowledgments. This paper is derived from my PhD thesis at Johns Hopkins University. I am grateful to my advisor C. Consani for many useful discussions. I am also grateful to M. Khalkhali for important suggestions and comments.

2. The Steenbrink Complex and the Cyclic Homology of the Sheaf of Differential Operators

In this section, we consider the following situation. Let $f : X \longrightarrow S$ be a proper morphism from an $n+1$-dimensional variety X to the complex unit disc S. We assume that the fibers $f^{-1}(t)$, $t \in S$ are smooth whenever $t \neq 0$, while the divisor $Y = f^{-1}(0)$ is such that Y_{red} is a divisor with normal crossings, having smooth n-dimensional components $Y_1, Y_2, \ldots, Y_m$. Let $X^* = X - Y$, $S^* = S - \{0\}$ and let $p : \tilde{S}^* \longrightarrow S^*$ be a universal covering of S^*. Set $\tilde{X}^* = X^* \times_S \tilde{S}^*$. Let $\tilde{j}_* : \tilde{X}^* \longrightarrow X$, $j : X^* \longrightarrow X$ and $i : Y \hookrightarrow X$ denote the natural morphisms. Then we have the following diagram:

$$(2.1) \quad \begin{array}{ccccccc} \tilde{X}^* & \longrightarrow & X^* & \overset{j}{\longrightarrow} & X & \overset{i}{\longleftarrow} & Y \\ \downarrow & & \downarrow & & f\downarrow & & \downarrow \\ \tilde{S}^* & \overset{p}{\longrightarrow} & S^* & \longrightarrow & S & \longleftarrow & \{0\} \end{array}$$

Then $\mathbf{R}\psi^*(\mathbb{C})$ is the the complex of nearby cycles, the functor $\mathbf{R}\psi^*$ being defined by $\mathbf{R}\psi^*(\mathcal{F}) := i^* \mathbf{R}\tilde{j}_* \tilde{j}^* \mathcal{F}$ for a sheaf $\mathcal{F}$ of abelian groups on X. If $\tilde{S}^*$ is the upper half plane, then p is defined by $p(u) = e^u$ for each $u \in \tilde{S}^*$. The complex $\mathbf{R}\psi^*(\mathbb{C})$ is provided with a monodromy T which is induced by the translation $u \mapsto u + 2\pi\sqrt{-1}$ on $\tilde{S}^*$.

Now, if t denotes the uniform coordinate on the disc S, let us denote by $\theta = f^*(dt/t)$ the pullback of the logarithmic differential form dt/t to X. It may be verified (see [**St2**]) that, if the component Y_i, $1 \leq i \leq m$ of the divisor $Y = f^{-1}(0)$ has multiplicity m_i, $1 \leq i \leq m$, then given a local coordinate system $(z_1, z_2, \ldots, z_{n+1})$ on X, we have

$$(2.2) \quad \theta := f^* \left(\frac{dt}{t} \right) = \sum_{i=1}^{m} m_i \frac{dz_i}{z_i}$$

Y_i being defined in the local coordinate system by $\{z_i = 0\}$ for $1 \leq i \leq m$. Following [**GN**, §2.2], let us define the following complex of sheaves on X, which is yet another resolution of the nearby cycles complex

$$(2.3) \quad \psi^*(\mathbb{C}) = \mathbb{C}[u] \otimes_{\mathbb{C}} \Omega_X^*(\log Y) = \sum_{p \geq 0} \mathbb{C}u^p \otimes_{\mathbb{C}} \Omega_X^*(\log Y)$$

with differential d defined by $d(u \otimes 1) = 1 \otimes \theta$ and $d(1 \otimes \omega) = 1 \otimes d(\omega)$ and extended so that this complex forms a differential graded subalgebra of $\tilde{j}_* \Omega_{\tilde{X}^*}^*$ generated by $\Omega_X^*(\log Y)$ and u. Then the bicomplex $\psi^{**}(\mathbb{C})$ (see Figure 1) defined by

$$(2.4) \quad \psi(\mathbb{C})^{-p,q} = \mathbb{C}u^p \otimes_{\mathbb{C}} \Omega_X^{q-p}(\log Y) \qquad p \geq 0$$

with the differentials

(2.5)
$$d' : \psi(\mathbb{C})^{-p,q} \longrightarrow \psi(\mathbb{C})^{-p+1,q} d'(u^p \otimes \omega) = (pu^{p-1} \otimes \theta \wedge \omega)$$
$$d'' : \psi(\mathbb{C})^{-p,q} \longrightarrow \psi(\mathbb{C})^{-p,q+1} d''(u^p \otimes \omega) = (u^p \otimes d\omega)$$

is a resolution of the complex $\psi^*(\mathbb{C})$. On $\psi^*(\mathbb{C})$, the monodromy is expressed by the operator

(2.6)
$$T : \psi^*(\mathbb{C}) \longrightarrow \psi^*(\mathbb{C})$$

given by

(2.7)
$$T(u^p \otimes \omega) = (u + 2\pi\sqrt{-1})^p \otimes \omega$$

From now on, we shall denote $u^{[p]} = u^p/p!$ and $u_{[p]} = (-1)^{p-1}(p-1)!u^{-p}$, for $p \geq 0$. If we let $N = \log T/2\pi i$ as before, we note that N has the following description on $\psi^{**}(\mathbb{C})$ (see [**GN**, §**2.2**] for details)

(2.8)
$$N(u^{[p]} \otimes \omega) = u^{[p-1]} \otimes \omega \qquad p \geq 0$$

FIGURE 1 The bicomplex $\psi^{**}(\mathbb{C})$.

By abuse of notation, we shall often refer to $\psi^{**}(\mathbb{C})$ itself as the nearby cycles complex. The nearby cycles complex can be used to compute the cohomology of the "universal fiber" $\tilde{X}^*$, i.e., we have the following proposition.

PROPOSITION 2.1. *With the notation and setup as in (2.1)–(2.8), there is an isomorphism of hypercohomologies*

$$\mathbb{H}^*(Y, \psi^*(\mathbb{C})) \simeq \mathbb{H}^*(Y, \mathbf{R}\psi^*(\mathbb{C})) \simeq H^*(\tilde{X}^*, \mathbb{C})$$

PROOF. We have mentioned that $\psi^{**}(\mathbb{C})$ is a resolution of the nearby cycles complex and hence it induces the isomorphism $\mathbb{H}^*(Y, \psi^*(\mathbb{C})) \approx \mathbb{H}^*(Y, \mathbf{R}\psi^*(\mathbb{C}))$. For the proof of the isomorphism $\mathbb{H}^*(Y, \mathbf{R}\psi^*(\mathbb{C})) \simeq H^*(\tilde{X}^*, \mathbb{C})$, see [**Del1**, §4].

The complex $(\Omega_X^*(\log Y), d)$ (d being differentiation extended to differential forms with log poles along Y) carries a weight filtration given by

$$W_k\Omega_X^*(\log Y) = \wedge^k\Omega_X^1(\log Y) \wedge \Omega_X^{*-k} \qquad k \in \mathbb{Z}$$

Furthermore, by the Poincaré residue theorem (see [**St2**, §6]), we have an isomorphism of complexes

$$(2.9) \qquad Gr_k^W \Omega_X^* (\log Y) \xrightarrow{\sim} (a_k)_* \Omega_{\tilde{Y}^{(k)}}^* [-k]$$

where the map a_k is as follows. In (2.9), $\tilde{Y}^{(k)}$ denotes the disjoint union of all "k-intersections" $Y_{i_1} \cap Y_{i_2} \cap \ldots \cap Y_{i_k}$ of components of Y as we vary over all tuples $1 \leq i_1 < i_2 < \ldots < i_k \leq m$. Then $\tilde{Y}^{(k)}$ is equipped with an obvious map $a_k : \tilde{Y}^{(k)} \longrightarrow Y$.

In this context, Steenbrink (see [**St1**] or [**St2**, §13]) has introduced the following bicomplex A^{**} of sheaves

$$A^{pq} = \mathbb{C}u_{[p+1]} \otimes_{\mathbb{C}} \Omega_X^{p+q+1} (\log Y) / W_p \Omega_X (\log Y)^{p+q+1}$$

along with differentials

$$d' : A^{pq} \longrightarrow A^{p+1,q} \qquad d'(u_{[p+1]} \otimes \omega) = u_{[p+2]} \otimes \theta \wedge \omega$$
$$d'' : A^{pq} \longrightarrow A^{p,q+1} \qquad d''(u_{[p+1]} \otimes \omega) = u_{[p+1]} \otimes d\omega$$

Following [**St2**], on the terms A^{pq}, we have two filtrations W and L, given by

$$(2.10) \qquad \begin{aligned} W_r A^{pq} &= \mathbb{C}u_{[p+1]} \otimes_{\mathbb{C}} W_{r+p+1} \Omega_X^{p+q+1} (\log Y) / W_p \Omega_X^{p+q+1} (\log Y) \\ L_r A^{pq} &= \mathbb{C}u_{[p+1]} \otimes_{\mathbb{C}} W_{r+2p+1} \Omega_X^{p+q+1} (\log Y) / W_p \Omega_X^{p+q+1} (\log Y) \end{aligned}$$

It is important to note that the differentials d' and d'' respect the latter filtration L (usually called the monodromy filtration) and that we have (see [**St2**])

$$(2.11) \qquad Gr_r^L A^* \simeq \bigoplus_{k \geq 0, -r} \mathbb{C}u_{[k+1]} \otimes_{\mathbb{C}} Gr_{r+2k+1}^W \Omega_X^{*+1} (\log Y)$$

Using the Poincaré residue theorem described in (2.9), the right hand side of (2.11) is isomorphic to

$$Gr_r^L A^* \simeq \bigoplus_{k \geq 0, -r} \Omega_{\tilde{Y}^{r+2k+1}}^{*+1} [-r - 2k - 1]$$

Finally, the monodromy on Steenbrink's complex is described by

$$N : A^{pq} \to A^{p+1,q-1} \qquad N(u_{[p+1]} \otimes \omega) := u_{[p+2]} \otimes \omega$$

We note that the operator N is explicitly nilpotent on the bicomplex A^{**} (as opposed to simply on the hypercohomology, as in the case of $\psi^{**}(\mathbb{C})$). It may be easily checked that $[N, d'] = [N, d''] = 0$. Furthermore, we have the following proposition.

PROPOSITION 2.2. (see [**GN**, §**2.5**]) *The morphism* $\mu : \psi^*(\mathbb{C}) \longrightarrow A^*$ *defined by*

$$(2.12) \qquad \mu(u^{[p]} \otimes \omega_p) = \begin{cases} 0 & \text{if } p \neq 0 \\ (-1)^{|\omega_0|} u_{[1]} \otimes \theta \wedge \omega_0 & \text{if } p = 0 \end{cases}$$

is a quasi-isomorphism of complexes. Further, if $N = \frac{\log T}{2\pi i}$, *where* T *is the monodromy operator, the morphisms* $N \circ \mu$ *and* $\mu \circ N$ *are homotopic. The homotopy is* *induced by the map h of bidegree $(0, -1)$ defined as*

$$(2.13) \qquad h(u^{[p]} \otimes \omega_p) = \begin{cases} 0 & \text{if } p \neq 0 \\ (-1)^{|\omega_0|} u_{[1]} \otimes \omega_0 & \text{if } p = 0 \end{cases}$$

PROOF. It is clear that $[\mu, d''] = [\mu, d'] = 0$ and hence μ induces a morphism of complexes. The proof that μ is actually a quasi-isomorphism may be found in [**N**]. Further, we note that $h \circ d'' + d'' \circ h = 0$ and $N \circ \mu - \mu \circ N = h \circ d' + d' \circ h$, whence h defines a homotopy from $N \circ \mu$ to $\mu \circ N$.

As mentioned above, the maps N and μ commute only up to homotopy. Consider the subcomplex $Ker(N)^*$ of $\psi^*(\mathbb{C})$ and the subcomplex $Im(N)^*$ of A^*. The following theorem expresses the idea that this "commutation up to homotopy" entails an interchange between the kernel and the cokernel of the monodromy on $\psi^{**}(\mathbb{C})$ and A^{**}. This interchange will be key to our understanding of the results of Section 3.

THEOREM 2.3. *Let us denote the monodromy on $\psi^{**}(\mathbb{C})$ by N_ψ and the monodromy operator on A^{**} by N_A. Let us consider the complexes $(Ker(N_\psi)^*, d'')$ and $(Coker(N_A)^*, d'')$. Then we have*

$$(2.14) \qquad (Ker(N_\psi)^*, d'')/W_0(Ker(N_\psi)^*, d'') = (Coker(N_A)^*, d'')[-1]$$

where $W_0(Ker(N_\psi)^*, d'')$ denotes the subcomplex of $(Ker(N_\psi)^*, d'')$ of logarithmic differential forms of weight 0.

PROOF. The map N_ψ acts by $N_\psi(u^{[p]} \otimes \omega) = u^{[p-1]} \otimes \omega$ and hence it is clear that the kernel of N_ψ consists of the terms $u^{[0]} \otimes \omega$, which form the column $\psi^{0*}(\mathbb{C})$ of $\psi^{**}(\mathbb{C})$. Hence $Ker(N_\psi)^i = \mathbb{C}u^0 \Omega_X^i(\log Y)$.

Similarly, since N_A acts as $N_A(u_{[p]} \otimes \omega) = u_{[p+1]} \otimes \omega$, the cokernel of N_A consists of the terms $u_{[1]} \otimes \omega$ which form the column $A^{0*} = \mathbb{C}u_{[1]} \otimes \Omega_X^{*+1}(\log Y)/W_0\Omega_X^{*+1}(\log Y)$ of A^{**}. Hence it follows that

$$(Ker(N_\psi)^*, d'')/W_0(Ker(N_\psi)^*, d'') = (\mathbb{C}u^0 \otimes \Omega_X^*(\log Y)/W_0\Omega_X^*(\log Y), d'') \xrightarrow{\sim}$$
$$(\mathbb{C}u^0 \otimes \Omega_X^{*+1}(\log Y)/W_0\Omega_X^{*+1}(\log Y), d'')[-1] = (A^{0*}, d'')[-1]$$
$$= (Coker(N_A)^*, d'')[-1]$$

2.1. Cyclic Homology of the Sheaf of Differential Operators

In the notation above, let U be an open subset of X^* and let $D(U)$ denote the ring of holomorphic differential operators on U. Denote by D_{X^*} the sheaf of differential operators on X^*, i.e., D_{X^*} is the sheaf associated to the presheaf $U \mapsto D(U)$ for every open set $U \subseteq X^*$. For an algebra A over $\mathbb{C}$, let $C_*^h(A)$ denote the complex of [**Lod**, §1.1.3] defining the Hochschild homology of A and let $CC_{**}(A)$ denote the bicomplex of [**Lod**, §2.1.2] defining the cyclic homology of A.

DEFINITION 2.4. Let $C^h_*(D_{X^*})$ (resp. $CC_{**}(D_{X^*})$) denote the sheafification of the complexes defining the Hochschild (resp. cyclic) homologies of the algebras $D(U)$ (for each open $U \subseteq X^*$). The Hochschild homology $HH_*(D_{X^*})$ and cyclic homology $HC_*(D_{X^*})$ of D_{X^*} are defined to be the respective hypercohomologies, i.e

$$HH_*(D_{X^*}) = \mathbb{H}^{-*}(C^h_*(D_{X^*})) \qquad HC_*(D_{X^*}) = \mathbb{H}^{-*}(Tot(CC_{**}(D_{X^*})))$$

We recall here that the complexes $C^h_*(D_{X^*})$ and $Tot(CC_{**}(D_{X^*}))$ are unbounded below and therefore the hypercohomology groups are defined in the sense of Weibel [**Wei1, Appendix**]. We have the following result of Wodzicki describing the Hochschild and cyclic homologies of an algebra $D(U)$ of holomorphic differential operators on a Stein manifold.

PROPOSITION 2.5. *(Wodzicki) If U is a Stein manifold and $dim(U) = d$, then*

(1) $HC_q(D(U)) \simeq H^{2d-q}_{dR}(U) \oplus H^{2d-q+2}_{dR}(U) \oplus H^{2d-q+4}_{dR}(U) \oplus \cdots$

(2) $HH_q(D(U)) \simeq H^{2d-q}_{dR}(U)$

Also, the analogous result holds for the ring of algebraic differential operators on an affine variety and for C^∞-differential operators on a compact C^∞-manifold.

PROOF. See [**W, Theorem 2**] and [**W, Theorem 3**].

In the next theorem, we will prove the analogous result in the setting of hypercohomology, which we shall then apply in Sections 2.2 and 2.3. Recall that we assumed that; in the degeneration $f : X \longrightarrow S$ over the unit disc, X has dimension $n + 1$. Hence, it follows that $X^* = X \backslash Y$ is a locally Stein manifold of dimension $n + 1$.

THEOREM 2.6. *Let X^* be a locally Stein complex manifold with $dim(X^*) = n + 1$ and let D_{X^*} be the sheaf of holomorphic differential operators on X^*. Then, there are isomorphisms*

(1) $HC_q(D_{X^*}) \simeq H^{2n+2-q}_{dR}(X^*) \oplus H^{(2n+2)-q+2}_{dR}(X^*) \oplus H^{(2n+2)-q+4}_{dR}(X^*) \oplus \cdots$

(2) $HH_q(D_{X^*}) \simeq H^{(2n+2)-q}_{dR}(X^*)$

PROOF. Let $C^h_*(D_{X^*})$ (resp. $CC_{**}(D_{X^*})$) denote the sheafification of the complex defining the Hochschild (resp. cyclic) homologies of the algebras $D(U)$ for open subsets $U \subseteq X^*$ as in Definition 2.4.

The fact that the Hochschild complex $C^h_*(D_{X^*})$ of the sheaf of differential operators D_{X^*} is quasi-isomorphic to $\mathbb{C}[2n+2]$ (the constant sheaf $\mathbb{C}$ shifted $(2n + 2)$-places, where $n + 1 = dim(X^*)$) is already known from [**Bry1**] and [**Bry2**]. For any open subset $U \subseteq X^*$, consider the complex $CC_{**}(D(U))$ defining the cyclic homology of the algebra $D(U)$. The periodicity operator on $CC_{**}(D(U))$ is defined by taking the quotient over the terms coming from the first two columns of the bicomplex $CC_{**}(D(U))$ (see [**Lod, §2.1.2, §2.2.1**]) and considering the corresponding long exact sequence. One of the two columns dropped is a Hochschild complex and the other column is made up of differentials b' which, following [**Lod, §1.1.12**] make this latter column chain homotopic to 0. Since every alternate column of $CC_{**}(D(U))$ is chain homotopic to zero, it follows that (the total complex of) the sheafification of $CC_{**}(D(U))$ is quasi-isomorphic to a direct sum

$$\mathbb{C}[(2n+2)] \oplus \mathbb{C}[(2n+2)+2)] \oplus \cdots$$

and that the periodicity operator acts by dropping the first summand $\mathbb{C}[(2n+2)]$ which is quasi-isomorphic to a Hochschild complex. Hence we have the isomorphism of hypercohomologies

$$HC_q(D_{X^*}) \simeq H_{dR}^{(2n+2)-q}(X^*) \oplus H_{dR}^{(2n+2)-q+2}(X^*) \oplus H_{dR}^{(2n+2)-q+4}(X^*) \oplus \cdots$$

COROLLARY 2.7. *There exists a long exact sequence*

$$\cdots \to HH_q(D_{X^*}) \to HC_q(D_{X^*}) \xrightarrow{S} HC_{q-2}(D_{X^*}) \to \cdots$$

The periodicity operator

$$S : HC_q(D_{X^*}) \longrightarrow HC_{q-2}(D_{X^*})$$

acts by dropping the summand $HH_q(D_{X^*}) \simeq H_{DR}^{(2n+2)-q}(X)$.

PROOF. This follows directly from the proof of Theorem 2.6 in which we showed that the sheafification of the double complex $CC(D(U))$ is quasi-isomorphic to the direct sum $\mathbb{C}[(2n+2)] \oplus \mathbb{C}[(2n+2)+2] \oplus \cdots$ and that the periodicity operator acts by dropping the first summand $\mathbb{C}[(2n+2)]$.

2.2. The Complex Ker(N)* and the Hochschild Complex

As mentioned before, the bicomplex $\psi^{**}(\mathbb{C})$ is a resolution of the nearby cycles complex $\mathbf{R}\psi^*(\mathbb{C}) = i^* \mathbf{R}\tilde{j}_* \tilde{j}^* \mathbb{C}$. The monodromy N, which is given by $N = \frac{\log T}{2\pi i}$, acts on $\psi^{**}(\mathbb{C})$ as follows; given $u^{[p]} \otimes \omega \in \psi^{-p,q}(\mathbb{C})$, we have

$$N(u^{[p]} \otimes \omega) = (u^{[p-1]} \otimes \omega) \in \psi^{-p+1,q-1}(\mathbb{C}) \qquad p \geq 0$$

Moreover, it is easy to check that $N : \psi^{-p,q}(\mathbb{C}) \to \psi^{-p+1,q-1}(\mathbb{C})$ commutes with the differentials d' and d'' of $\psi^{**}(\mathbb{C})$, i.e.,

$$[N, d'] = [N, d''] = 0$$

Therefore, we can consider the subcomplex of $Tot(\psi^{**}(\mathbb{C}))$ given by $(Ker(N)^*, d' + d'')$, which we shall denote by $Ker(N)^*$. We consider the increasing filtration F_k on $Tot(\psi^{**}(\mathbb{C}))$ by subcomplexes

$$F_k(Tot(\psi^{**}(\mathbb{C}))) = Ker(N^k)^* \qquad k \geq 0$$

and the graded pieces $F_{k+1}/F_k = Ker(N^{k+1})^*/Ker(N^k)^*$. Note that the filtration F_k is bounded below, i.e., $F_0(Tot(\psi^{**}(\mathbb{C}))) = 0$ and that this filtration is exhaustive, i.e., $Tot(\psi^{**}(\mathbb{C})) = \bigcup_{k \geq 0} F_k(Tot(\psi^{**}(\mathbb{C})))$.

PROPOSITION 2.8. *Let D_{X^*} denote the sheaf of differential operators on X^* and let $j : X^* \longrightarrow X$ denote the inclusion. Then there are canonical isomorphisms*

$$(2.15) \qquad HH_q(D_{X^*}) \xrightarrow{\sim} \mathbb{H}^{(2n+2)-q}(Ker(N)^*)$$

where $HH_*(D_{X^*})$ denotes the Hochschild homology of the sheaf of differential operators on X^* as defined in Definition 2.4.

PROOF. Since $N(u^{[p]} \otimes \omega) = (u^{[p-1]} \otimes \omega)$, $u^{[p]} \otimes \omega$ lies in $Ker(N)^*$ if and only if $p = 0$. Hence, we have

$$(2.16) \qquad Ker(N)^* = \psi^{0,*}(\mathbb{C})$$

From the description of the terms $\psi^{0,*}(\mathbb{C})$, the complex $Ker(N)^*$ can be written explicitly as

$$(2.17) \qquad Ker(N)^* = \mathbb{C}u^0 \otimes \Omega_X^*(\log Y)$$

Let $j : X^* \to X$ denote the open immersion. From [**Del2**, §**3.1.8**], we know that there is a quasi-isomorphism of filtered complexes

$$(2.18) \qquad (\Omega_X^*(\log Y), W) \leftarrow (\Omega_X^*(\log Y), \tau) \hookrightarrow (j_*\Omega_{X^*}^*, \tau)$$

where W is the weight filtration on the de Rham complex with logarithmic poles and τ is the canonical filtration (see [**Del2**, §**3.1.7**] for details). Both W and τ are finite filtrations (on each term of the complex) and hence, combining with (2.17), we have a quasi-isomorphism of complexes

$$(2.19) \qquad Ker(N)^* = \mathbb{C}u^0 \otimes \Omega_X^*(\log Y) \xrightarrow{\sim} \Omega_X^*(\log Y) \xrightarrow{\sim} j_*\Omega_{X^*}^*$$

On the other hand, from the proof of Theorem 2.6, we have a quasi-isomorphism

$$(2.20) \qquad C_*^h(D_{X^*}) \xrightarrow{\sim} \mathbb{C}_{X^*}[(2n+2)]$$

Since $\mathbb{C}_{X^*}$ has a resolution by sheaves $\Omega_{X^*}^*$ that are j_*-acyclic, we have canonical isomorphisms of hypercohomologies (combining (2.19) and (2.20) and taking hypercohomologies)

$$HH_q(D_{X^*}) \simeq \mathbb{H}^{(2n+2)-q}(\mathbb{C}_{X^*}) \simeq \mathbb{H}^{(2n+2)-q}(\Omega_{X^*}^*)$$
$$\simeq \mathbb{H}^{(2n+2)-q}(j_*\Omega_{X^*}^*) \simeq \mathbb{H}^{(2n+2)-q}(Ker(N)^*)$$

COROLLARY 2.9. *The increasing filtration by columns on the total complex $Tot(\psi^{**}(\mathbb{C}))$ coincides with the filtration by $F_j(Tot(\psi^{**}(\mathbb{C}))) = Ker(N^j)^*$, $j \geq 0$ on $Tot(\psi^{**}(\mathbb{C}))$ and moreover, the Hochschild homology of D_{X^*} can be computed as (for any $k \geq 0$)*

$$HH_q(D_{X^*}) \simeq \mathbb{H}^{(2n+2)-q}((Ker(N^{k+1})/Ker(N^k))^*) \simeq \mathbb{H}^{(2n+2)-q}(gr_I^k Tot(\psi^{**}(\mathbb{C})))$$

where $gr_I^k Tot(\psi^{**}(\mathbb{C}))$ denote the graded pieces corresponding to the filtration of $Tot(\psi^{**}(\mathbb{C}))$ by columns.

PROOF. For any $k \geq 0$, we have that $F_k(Tot(\psi^{**}(\mathbb{C}))) = Ker(N^k)^*$ comprises the terms $\psi^{p,q}(\mathbb{C})$ where $p \geq -k + 1$. Consequently, the quotient complex $Ker(N^{k+1})^*/Ker(N^k)^*$ consists of all those terms of $(Tot(\psi^{**}(\mathbb{C})))$ that come from $\psi^{-k,*}(\mathbb{C})$. As a subquotient of $Tot(\psi^{**}(\mathbb{C}))$, the complex $(Ker(N^{k+1})/Ker(N^k))^*$ may therefore be described as

$$(Ker(N^{k+1})/Ker(N^k))^* = \mathbb{C}u^{[k]} \otimes \Omega_X^*(\log Y) \simeq \Omega_X^*(\log Y)$$

Thus, exactly as in the proof of Proposition 2.8, it follows that we have isomorphisms $HH_q(D_{X^*}) \simeq \mathbb{H}^{(2n+2)-q}((Ker(N^{k+1})/Ker(N^k))^*)$. Further, we see that the complex $Ker(N^k)^*$ consists of all terms of $Tot(\psi^{**}(\mathbb{C}))$ that come from columns $0, 1, \ldots, k$ of $\psi^{**}(\mathbb{C})$. Therefore, F_k coincides with the increasing filtration by columns on $Tot(\psi^{**}(\mathbb{C}))$ and, using Proposition 2.8, we have

$$HH_q(D_{X^*}) \simeq \mathbb{H}^{(2n+2)-q}((Ker(N^{k+1})/Ker(N^k))^*) \simeq \mathbb{H}^{(2n+2)-q}(gr_I^k Tot(\psi^{**}(\mathbb{C})))$$

2.3. The Connes Periodicity Operator and the Monodromy Operator

In the previous section, we have shown in Proposition 2.8 that the graded pieces of the filtration on $Tot(\psi^{**}(\mathbb{C}))$ by $F_j Tot(\psi^{**}(\mathbb{C})) = Ker(N^j)$, $j \geq 0$ compute the Hochschild homology of D_{X^*}. Furthermore, we have shown in Corollary 2.9 that the filtration by $Ker(N^j)$, $j \geq 0$ coincides with the filtration by columns on $Tot(\psi^{**}(\mathbb{C}))$. Moreover, this is an increasing filtration on the total complex ψ^*, which is also bounded below and exhaustive. Consequently, we have a converging spectral sequence in hypercohomology

$$(2.21) \qquad HH_q(D_{X^*}) \simeq \mathbb{H}^{(2n+2)-q}((Ker(N^p)/Ker(N^{p-1}))^*) = E_1^{-p,(2n+2)+p-q}$$
$$\Rightarrow \mathbb{H}^{(2n+2)-q}(\psi^*)$$

We will now define a triple complex $\mathcal{BC}^{***}$ by "vertically stacking" copies of the bicomplexes $\psi^{**}(\mathbb{C})$ in a certain way. Then a "column" of the triple complex $\mathcal{BC}^{***}$ is a bicomplex $\mathcal{BC}^{p**}$ obtained by fixing the first index p. The "column" $\mathcal{BC}^{p**}$ therefore consists of a column from different copies of $\psi^{**}(\mathbb{C})$ that have been stacked to form the triple complex $\mathcal{BC}^{***}$.

Thereafter, in Theorem 2.10, we shall show that the hypercohomology of each of the "columns" $\mathcal{BC}^{p**}$ is canonically isomorphic to the cyclic homology of D_{X^*}. This is a counterpart to Corollary 2.9, where we showed that the hypercohomology of the columns of $\psi^{**}(\mathbb{C})$ is canonically isomorphic to Hochschild homology of D_{X^*}.

The $\psi^{**}(\mathbb{C})$ complexes stacked vertically to form the complex $\mathcal{BC}^{***}$ induce a monodromy operator N on $\mathcal{BC}^{***}$ and hence on each column $\mathcal{BC}^{p**}$. We have just mentioned that the hypercohomology of the column $\mathcal{BC}^{p**}$ is canonically isomorphic to the cyclic homology of D_{X^*}. Therefore, the hypercohomology of each column $\mathcal{BC}^{p**}$ carries a periodicity operator S. We will also show in Theorem 2.10 that, in this formulation, the monodromy operator N coincides with the Connes periodicity operator S in each column $\mathcal{BC}^{p**}$.

Finally, in Theorem 2.11, we show that the filtration by columns $\mathcal{BC}^{p**}$ on the total complex $Tot(\mathcal{BC}^{***})$ yields a spectral sequence converging to the hypercohomology of $\mathcal{BC}^{***}$ whose E_1-terms are cyclic homologies of D_{X^*}. This is a counterpart to (2.21) above.

Consider the following triple complex of sheaves

$$\mathcal{BC}^{p,q,r} = \psi^{p+r,q-2r}(\mathbb{C}) = \mathbb{C}u^{[-p-r]} \otimes \Omega_X^{q+p-r}(\log Y) \qquad p,q,r \in \mathbb{Z}, p+r \leq 0$$

and $\mathcal{BC}^{p,q,r} = 0$ otherwise; along with differentials

$$d' : \mathcal{BC}^{p,q,r} \to \mathcal{BC}^{p+1,q,r} \quad d'(u^{[p]} \otimes \omega) = u^{[p-1]} \otimes \theta \wedge \omega \quad \text{(projected onto } \mathcal{BC}^{p+1,q,r})$$
$$d'' : \mathcal{BC}^{p,q,r} \to \mathcal{BC}^{p,q+1,r} \quad d''(u^{[p]} \otimes \omega) = u^{[p]} \otimes d\omega \quad \text{(projected onto } \mathcal{BC}^{p,q+1,r})$$

The third differential of the triple complex, i.e., the map from $\mathcal{BC}^{p,q,r} \to \mathcal{BC}^{p,q,r+1}$ is taken to be 0. This is supposed to reflect the fact that the cyclic homology of D_{X^*} splits as a direct sum of Hochschild homologies of D_{X^*}. The monodromy on the complex $\psi^{**}(\mathbb{C})$ extends to a monodromy operator on $\mathcal{BC}^{***}$ defined as

$$N : \mathcal{BC}^{p,q,r} \longrightarrow \mathcal{BC}^{p+1,q-1,r} \qquad N(u^{[t]} \otimes \omega) = u^{[t-1]} \otimes \omega \tag{2.22}$$

For any fixed $p \in \mathbb{Z}$, the bicomplex $\mathcal{BC}^{p,*,*}$ may be described explicitly as in Figure 2.

$$\begin{array}{lccccccc}
r{=}{-p}: & & & & & \overbrace{\mathbb{C}u^0\Omega^0_X(\log Y)}^{\mathcal{BC}^{p,-2p,-p}} & \xrightarrow{d''} \\
& & & & & \uparrow{\scriptstyle 0} & \\
r{=}{-p-1}: & & & \overbrace{\mathbb{C}u^1\Omega^0_X(\log Y)}^{\mathcal{BC}^{p,-2p-1,-p-1}} & \xrightarrow{d''} & \mathbb{C}u^1\Omega^1_X(\log Y) & \xrightarrow{d''} \\
& & & \uparrow{\scriptstyle 0} & & \uparrow{\scriptstyle 0} & \\
r{=}{-p-2}: & \overbrace{\mathbb{C}u^2\Omega^0_X(\log Y)}^{\mathcal{BC}^{p,-2p-2,-p-2}} & \xrightarrow{d''} & \mathbb{C}u^2\Omega^1_X(\log Y) & \xrightarrow{d''} & \mathbb{C}u^2\Omega^2_X(\log Y) & \xrightarrow{d''}
\end{array}$$

FIGURE 2 The bicomplex $\mathcal{BC}^{p,*,*}$.

While dealing with the Hochschild homology of D_{X^*} in Section 2.2, we filtered the complex $\psi^{p,q}(\mathbb{C})$ by its columns fixing the value of p. We will now deal with one "column" of the triple complex $\mathcal{BC}^{***}$ at a time, again by considering only the terms $\mathcal{BC}^{p,q,r}$ with a fixed value of p.

THEOREM 2.10. *Let D_{X^*} denote the sheaf of differential operators on X^* and let $j : X^* \longrightarrow X$ denote the open immersion. Then*

(a) For each fixed $p \in \mathbb{Z}$, the bicomplex $\mathcal{BC}^{p,*,*}$ computes the cyclic homology of D_{X^*}, i.e., there are canonical isomorphisms

$$HC_k(D_{X^*}) \simeq \mathbb{H}^{(2n+2)-k}(Tot(\mathcal{BC}^{p,*,*})) \qquad k \in \mathbb{N}, \quad n = dim(X^*) \tag{2.23}$$

where the total complex $Tot(\mathcal{BC}^{p,*,*})$ is described by

$$Tot(\mathcal{BC}^{p,*,*})^l = \bigoplus_{q+r=-p+l} \mathcal{BC}^{p,q,r} = \bigoplus_{q+r=-p+l} \mathbb{C}u^{[-p-r]} \otimes \Omega_X^{q+p-r}(\log Y) \tag{2.24}$$

(b) Using the isomorphism in (2.23), the monodromy N on the hypercohomology of the complex $Tot(\mathcal{BC}^{p,*,*})$ coincides with the periodicity operator S on the cyclic homology of D_{X^*}, i.e.

$$(2.25) \qquad N : \mathbb{H}^{(2n+2)-k}(Tot(\mathcal{BC}^{p,*,*})) \simeq HC_k(D_{X^*}) \xrightarrow{S}$$
$$HC_{k-2}(D_{X^*}) \simeq \mathbb{H}^{(2n+2)-k+2}(Tot(\mathcal{BC}^{p,*,*}))$$

PROOF. (a) The rows of $\mathcal{BC}^{p**}$ with d''-differentials are logarithmic de Rham complexes and hence their hypercohomologies are canonically isomorphic to de Rham cohomologies of X^*. We have the following quasi-isomorphism of complexes

$$(Tot(\mathcal{BC}^{p**})) = \left(\bigoplus_{l=0}^{\infty}(\Omega_X^*(\log Y)[2l]) \right) \xrightarrow{q.i.} \bigoplus_{l=0}^{\infty}(j_*\Omega_{X^*}^*[2l])$$

Further, from the proof of Theorem 2.6, it follows that:

$$Tot(CC(D_{X^*})) \xrightarrow{q.i.} \left(\bigoplus_{l=0}^{\infty}\mathbb{C}_{X^*}[2l] \right) [(2n+2)] \qquad (n+1 = dim(X^*))$$

Hence we have canonical isomorphisms

$$HC_k(D_{X^*}) \simeq \mathbb{H}^{(2n+2)-k}(\bigoplus_{l=0}^{\infty}\mathbb{C}_{X^*}[2l]) \simeq \mathbb{H}^{(2n+2)-k}(\bigoplus_{l=0}^{\infty}j_*\Omega_{X^*}^*[2l])$$
$$\simeq \mathbb{H}^{(2n+2)-k}(Tot(\mathcal{BC}^{p,*,*}))$$

(b) Since $N(u^{[p]}\otimes\omega) = u^{[p-1]}\otimes\omega$, it follows that N acts on $\mathcal{BC}^{p**}$ by dropping the terms in Figure 2 with u^0 coefficients, i.e. the terms $\mathcal{BC}^{p,*,-p}$ from the bicomplex $\mathcal{BC}^{p**}$. We also note that $N(\mathcal{BC}^{p,*,*}) \simeq \mathcal{BC}^{p,*,*}[0,1,1]$. Hence

$$\mathbb{H}^{(2n+2)-k}(N(\mathcal{BC}^{p,*,*})) = \mathbb{H}^{(2n+2)-k+2}(Tot(\mathcal{BC}^{p,*,*}))$$

and N coincides with the projection

$$HC_k(D_{X^*}) \simeq \bigoplus_{l=0}^{\infty} H_{DR}^{(2n+2)-k+l}(X^*) \simeq \mathbb{H}^{(2n+2)-k}(Tot(\mathcal{BC}^{p,*,*}))$$
$$\simeq \mathbb{H}^{(2n+2)-k}\left(\bigoplus_{l=0}^{\infty}(\mathbb{C}u^{[l]}\otimes\Omega_X^*(\log Y)[2l]) \right) \xrightarrow{N}$$
$$\mathbb{H}^{(2n+2)-k}\left(\bigoplus_{l=1}^{\infty}(\mathbb{C}u^{[l]}\otimes\Omega_X^*(\log Y)[2l]) \right)$$
$$\simeq \mathbb{H}^{(2n+2)-k}(N(\mathcal{BC}^{p,*,*})) \simeq \mathbb{H}^{(2n+2)-k+2}(Tot(\mathcal{BC}^{p,*,*}))$$
$$\simeq \bigoplus_{l=1}^{\infty} H_{DR}^{(2n+2)-k+l} \simeq HC_{k-2}(D_{X^*})$$

The operator S can be identified with the same projection, as we have shown in Corollary 2.7.

We also have the following result, which is a cyclic counterpart of (2.21).

THEOREM 2.11. *Let $\mathcal{BC}^*$ denote the total complex associated to the triple complex $\mathcal{BC}^{***}$. Let F'_k, $k \in \mathbb{Z}$ denote the filtration induced on the total complex $\mathcal{BC}^*$ by "columns" $\mathcal{BC}^{p**}$, i.e.*

$$F'_k \mathcal{BC}^* = \left(\bigoplus_{q+r=-p+*,p\geq -k} \mathcal{BC}^{p,q,r}, d' + d'' + d''' \right)$$

Then, there is a converging spectral sequence in hypercohomology

$$E_1^{-p,(2n+2)+p-q} := \mathbb{H}^{(2n+2)-q}(F'_p/F'_{p-1}) \Rightarrow \mathbb{H}^{(2n+2)-q}(\mathcal{BC}^*)$$

The E_1-terms are, moreover, canonically isomorphic to cyclic homologies of D_{X^*}, i.e.

$$HC_q(D_{X^*}) \simeq E_1^{-p,(2n+2)+p-q}$$

PROOF. For each fixed $p \in \mathbb{Z}$, the bicomplexes $\mathcal{BC}^{p**}$ form the graded pieces of a "filtration by columns" on the triple complex $\mathcal{BC}^{***}$. Hence the complexes $Tot(\mathcal{BC}^{p**})$ form the graded pieces of a filtration on the total complex $\mathcal{BC}^*$, i.e.

$$F'_p/F'_{p-1} = Tot(\mathcal{BC}^{p**})$$

Further, this filtration is bounded below (separately on each term of $\mathcal{BC}^*$) and exhaustive (see [**Wei2**, §**5.4.2**]). From Theorem 2.10, we know that

$$HC_k(D_{X^*}) \simeq \mathbb{H}^{(2n+2)-k}(Tot(\mathcal{BC}^{p,*,*})) \qquad k \in \mathbb{N}$$

From the spectral sequence associated to the filtration by columns $\mathcal{BC}^{p**}$ on the complex $Tot(\mathcal{BC}^{p**})$, $p \in \mathbb{Z}$, we get (see [**Wei2**, §**5.51**])

$$HC_q(D_{X^*}) \simeq \mathbb{H}^{(2n+2)-q}(Tot(\mathcal{BC}^{p,*,*})) \simeq \mathbb{H}^{(2n+2)-q}(F'_p/F'_{p-1}) = E_1^{-p,(2n+2)+p-q}$$
$$\Rightarrow \mathbb{H}^{(2n+2)-q}(\mathcal{BC}^*)$$

For each fixed p, the kernel of the map N

$$(2.26) \qquad\qquad N : \mathcal{BC}^{p**} \longrightarrow \mathcal{BC}^{p+1,**}$$

is the complex $\mathcal{BC}^{p,*,-p}$, which is a logarithmic de Rham complex. Hence, it is clear, following the same reasoning as the proof of Proposition 2.8 that

$$(2.27) \qquad HH_k(D_{X^*}) \simeq \mathbb{H}^{(2n+2)-k}(Ker(N)^*) \qquad k \in \mathbb{Z}$$

Finally, we have the following proposition.

PROPOSITION 2.12. *We have the following commutative diagram of long exact sequences in which the vertical maps are isomorphisms*

$$\mathbb{H}^{(2n+2)-k}(Ker(N)^*) \xrightarrow{I} \mathbb{H}^{(2n+2)-k}(Tot(\mathcal{BC}^{p**})) \xrightarrow{N} \mathbb{H}^{(2n+2)-k+2}(Tot(\mathcal{BC}^{p**})) \rightarrow$$
$$\simeq\downarrow \qquad\qquad\qquad\qquad \simeq\downarrow \qquad\qquad\qquad\qquad\qquad \simeq\downarrow$$
$$HH_k(D_{X^*}) \qquad\xrightarrow{I}\qquad HC_k(D_{X^*}) \qquad\xrightarrow{S}\qquad HC_{k-2}(D_{X^*}) \longrightarrow$$

PROOF. Let $j : X^* \longrightarrow X$ be the open immersion. We have the following diagrams of quasi-isomorphisms

$$0 \to Ker(N)^*[2n+2] \to Tot(\mathcal{BC}^{p**})[2n+2] \xrightarrow{N} N(\mathcal{BC}^{p**})[2n+2] \to 0$$

$$q.i. \downarrow \qquad\qquad q.i. \downarrow \qquad\qquad q.i. \downarrow$$

$$0 \to j_*\Omega^*_{X^*}[2n+2] \to \left(\bigoplus_{k=0}^{\infty} j_*\Omega^*_{X^*}[2k]\right)[2n+2] \to \left(\bigoplus_{k=1}^{\infty} j_*\Omega^*_{X^*}[2k]\right)[2n+2] \to 0$$

$$0 \to \Omega^*_{X^*}[2n+2] \to \left(\bigoplus_{k=0}^{\infty} \Omega^*_{X^*}[2k]\right)[2n+2] \to \left(\bigoplus_{k=1}^{\infty} \Omega^*_{X^*}[2k]\right)[2n+2] \to 0$$

$$q.i. \uparrow \qquad\qquad q.i. \uparrow \qquad\qquad q.i. \uparrow$$

$$0 \to \mathbb{C}_{X^*}[2n+2] \to \left(\bigoplus_{k=0}^{\infty} \mathbb{C}_{X^*}[2k]\right)[2n+2] \to \left(\bigoplus_{k=1}^{\infty} \mathbb{C}_{X^*}[2k]\right)[2n+2] \to 0$$

$$q.i. \uparrow \qquad\qquad q.i. \uparrow \qquad\qquad q.i. \uparrow$$

$$0 \to C^h(D_{X^*}) \to \qquad Tot(CC(D_{X^*})) \qquad \to \qquad Tot(CC(D_{X^*})[2]) \to 0$$

As in the proof of Proposition 2.8, we know that there are canonical isomorphisms $\mathbb{H}^l(\Omega^*_{X^*}) \simeq \mathbb{H}^l(j_*\Omega^*_{X^*})$, $\forall\ l \in \mathbb{Z}$. Using (2.23) and (2.27), the result now follows immediately by taking hypercohomologies and from the observation that $N(\mathcal{BC}^{p**}) \simeq \mathcal{BC}^{p**}[0,1,1]$.

3. Consani's Complex and the Cyclic Cohomology of the Ring of Differential Operators

The formalism of nearby cycles associated to an algebraic degeneration over a disc can be extended to the case of the reduced and irreducible fiber at infinity of an arithmetic variety. More precisely, there exist archimedean analogues for the direct summands of the E_1-terms of the spectral sequence associated to the L-filtration on Steenbrink's complex A^{**}. From (2.10), we recall that the L-filtration on Steenbrink's complex A^{**} is given by

$$L_r A^{pq} = \mathbb{C}u_{[p+1]} \otimes_{\mathbb{C}} W_{r+2p+1}\Omega^{p+q+1}_X(\log Y)/W_p\Omega^{p+q+1}_X(\log Y)$$

We define, in the notation of (2.9), the following terms

$$(3.1) \qquad K_S^{i,j,k} = \begin{cases} H^{i+j-2k+n}(\tilde{Y}^{(2k-i+1)}, \mathbb{C}) & \text{if } k \geq 0, i \\ 0 & \text{otherwise} \end{cases}$$

(recall that $n + 1 = dim(X)$) and note that the E_1-terms of the spectral sequence associated to the L-filtration can be made explicit in terms of the objects $K_S^{i,j,k}$ as follows (see [**GN**, §**2.6**])

$$(3.2)\ \ E_1^{-r,q+r} := \mathbb{H}^q(Y, Gr_r^W A^*) = \bigoplus_{k\geq 0, -r} H^{q-r-2k}(\tilde{Y}^{(r+2k+1)}, \mathbb{C}) = \bigoplus_k K_S^{r,q-n,k}$$

where the middle equality follows from the result (2.9) of Poincaré residue theorem. If we let $K_S^{i,j} = \bigoplus_k K_S^{i,j,k}$, we can form a bicomplex (K^{**}, d', d'') whose differentials d' and d'' are induced by the differentials d' and d'' on A^{**}. The bicomplex

(K_S^{**}, d', d'') also carries a monodromy operator $N : K_S^{i,j,k} \longrightarrow K_S^{i+2,j,k+1}$ that commutes with the differentials d' and d''. The modules $K_S^{i,j}$ have a number of key properties, the most interesting being that for any $i, j, \in \mathbb{Z}$, $K^{i,j}$ carries a pure Hodge structure of weight $j - i + n$. Further, it may be shown that

$$N : K_S^{i,j} \longrightarrow K_S^{i+2,j}(-1)$$

is a morphism of pure Hodge structures of weight $j - i + n$, where $K_S^{i+2,j}(-1)$ denotes the module $K_S^{i+2,j}$ (with Hodge structure of weight $j - i - 2 + n$) "twisted" by tensoring with the "Tate object", i.e., the module $(2\pi i)^{-1}\mathbb{Z}$ (see, for instance, [St2, §11.4]) which carries a pure Hodge structure of weight 2 concentrated in bidegree $(1, 1)$. Hence $K_S^{i+2,j}(-1)$ carries a pure Hodge structure of weight $j - i + n$. For each $i \in \mathbb{Z}$, there is the following isomorphism of Hodge structures

$$N^i : K^{-i,j} \longrightarrow K^{i,j}(-i)$$

which is an analogue of hard Lefschetz theorem. We refer the reader to [GN, §2.6-2.7] for further details.

The archimedean analogue of the above construction was carried out by Consani who, in her PhD thesis (see [Con2], [Con3]), introduced a complex $K^{i,j,k}$, again with monodromy N, that is the "archimedean" analogue of the Steenbrink complex at archimedean infinity. In this new setup, the complex (or real) points of an arithmetic variety over a number field play the role of the nearby fiber over infinity. The complex points of the nearby fiber at infinity form a smooth complex manifold $X(\mathbb{C})$ of dimension, say, m, over $\mathbb{C}$. The manifold $X(\mathbb{C})$ may also be extended from $\mathbb{R}$, i.e., there exists a real manifold $X(\mathbb{R})$ of dimension m such that $X(\mathbb{C}) = X(\mathbb{R}) \otimes_{\mathbb{R}} \mathbb{C}$. The complex carries differentials d' and d'' and we set $K^{i,j} = \bigoplus_k K^{i,j,k}$ to define the bicomplex (K^{**}, d', d'').

The objects $K^{i,j,k}$ defined by Consani are modules of real differentials twisted by powers of $(2\pi i)$, with the "monodromy" N acting by twisting the differential forms by -1, i.e., multiplying by $(2\pi i)^{-1}$. For any given $t \in \mathbb{Z}$, we consider the terms $K^{i,j,k}$ twisted by $(2\pi i)^t$ and these terms form a bicomplex (K_t^{**}, d', d''). Then, in Proposition 3.8, we show that there are natural surjections from the "real Hochschild cohomology" (see Definition 3.4) of the ring of C^{∞}-differential operators $D(X)$ on X to the cohomology of the graded pieces of the filtration defined on $Tot(K_t^{**})$ by $Im(N^j)$, $j \geq 0$. We also note that the filtration by $Im(N^j)$ coincides with the filtration by rows. Thereafter, we construct a triple complex $\mathcal{K}^{***}$ such that there are natural surjections from the "real cyclic cohomology" (see Definition 3.4) of $D(X)$ to the cohomology of the "rows" $\mathcal{K}^{*q*}$ (for fixed $q \in \mathbb{Z}$) of the triple complex $\mathcal{K}^{***}$. The complex $\mathcal{K}^{***}$ is, in fact, a deck of complexes K_t^{**} for different values of t and hence carries an induced operator N. Finally, Proposition 3.10 shows that, in this formalism, the induced monodromy operator N on the rows $\mathcal{K}^{*q*}$ of $\mathcal{K}^{***}$ corresponds to the periodicity operator S on the real cyclic cohomology of $D(X)$.

We notice, therefore, that the results in this section are essentially the "dual" of the results of Section 2, with the kernel filtration $Ker(N^j)$, $j \geq 0$ replaced by the image filtration $Im(N^j)$, $j \geq 0$ and the Hochschild (resp. cyclic) homology of the differential operators replaced by the Hochschild (resp. cyclic) cohomology.

We understand this "interchange" of roles between the framework of cyclic homology and that of cyclic cohomology between the two sections as follows. The

$K^{i,j,k}$'s appearing in Consani's complex are an analogue of the direct summands of the E_1-terms for the L-filtration on Steenbrink's complex A^*, which appears as the target of the quasi-isomorphism $\mu : \psi^*(\mathbb{C}) \longrightarrow A^*$ in Proposition 2.2. Then the interchange between the kernel of the monodromy N appearing on $\psi^*(\mathbb{C})$ and the cokernel of the monodromy N appearing on A^*, which was already made explicit in Theorem 2.3, will be understood as an interchange between cyclic homology and cyclic cohomology. In the archimedean case, we are working with the $K^{i,j,k}$'s, which are analogous to the $K_S^{i,j,k}$'s defined in (3.1) that come out of a filtration on Steenbrink's complex. This suggests that the archimedean analogue of the results of Section 2 should be carried out by using cyclic cohomology and the Steenbrink complex A^{**}, instead of cyclic homology and the complex $\psi^{**}(\mathbb{C})$.

Finally, in Section 3.4, we will show that it is possible to "recover" an archimedean version of the original theory of $\psi^{**}(\mathbb{C})$ using the objects $K^{i,j,k}$. More precisely, we will assemble the terms $K^{i,j,k}$ to construct a complex φ^{**} which serves as the analogue of $\psi^{**}(\mathbb{C})$ at archimedean infinity. Correspondingly, we will also assemble the objects $K^{i,j,k}$ to form a complex B^{**} which replaces the Steenbrink complex A^{**} of Proposition 2.2. Thereafter, we define a morphism $\mu : \varphi^* \to B^*$ of complexes. Further, both the complexes φ^{**} and B^{**} are equipped with an endomorphism N, and, as in Proposition 2.2, $\mu \circ N$ is chain homotopic to the operator $N \circ \mu$.

3.1. Consani's Complex and the Ring of Differential Operators

Let X be a smooth compact manifold of dimension m over $\mathbb{C}$ or $\mathbb{R}$. If X is defined over $\mathbb{R}$, then we let $X(\mathbb{C})$ denote the complex manifold $X(\mathbb{C}) = X \otimes_{\mathbb{R}} \mathbb{C}$. Then in either case, $dim_{\mathbb{C}}(X(\mathbb{C})) = m$ and, by abuse of notation, we will continue to refer to $X(\mathbb{C})$ simply as X, whenever X is defined over $\mathbb{R}$. For any $a, b \in \mathbb{Z}$, we let $\Omega_{X,\mathbb{R}}^{a,b} := (\Omega_X^{a,b} + \Omega_X^{b,a})_{\mathbb{R}}$ denote the submodule of $(\Omega_X^{a,b} + \Omega_X^{b,a})$ consisting of real differentials, i.e., those fixed under conjugation. Then following [**Con2**, §4], we define the following. For all $i, j, k \in \mathbb{Z}$, let

$$
K^{i,j,k} = \begin{cases} \displaystyle\bigoplus_{a \leq b, a+b=j+m, |a-b| \leq 2k-i} \Omega_{X,\mathbb{R}}^{a,b}\left(\frac{m+j-i}{2}\right) & \text{if } k \geq max\{0, i\} \\ 0 & \text{otherwise} \end{cases}
$$

where the term $\Omega_{X,\mathbb{R}}^{a,b}(r)$ refers to the module of real differentials $\Omega_{X,\mathbb{R}}^{a,b}$ multiplied by a factor of $(2\pi i)^r$. We refer to the number r in $\Omega_{X,\mathbb{R}}^{a,b}(r)$ as the (Tate) twist of the real differential forms.

We will now define the maps d', d'' and N as follows. Denote by ∂ and $\overline{\partial}$ the usual partial differential operators on $\Omega_{X,\mathbb{R}}^{a,b}$. We define

(3.3)
$$
\begin{aligned}
d' &: K^{i,j,k} \to K^{i+1,j+1,k+1} & d' &= \partial + \overline{\partial} \\
d'' &: K^{i,j,k} \to K^{i+1,j+1,k} & d'' &= i(\partial - \overline{\partial}) \\
N &: K^{i,j,k} \to K^{i+2,j,k+1} & N(a) &= (2\pi i)^{-1} a
\end{aligned}
$$

where the differential d'' is considered as composed with the projection onto its image $K^{i+1,j+1,k}$. We will also maintain

$$
d = d' + d''
$$

For $i, j \in \mathbb{Z}$, write $K^{i,j} = \bigoplus_{k \in \mathbb{Z}} K^{i,j,k}$. We will consider the bicomplex $(K^{\cdot \cdot}, d', d'')$ and the associated total complex $K^* = \bigoplus_{i+j=*} K^{i,j}$. It is easy to check that $[d', N] = [d'', N] = 0$.

Given X as above, we can consider X as a smooth compact $\mathbb{C}$-manifold X^{an} of complex dimension m (and real dimension $n = 2m$), which we shall continue to denote by X. Let $D(X)$ denote the ring of C^∞-differential operators on X. We have the following result of Wodzicki [**W**], which is the counterpart of Proposition 2.5.

PROPOSITION 3.1. *If X is a compact complex manifold of real dimension n, the cyclic homology of the ring $D(X)$ of C^∞-differential operators on X decomposes canonically as*

$$HC_q(D(X)) \simeq H_{dR}^{2n-q}(X) \bigoplus H_{dR}^{2n-q+2}(X) \bigoplus H_{dR}^{2n-q+4}(X) \bigoplus \cdots$$

The Hochschild homology of $D(X)$ is given by

$$HH_q(D(X)) \simeq H_{dR}^{2n-q}(X)$$

Moreover, the isomorphisms are functorial with respect to embeddings of codimension zero.

PROOF. See [**W, Theorem 3**].

Now, suppose that $(C^\cdot, d_1)$ and $(D^\cdot, d_2)$ are two complexes of $\mathbb{C}$-modules and let $f : C^\cdot \to D^\cdot$ be a morphism of complexes (of degree 0). We assume that each of the modules $C^\cdot$ and $D^\cdot$ is provided with a natural conjugation (denoted $x \mapsto x^c$ for x belonging to any C^n or D^m) compatible with complex conjugation, i.e., $(\alpha x)^c = \bar{\alpha} x^c$ for any $\alpha \in \mathbb{C}$ and x belonging to any C^m or D^n and let $C_{\mathbb{R}}^\cdot$ and $D_{\mathbb{R}}^\cdot$ denote the $\mathbb{R}$-submodule of $C^\cdot$ and $D^\cdot$ resp. elements fixed under this conjugation.

Assume that the differentials d_i, $i = 1, 2$ on the complexes $C^\cdot$ and $D^\cdot$ respectively, commute with the conjugation and so do the maps f^j, $j \in \mathbb{Z}$ of the morphism $f : C^\cdot \to D^\cdot$. This means we have the equalities

$$d_i(x^c) = d_i(x)^c \qquad f^j(x^c) = f^j(x)^c$$

Then we have two new complexes $(C_{\mathbb{R}}^\cdot, d_1)$ and $(D_{\mathbb{R}}^\cdot, d_2)$ and an induced morphism connecting them, which we also denote by $f : (C_{\mathbb{R}}^\cdot, d_1) \to (D_{\mathbb{R}}^\cdot, d_2)$. Finally, we set $H^*(C_{\mathbb{R}}^\cdot) = H^*(C_{\mathbb{R}}^\cdot, d_1)$ and similarly for $(D_{\mathbb{R}}^\cdot, d_2)$. It is also clear that if the $f_* : H^*(C^\cdot) \longrightarrow H^*(D^\cdot)$ are isomorphisms, so are $f_* : H^*(C^\cdot)_{\mathbb{R}} \longrightarrow H^*(D^\cdot)_{\mathbb{R}}$.

LEMMA 3.2. *Let $(C^\cdot, d_1)$ be a complex of $\mathbb{C}$-modules where each module C^m, $m \in \mathbb{Z}$ is equipped with a conjugation $x \mapsto x^c$ compatible with the complex conjugation and with the differential d_1, (i.e., $d_1(x^c) = (d_1(x))^c$). Then the homology groups $H^*(C^\cdot)$ carry a conjugation compatible with complex conjugation. Let $H^*(C^\cdot)_{\mathbb{R}}$ denote the submodule of $H^*(C^\cdot)$ invariant under this conjugation. Then*

$$H^*(C_{\mathbb{R}}^\cdot) \approx H^*(C^\cdot)_{\mathbb{R}}.$$

PROOF. Let $d_1^m : C^m \to C^{m+1}$, $(m \in \mathbb{Z})$ be the differential. If $z \in C^m$ is such that $d_1^m(z) = 0$, we define the conjugate of its class in homology by

$$(z + Im(d_1^{m-1}))^c = z^c + Im(d_1^{m-1})$$

This class is well defined since d_1^{m-1} is compatible with the conjugation $z \mapsto z^c$. Let $w \in C_{\mathbb{R}}^m$ (hence $w = w^c$) be such that $d_1^m(w) = 0$. Then define the homomorphism

$$\phi_m : H^m(C_{\dot{\mathbb{R}}}) \longrightarrow H^m(C^{\cdot})_{\mathbb{R}} \qquad \phi_m(w + Im(d_1^{m-1})) = w + Im(d_1^{m-1}) \in H^m(C^{\cdot})_{\mathbb{R}}$$

Suppose that $\phi_m(w + Im(d_1^{m-1})) = 0$. Then $w = d_1^{m-1}(x)$ for some $x \in C^{m-1}$. But then $w = w^c = d_1^{m-1}(x^c)$ and hence $w = d_1^{m-1}((x + x^c)/2)$. However, since $(x + x^c)/2 \in C_{\mathbb{R}}^{m-1}$ we obtain $w + Im(d_1^{m-1}) = 0$ in $H^m(C_{\dot{\mathbb{R}}})$. This proves that ϕ_m is one-one.

We now choose $z + Im(d_1^{m-1}) \in H^m(C^{\cdot})_{\mathbb{R}}$ such that $(z + Im(d_1^{m-1}))^c = z + Im(d_1^{m-1})$. Then there exists $x \in C^{m-1}$ such that $z - z^c = d_1^{m-1}(x)$. Consider $w = z - d_1^{m-1}(x/2) = z - (z - z^c)/2 = (z + z^c)/2$. Hence $w^c = w$ and it is clear that $\phi_m(w + Im(d_1^{m-1})) = z + Im(d_1^{m-1})$. This proves that ϕ_m is also onto and hence an isomorphism.

Hence if $f : (C^{\cdot}, d_1) \longrightarrow (D^{\cdot}, d_2)$ induces a quasi-isomorphism, we have

$$H^*(C_{\dot{\mathbb{R}}}) \simeq H^*(C^{\cdot})_{\mathbb{R}} \xrightarrow[\simeq]{f_*} H^*(D^{\cdot})_{\mathbb{R}} \simeq H^*(D_{\dot{\mathbb{R}}})$$

i.e., an induced quasi-isomorphism $f_{\mathbb{R}} : (C_{\dot{\mathbb{R}}}, d_1) \longrightarrow (D_{\dot{\mathbb{R}}}, d_2)$.

LEMMA 3.3. *Let $(E^{\cdot}, d_1)$ be an exact sequence of $\mathbb{C}$-modules where each module E^m, $m \in \mathbb{Z}$ is equipped with a conjugation $x \mapsto x^c$ compatible with the complex conjugation and with the differential d_1, i.e., $d_1(x^c) = (d_1(x))^c$. Then the sequence $(E_{\dot{\mathbb{R}}}, d_1)$, where d_1 denotes the induced differential, is also exact.*

PROOF. This follows immediately from the statement above and from the fact that a complex is exact if and only if it is quasi-isomorphic to the zero complex.

DEFINITION 3.4. The complexes defining the cyclic homology $HC_*(D(X))$ (resp. cyclic cohomology $HC^*(D(X))$) of the ring $D(X)$ of C^{∞}-differential operators on X and the de Rham cohomology of X carry natural conjugations. We shall denote the homologies of their subcomplexes fixed under conjugation by $HC_*(D(X), \mathbb{R})$ (resp. by $HC^*(D(X), \mathbb{R})$) and $H_{dR}^*(X, \mathbb{R})$ respectively. We shall refer to $HC_*(D(X), \mathbb{R})$ (resp. $HC^*(D(X), \mathbb{R})$) as the real cyclic homology of (resp. real cyclic cohomology) of $D(X)$, while we shall refer to $H_{dR}^*(X, \mathbb{R})$ as the real de Rham cohomology.

Similarly, we will refer to the homology of the subcomplex of the complex defining the Hochschild homology (resp. Hochschild cohomology) of $D(X)$ fixed under conjugation as the real Hochschild homology (resp. real Hochschild cohomology) of $D(X)$ and denote it by $HH_*(D(X), \mathbb{R})$ (resp. by $HH^*(D(X), \mathbb{R})$).

We note moreover, that, if M is a $\mathbb{C}$-module equipped with a complex conjugation $m \mapsto m^c$ for all $m \in M$, compatible with complex conjugation, there is a conjugation $f \mapsto f^c$ for $f \in Hom(M, \mathbb{C})$ which is

$$f^c(m) = \overline{f(m^c)} \qquad m \in M, f \in Hom(M, \mathbb{C})$$

In particular, we will apply this to the modules $Hom(HC_q(D(X), \mathbb{C}))$.

PROPOSITION 3.5. *If the complex manifold X has real dimension n, we have canonical isomorphisms of $\mathbb{R}$-vector spaces*

$$HC_q(D(X), \mathbb{R}) \simeq H_{dR}^{2n-q}(X, \mathbb{R}) \bigoplus H_{dR}^{2n-q+2}(X, \mathbb{R}) \bigoplus H_{dR}^{2n-q+4}(X, \mathbb{R}) \bigoplus \cdots$$

$$HH_q(D(X), \mathbb{R}) \simeq H_{dR}^{2n-q}(X, \mathbb{R})$$

PROOF. From Proposition 3.1, we have canonical isomorphisms

$$HH_q(D(X)) \simeq H_{dR}^{2n-q}(X)$$

Further, these isomorphisms are compatible with embeddings of codimension 0, which, in particular, implies that they are compatible with conjugation. From Lemma 3.2, it follows that the operation of taking homologies commutes with conjugation and hence we have

$$HH_q(D(X), \mathbb{R}) \simeq H_{dR}^{2n-q}(X, \mathbb{R})$$

This proves the second isomorphism in the proposition. Again, for cyclic homology, from Proposition 3.1, we have the canonical isomorphism

$$HC_q(D(X)) \simeq H_{dR}^{2n-q}(X) \bigoplus H_{dR}^{2n-q+2}(X) \bigoplus H_{dR}^{2n-q+4}(X) \bigoplus \cdots$$

compatible with embeddings of codimension 0. The bicomplex $CC(D(X))$ that is used to compute cyclic homology also carries a natural conjugation and by similar reasoning, we have the isomorphism

$$HC_q(D(X), \mathbb{R}) \simeq H_{dR}^{2n-q}(X, \mathbb{R}) \bigoplus H_{dR}^{2n-q+2}(X, \mathbb{R}) \bigoplus H_{dR}^{2n-q+4}(X, \mathbb{R}) \bigoplus \cdots$$

We now consider the cohomological viewpoint. By definition, the Hochschild (resp. cyclic) cohomology of the ring $D(X)$ is the cohomology of the complex formed by taking the linear duals of the terms in the Hochschild (resp. cyclic) complex of the ring $D(X)$. Since all the terms involved are finite dimensional vector spaces over the field $\mathbb{C}$, we have isomorphisms

$$HC^q(D(X)) \xrightarrow{\simeq} Hom(HC_q(D(X)), \mathbb{C})$$

On the other hand, the Poincaré duality isomorphism for the compact manifold X gives us:

$$Hom(H_{dR}^{n-q}(X), \mathbb{C}) \xrightarrow{\simeq} H_{dR}^q(X) \qquad \forall\, q \geq 0$$

PROPOSITION 3.6. *(a) There is a commutative diagram of long exact sequences*

$$
\begin{array}{ccccc}
HC^{q+n-2}(D(X)) & \xrightarrow{\ S\ } & HC^{q+n}(D(X)) & \xrightarrow{\ I\ } & HH^{q+n}(D(X)) \to \\
\simeq \downarrow & & \simeq \downarrow & & \simeq \downarrow \\
Hom(HC_{q+n-2}(D(X)), \mathbb{C}) & \xrightarrow{\ S\ } & Hom(HC_{q+n}(D(X)), \mathbb{C}) & \longrightarrow & Hom(HH_{q+n}(D(X)), \mathbb{C}) \to \\
\simeq \downarrow & & \simeq \downarrow & & \simeq \downarrow \\
H_{dR}^{q-2}(X) \oplus H_{dR}^{q-4}(X) \oplus \cdots & \xrightarrow{\ S\ } & H_{dR}^q(X) \oplus H_{dR}^{q-2}(X) \oplus \cdots & \longrightarrow & H_{dR}^q(X) \to
\end{array}
$$

(b) By conjugating and taking invariants, the vertical maps in the diagram above give us isomorphisms

$$HC^{q+n}(D(X), \mathbb{R}) \xrightarrow{\simeq} H^q_{dR}(X, \mathbb{R}) \oplus H^{q-2}_{dR}(X, \mathbb{R}) \oplus \cdots$$

$$HH^{q+n}(D(X), \mathbb{R}) \xrightarrow{\simeq} H^q_{dR}(X, \mathbb{R})$$

Further, we have a long exact periodicity sequence

$$\cdots \to HC^{q+n-2}(D(X), \mathbb{R}) \xrightarrow{S} HC^{q+n}(D(X), \mathbb{R}) \xrightarrow{I} HH^{q+n}(D(X), \mathbb{R}) \to \cdots$$

PROOF. (a) Since the pairing of HC^q and HC_q extends the pairing on Hochschild homology, the natural isomorphisms induced by the pairings:

$$HC^{q+n}(D(X)) \xrightarrow{\simeq} Hom(HC_{q+n}(D(X)), \mathbb{C})$$

and

$$HH^{q+n}(D(X)) \xrightarrow{\simeq} Hom(HH_{q+n}(D(X)), \mathbb{C})$$

commute with the maps in the periodicity sequence. The isomorphisms

$$Hom(HH_{q+n}(D(X)), \mathbb{C}) \xrightarrow{\simeq} Hom(H^{n-q}_{dR}(X), \mathbb{C}) \xrightarrow{\simeq} H^q_{dR}(X)$$

$$Hom(HC_{q+n}(D(X)), \mathbb{C}) \xrightarrow{\simeq} \bigoplus_{l=0}^{\infty} Hom(H^{n-q+2l}_{dR}, \mathbb{C}) \xrightarrow{\simeq} \bigoplus_{l=0}^{\infty} H^{q-2l}_{dR}(X)$$

follow from Poincaré duality.
(b) We consider the sequence of isomorphisms:

$$HH^q(D(X)) \xrightarrow{\simeq} Hom(HH_q(D(X)), \mathbb{C}) \xrightarrow{\simeq} H^q_{dR}(X)$$

All three homology modules $HH^{q+n}(D(X))$, $Hom(HH_{q+n}(D(X)), \mathbb{C})$ and $H^q_{dR}(X)$ carry a natural conjugation and we have the sequence of induced isomorphisms between the invariants

$$HH^{q+n}(D(X))^{c=id} \xrightarrow{\simeq} Hom(HH_{q+n}(D(X)), \mathbb{C})^{c=id} \xrightarrow{\simeq} H^q_{dR}(X)^{c=id}$$

By applying Lemma 3.2, it follows that $HH^q(D(X))^{c=id} \simeq HH^q(D(X), \mathbb{R})$ and $H^q_{dR}(X)^{c=id} \simeq H^q_{dR}(X, \mathbb{R})$. Therefore, we have isomorphisms

$$HH^{q+n}(D(X), \mathbb{R}) \xrightarrow{\simeq} H^q_{dR}(X, \mathbb{R})$$

and the same argument shows that $HC^{q+n}(D(X), \mathbb{R}) \xrightarrow{\simeq} \oplus_{l=0}^{\infty} H^{q-2l}_{dR}(X, \mathbb{R})$. Finally, the fact that the complex

$$\to HC^{q+n-2}(D(X), \mathbb{R}) \xrightarrow{S} HC^{q+n}(D(X), \mathbb{R}) \xrightarrow{I} HH^{q+n}(D(X), \mathbb{R}) \to$$

is exact follows from Lemma 3.3.

3.2. The Image Filtration and Hochschild Cohomology

In this section, we show how the graded pieces of the decreasing filtration by $Im(N^l)$, $l \geq 0$, N being the monodromy, are connected with Hochschild cohomology. This is intuitively dual to using the kernel filtration, the graded pieces of which were used in the last section to describe Hochschild homology. The motivation for this comes from Theorem 2.3 where we see the "switch" from kernel to cokernel, the cokernel being the first graded piece of the image filtration. This "duality" is also the motivation behind using Hochscild (and cyclic) cohomologies in Section 3, instead of the Hochschild (and cyclic) homologies used in Section 2. From Section 3.1, we recall that

$$(3.4) \quad K^{i,j,k} = \begin{cases} \displaystyle\bigoplus_{a \leq b, a+b=j+n/2, |a-b| \leq 2k-i} \Omega^{a,b}_{X,\mathbb{R}}\left(\frac{n/2+j-i}{2}\right) & \text{if } k \geq max\{0,i\} \\ 0 & \text{otherwise} \end{cases}$$

$n = 2m$ being the real dimension of X, which we now see as a compact C^∞-manifold. We start with the following lemma

LEMMA 3.7. *For any given $t \in \mathbb{Z}$ and $k \geq 0$ and $j \in \mathbb{Z}$, the projection maps $p^t_{j,k} : \Omega^{j+n/2}_{\mathbb{R}} \to K^{j-t',j,k}$ induce a morphism of complexes (where $t' = 2t - n/2$),*

$$(3.5) \quad p^t_{*,k} : (\Omega^{*+n/2}_{\mathbb{R}}, -d^c) \to (K^{*-t',*,k}, d'')$$

PROOF. First, we suppose that $j - t' < k$. In order to check that the projections $p^t_{j,k}$ induce a map of complexes, it suffices to show that, for $x \neq 0$, $x \in \Omega^{j+n/2}_{\mathbb{R}}$, if $p^t_{j,k}(x) = 0$, then $p^t_{j+1,k}(d^c(x)) = 0$. For sake of convenience, we can assume that x lies in some $(\Omega^{a,b} + \Omega^{b,a})_{\mathbb{R}}$ where $a + b = j + n/2$. From (3.4) it follows that if $p^t_{j,k}(x) = 0$, we must have $|a - b| > 2k - i$, where $i = j - t' = j - 2t + n/2$. On the other hand, if $p^t_{j+1,k}(d^c(x)) \neq 0$, it must follow that $|a - b| - 1 \leq 2k - i - 1$, which is a contradiction.

Finally, we note that if $j - t' \geq k$, then the target $K^{j+1-t',j+1,k}$ of $p^t_{j+1,k}$ is zero by definition and there is nothing to prove.

We now consider all terms $K^{i,j,k}$ "having a fixed level," i.e., for each fixed value of the Tate twist $t \in \mathbb{Z}$, we define the following bicomplex (K^{**}_t, d', d'') as

$$(3.6) \quad K^{p,q}_t = \mathbb{R}u^{-q-1} \otimes K^{p+q,p+q+t',q} \qquad p \leq 0, q \geq 0$$

where $t' = 2t - n/2$ and $K^{p,q}_t = 0$ otherwise. We introduce the differentials

$$d' : K^{p,q}_t \to K^{p,q+1}_t \qquad d'(u_{[q+1]} \otimes (2\pi i)^t \omega) = u_{[q+2]} \otimes (2\pi i)^t (\partial + \bar{\partial})\omega$$
$$\text{(if } K^{p,q+1}_t \neq 0)$$

$$d'' : K^{p,q}_t \to K^{p+1,q}_t \qquad d''(u_{[q+1]} \otimes (2\pi i)^t \omega) = u_{[q+1]} \otimes (2\pi i)^t i(\partial - \bar{\partial})\omega$$
$$\text{(if } K^{p+1,q}_t \neq 0)$$

The monodromy N acts between bicomplexes (K^{**}_t, d', d'') at different levels as follows

$$(3.7) \quad N : K^{pq}_t \longrightarrow K^{p+1,q+1}_{t-1} \qquad N(u_{[q+1]} \otimes (2\pi i)^t \omega) = u_{[q+2]} \otimes (2\pi i)^{t-1} \omega$$

In fact, we shall also consider powers of N; for any $l \geq 0$, we have maps

$$(3.8) \qquad N^{l+1} : K_{t+l+1}^{**} \to K_t^{*+l+1,*+l+1}$$

$$N^l : K_{t+l}^{**} \to K_t^{*+l,*+l}$$

Here it is understood that N^0 refers to the identity map.

We consider the decreasing filtration $\{G^l\}_{l \geq 0}$ on $Tot(K_t^{**})$ defined by the sub-complexes $G^l(Tot(K_t^{**})) = Im(N^l)^*$ for each $l \geq 0$ and also the graded pieces $gr_G^l(Tot(K_t^{**})) := G^l(Tot(K_t^{**}))/G^{l+1}(Tot(K_t^{**}))$ of this filtration.

PROPOSITION 3.8. *For each fixed $l \in \mathbb{Z}_{\geq 0}$ and for each fixed value $t \in \mathbb{Z}$ of the Tate twist, there are natural surjections from the real Hochschild cohomology of $D(X)$ to the cohomology of the quotient complexes $(Im(N^l)/Im(N^{l+1}))^*$, i.e.,*

$$HH^{q+n}(D(X), \mathbb{R}) \twoheadrightarrow H^{q-2t}(gr_G^l(Tot(K_t^{**}))) := H^{q-2t}((Im(N^l)/Im(N^{l+1}))^*)$$

PROOF. We let $t' = 2t - n/2$. Then, by definition, we have the map N

$$N : K_{t+1}^{pq} = \mathbb{R}u^{-q-1} \otimes K^{p+q,p+q+t'+2,q} \to \mathbb{R}u^{-q-2} \otimes K^{p+q+2,p+q+t'+2,q+1} = K_t^{p+1,q+1}$$

$$N(u_{[q+1]} \otimes (2\pi i)^t \omega) = u_{[q+2]} \otimes (2\pi i)^{t-1} \omega$$

If $K^{p+q+2,p+q+t'+2,q+1} \neq 0$, then $p+q+2 \leq q+1$ and hence $p+q < q$. Thus, $K^{p+q,p+q+(t'+2),q} \neq 0$ is mapped by N onto $K^{p+q+2,p+q+t'+2,q+1}$, unless $q+1 = 0$. It follows that the cokernel of N consists solely of the terms $K_t^{p,0} = \mathbb{R}u^{-1} \otimes K^{p,p+t',0}$ whereas the image of N consists of the terms K_t^{pq} with $q > 0$.

From these considerations, we deduce that the graded piece $Im(N^l)/Im(N^{l+1})$, $(l \geq 0)$ of the filtration by $Im(N^l)$, $l \geq 0$ on $Tot(K_t^{**})$ consists of the terms coming from $K_t^{*,l}$. By applying Lemma 3.7, we obtain the surjections

$$(3.9) \quad p_{*-n/2,l}^t : (\Omega_{\mathbb{R}}^*, -d^c) \twoheadrightarrow (K^{*-2t,*-n/2,l}, d'') \simeq ((Im(N^l)/Im(N^{l+1}))^{*-2t}, d'')$$

From the definitions in (3.4), it is clear that the surjections in (3.9) are actually projections onto direct summands and therefore induce surjections in cohomology. By composing with the isomorphisms of Proposition 3.6, we get a surjection of cohomologies (for each $q \in \mathbb{Z}$)

$$(3.10) \qquad HH^{q+n}(D(X), \mathbb{R}) \xrightarrow{\cong} H_{dR}^q(X, \mathbb{R}) \xrightarrow{p_{q-n/2,l}^t}$$

$$H^{q-2t}((Im(N^l)/Im(N^{l+1}))^*) = H^{q-2t}(gr_G^l(Tot(K_t^{**})))$$

3.3. The Connes Periodicity Operator and the Monodromy Operator

In Proposition 3.8 (for each fixed $t \in \mathbb{Z}$) we have shown that there exist natural surjections from the real Hochschild cohomology of the ring $D(X)$ to the cohomology of the graded pieces $gr_G^l(Tot(K_t^{**})) := Im(N^l)/Im(N^{l+1})$ of the filtration $G^l(Tot(K_t^{**})) := Im(N^l)$, $l \geq 0$ on the total complex of K_t^{**}. From the proof of Proposition 3.8 (also see Figure 3), we also note that the decreasing filtration on the complex $Tot(K_t^{**})$ by rows of K_t^{**} coincides with the filtration by the images $Im(N^l)$, $l \geq 0$.

$$
\left.
\begin{array}{c}
\left.
\begin{array}{ccc}
\uparrow & d'\uparrow & \uparrow \\
\to u^{-3}K^{0,t',2} \xrightarrow{\ d''\ } u^{-3}K^{1,t'+1,2} \xrightarrow{\ d''\ } u^{-3}K^{2,t'+2,2}
\end{array}
\right\} = Im(N^2) \\[2mm]
\begin{array}{ccc}
d'\uparrow & d'\uparrow & \uparrow \\
\to u^{-2}K^{-1,t'-1,1} \xrightarrow{\ d''\ } u^{-2}K^{0,t',1} \xrightarrow{\ d''\ } u^{-2}K^{1,t'+1,1}
\end{array} \\[2mm]
\begin{array}{ccc}
d'\uparrow & d'\uparrow & \uparrow \\
\to \quad u^{-1}K^{-2,t'-2,0} \xrightarrow{\ d''\ } u^{-1}K^{-1,t'-1,0} \xrightarrow{\ d''\ } u^{-1}K^{0,t',0}
\end{array}
\end{array}
\right\} = Im(N)
$$

FIGURE 3 The bicomplex K_t^{**}.

We will now define a triple complex $\mathcal{K}^{***}$ such that there exist natural surjections from the real cyclic cohomology of $D(X)$ to the cohomology of the associated graded pieces of the decreasing filtration by "rows" on $Tot(\mathcal{K}^{***})$ (a "row" of $\mathcal{K}^{***}$ is a bicomplex $\mathcal{K}^{*q*}$ for a fixed $q \in \mathbb{Z}$). This is the counterpart at infinity (though only in a "dual" sense) of the triple complex $\mathcal{BC}^{***}$ constructed in Section 2. Consider, therefore, the following complex. For $p, q, r \in \mathbb{Z}$, set $\mathcal{K}^{***}$ to be

$$\mathcal{K}^{p,q,r} = \mathbb{R}u_{[q+r+1]} \otimes K^{p+q+r,p+q-n/2-r,q+r} \qquad p \le 0, \ q + r \ge 0$$

and $\mathcal{K}^{p,q,r} = 0$ otherwise; with differentials (here the twist $t = -r$)

$$d' : \mathcal{K}^{p,q,r} \to \mathcal{K}^{p,q+1,r} \qquad d'(u_{[q+r+1]} \otimes (2\pi i)^t \omega) = u_{[q+r+2]} \otimes (2\pi i)^t (\partial + \bar\partial)\omega$$
$$(\text{if } \mathcal{K}^{p,q+1,r} \ne 0)$$

$$d'' : \mathcal{K}^{p,q,r} \to \mathcal{K}^{p+1,q,r} \qquad d''(u_{[q+r+1]} \otimes (2\pi i)^t \omega) = u_{[q+r+1]} \otimes (2\pi i)^t i(\partial - \bar\partial)\omega$$
$$(\text{if } \mathcal{K}^{p+1,q,r} \ne 0)$$

The third differential $\mathcal{K}^{p,q,r} \to \mathcal{K}^{p,q,r+1}$ is taken to be zero. The triple complex $\mathcal{K}^{***}$ is actually a deck of complexes K_t^{**} for different values of $t \in \mathbb{Z}$. The complexes K_t^{**} have been related to Hochschild complexes by Proposition 3.8. (Recall here that the complex $\mathcal{BC}^{***}$ of Section 2.3 is a deck of complexes $\psi^{**}(\mathbb{C})$ which are also related to Hochschild complexes by Corollary 2.9.) The bicomplexes K_t^{**} stacked to form the complex $\mathcal{K}^{***}$ induce an operator N described on $\mathcal{K}^{***}$ as

$$N : \mathcal{K}^{p,q,r} \to \mathcal{K}^{p+1,q,r+1} \qquad N(u_{[q+r+1]} \otimes (2\pi i)^t \omega) = u_{[q+r+2]} \otimes (2\pi i)^{t-1}\omega$$

PROPOSITION 3.9. *For each fixed $q \in \mathbb{Z}$, there are natural surjections from the real cyclic cohomology of the ring $D(X)$ to the cohomology of the bicomplex $\mathcal{K}^{*q*}$*

$$HC^{j+n}(D(X), \mathbb{R}) \twoheadrightarrow H^j(\mathcal{K}^{*q*}) \qquad j \ge 0$$

which are induced by the surjections of Lemma 3.7.

PROOF. From Lemma 3.7, we know that each K_t^{**} is a quotient of the complex $(\Omega_{\mathbb{R}}^*, -d^c)$. As mentioned before, the triple complex $\mathcal{K}^{***}$ is obtained by stacking the bicomplexes K_t^{**} for different values of $t \in \mathbb{Z}$ in a specific manner and hence each "row" of $\mathcal{K}^{***}$ consists of rows from different bicomplexes K_t^{**}. It follows that, for a fixed $q \in \mathbb{Z}$, the "row" $\mathcal{K}^{*q*}$ of the triple complex $\mathcal{K}^{***}$ is a quotient of the following complex (where $d^c = i(\bar{\partial} - \partial)$)

(3.11)

$$
\begin{array}{ccccccc}
& \vdots & & \vdots & & \\
& \uparrow & & \uparrow & & \\
u^{-3}\Omega_{\mathbb{R}}^0 & \xrightarrow{-d^c} & u^{-3}\Omega_{\mathbb{R}}^1 & \xrightarrow{-d^c} & u^{-3}\Omega_{\mathbb{R}}^2 & \longrightarrow & \cdots \\
\quad \uparrow 0 & & \quad \uparrow 0 & & \quad \uparrow 0 & & \\
u^{-2}\Omega_{\mathbb{R}}^0 & \xrightarrow{-d^c} & u^{-2}\Omega_{\mathbb{R}}^1 & \xrightarrow{-d^c} & u^{-2}\Omega_{\mathbb{R}}^2 & \xrightarrow{-d^c} & u^{-2}\Omega_{\mathbb{R}}^3 & \longrightarrow & \cdots \\
\quad \uparrow 0 & & \quad \uparrow 0 & & \quad \uparrow 0 & & \quad \uparrow 0 \\
u^{-1}\Omega_{\mathbb{R}}^0 \xrightarrow{-d^c} u^{-1}\Omega_{\mathbb{R}}^1 & \xrightarrow{-d^c} & u^{-1}\Omega_{\mathbb{R}}^2 & \xrightarrow{-d^c} & u^{-1}\Omega_{\mathbb{R}}^3 & \xrightarrow{-d^c} & u^{-1}\Omega_{\mathbb{R}}^4 & \longrightarrow & \cdots
\end{array}
$$

(here the projections p_{**}^{-1} of Lemma 3.7 map the lowest row in the diagram above to the terms $\mathcal{K}^{*,q,-q}$ (note that we impose $r + q \geq 0$)). For the sake of convenience, we shall denote this bicomplex by $\Omega_{\mathbb{R}}^{**}[u^{-1}]$. Moreover, as in (3.9), the quotient map from $\Omega_{\mathbb{R}}[u^{-1}]$ to $\mathcal{K}^{*q*}$ is actually a projection onto a direct summand, thus inducing a surjection in cohomology.

The jth total cohomology of the bicomplex $\Omega_{\mathbb{R}}^{**}[u^{-1}]$ is given by the direct sum

$$H^j(\Omega_{\mathbb{R}}^{**}[u^{-1}]) = H_{dR}^j(X, \mathbb{R}) \oplus H_{dR}^{j-2}(X, \mathbb{R}) \oplus \cdots$$

Hence we have the following composition of maps, forming a surjection
(3.12)
$$HC^{j+n}(D(X), \mathbb{R}) \xrightarrow{\sim} H_{dR}^j(X, \mathbb{R}) \oplus H_{dR}^{j-2}(X, \mathbb{R}) \oplus \cdots = H^j(\Omega_{\mathbb{R}}^{**}[u^{-1}])$$
$$\twoheadrightarrow H^j(\mathcal{K}^{*q*})$$

The first map in (3.12), which is an isomorphism, comes from Proposition 3.6, while the last, which is a surjection, follows from the fact that $\mathcal{K}^{*q*}$ is a direct summand of $\Omega_{\mathbb{R}}^{**}[u^{-1}]$.

For each fixed $q \in \mathbb{Z}$, the cokernel of the map

$$N : \mathcal{K}^{p,q,r} \to \mathcal{K}^{p+1,q,r+1} \qquad N(u_{[q+r+1]} \otimes (2\pi i)^t \omega) = u_{[q+r+2]} \otimes (2\pi i)^{t-1} \omega$$

is the complex $\mathcal{K}^{*,q,-q}$. Hence, following the same reasoning as in the proof of Proposition 3.9, we see that there are natural surjections

(3.13) $\qquad HH^{j+n}(D(X), \mathbb{R}) \xrightarrow{\sim} H_{dR}^j(X, \mathbb{R}) \twoheadrightarrow H^j(Coker(N)^*)$

In what follows, we shall continue to use $\Omega_{\mathbb{R}}^{**}[u^{-1}]$ to denote the bicomplex of (3.11). The next proposition now shows how the periodicity operator corresponds to the monodromy

PROPOSITION 3.10. *For any $q \in \mathbb{Z}$, the periodicity operator S appearing on the real cyclic cohomology of $D(X)$ (see Proposition 3.5) can be identified with the monodromy operator N appearing on the complex $\mathcal{K}^{*q*}$, in other words, for any $j \in \mathbb{Z}$ and $n = dim(X)$, we have a commutative diagram*

$$
\begin{array}{ccccc}
HC^{j+n-2}(D(X)) & \xrightarrow{S} & HC^{j+n}(D(X)) & \longrightarrow & HH^{j+n}(D(X)) \\
\downarrow & & \downarrow & & \downarrow \\
H^{j-2}(\mathcal{K}^{*q*}) & \xrightarrow{N} & H^j(\mathcal{K}^{*q*}) & \longrightarrow & H^j(Coker(N)^*)
\end{array}
$$

where the vertical maps are all surjections.

PROOF. Consider the bicomplex $\Omega_{\mathbb{R}}^{**}[u^{-1}]$ of (3.11). For sake of convenience, we introduce a map $\tilde{N}$ on $\Omega_{\mathbb{R}}^{**}[u^{-1}]$ that acts as

$$
\tilde{N}(u_{[q+1]} \otimes \omega) = u_{[q+2]} \otimes \omega
$$

Notice that $\tilde{N}$ lies above the monodromy operator N on $\mathcal{K}^{*q*}$, since the latter is a quotient of $\Omega_{\mathbb{R}}^{**}[u^{-1}]$. Furthermore, we note that $\tilde{N}$ is injective on $\Omega_{\mathbb{R}}^{**}[u^{-1}]$.

We therefore have a short exact sequence of bicomplexes

$$
0 \to \Omega_{\mathbb{R}}^{**}[u^{-1}][-1,-1] \xrightarrow{\tilde{N}} \Omega_{\mathbb{R}}^{**}[u^{-1}] \to Coker(\tilde{N})^* \to 0
$$

which gives rise to a long exact sequence of associated homologies

$$
\cdots \longrightarrow H^{l-2}(\Omega_{\mathbb{R}}^{**}[u^{-1}]) \xrightarrow{\tilde{N}} H^l(\Omega_{\mathbb{R}}^{**}[u^{-1}]) \longrightarrow
$$
$$
H^j(Coker(\tilde{N})^*) \longrightarrow \cdots
$$

The isomorphisms in Proposition 3.5 and Proposition 3.6 are canonical and hence the periodicity operator on $HC^{j+n}(D(X), \mathbb{R})$ acts by dropping the top summand as follows

(3.14)
$$
\begin{array}{ccc}
HC^{j+n}(D(X), \mathbb{R}) & \xrightarrow{\simeq} & H_{dR}^j(X, \mathbb{R}) \oplus H_{dR}^{j-2}(X, \mathbb{R}) \oplus H_{dR}^{j-4}(X, \mathbb{R}) \oplus \cdots \\
S \downarrow & & p \downarrow \\
HC^{j+n-2}(D(X), \mathbb{R}) & \xrightarrow{\simeq} & H_{dR}^{j-2}(X, \mathbb{R}) \oplus H_{dR}^{j-4}(X, \mathbb{R}) \oplus \cdots
\end{array}
$$

where the horizontal arrows are the isomorphisms of Proposition 3.6 and p is the obvious projection. Combining (3.14) with the fact that $\mathcal{K}^{*q*}$ is a quotient of $\Omega_{\mathbb{R}}^{**}[u^{-1}]$ and considering the surjections $HC^{j+n}(D(X), \mathbb{R}) \to H^j(\mathcal{K}^{*q*})$ described in Proposition 3.9 along with the surjections (from (3.13))

$$
HH^{j+n}(D(X), \mathbb{R}) \xrightarrow{\simeq} H_{dR}^j(X, \mathbb{R}) = H^j(Coker(\tilde{N})^*) \twoheadrightarrow H^j(Coker(N)^*)
$$

we have the commutative diagram

$$
\begin{array}{ccccc}
HC^{j+n-2}(D(X), \mathbb{R}) & \xrightarrow{S} & HC^{j+n}(D(X), \mathbb{R}) & \longrightarrow & HH^{j+n}(D(X), \mathbb{R}) \\
\downarrow & & \downarrow & & \downarrow \\
H^{j-2}(\mathcal{K}^{*q*}) & \xrightarrow{N} & H^j(\mathcal{K}^{*q*}) & \longrightarrow & H^j(Coker(N)^*)
\end{array}
$$

wherein all the vertical maps are surjective.

3.4. An Analogue of the Complex of Nearby Cycles at Archimedean Infinity

In this section, our objective is to define a complex φ^{**} endowed with an operator N that is in direct analogy to the complex of nearby cycles $\psi^{**}(\mathbb{C})$ of Section 2. Thereafter, in Proposition 3.11, we prove an analogue of the homotopy result in Proposition 2.2. In fact, our proof of Proposition 3.11 is also formally similar to that of Proposition 2.2.

Let X be a compact manifold defined over $\mathbb{C}$ or $\mathbb{R}$. Again, if X has total dimension n, $dim_{\mathbb{C}}(X(\mathbb{C})) = n/2$. The idea is as follows: from the discussion at the beginning of Section 3, we know that the terms $K_S^{i,j,k}$ defined by Steenbrink are direct summands of the E_1-term of the spectral sequence associated to the L-filtration on Steenbrink's complex A^{**} (see (3.1) and (3.2)). Further, Proposition 2.2 exhibits a quasi-isomorphism $\mu : \psi^* \longrightarrow A^*$.

This suggests that the terms $K^{i,j,k}$ appearing in Consani's complex (3.3) may be assembled, in place of the terms $K_S^{i,j,k}$, to define a complex B^{**} that plays the role, at archimedean infinity, of Steenbrink's complex A^{**}

$$A^{pq} = \mathbb{C}u_{[p+1]} \otimes \Omega_X^{p+q+1}(\log Y)/W_p\Omega_X^{p+q+1}(\log Y) \qquad p \geq 0$$

Similarly, we will assemble the terms $K^{i,j,k}$ to form a complex φ^{**} that plays the role, in this context, of the nearby cycles complex.

Both complexes φ^{**} and B^{**} are equipped with an operator N and we will show in Proposition 3.11 that there exists a morphism $\mu : \varphi^{**} \to B^{**}$ of bicomplexes which is compatible with N up to homotopy, in other words, $\mu \circ N - N \circ \mu$ is homotopic to 0.

For easy comparison, we present the complexes $\psi^{**}(\mathbb{C})$ and A^{**} side by side in Table 1.

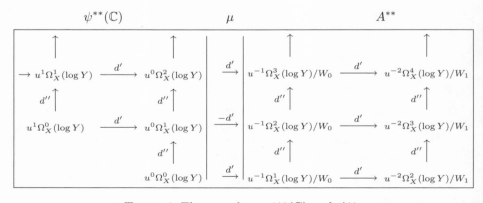

TABLE 1 The complexes $\psi^{**}(\mathbb{C})$ and A^{**}.

From Table 1, we notice the following

(1) Along each row with d'-differentials, the terms on the A^{**} side are essentially a continuation of the d'-complex from the $\psi^{**}(\mathbb{C})$ side, except that, on the A^{**} side, we have removed the differential forms of weight "at most p" by considering the quotient $A^{pq} = \mathbb{C}u_{[p+1]} \otimes \Omega_X^{p+q+1}(\log Y)/W_p\Omega_X^{p+q+1}(\log Y)$.

However, this apparent difference between the $\psi^{**}(\mathbb{C})$ side and the A^{**} side is removed if we simply rewrite the expression for the bicomplex $\psi^{**}(\mathbb{C})$ as

$$\psi(\mathbb{C})^{-p,q} = \mathbb{C}u^p \otimes_{\mathbb{C}} \Omega_X^{q-p}(\log Y)/W_{-p-1}\Omega_X^{q-p} \qquad p \geq 0$$

where the terms of weight $\leq -p-1$ are differential forms of negative degree and hence zero. Therefore, we see in Table 1 that, as we proceed along any d'-row of $\psi^{**}(\mathbb{C})$, we have quotients over terms of negative weight all the way up to the column ψ^{0*} where we take quotient over terms of weight at most -1. Thereafter, following the same row into the A^{**} side, we have quotients over terms of weight at most 0, then at most 1 and so on.

(2) The connecting map μ between the nearby cycles complex $\psi^{**}(\mathbb{C})$ and Steenbrink's complex A^{**} has (up to powers of u) the same expression (see (2.12)) as the differential d' with alternating signs.

It is now clear that we should view $\psi^{**}(\mathbb{C})$ and A^{**} as the two complementary pieces of the same bicomplex with differentials (d', d''), "chopped" at the 0th column.

If we apply the weight filtration to both of these complexes and consider the associated graded objects, then using the isomorphism from Poincaré lemma $((Gr_k^W \Omega_X^*(\log Y) \xrightarrow{\sim} \Omega_{\tilde{Y}^k}^*[-k])$ as in the notation of (2.9)) we obtain the results shown in Table 2.

$$\begin{array}{cccc} \psi^{**} & & \mu & A^{**} \end{array}$$

$$\begin{array}{ccc}
\uparrow & \uparrow & & \uparrow \qquad\qquad \uparrow \\
\Omega_{\tilde{Y}^0}^1 \oplus \Omega_{\tilde{Y}^1}^0 \xrightarrow{d'} \Omega_{\tilde{Y}^0}^2 \oplus \Omega_{\tilde{Y}^1}^1 \oplus \Omega_{\tilde{Y}^2}^0 & \xrightarrow{d'} & \Omega_{\tilde{Y}^1}^2 \oplus \Omega_{\tilde{Y}^2}^1 \oplus \Omega_{\tilde{Y}^3}^0 \xrightarrow{d'} \Omega_{\tilde{Y}^2}^2 \oplus \Omega_{\tilde{Y}^3}^1 \oplus \Omega_{\tilde{Y}^4}^0 \\
d'' \uparrow \qquad d'' \uparrow & & d'' \uparrow \qquad\qquad d'' \uparrow \\
\Omega_{\tilde{Y}^0}^0 \xrightarrow{d'} \Omega_{\tilde{Y}^0}^1 \oplus \Omega_{\tilde{Y}^1}^0 & \xrightarrow{-d'} & \Omega_{\tilde{Y}^1}^1 \oplus \Omega_{\tilde{Y}^2}^0 \xrightarrow{d'} \Omega_{\tilde{Y}^2}^1 \oplus \Omega_{\tilde{Y}^3}^0 \\
d'' \uparrow & & d'' \uparrow \qquad\qquad d'' \uparrow \\
\Omega_{\tilde{Y}^0}^0 & \xrightarrow{d'} & \Omega_{\tilde{Y}^1}^0 \xrightarrow{d'} \Omega_{\tilde{Y}^2}^0
\end{array}$$

TABLE 2 ψ^{**} and A^{**}.

The definition of the K^{***} is as in (3.3). Fix any initial $i, j, k \in \mathbb{Z}$ (the natural choice would be to fix $i = 0$, $j = -n/2$ and $k = 0$, where n is the total dimension of $X(\mathbb{C})$). We define the bicomplex

$$\varphi(i,j,k)^{-p,q} = \begin{cases} \mathbb{R}u^{[p]} \otimes \displaystyle\bigoplus_{l=0}^{q-p} K^{i+q-3p-2l,\, j+q-p,\, k-p} & \text{if } q \geq p \geq 0 \\[2em] = \mathbb{R}u^{[p]} \otimes \displaystyle\bigoplus_{l=0}^{q-p} \bigoplus_{\substack{|a-b| \leq 2k-i+p-q+1 \\ a+b=j+q-p+n/2,\, a\leq b}} (\Omega^{a,b} + \Omega^{b,a})_{\mathbb{R}}\left(\frac{n+j-i}{2}+p+l\right) \\[2em] 0 & \text{otherwise} \end{cases}$$

For the sake of convenience, we will denote the bicomplex $\varphi(i,j,k)^{**}$ by φ^{**}. Then the term $\varphi^{-p,q}$ consists of sums of terms of the form $u^{[p]} \otimes (2\pi i)^t \omega$, where $(2\pi i)^t \omega \in K^{i+q-3p-2l,j+q-p,k-p}$, $0 \leq l \leq q-p$ (from which, by definition of the term $K^{i+q-3p-2l,j+q-p,k-p}$, it follows that $t = \left(\frac{n/2+j-i}{2} + p + l\right)$). On the bicomplex φ^{**}, we introduce the two differentials

$$d' : \varphi^{-p,q} \to \varphi^{-p+1,q} \qquad d'(u^{[p]} \otimes (2\pi i)^t \omega) = u^{[p-1]} \otimes (2\pi i)^t (\partial + \bar{\partial})\omega$$
$$d'' : \varphi^{-p,q} \to \varphi^{-p,q+1} \qquad d''(u^{[p]} \otimes (2\pi i)^t \omega) = u^{[p]} \otimes (2\pi i)^t i(\partial - \bar{\partial})\omega$$

and the operator

$$N_\varphi : \varphi^{-p,q} \to \varphi^{-p+1,q-1} \qquad N_\varphi(u^{[p]} \otimes (2\pi i)^t \omega) = u^{[p-1]} \otimes (2\pi i)^{t-1} \omega$$

For the same fixed values i,j,k, we define the "Steenbrink Complex" B^{**} as

$$B(i,j,k)^{pq} = \begin{cases} \mathbb{R}u^{-p-1} \otimes \displaystyle\bigoplus_{l=0}^{q} K^{i+p+q+1-2l,j+p+q+1,k+p+1} & \text{if } p \geq 0 \\ \\ 0 & \text{otherwise} \end{cases}$$

For the sake of convenience, we will denote $B(i,j,k)^{**}$ simply by B^{**}. On the bicomplex B^{**} we introduce the differentials

$$d' : B^{pq} \to B^{p+1,q} \qquad d'(u_{[p+1]} \otimes (2\pi i)^t \omega) = u_{[p+2]} \otimes (2\pi i)^t (\partial + \bar{\partial})\omega$$
$$d'' : B^{pq} \to B^{p,q+1} \qquad d''(u_{[p+1]} \otimes (2\pi i)^t \omega) = u_{[p+1]} \otimes (2\pi i)^t i(\partial - \bar{\partial})\omega$$

and the operator

$$N_B : B^{pq} \to B^{p+1,q-1} \qquad N_B(u_{[p+1]} \otimes (2\pi i)^t \omega) = u_{[p+2]} \otimes (2\pi i)^{t-1} \omega$$

To illustrate the analogy, compare the following tables with Table 2 (compare Table 3(1) to the left side of Table 2 and Table 3(2) to the right side of Table 2).

TABLE 3(1) $\varphi(i,j,k)^{**}$.

$$
\begin{array}{ccc}
\begin{array}{c} K^{i+3,j+3,k+1} \oplus K^{i+1,j+3,k+1} \\ \oplus K^{i-1,j+3,k+1} \end{array} & \xrightarrow{\ d'\ } & \begin{array}{c} K^{i+4,j+4,k+2} \oplus K^{i+2,j+4,k+2} \\ \oplus K^{i,j+4,k+2} \end{array} \\[4pt]
\Big\uparrow d'' & & \Big\uparrow d'' \\[4pt]
K^{i+2,j+2,k+1} \oplus K^{i,j+2,k+1} & \xrightarrow{\ d'\ } & K^{i+3,j+3,k+2} \oplus K^{i+1,j+3,k+2} \\[4pt]
\Big\uparrow d'' & & \Big\uparrow d'' \\[4pt]
\underset{\underset{B^{0,0}}{\|}}{K^{i+1,j+1,k+1}} & \xrightarrow{\ d'\ } & \underset{\underset{B^{1,0}}{\|}}{K^{i+2,j+2,k+2}}
\end{array}
$$

TABLE 3(2) $B(i,j,k)^{**}$.

The following proposition now shows that the maps N appearing on the bicomplexes φ^{**} and B^{**} are homotopic.

PROPOSITION 3.11. *Consider the morphism* $\mu : \varphi^* \to B^*$ *defined as*

$$
\mu(u^{[p]} \otimes (2\pi i)^t \omega) = \begin{cases} (-1)^{|\omega|} u_{[1]} \otimes (2\pi i)^t (\partial + \bar{\partial})\omega & \text{if } p = 0 \\ 0 & \text{if } p \neq 0 \end{cases}
$$

Then the morphisms $\mu \circ N_\varphi$ *and* $N_B \circ \mu$ *are homotopic, by means of the homotopy* $h : \varphi^{**} \to B^{**}$ *of bidegree* $(0,-1)$ *defined by*

$$
h(u^{[p]} \otimes (2\pi i)^t \omega) = \begin{cases} 0 & \text{if } p \neq 0 \\ (-1)^{|\omega|} u_{[1]} \otimes (2\pi i)^{t-1} \omega & \text{if } p = 0 \end{cases}
$$

PROOF. The following diagram describes the setup.

$$
\begin{array}{ccccccccc}
& & & & \varphi^{0,0} & \xrightarrow{\mu} & B^{0,0} & \xrightarrow{d'} & B^{1,0} & \longrightarrow & \ldots \\
& & & & \downarrow d'' & & \downarrow d'' & & \downarrow d'' \\
& & \varphi^{-1,1} & \xrightarrow{d'} & \varphi^{0,1} & \xrightarrow{\mu} & B^{0,1} & \xrightarrow{d'} & B^{1,1} & \longrightarrow & \ldots \\
& & \downarrow d'' & & \downarrow d'' & & \downarrow d'' & & \downarrow d'' \\
\varphi^{-2,2} & \xrightarrow{d'} & \varphi^{-1,2} & \xrightarrow{d'} & \varphi^{0,2} & \xrightarrow{\mu} & B^{0,2} & \xrightarrow{d'} & B^{1,2} & \longrightarrow & \ldots \\
\downarrow & & \downarrow & & \downarrow & & \downarrow & & \downarrow \\
\vdots & & \vdots & & \vdots & & \vdots & & \vdots
\end{array}
$$

The maps d' and d'' describe the differentials of the bicomplexes φ^{**} and B^{**} and the map μ induces a morphism of bicomplexes. This means that while the (d', d'') squares anti-commute, the (μ, d'') squares commute. For any $(u^{[0]} \otimes (2\pi i)^t \omega)$ in $\varphi^{0,q}$, we have

$$
d'' h(u^{[0]} \otimes (2\pi i)^t \omega) = d''((-1)^{|\omega|} u_{[1]} \otimes (2\pi i)^{t-1} \omega) = (-1)^{|\omega|} u_{[1]} \otimes (2\pi i)^{t-1} d''(\omega)
$$

$$
h d''(u^{[0]} \otimes (2\pi i)^t \omega) = h(u^{[0]} \otimes (2\pi i)^t d''(\omega)) = (-1)^{|d''(\omega)|} u_{[1]} \otimes (2\pi i)^{t-1} d''(\omega)
$$

It follows that $d'' \circ h + h \circ d'' = 0$.

Now, note that given any $\varphi^{-p,q}$, the morphism μ is zero (by definition) unless $p = 0$. Hence $N_B \circ \mu - \mu \circ N_\varphi$ is nontrivial only on terms $\varphi^{-1,*}$ and $\varphi^{0,*}$. Given $u^{[1]} \otimes (2\pi i)^t \omega_1 \in \varphi^{-1,q}$ and $u^{[0]} \otimes (2\pi i)^{t'} \omega_0 \in \varphi^{0,q'-1}$, we get

$$(N_B \circ \mu - \mu \circ N_\varphi)(u^{[0]} \otimes (2\pi i)^{t'} \omega_0 + u^{[1]} \otimes (2\pi i)^t \omega_1)$$

$$= (-1)^{|\omega_0|} u_{[2]} \otimes (2\pi i)^{t'-1}(\partial + \bar\partial)\omega_0 - (-1)^{|\omega_1|} u_{[1]} \otimes (2\pi i)^{t-1}(\partial + \bar\partial)\omega_1$$

$$= (d' \circ h + h \circ d')(u^{[0]} \otimes (2\pi i)^{t'} \omega_0 + u^{[1]} \otimes (2\pi i)^t \omega_1)$$

Therefore the morphisms $N_B \circ \mu$ and $\mu \circ N_\varphi$ are homotopic.

References

[BGS]   S. Bloch, H. Gillet, C. Soulé, *Algebraic cycles on degenerate fibers.*, Arithmetic geometry (Cortona, 1994), Sympos. Math., XXXVII, Cambridge Univ. Press, Cambridge, 1997, pp. 45–69.

[Bry1] J. L. Brylinski, *A differential complex for Poisson manifolds*, J. Differential Geom., **28, no. 1** (1988), 93–114.

[Bry2] J. L. Brylinski, *Some examples of Hochschild and cyclic homology*, Lecture Notes in Math., 1271, Springer, Berlin, 1987, pp. 33–72.

[Co1]   A. Connes, *Cohomologie cyclique et foncteurs Ext^n*, C. R. Acad. Sci. Paris Sér. I Math., **296, no. 23** (1983), 953–988.

[Con2] C. Consani, *Double complexes and Euler L-factors*, Compositio Math., **111, no. 3,** (1998), 323–358.

[Con3] C. Consani, *Double complexes and Euler L-factors*, PhD Thesis, University of Chicago, 1996.

[Del1] P. Deligne, *Comparison avec la théorie transcedente, Exp XIV*, Groupes de Monodromie en Géométrie Algébrique (SGA 7 II). Lecture Notes in Math. **340**, Springer-Verlag, Berlin-Heidelberg, 1973.

[Del2] P. Deligne, *Théorie de Hodge. II.*, Inst. Hautes Études Sci. Publ. Math. 40, 1971, pp. 5–57.

[GN]   F. Guillén, V. Navarro Aznar, *Sur le théorème local des cycles invariants*, Duke Math. J., **61, no. 1** (1990), 323–358.

[Lod]   J. L. Loday, *Cylic Homology, Grundlehren der Mathematischen Wissenschaften, 301*, Springer-Verlag, Berlin, 1998.

[N]   V. Navarro Aznar, *Sur la théorie de Hodge-Deligne*, Invent. Math., **90, no. 1** (1987), 11–76.

[St1]   J. Steenbrink, *Limits of Hodge structures*, Invent. Math., **31, no. 3** (1975/76), 229–257.

[St2]   J. Steenbrink, *Double complexes and Euler L-factors*, PhD Thesis, Universiteit van Amsterdam, 1974.

[Tsy]   B. L. Tsygan, *Homology of matrix Lie algebras over rings and the Hochschild homology.*, Russian Math. Surveys, **38, no. 2** (1983), 198–199.

[Wei1] C. Weibel, *Cyclic homology for schemes*, Proc. Amer. Math. Soc., **124, no. 6** (1996), 1655–1662.

[Wei2] C. Weibel, *An introduction to homological algebra. Cambridge Studies in Advanced Mathematics, 38*, Cambridge Univ. Press, Cambridge, MA, 1994.

[W]   M. Wodzicki, *Cyclic homology of differential operators*, Duke Math. J., **54, no. 2** (1987), 641–647.

DEPARTMENT OF MATHEMATICS, JOHNS HOPKINS UNIVERSITY, 3400 N CHARLES ST., 404 KRIEGER HALL, BALTIMORE, MD 21218, USA.

Current address: Department of Mathematics, Ohio State University, 231 W 18th Avenue, 100 Math Tower, Columbus, OH 43210, USA.

E-mail address: `abhishekbanerjee1313@gmail.com`

Modular Index Invariants of Mumford Curves

Alan Carey, Matilde Marcolli, and Adam Rennie

ABSTRACT. We continue an investigation initiated by Consani–Marcolli of the relation between the algebraic geometry of p-adic Mumford curves and the noncommutative geometry of graph C^*-algebras associated to the action of the uniformizing p-adic Schottky group on the Bruhat–Tits tree. We reconstruct invariants of Mumford curves related to valuations of generators of the associated Schottky group by developing a graphical theory for KMS weights on the associated graph C^*-algebra and using modular index theory for KMS weights. We give explicit examples of the construction of graph weights for low genus Mumford curves. We then show that the theta functions of Mumford curves, and the induced currents on the Bruhat–Tits tree, define functions that generalize the graph weights. We show that such inhomogeneous graph weights can be constructed from spectral flows, and that one can reconstruct theta functions from such graphical data.

1. Introduction

Mumford curves generalize the Tate uniformization of elliptic curves and provide p-adic analogues of the uniformization of Riemann surfaces [**Mum**]. The type of p-adic uniformization considered for these curves is a close analogue of the Schottky uniformization of complex Riemann surfaces, where instead of a Schottky group $\Gamma \subset \mathrm{PSL}_2(\mathbb{C})$ acting on the Riemann sphere $\mathbb{P}^1(\mathbb{C})$, one has a p-adic Schottky group acting on the boundary of the Bruhat–Tits tree and on the Drinfeld p-adic upper half plane.

The analogy between Mumford curves and Schottky uniformization of Riemann surfaces was a key ingredient in the results of Manin on Green functions of Arakelov geometry in terms of hyperbolic geometry [**Man**], motivated by the analogy with earlier results of Drinfeld–Manin for the case of p-adic Schottky groups [**DriMan**]. Manin's computation of the Green function for a Schottky-uniformized Riemann surface in terms of geodesic lengths in the hyperbolic handlebody uniformized by the same Schottky group provides a geometric interpretation of the missing "fiber at infinity" in Arakelov geometry in terms the tangle of bounded geodesics inside the hyperbolic 3-manifold. In order to make this result compatible with Deninger's description of the Gamma factors of L-functions as regularized determinants and with Consani's archimedean cohomology [**Cons**], this formulation of Manin was reinterpreted in terms of noncommutative geometry by Consani–Marcolli [**CM**]. In

particular, the model proposed in [**CM**] for the "fiber at infinity" uses a noncommutative space which describes the action of the Schottky group Γ on its limit set
$\Lambda_\Gamma \subset \mathbb{P}^1(\mathbb{C})$ via the crossed product C^*-algebra $C(\Lambda_\Gamma) \rtimes \Gamma$. This is a particular
case of a Cuntz–Krieger algebra given by the graph C^*-algebra of the finite graph
Δ_Γ/Γ, with Δ_Γ the Cayley graph of $\Gamma \simeq \mathbb{Z}^{*g}$.

Following the same analogy between Schottky uniformization of Riemann surfaces and p-adic uniformization of Mumford curves, Consani and Marcolli extended
their construction [**CM**] to the case of Mumford curves, ([**CM1**], [**CM2**]). More
interesting graph C^*-algebras appear in the p-adic case than in the archimedean
setting, namely the ones associated to the graph Δ_Γ/Γ, which is the dual graph of
the specialization of the Mumford curve and to Δ'_Γ/Γ, which is the dual graph of
the closed fiber of the minimal smooth model of the curve. After these results of
Consani–Marcolli, the construction was further refined in [**CMRV**] and extended
to some classes of higher rank buildings generalizing the rank-one case of Schottky
groups acting on Bruhat–Tits trees. The relation between Schottky uniformiza-
tions, noncommutative geometry, and graph C^*-algebras was further analyzed in
([**CLM**], [**CorMa**]). More recently, it was shown in [**deJong**] that, for a free action
of a discrete group of isometries on a tree, one can make the algebra of functions
on the boundary of the quotient graph into a commutative spectral triple in such
a way that the family of zeta functions of this finitely summable spectral triple
determines the graph.

The main question in this approach is how much of the algebraic geometry of
Mumford curves can be recovered by means of the noncommutative geometry of
certain C^*-algebras associated to the action of the Schottky group on the Bruhat–
Tits tree, on its limit set, and on the Drinfeld upper half plane, and conversely how
much the noncommutative geometry is determined by algebro-geometric information coming from the Mumford curve.

In this paper we analyze another aspect of this question, based on recent results
on circle actions on graph C^*-algebras and associated KMS states and modular index theory [**CNNR**]. In particular, we first show how numerical information like
the Schottky invariants given by the translation lengths of a given set of generators
of the Schottky group can be recovered from the modular index invariants of the
graph C^*-algebra determined by the action of the p-adic Schottky group on the
Bruhat–Tits tree. We then analyze the relation between graph weights (see Section
4) for this same graph C^*-algebra and theta functions of the Mumford curve. Unlike
the previous results on Mumford curves and noncommutative geometry, which concentrated on the use of the graph C^*-algebra associated to the *finite* graph Δ'_Γ/Γ or
Δ_Γ/Γ, here we use the full infinite graph $\Delta_\mathbb{K}/\Gamma$, where $\Delta_\mathbb{K}$ is the Bruhat–Tits tree,
with boundary at infinity $\partial\Delta_\mathbb{K}/\Gamma = X_\Gamma(\mathbb{K})$, the $\mathbb{K}$-points of the Mumford curve,
with $\mathbb{K}$ a finite extension of $\mathbb{Q}_p$. Working with an infinite graph requires a more
subtle analysis of the modular index theory and a setting for the graph weights
(which we recall in Section 4), where the main information is located inside the
finite graph Δ'_Γ/Γ and is propagated along the infinite trees in $\Delta_\mathbb{K}/\Gamma$ attached to
the vertices of Δ'_Γ/Γ, towards the conformal boundary $X_\Gamma(\mathbb{K})$. The graph weights
are solutions of a combinatorial equation at the vertices of the graph, which can
be thought of as governing a momentum flow through the graph. We prove that
for graphs such as $\Delta_\mathbb{K}/\Gamma$ faithful graph weights are the same as gauge invariant,
norm lower semicontinuous faithful semifinite functionals on the graph C^*-algebra.

This is the basis for constructing KMS states associated to graph weights, which are then used to compute modular index invariants for these type III geometries.

The main result, which we obtain in several steps in Sections 4 and 5 states that one can recover the translation lengths of the generators of the Schottky group from the modular index theorem for the graph C^*-algebra endowed with a suitable periodic time evolution. We formulate the result in terms of Mumford curves, because that is our main motivating example, but in fact it can be stated purely in terms of actions of discrete free subgroups of isometry groups of finite valence trees. Namely, the main result we obtain in Sections 4 and 5 is the following.

THEOREM 1.1. *Let* Γ *be a discrete free subgroup of isometries of a finite valence tree* Δ. *If the graph* Δ/Γ *admits a faithful graph weight and has a finite subgraph satisfying the conditions of Definition 2.3 (zhyvot), then the modular index theorem for the graph* C^*-algebra $C^*(\Delta/\Gamma)$, *with the zhyvot circle action (see Definition 2.3) determines the translation lengths of (generators of) the group.*

What refers specifically to the case of Mumford curves in this result is the fact that the zhyvot circle action used in the modular index theorem is adapted to the structure of Δ/Γ consisting of a finite graph (the zhyvot of Definition 2.3) with infinite trees attached to its vertices. This is the typical form of graphs obtained as quotients of the Bruhat–Tits tree by a Schottky group in the theory of Mumford curves. The circle action adapted to this particular structure of the graph ensures that the necessary "spectral subspace condition" for the modular index theorem is satisfied. However, it is important to notice that, in terms of the geometry of Mumford curves, this construction captures only invariants up to conjugacy by tree isometries, not by isometries in PGL(2), which is the rigidity group for isomorphisms of Mumford curves. It is well known [**Lubo**] that the Schottky group Γ has countably many conjugacy classes in the isometry group of the tree, but uncountably many in PGL(2). In the specific case of translation lengths, one sees that these are in fact invariant under the full isometry group of the tree and not only under PGL(2).

About the existence of faithful graph weights, we prove in Theorem 4.5 that these are equivalent to gauge invariant norm lower semicontinuous faithful semifinite functionals on the graph C^*-algebra. We concentrate in Section 4 on a particular class of "special graph weights" and we give in Lemma 4.6 a general existence result for such weights in terms of a condition on a matrix associated to the graph. It is likely that restriction to "special graph weights" is artificial, and in work in progress the modular index theory is being extended to KMS weights for quasi-periodic actions of $\mathbb{R}$.

In Section 7 we work specifically in the setting of Mumford curves and we establish a relation between the construction of theta functions and a modified version of the graph weight equation. The theta functions of Mumford curves in turn can be described as in [**vdP**] in terms of Γ-invariant currents on the Bruhat–Tits tree $\Delta_{\mathbb{K}}$ and corresponding signed measures of total mass zero on the boundary. We show that this description of theta functions leads to an inhomogeneous version of the equation defining graph weights, or equivalently to a homogeneous version, but where the weights are allowed to have a sign instead of being positive and are also required to be integer valued. We also show that there is an isomorphism between the abelian group of Γ-invariant currents on the Bruhat-Tits tree and linear functionals on the K_0 of the graph C^*-algebra of the quotient graph $C^*(\Delta_{\mathbb{K}}/\Gamma)$.

This implies that theta functions of the Mumford curve define functionals on the K-theory of the graph algebra, with the only ambiguity given by the action of $\mathbb{K}^*$.

Finally, we discuss how to use the spectral flow to construct solutions of the inhomogeneous graph weight equations and how to use these to construct theta functions in the case where one has more than one (positive) graph weight on $\Delta_{\mathbb{K}}/\Gamma$.

It would be interesting, in a similar manner, to explore how other invariants associated to type III noncommutative geometries, such as the approach followed by Connes–Moscovici in [**CoMo**] and by Moscovici in [**Mosc**], may relate to the algebraic geometry of Mumford curves in the specific case of the algebras $C^*(\Delta_{\mathbb{K}}/\Gamma)$.

2. Mumford Curves

We recall here some well-known facts from the theory of Mumford curves, which we need in the rest of the paper. The results mentioned in this brief introduction can be found for instance in [**Mum**], [**Ma**], [**GvP**] and were also reviewed in more detail in [**CM1**].

2.1. The Bruhat–Tits tree

Let $\mathbb{K}$ denote a finite extension of $\mathbb{Q}_p$ and let $\mathcal{O} = \mathcal{O}_{\mathbb{K}} \subset \mathbb{K}$ be its ring of integers, with $\mathfrak{m} \subset \mathcal{O}$ the maximal ideal. The finite field $k = \mathcal{O}/\mathfrak{m}$ of cardinality $q = \#\mathcal{O}/\mathfrak{m}$ is called the residue field.

Let $\Delta_{\mathbb{K}}^0$ denote the set of equivalence classes of free rank 2 $\mathcal{O}$-modules, with the equivalence relation

$$M_1 \sim M_2 \Leftrightarrow \exists \lambda \in \mathbb{K}^*, \quad M_1 = \lambda M_2.$$

The group $\mathrm{GL}_2(\mathbb{K})$ acts on $\Delta_{\mathbb{K}}^0$ by $gM = \{gm \mid m \in M\}$. This descends to an action of $\mathrm{PGL}_2(\mathbb{K})$, since $M_1 \sim M_2$ for M_1 and M_2 in the same $\mathbb{K}^*$-orbit. Given $M_2 \subset M_1$, one has $M_1/M_2 \simeq \mathcal{O}/\mathfrak{m}^l \oplus \mathcal{O}/\mathfrak{m}^k$, for some $l, k \in \mathbb{N}$. The action of $\mathbb{K}^*$ preserves the inclusion $M_2 \subset M_1$, hence one has a well-defined metric

$$(2.1) \qquad\qquad d(M_1, M_2) = |l - k|.$$

The Bruhat–Tits tree of $\mathrm{PGL}_2(\mathbb{K})$ is the infinite graph with set of vertices $\Delta_{\mathbb{K}}^0$, and an edge connecting two vertices M_1, M_2 whenever $d(M_1, M_2) = 1$. It is an infinite tree with vertices of valence $q + 1$ where $q = \#\mathcal{O}/\mathfrak{m}$. The group $\mathrm{PGL}_2(\mathbb{K})$ acts on $\Delta_{\mathbb{K}}$ by isometries. The boundary $\partial\Delta_{\mathbb{K}}$ is naturally identified with $\mathbb{P}^1(\mathbb{K})$.

2.2. p-adic Schottky Groups and Mumford Curves

A Schottky group Γ is a finitely generated, discrete, torsion-free subgroup of $\mathrm{PGL}_2(\mathbb{K})$ whose nontrivial elements $\gamma \neq 1$ are all *hyperbolic*, *i.e.* such that the eigenvalues of γ in $\mathbb{K}$ have different valuation. The group Γ acts freely on the tree $\Delta_{\mathbb{K}}$. Hyperbolic elements γ have two fixed points $z^{\pm}(\gamma)$ on the boundary $\mathbb{P}^1(\mathbb{K})$. For an element $\gamma \neq 1$ in Γ the *axis* $L(\gamma)$ of γ is the unique geodesic in the Bruhat–Tits tree $\Delta_{\mathbb{K}}$ with endpoints consisting of the two fixed points $z^{\pm}(\gamma) \in \mathbb{P}^1(\mathbb{K}) = \partial\Delta_{\mathbb{K}}$.

Let $\Lambda_{\Gamma} \subset \mathbb{P}^1(\mathbb{K})$ be the closure in $\mathbb{P}^1(\mathbb{K})$ of the set of fixed points of the elements $\gamma \in \Gamma \backslash \{1\}$. This is called the *limit set* of Γ. Only in the case of $\Gamma = (\gamma)^{\mathbb{Z}} \simeq \mathbb{Z}$, with a single hyperbolic generator γ, one has $\#\Lambda_{\Gamma} < \infty$. This special case, as we see below, correponds to Mumford curves of genus one. In general, for higher genus (higher rank Schottky groups), the limit set is uncountable (typically a fractal). The *domain of discontinuity* for the Schottky group Γ is the complement $\Omega_{\Gamma}(\mathbb{K}) = \mathbb{P}^1(\mathbb{K}) \smallsetminus \Lambda_{\Gamma}$.

The quotient $X_\Gamma := \Omega_\Gamma/\Gamma$ gives the analytic model (via uniformization) of an algebraic curve X defined over $\mathbb{K}$ (*cf.* [**Mum**, p. 163]).

For a p-adic Schottky group $\Gamma \subset \mathrm{PGL}(2, \mathbb{K})$ there is a smallest subtree $\Delta'_\Gamma \subset \Delta_\mathbb{K}$ that contains the axes $L(\gamma)$ of all the elements $\gamma \neq 1$ of Γ. The set of ends of Δ'_Γ in $\mathbb{P}^1(\mathbb{K})$ is the limit set Λ_Γ of Γ. The group Γ acts on Δ'_Γ with quotient Δ'_Γ/Γ a finite graph. There is also a smallest tree Δ_Γ on which Γ acts, with vertices a subset of vertices of the Bruhat–Tits tree. The tree Δ'_Γ contains extra vertices with respect to Δ_Γ. These come from vertices of $\Delta_\mathbb{K}$ that are not vertices of Δ_Γ, but which lie on paths in Δ_Γ. (Δ_Γ is not a subtree of $\Delta_\mathbb{K}$, while Δ'_Γ is.) The quotient Δ_Γ/Γ is also a finite graph. Both the graphs Δ'_Γ/Γ and Δ_Γ/Γ have algebro-geometric significance: Δ'_Γ/Γ is the dual graph of the closed fiber of the minimal smooth model of the algebraic curve X over $\mathbb{K}$; Δ_Γ/Γ is the dual graph of the specialization of the curve X. The latter is a k-split degenerate, stable curve, with k the residue field of $\mathbb{K}$.

The set of $\mathbb{K}$-points $X_\Gamma(\mathbb{K})$ of the Mumford curve is identified with the ends of the graph $\Delta_\mathbb{K}/\Gamma$. With a slight abuse of notation we sometimes say that X_Γ is the boundary of $\Delta_\mathbb{K}/\Gamma$ and write

$$(2.2) \qquad \partial \Delta_\mathbb{K}/\Gamma = X_\Gamma.$$

The graph $\Delta_\mathbb{K}/\Gamma$ contains the finite subgraph Δ'_Γ/Γ. Infinite trees depart from the vertices of the subgraph Δ'_Γ/Γ with ends on the boundary at infinity X_Γ. We assume that the base point v belongs to Δ'_Γ so that all these trees are oriented outward from the finite graph Δ'_Γ/Γ. All the nontrivial topology resides in the graph Δ'_Γ/Γ from which one can read off the genus of the curve.

So far, the use of methods of noncommutative geometry in the context of Mumford curves and Schottky uniformization (*cf.* [**CM**], [**CM1**], [**CM2**], [**CLM**], [**CMRV**]) concentrated on the finite graphs Δ'_Γ/Γ and Δ_Γ/Γ and noncommutative spaces associated to the dynamics of the action of the Schottky group on its limit set. Here we consider the full infinite graph $\Delta_\mathbb{K}/\Gamma$ of (2.2). In fact, we will show that it is precisely the presence in $\Delta_\mathbb{K}/\Gamma$ of the infinite trees attached to the vertices of the finite subgraph Δ'_Γ/Γ that makes it possible to construct interesting KMS states on the associated graph C^*-algebra and hence to apply the techniques of modular index theory to obtain new invariants of a K-theoretic nature for Mumford curves.

2.3. Directed Graphs and Their Algebras

There are different ways to introduce a structure of directed graph on the finite graphs Δ_Γ/Γ, Δ'_Γ/Γ and on the infinite graph $\Delta_\mathbb{K}/\Gamma$.

One possibility, considered for instance in [**CLM**], is not to prescribe an orientation on the graphs. This means that one keeps for each edge the choice of both possible orientations. The associated directed graph has then double the number of edges to account for the two possible orientations. This approach is very helpful in the K-theoretic computations performed in [**CLM**], where it actually simplifies the approach, while in the setting we describe here, where we will be concerned with constructing explicit solutions of the graph weights equation, it has the problem that it makes the graph C^*-algebras more complicated and the combinatorics correspondingly more involved than strictly necessary, so we will not follow it here.

Another way to make the graphs of Mumford curves into directed graphs is by the choice of a projective coordinate $z \in \mathbb{P}^1(\mathbb{K})$ (*cf.* [**CM1**], [**CM2**]). The choice of the coordinate z determines uniquely a base point $v \in \Delta^0_\mathbb{K}$, given by the origin of three non-overlapping paths with ends the points 0, 1 and ∞ in $\mathbb{P}^1(\mathbb{K})$. The choice of v gives an orientation to the tree $\Delta_{\mathbb{K}}$ given by the outward direction from v. This gives an induced orientation to any fundamental domains of the Γ-action in $\Delta_{\mathbb{K}}$, Δ'_{Γ} and Δ_{Γ} containing the base vertex v, which one can use to obtain all the possible induced orientations on the quotient graphs.

There is still another possibility of orienting the tree $\Delta_{\mathbb{K}}$ in a way that is adapted to the action of Γ, and this is the one we adopt here. It is described in Lemma 2.1

Suppose we are given the choice of a projective coordinate $z \in \mathbb{P}^1(\mathbb{K})$ and assume that the corresponding vertex v is in fact a vertex of Δ'_{Γ}. Let $\{\gamma_1, \ldots, \gamma_g\}$ be a set of generators for Γ. An orientation of $\Gamma \backslash \Delta_{\mathbb{K}}$ is then obtained from a Γ-invariant orientation of $\Delta_{\mathbb{K}}$ as follows.

LEMMA 2.1. *Consider the chain of edges in $\Delta_{\mathbb{K}}$ connecting the base vertex v to $\gamma_i v$. Then there is a choice of orientation of these edges that induces an orientation on the quotient graph and that extends to a Γ-invariant orientation of $\Delta_{\mathbb{K}}$.*

PROOF. Consider the chain of edges between v and $\gamma_i v$. If all of them have distinct images in the quotient graph, orient them all in the direction away from v and towards $\gamma_i v$. If there is more than one edge in the path from v to $\gamma_i v$ that maps to the same edge in the quotient graph, orient the first one that occurs from v to $\gamma_i v$ and the others consistently with the induced orientation of the corresponding edge in the quotient graph. Similarly, orient the edges between v and $\gamma_i^{-1} v$ in the direction pointing towards v, with the same caveat for edges with the same image in the quotient. Propagate this orientation across the tree Δ'_{Γ} by repeating the same procedure with the edges between $\gamma_i^{\pm 1} v$ and $\gamma_j \gamma_i^{\pm 1} v$ and between $\gamma_i^{\pm 1} v$ and $\gamma_j^{-1} \gamma_i^{\pm 1} v$ and so on. Continuing in this way, one obtains an orientation of the tree Δ'_{Γ} compatible with the induced orientation on the quotient graph $\Delta'_{\Gamma} / \Gamma$. One then orients the rest of the tree $\Delta_{\mathbb{K}}$ away from the subtree Δ'_{Γ}.

An example of the orientations obtained in this way on the tree Δ_{Γ} and on the quotient graph for the genus two case is given in Figures 7, 8, 9 below.

2.4. Graph Algebras for Mumford Curves

For a more detailed introduction to graph C^*-algebras we refer the reader to [**BPRS**], [**kpr**], [**R**] and the references therein. A directed graph $E = (E^0, E^1, r, s)$ consists of countable sets E^0 of vertices and E^1 of edges, and maps $r, s : E^1 \to E^0$ identifying the range and source of each edge. *We will always assume that the graph is row-finite*, which means that each vertex emits at most finitely many edges. Later we will also assume that the graph is *locally finite* which means it is row-finite and each vertex receives at most finitely many edges. We write E^n for the set of paths $\mu = \mu_1 \mu_2 \cdots \mu_n$ of length $|\mu| := n$; that is, sequences of edges μ_i such that $r(\mu_i) = s(\mu_{i+1})$ for $1 \leq i < n$. The maps r, s extend to $E^* := \bigcup_{n \geq 0} E^n$ in an obvious way. A *loop* in E is a path $L \in E^*$ with $s(L) = r(L)$, we say that a loop L has an exit if there is $v = s(L_i)$ for some i which emits more than one edge. If $V \subseteq E^0$ then we write $V \geq w$ if there is a path $\mu \in E^*$ with $s(\mu) \in V$ and $r(\mu) = w$ (we also sometimes say that w is downstream from V). A *sink* is a vertex $v \in E^0$ with $s^{-1}(v) = \emptyset$, a *source* is a vertex $w \in E^0$ with $r^{-1}(w) = \emptyset$.

A *Cuntz–Krieger E-family* in a C^*-algebra B consists of mutually orthogonal projections $\{p_v : v \in E^0\}$ and partial isometries $\{S_e : e \in E^1\}$ satisfying the

Cuntz–Krieger relations

$$S_e^* S_e = p_{r(e)} \text{ for } e \in E^1 \text{ and } p_v = \sum_{\{e : s(e) = v\}} S_e S_e^* \text{ whenever } v \text{ is not a sink.}$$

It is proved in Theorem 1.2 of [**kpr**] that there exists a universal C^*-algebra $C^*(E)$ generated by a non-zero Cuntz–Krieger E-family $\{S_e, p_v\}$. A product $S_\mu := S_{\mu_1} S_{\mu_2} \ldots S_{\mu_n}$ is non-zero precisely when $\mu = \mu_1 \mu_2 \cdots \mu_n$ is a path in E^n. Since the Cuntz–Krieger relations imply that the projections $S_e S_e^*$ are also mutually orthogonal, we have $S_e^* S_f = 0$ unless $e = f$, and words in $\{S_e, S_f^*\}$ collapse to products of the form $S_\mu S_\nu^*$ for $\mu, \nu \in E^*$ satisfying $r(\mu) = r(\nu)$ (*cf.* Lemma 1.1 of [**kpr**]) Indeed, because the family $\{S_\mu S_\nu^*\}$ is closed under multiplication and involution, we have

$$(2.3) \qquad C^*(E) = \overline{\text{span}}\{S_\mu S_\nu^* : \mu, \nu \in E^* \text{ and } r(\mu) = r(\nu)\}.$$

The algebraic relations and the density of span$\{S_\mu S_\nu^*\}$ in $C^*(E)$ play a critical role throughout the paper. We adopt the conventions that vertices are paths of length 0, that $S_v := p_v$ for $v \in E^0$, and that all paths μ, ν appearing in (2.3) are non-empty; we recover S_μ, for example, by taking $\nu = r(\mu)$, so that $S_\mu S_\nu^* = S_\mu p_{r(\mu)} = S_\mu$.

If $z \in \mathbb{T} = S^1$, then the family $\{z S_e, p_v\}$ is another Cuntz–Krieger E-family which generates $C^*(E)$, and the universal property gives a homomorphism $\gamma_z : C^*(E) \to C^*(E)$ such that $\gamma_z(S_e) = z S_e$ and $\gamma_z(p_v) = p_v$. The homomorphism $\gamma_{\bar{z}}$ is an inverse for γ_z, so $\gamma_z \in \text{Aut } C^*(E)$, and a routine $\epsilon/3$ argument using (2.3) shows that γ is a strongly continuous action of $\mathbb{T}$ on $C^*(E)$. It is called the *gauge action*. Because $\mathbb{T} = S^1$ is compact, averaging over γ with respect to normalized Haar measure gives an expectation Φ of $C^*(E)$ onto the fixed-point algebra $C^*(E)^\gamma$:

$$\Phi(a) := \frac{1}{2\pi} \int_{S^1} \gamma_z(a)\, d\theta \text{ for } a \in C^*(E), \quad z = e^{i\theta}.$$

The map Φ is positive, has norm 1, and is faithful in the sense that $\Phi(a^*a) = 0$ implies $a = 0$.

From (2.3), it is easy to see that a graph C^*-algebra is unital if and only if the underlying graph is finite. When we consider infinite graphs, formulas which involve sums of projections may contain infinite sums. To interpret these, we use strict convergence in the multiplier algebra of $C^*(E)$.

LEMMA 2.2 ([**kpr**]). *Let E be a row-finite graph, let A be a C^*-algebra generated by a Cuntz–Krieger E-family $\{T_e, q_v\}$, and let $\{p_n\}$ be a sequence of projections in A. If $p_n T_\mu T_\nu^*$ converges for every $\mu, \nu \in E^*$, then $\{p_n\}$ converges strictly to a projection $p \in M(A)$.*

The directed graph $\Delta_{\mathbb{K}}/\Gamma$ we obtain from a Mumford curve, with the orientation of Lemma 2.1, is locally finite, has no sources and contains a subgraph Δ_Γ'/Γ with no sources and with the following two properties. If v is any vertex in $\Delta_{\mathbb{K}}/\Gamma$ there exists a path in $\Delta_{\mathbb{K}}/\Gamma$ with range v and source contained in Δ_Γ'/Γ, and for any path with source outside M, the range is outside M. For such a graph we can define a new circle action by restricting the gauge action to the subgraph. The properties of this action turn out to be crucial for us.

The reason may be found in [**CNNR**], where the existence of a Kasparov A-A^σ module for a circle action σ on A was found to be equivalent to a condition on the spectral subspaces $A_k = \{a \in A : \sigma_z(a) = z^k a\}$. The condition, called the *spectral subspace condition* in [**CNNR**], states that for all $k \in \mathbb{Z}$, $A_k A_k^*$, always an ideal in A^σ, is in fact a complemented ideal in A^σ. Thus we must have $A^\sigma = A_k A_k^* \oplus G_k$ for some other ideal G_k. It turns out that the graphs arising from Mumford curves allow us to define a circle action for which the spectral subspaces satisfy the spectral subspace condition.

In the following, we consider a circle action σ_z which is closely related to the gauge action γ_z described above, but which accounts for the specific structure of the graphs of Mumford curves, consisting of a central finite graph (the zhyvot defined below) and infinite trees emanating from its vertices.

DEFINITION 2.3. Let E be a locally finite directed graph with no sources, $M \subset E$ a subgraph with no sources and such that
  (1) for any $v \in E^0$ there is a path μ with $s(\mu) \in M$ and $r(\mu) = v$
  (2) for all paths ρ with $s(\rho) \notin M$ we have $r(\rho) \notin M$.
Then we say that E has zhyvot M, and that M is a zhyvot of E.

The zhyvot action $\sigma : \mathbb{T} \to Aut(C^*(E))$ is defined by

$$\sigma_z(S_e) = \begin{cases} \gamma_z(S_e) & e \in M^1 \\ S_e & e \notin M^1 \end{cases} \qquad \sigma_z(p_v) = p_v, \quad v \in E^0,$$

where γ_z is the usual gauge action. If μ is a path in E, let $|\mu|_\sigma$ be the non-negative integer such that $\sigma_z(S_\mu) = z^{|\mu|_\sigma} S_\mu$.

REMARK. The zhyvot of a graph need not be unique.

EXAMPLE. In the case of Mumford curves, the finite graph Δ'_Γ/Γ gives a zhyvot for the infinite graph $\Delta_\mathbb{K}/\Gamma$. There are other possible choices of a zhyvot for the same graph $\Delta_\mathbb{K}/\Gamma$, which are interesting from the point of view of the geometry of Mumford curves. In particular, in the theory of Mumford curves, one considers the reduction modulo powers $\mathfrak{m}^n$ of the maximal ideal $\mathfrak{m} \subset \mathcal{O}_\mathbb{K}$, which provides infinitesimal neighborhoods of order n of the closed fiber. For each $n \geq 0$, we consider a subgraph $\Delta_{\mathbb{K},n}$ of the Bruhat–Tits tree $\Delta_\mathbb{K}$ defined by setting

$$\Delta^0_{\mathbb{K},n} := \{v \in \Delta^0_\mathbb{K} : d(v, \Delta'_\Gamma) \leq n\},$$

with respect to the distance (2.1), with $d(v, \Delta'_\Gamma) := \inf\{d(v, \tilde{v}) : \tilde{v} \in (\Delta'_\Gamma)^0\}$, and

$$\Delta^1_{\mathbb{K},n} := \{w \in \Delta^1_\mathbb{K} : s(w), r(w) \in \Delta^0_{\mathbb{K},n}\}.$$

Thus, we have $\Delta_{\mathbb{K},0} = \Delta'_\Gamma$ and $\Delta_\mathbb{K} = \cup_n \Delta_{\mathbb{K},n}$. For all $n \in \mathbb{N}$, the graph $\Delta_{\mathbb{K},n}$ is invariant under the action of the Schottky group Γ on Δ, and the finite graph $\Delta_{\mathbb{K},n}/\Gamma$ gives the dual graph of the reduction $X_\mathbb{K} \otimes \mathcal{O}/\mathfrak{m}^{n+1}$. Thus, we refer to the $\Delta_{\mathbb{K},n}$ as *reduction graphs*. They form a directed family with inclusions $j_{n,m} : \Delta_{\mathbb{K},n} \hookrightarrow \Delta_{\mathbb{K},m}$, for all $m \geq n$, with all the inclusions compatible with the action of Γ. Each of the quotient graphs $\Delta_{\mathbb{K},n}/\Gamma$ also gives a zhyvot for $\Delta_\mathbb{K}/\Gamma$. In the following we will concentrate on the case where $M = \Delta'_\Gamma/\Gamma$ but one can equivalently work with the reduction graphs.

Given a graph E with zhyvot M and $k \geq 0$ define

$$F_k := \overline{\text{span}}\{S_\mu S_\nu^* : |\mu|_\sigma = |\nu|_\sigma \geq k\},$$

and

$$G_k := \overline{\mathrm{span}} \left\{ S_\mu S_\nu^* : \begin{array}{c} 0 \leq |\mu|_\sigma = |\nu|_\sigma < k, \\ \text{and} \\ \text{either} \quad r(\mu) = r(\nu) \notin M \quad \text{or} \quad r(\mu) = r(\nu) \quad \text{is a sink in } M \end{array} \right\}.$$

Observe that in the definition of G_k, the sinks need not be sinks of the full graph E, just sinks of the subgraph M.

NOTATION. Given a path $\rho \in E^*$, we let $\underline{\rho}$ denote the initial segment of ρ and let $\overline{\rho}$ denote the final segment; in all cases the length of these segments will be clear from context. We always have $\rho = \underline{\rho}\overline{\rho}$.

LEMMA 2.4. *Let E be a locally finite directed graph with no sources and zhyvot $M \subset E$. Let $F = C^*(E)^\sigma$ be the fixed point algebra for the zhyvot action. Then*

$$F = F_k \oplus G_k, \qquad k = 1, 2, 3, \ldots .$$

PROOF. We first check using generators that $F_k G_k = G_k F_k = \{0\}$; once we have shown that $F_k + G_k = F$ this will also show that F_k and G_k are both ideals (that they are subalgebras follows from similar, but simpler, calculations to those below).

Fix $k \geq 1$. Let $S_\mu S_\nu^* \in G_k$ so that $0 \leq |\mu|_\sigma = |\nu|_\sigma < k$ and either $r(\mu) = r(\nu) \notin M$ or is a sink of M. Let $S_\rho S_\tau^* \in F_k$ so that $|\rho|_\sigma = |\tau|_\sigma \geq k$. Then

$$S_\mu S_\nu^* S_\rho S_\tau^* = \begin{cases} S_\mu S_{\overline{\rho}}^- S_\tau^* \delta_{\nu, \underline{\rho}} & |\nu| \leq |\rho| \\ S_\mu S_{\overline{\nu}}^* S_\tau^* \delta_{\underline{\nu}, \rho} & |\nu| \geq |\rho| \end{cases},$$

where $|\cdot|$ denotes the usual length of paths. When $|\nu| \geq |\rho|$, the product is non-zero if and only if $\underline{\nu} = \rho$ but

$$|\underline{\nu}|_\sigma \leq |\nu|_\sigma < |\rho|_\sigma,$$

so this can not happen. When $|\nu| \leq |\rho|$, the product is non-zero if and only if $\nu = \underline{\rho}$, but the range of $\nu \notin M$ or is a sink of M while $|\nu|_\sigma = |\underline{\rho}|_\sigma$ implies that $|\underline{\rho}|_\sigma < |\rho|_\sigma$, and so $r(\underline{\rho}) \in M$ and is not a sink of M. Hence the product is zero, and $G_k F_k = \{0\}$. The computation $F_k G_k = \{0\}$ is entirely analogous, so we omit it.

To see that $F_k + G_k = F$, we need only show that the generators $S_\mu S_\nu^*$ with $0 \leq |\mu|_\sigma = |\nu|_\sigma < k$ and $r(\mu) = r(\nu) \in M$ not a sink, are sums of elements from F_k and G_k, all other generators having been accounted for.

So let $0 \leq n = |\mu|_\sigma = |\nu|_\sigma < k$. The notation $|\rho| \preceq k$ means that $|\rho| = k$ or $|\rho| < k$ and $r(\rho)$ is a sink. Then

$$S_\mu S_\nu^* = \sum_{\rho \in E^*, \ s(\rho) = r(\mu), \ |\rho| \preceq k-n+1} S_\mu S_\rho S_\rho^* S_\nu^*.$$

If $0 \leq |\rho|_\sigma < k-n$, then we must have $r(\rho) \notin M$ or $r(\rho)$ a sink of M. This is because $|\rho|_\sigma \leq |\rho|$, and if $r(\rho)$ is not a sink, we have strict inequality since $|\rho| = k - n + 1$. Hence if $r(\rho)$ is not a sink, $r(\rho) \notin M$. On the other hand if $r(\rho)$ is a sink of E, then either $r(\rho) \notin M$ or $r(\rho)$ is a sink of M.

Thus for $0 \leq |\rho|_\sigma < k - n$, we have $S_\mu S_\rho S_\rho^* S_\nu^* \in G_k$, while if $k - n \leq |\rho|_\sigma \leq k - n + 1$, we have $S_\mu S_\rho S_\rho^* S_\nu^* \in F_k$.

Finally, to see that $F = F_k \oplus G_k$ for each $k \geq 0$, observe that we can split the sequence

$$0 \to F_k \xrightarrow{i} F \to G_k \to 0$$

using the homomorphism $\phi_k : F \to F_k$ defined by

$$\phi_k(f) = P_k f P_k, \qquad P_k = \sum_{|\mu|_\sigma = k} S_\mu S_\mu^*.$$

Checking that ϕ_k is a homomorphism and has range F_k is an exercise with the generators.

PROPOSITION 2.5. *Let E be a locally finite directed graph without sources and with zhyvot M. For $k \in \mathbb{Z}$ let $A_k = \{a \in C^*(E) : \sigma_z(a) = z^k a\}$ denote the spectral subspaces for the zhyvot action. Then*

$$A_k A_k^* = \begin{cases} F_k & k \geq 0 \\ F & k \leq 0 \end{cases}.$$

REMARK. In particular, the spectral subspace assumptions of [**CNNR**] are satisfied for the zhyvot action on a graph with a zhyvot.

PROOF. With $|\mu|$ denoting the ordinary length of paths in E, we have the product formula

$$(2.4) \qquad (S_\mu S_\nu^*)(S_\sigma S_\rho^*)^* = S_\mu S_\nu^* S_\rho S_\sigma^* = \begin{cases} S_\mu S_{\overline{\rho}} S_\sigma^* \delta_{\nu, \underline{\rho}} & |\nu| \leq |\rho| \\ S_\mu S_{\overline{\nu}}^* S_\sigma^* \delta_{\underline{\nu}, \rho} & |\nu| \geq |\rho| \end{cases},$$

where $\underline{\rho}$ is the initial segment of ρ of appropriate length, and $\overline{\rho}$ is the final segment. If $S_\mu S_\nu^*$, $S_\sigma S_\rho^*$ are in A_k, $k \geq 0$, then

$$|\mu|_\sigma - |\nu|_\sigma = k = |\gamma|_\sigma - |\rho|_\sigma,$$

so that $|\gamma|_\sigma \geq k$ and $|\mu|_\sigma \geq k$. Together with (2.4), this shows that for $k \geq 0$ we have $A_k A_k^* \in F_k$. Conversely, if $S_\alpha S_\beta^* \in F_k$, so $|\alpha|_\sigma = |\beta|_\sigma \geq k$, we can factor

$$S_\alpha S_\beta^* = S_{\underline{\alpha}} S_{\overline{\alpha}} S_\beta^* = S_{\underline{\alpha}} (S_\beta S_{\overline{\alpha}}^*)^* \in A_k A_k^*.$$

For $k \leq 0$ we of course have $A_k A_k^* \in F$, and so we need only show that for any $S_\alpha S_\beta^* \in F$, $S_\alpha S_\beta^* \in A_k A_k^*$.

Here we use the final property of zhyvot graphs, namely that we can find a path $\lambda \in E^*$ with $s(\lambda) \in M$ and $r(\lambda) = r(\alpha) = r(\beta)$. Moreover, because M has no sources, we can take $|\lambda|_\sigma$ as large as we like. Thus we can write

$$S_\alpha S_\beta^* = S_\alpha S_\lambda^* S_\lambda S_\beta^* = (S_\alpha S_\lambda^*)(S_\beta S_\lambda^*)^*.$$

Choosing $|\lambda|_\sigma = |\alpha|_\sigma + |k|$ shows that $S_\alpha S_\beta^* \in A_k A_k^*$.

This allows us to recover some known structure of the fixed point algebra of a graph algebra for the usual gauge action and to understand via the spectral subspace condition of [**CNNR**] exactly why the assumptions of [**PRen**] were required to construct a Kasparov module.

COROLLARY 2.6. *Let E be a locally finite directed graph without sources. Then the fixed point algebra for the usual gauge action decomposes as*

$$F = F_k \oplus G_k, \quad k = 1, 2, 3, \ldots,$$

where

$$F_k = \overline{\text{span}}\{S_\mu S_\nu^* : |\mu| = |\nu| \geq k\},$$
$$G_k = \overline{\text{span}}\{S_\mu S_\nu^* : 0 \leq |\mu| = |\nu| < k, \ r(\mu) = r(\nu) \text{ is a sink}\}.$$

PROOF. This follows from Proposition 2.5 since E is a graph with zhyvot E.

3. Schottky Invariants of Mumford Curves and Field Extensions

3.1. Schottky Lengths and Valuation

Let $\Gamma \subset \text{PGL}_2(\mathbb{K})$ be a p-adic Schottky group acting by isometries on the Bruhat–Tits tree $\Delta_{\mathbb{K}}$. As recalled in Section 2.2 a hyperbolic element $\gamma \in \Gamma$ determines a unique axis $L(\gamma)$ in $\Delta_{\mathbb{K}}$, which is the infinite path of edges connecting the two fixed points $z^\pm(\gamma) \in \Lambda_\Gamma \subset \mathbb{P}^1(\mathbb{K}) = \partial \Delta_{\mathbb{K}}$. The element γ acts on $L(\gamma)$ by a translation of length $\ell(\gamma)$.

To a given set of generators $\{\gamma_1, \cdots, \gamma_g\}$ of Γ one can associate the translation lengths $\ell(\gamma_i)$. We refer to the collection of values $\{\ell(\gamma_i)\}$ as the *Schottky invariants* of $(\Gamma, \{\gamma_i\})$. For example, in the case of genus one, one can assume the generator of Γ is given by a matrix for the form

$$\gamma = \begin{pmatrix} q & 0 \\ 0 & 1 \end{pmatrix}$$

with $|q| < 1$ so that the fixed points are $z^+(\gamma) = 0$ and $z^-(\gamma) = \infty$. The element γ acts on the axis $L(\gamma)$ as a translation by a length $\ell(\gamma) = \log |q|^{-1} = v_{\text{m}}(q)$ equal to the number of vertices in the closed graph (topologically a circle) Δ'_Γ / Γ. We see clearly that, even in the simple genus one case, knowledge of the Schottky invariant $\ell(\gamma)$ does not suffice to recover the curve. This is clear from the fact that the Schottky length only sees the valuation of $q \in \mathbb{K}^*$. Nonetheless, the Schottky lengths give useful computable invariants.

It is important to stress that, when we consider translation lengths as invariants, we are in fact working with *marked Schottky space $\mathcal{S}_g$*, where we consider pairs of a Schottky group Γ of genus g and a marking given by the choice of a set of generators $\{\gamma_1, \ldots, \gamma_g\}$. In fact, the set of translation lengths of a minimal set of generators of a free group of tree isometries is not an invariant of the group action, but it depends on the specific choice of generators.

3.2. Field Extensions

In the following section, where we derive explicit KMS states associated to the infinite graphs given by the quotients $\Delta_{\mathbb{K}} / \Gamma$, we also discuss the issue of how the invariants we construct in this way for Mumford curves behave under field extensions of $\mathbb{K}$. For this purpose, we recall here briefly how the graphs $\Delta_{\mathbb{K}}$ and Δ'_Γ are affected when passing to a field extension (*cf.* [**Ma**]). This was also recalled in more detail in [**CM1**].

Let $\mathbb{L} \supset \mathbb{K}$ be a field extension with finite degree, $[\mathbb{L} : \mathbb{K}] < \infty$, and let $e_{\mathbb{L}/\mathbb{K}}$ be its ramification index. Let $\mathcal{O}_{\mathbb{L}}$ and $\mathcal{O}_{\mathbb{K}}$ denote the respective rings of integers. There is an embedding of the sets of vertices $\Delta_{\mathbb{K}}^0 \hookrightarrow \Delta_{\mathbb{L}}^0$ obtained by assigning to a free $\mathcal{O}_{\mathbb{K}}$-module M of rank 2 the free $\mathcal{O}_{\mathbb{L}}$-module of the same rank given by $M \otimes_{\mathcal{O}_{\mathbb{K}}} \mathcal{O}_{\mathbb{L}}$. This operation preserves the equivalence relation. However, the embedding $\Delta_{\mathbb{K}}^0 \hookrightarrow \Delta_{\mathbb{L}}^0$ obtained in this way is not isometric, as one can see from the isomorphism $(\mathcal{O}_{\mathbb{K}}/\mathfrak{m}^r) \otimes \mathcal{O}_{\mathbb{L}} \simeq \mathcal{O}_{\mathbb{L}}/\mathfrak{m}^{re_{\mathbb{L}/\mathbb{K}}}$. This can be corrected by modifying the metric on the graphs $\Delta_{\mathbb{L}}$, for all extensions $\mathbb{L} \supset \mathbb{K}$: if one uses the $\mathbb{K}$-normalized distance

$$(3.1) \qquad d_{\mathbb{K}}(M_1, M_2) := \frac{1}{e_{\mathbb{L}/\mathbb{K}}} d_{\mathbb{L}}(M_1, M_2)$$

on $\Delta_{\mathbb{L}}^0$, one obtains an isometric embedding $\Delta_{\mathbb{K}}^0 \hookrightarrow \Delta_{\mathbb{L}}^0$.

Geometrically, the relation between the Bruhat–Tits trees $\Delta_{\mathbb{K}}$ and $\Delta_{\mathbb{L}}$ is described by the following procedure that constructs $\Delta_{\mathbb{L}}$ from $\Delta_{\mathbb{K}}$ given the values of $e_{\mathbb{L}/\mathbb{K}}$ and $[\mathbb{L} : \mathbb{K}]$. The rule for inserting new vertices and edges when passing to a field extension $\mathbb{L} \supset \mathbb{K}$ is the following.

(1) $e_{\mathbb{L}/\mathbb{K}} - 1$ new vertices $\{v_1, \ldots, v_{e_{\mathbb{L}/\mathbb{K}}-1}\}$ are inserted between each pair of adjacent vertices in $\Delta_{\mathbb{K}}^0$. Let $\Delta_{\mathbb{L},\mathbb{K}}^0$ denote the set of all these additional vertices.

(2) $q^\zeta + 1$ edges depart from each vertex in $\Delta_{\mathbb{K}}^0 \cup \Delta_{\mathbb{L},\mathbb{K}}^0$, with $\zeta = \frac{1}{e_{\mathbb{L}/\mathbb{K}}}[\mathbb{L} : \mathbb{K}]$. Each such edge has length $\frac{1}{e_{\mathbb{L}/\mathbb{K}}}$.

(3) Each new edge attached to a vertex in $\Delta_{\mathbb{K}}^0 \cup \Delta_{\mathbb{L},\mathbb{K}}^0$ is the base of a number of homogeneous trees of valence $q^\zeta + 1$. The number is determined by the property that in the resulting graph the vertex from which the trees stem also has to have valence $q^\zeta + 1$. The Bruhat–Tits tree $\Delta_{\mathbb{L}}$ is the union of $\Delta_{\mathbb{K}}$ with the additional inserted vertices $\Delta_{\mathbb{L},\mathbb{K}}^0$ and the added trees stemming from each vertex.

This procedure is illustrated in Figure 1 which we report here from [**CM1**].

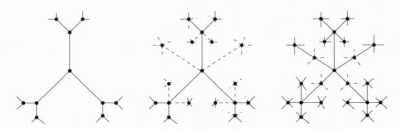

FIGURE 1 The tree $\Delta_{\mathbb{K}}$ for $\mathbb{K} = \mathbb{Q}_2$ and $\Delta_{\mathbb{L}}$ for a field extension with $f = 2$ and $e_{\mathbb{L}/\mathbb{K}} = 2$.

Suppose we are given a p-adic Schottky group $\Gamma \subset \mathrm{PGL}_2(\mathbb{K})$. Since all nontrivial elements of Γ are hyperbolic (the eigenvalues have different valuation), one can see that the two fixed points of any nontrivial element of Γ are in $\mathbb{P}^1(\mathbb{K}) = \partial\Delta_{\mathbb{K}}$. Thus, the limit set Λ_Γ is contained in $\mathbb{P}^1(\mathbb{K})$.

When one considers a finite extension $\mathbb{L} \supset \mathbb{K}$ and the corresponding Mumford curve $X_\Gamma(\mathbb{L}) = \Omega_\Gamma(\mathbb{L})/\Gamma$ with $\Omega_\Gamma(\mathbb{L}) = \mathbb{P}^1(\mathbb{L}) \smallsetminus \Lambda_\Gamma$, one can see this as the boundary of the graph $\Delta_{\mathbb{L}}/\Gamma$. Notice that the subtree $\Delta_{\Gamma,\mathbb{L}}'$ of $\Delta_{\mathbb{L}}$ and the subtree $\Delta_{\Gamma,\mathbb{K}}'$ of $\Delta_{\mathbb{K}}$, both of which have boundary Λ_Γ only differ by the presence of the additional $e_{\mathbb{L}/\mathbb{K}} - 1$ new vertices in between any two adjacent vertices of $\Delta_{\Gamma,\mathbb{K}}'$, while no new direction has been added (the limit points are the same). In particular, this means that the finite graph $\Delta'_{\Gamma,\mathbb{L}}/\Gamma$ is obtained from $\Delta'_{\Gamma,\mathbb{K}}/\Gamma$ by adding $e_{\mathbb{L}/\mathbb{K}} - 1$ vertices on each edge. The infinite graph $\Delta_{\mathbb{K}}/\Gamma$ is obtained by adding to each vertex of the finite graph $\Delta'_{\Gamma,\mathbb{K}}/\Gamma$ a finite number (possibly zero) of infinite homogeneous trees of valence $q + 1$ with base at that vertex. Given the finite graph $\Delta'_{\Gamma,\mathbb{K}}/\Gamma$, the number of such trees to be added at each vertex is determined by the requirement that the valence of each vertex of $\Delta_{\mathbb{K}}/\Gamma$ equals $q + 1$. The infinite graph $\Delta_{\mathbb{L}}/\Gamma$ is obtained from the graph $\Delta_{\mathbb{K}}/\Gamma$ by replacing the homogeneous trees of valence $q + 1$ starting from the vertices of $\Delta'_{\Gamma,\mathbb{K}}/\Gamma$ with homogeneous trees of valence $q^{\zeta} + 1$ stemming from the vertices of $\Delta'_{\Gamma,\mathbb{L}}/\Gamma$, so that each resulting vertex of $\Delta_{\mathbb{L}}/\Gamma$ has valence $q^{\zeta} + 1$.

We analyze the effect of field extensions from the point of view of KMS weights and modular index theory in Section 4.1.

4. Graph KMS Weights on Directed Graphs

Let E be a row-finite graph and $C^*(E)$ the associated graph C^*-algebra.

DEFINITION 4.1. A graph weight on E is a pair of functions $g : E^0 \to [0, \infty)$ and $\lambda : E^1 \to [0, \infty)$ such that for all vertices v

$$g(v) = \sum_{s(e)=v} \lambda(e)g(r(e)).$$

A graph weight is called faithful if $g(v) \neq 0$ for all $v \in E^0$. If $\sum_{v \in E^0} g(v) = 1$, we call (g, λ) a graph state.

REMARK. If $\lambda(e) = 1$ for all $e \in E^1$, we obtain the definition of a graph trace [**T**].

EXAMPLE. Suppose e is a simple loop in a graph, with exit f at the vertex v, and that there are no other loops and no other exits from v, as in Figure 2. Then the condition of Definition 4.1 becomes

$$g(v) = \lambda(e)g(v) + \lambda(f)g(r(f)),$$

so we find $g(v) = \frac{\lambda(f)}{1-\lambda(e)}g(r(f))$.

FIGURE 2 Loop with exit.

REMARK. A graph weight is in fact specified by a single function $h : E^* \to [0, \infty)$. For paths v of length zero, i.e. vertices, $h(v) = g(v)$ and for paths μ of length $k \geq 1$, $h(\mu) = \lambda(\mu_1)\lambda(\mu_2)\cdots\lambda(\mu_k)$. We retain the (g, λ) notation but extend the definition of λ by $\lambda(\mu) = \prod_{i=1}^{k} \lambda(\mu_i)$.

Recall that a path $|\mu|$ has length $|\mu| \preceq k$ if $|\mu| = k$ or $|\mu| < k$ and $r(\mu)$ is a sink. We then have the following result, which can be proved by induction.

LEMMA 4.2. *If* (g, λ) *is a graph weight on* E, *then*

$$g(v) = \sum_{s(\mu)=v, \ |\mu| \preceq k} \lambda(\mu) g(r(\mu)),$$

where for a path $\mu = e_1 \cdots e_j$, $j \leq k$, $\lambda(\mu) = \prod \lambda(e_j)$.

We then define a functional $\phi_{g,\lambda}$ associated to a graph weight (g, λ) as follows.

DEFINITION 4.3. Given (g, λ) on E a graph weight, define $\phi_{g,\lambda} : \mathrm{span}\{S_\mu S_\nu^* : \mu, \nu \in E^*\} \to \mathbb{C}$ by

$$\phi_{g,\lambda}(S_\mu S_\nu^*) := \delta_{\mu,\nu} \lambda(\nu) \phi_{g,\lambda}(p_{r(\nu)}) := \lambda(\nu) \delta_{\mu,\nu} g(r(\nu)).$$

This yields the following useful results.

PROPOSITION 4.4. *Let* $A_c = \mathrm{span}\{S_\mu S_\nu^* : \mu, \nu \in E^*\}$, *and let* (g, λ) *be a faithful graph weight on* E. *Then* A_c *with the inner product*

$$\langle a, b \rangle := \phi_{g,\lambda}(a^* b)$$

is a modular Hilbert algebra (or Tomita algebra).

PROOF. To complete the definition of modular Hilbert algebra, we must supply a complex one parameter group of algebra automorphisms σ_z and verify a number of conditions set out in [**Ta**]. So for $z \in \mathbb{C}$ define

$$\sigma_z(S_\mu S_\nu^*) = \left(\frac{\lambda(\mu)}{\lambda(\nu)} \right)^z S_\mu S_\nu^*.$$

Extending by linearity we can define σ_z on all of A_c. To verify the algebra automorphism property, it suffices to show that

$$\sigma_z(S_\mu S_\nu^* S_\rho S_\kappa^*) = \sigma_z(S_\mu S_\nu^*) \sigma_z(S_\rho S_\kappa^*).$$

First we compute the product on the left hand side.

$$S_\mu S_\nu^* S_\rho S_\kappa^* = \begin{cases} \delta_{\nu,\underline{\rho}} S_\mu S_{\overline{\rho}} S_\kappa^* & |\nu| \leq |\rho| \\ \delta_{\underline{\nu},\rho} S_\mu S_{\overline{\nu}}^* S_\kappa^* & |\nu| \geq |\rho| \end{cases}.$$

So

$$\sigma_z(S_\mu S_\nu^* S_\rho S_\kappa^*) = \begin{cases} \left(\frac{\lambda(\mu \overline{\rho})}{\lambda(\kappa)} \right)^z \delta_{\nu,\underline{\rho}} S_\mu S_{\overline{\rho}} S_\kappa^* & |\nu| \leq |\rho| \\ \left(\frac{\lambda(\mu)}{\lambda(\kappa \underline{\nu})} \right)^z \delta_{\underline{\nu},\rho} S_\mu S_{\overline{\nu}}^* S_\kappa^* & |\nu| \geq |\rho| \end{cases}.$$

On the right hand side we have

$$\sigma_z(S_\mu S_\nu^*) \sigma_z(S_\rho S_\kappa^*) = \left(\frac{\lambda(\mu) \lambda(\rho)}{\lambda(\nu) \lambda(\kappa)} \right)^z \begin{cases} \delta_{\nu,\underline{\rho}} S_\mu S_{\overline{\rho}} S_\kappa^* & |\nu| \leq |\rho| \\ \delta_{\underline{\nu},\rho} S_\mu S_{\overline{\nu}}^* S_\kappa^* & |\nu| \geq |\rho| \end{cases}$$

$$= \begin{cases} \left(\frac{\lambda(\mu) \lambda(\overline{\rho})}{\lambda(\kappa)} \right)^z \delta_{\nu,\underline{\rho}} S_\mu S_{\overline{\rho}} S_\kappa^* & |\nu| \leq |\rho| \\ \left(\frac{\lambda(\mu)}{\lambda(\overline{\nu}) \lambda(\kappa)} \right)^z \delta_{\underline{\nu},\rho} S_\mu S_{\overline{\nu}}^* S_\kappa^* & |\nu| \geq |\rho| \end{cases},$$

and this is easily seen to be the same as the left hand side whenever the product is non-zero. Observe we have used the fact that

$$\lambda(\rho) = \lambda(\underline{\rho}) \lambda(\overline{\rho}).$$

We need to show that $\langle a, b \rangle = \phi_{g,\lambda}(a^* b)$ does define an inner product. Let $a \in A_c$ and let $p \in A_c$ be a finite sum of vertex projections such that $pa = ap = a$ (p is a local unit for a). Then, since $g(v) > 0$ for all $v \in E^0$,

$$b \mapsto \frac{\phi_{g,\lambda}(pbp)}{\phi_{g,\lambda}(p)}$$

is a state on $pC^*(E)p$ and so is positive. Hence

$$\phi_{g,\lambda}(a^* a) \geq 0.$$

To show that the inner product is definite requires more care. First observe that if $\Psi : A_c \to \text{span}\{P_\mu = S_\mu S_\mu^*\}$ is the expectation on to the diagonal subalgebra, then $\phi_{g,\lambda} = \phi_{g,\lambda} \circ \Psi$. So we consider $a \in A_c$ and write $\Psi(a^* a) = \sum_\mu c_\mu P_\mu - \sum_\nu c_\nu P_\nu$. Here the c_μ, $c_\nu > 0$ and none of the paths μ is repeated in the sum. The average $\Psi(a^* a)$ is a positive operator, so if $\Psi(a^* a)$ is non-zero, all the P_ν in the negative part must be subprojections of $\sum_\mu P_\mu$ (otherwise $\Psi(a^* a)$ would have some negative spectrum). Since we are in a graph algebra, the Cuntz–Krieger relations tell us we can write

$$\sum_\mu P_\mu = \sum_\mu \sum_\rho P_{\mu\rho}$$

for some paths ρ extending the various μ, and that moreover all the P_ν appear as some $P_{\mu\rho}$. Thus

$$\Psi(a^* a) = \sum_\mu c_\mu P_\mu - \sum_\nu c_\nu P_\nu = \sum_\mu c_\mu \sum_\rho P_{\mu\rho} - \sum_\nu c_\nu P_\nu = \sum_\mu \sum_\rho d_{\mu\rho} P_{\mu\rho},$$

where the $d_{\mu\rho}$ are necessarily positive. Now we can compute

$$\phi_{g,\lambda}(a^* a) = \phi_{g,\lambda}\left(\sum_\mu \sum_\rho d_{\mu\rho} P_{\mu\rho}\right) = \sum_\mu \sum_\rho d_{\mu\rho} \lambda(\mu\rho) g(r(\mu\rho)) > 0.$$

So now we come to verifying the various conditions defining a modular Hilbert algebra. First, we need to consider the action of A_c on itself by left multiplication. This action is multiplicative,

$$\langle ba, a \rangle := \phi_{g,\lambda}(a^* b^* a) = \langle a, b^* a \rangle,$$

and continuous,

$$\langle ba, ba \rangle = \phi_{g,\lambda}(a^* b^* ba) \leq \|b^* b\| \langle a, a \rangle,$$

where $\|\cdot\|$ denotes the C^*-norm coming from $C^*(E)$. As $A_c^2 = A_c$, the density of A_c^2 in A_c is trivially fulfilled. Also for all real t, $(1+\sigma_t)(S_\mu S_\nu^*) = (1+(\lambda(\mu)/\lambda(\nu))^t)S_\mu S_\nu^*$, and so it is an easy check to see that $(1 + \sigma_t)(A_c)$ is dense in A_c for all real t. Also

$$\langle \sigma_{\bar{z}}(S_\mu S_\nu^*), S_\rho S_\kappa^* \rangle = \left(\frac{\lambda(\mu)}{\lambda(\nu)}\right)^z \langle S_\mu S_\nu^*, S_\rho S_\kappa^* \rangle$$

is plainly analytic in z (the reason for $\sigma_{\bar{z}}$ is that our inner product is conjugate linear in the first variable). Since a finite sum of analytic functions is analytic, $\langle \sigma_{\bar{z}}(a), b \rangle$ is analytic for all $a, b \in A_c$.

The remaining items to check are the compatibility of σ_z with the inner product and involution, and all of these we can check for monomials $S_\mu S_\nu^*$. The first item to check is

$$
\begin{aligned}
(\sigma_z(S_\mu S_\nu^*)^*) &= \left(\frac{\lambda(\mu)}{\lambda(\nu)}\right)^{\bar{z}} S_\nu S_\mu^* \\
&= \left(\frac{\lambda(\nu)}{\lambda(\mu)}\right)^{-\bar{z}} S_\nu S_\mu^* \\
&= \sigma_{-\bar{z}}((S_\mu S_\nu^*)^*).
\end{aligned}
$$

Next we require $\langle \sigma_z(a), b \rangle = \langle a, \sigma_{\bar{z}}(b) \rangle$. So we compute

$$
\begin{aligned}
\langle \sigma_z(S_\mu S_\nu^*), S_\rho S_\kappa^* \rangle &= \left(\frac{\lambda(\mu)}{\lambda(\nu)}\right)^{\bar{z}} g(r(\kappa)) \begin{cases} \delta_{\mu,\rho}\delta_{\nu,\kappa\overline{\mu}}\lambda(\kappa\overline{\mu}) & |\mu| \geq |\rho| \\ \delta_{\mu,\rho}\delta_{\nu\overline{\rho},\kappa}\lambda(\kappa) & |\mu| \leq |\rho| \end{cases} \\
&= g(r(\kappa)) \begin{cases} \left(\frac{\lambda(\mu)}{\lambda(\kappa\overline{\mu})}\right)^{\bar{z}} \delta_{\mu,\rho}\delta_{\nu,\kappa\overline{\mu}}\lambda(\kappa\overline{\mu}) & |\mu| \geq |\rho| \\ \left(\frac{\lambda(\rho)}{\lambda(\nu)}\right)^{\bar{z}} \delta_{\mu,\underline{\rho}}\delta_{\nu\overline{\rho},\kappa}\lambda(\kappa) & |\mu| \leq |\rho| \end{cases} \\
&= g(r(\kappa)) \begin{cases} \left(\frac{\lambda(\mu)}{\lambda(\kappa)}\right)^{\bar{z}} \delta_{\underline{\mu},\rho}\delta_{\nu,\kappa\overline{\mu}}\lambda(\kappa\overline{\mu}) & |\mu| \geq |\rho| \\ \left(\frac{\lambda(\rho)}{\lambda(\underline{\kappa})}\right)^{\bar{z}} \delta_{\mu,\underline{\rho}}\delta_{\nu\overline{\rho},\kappa}\lambda(\kappa) & |\mu| \leq |\rho| \end{cases} \\
&= g(r(\kappa)) \begin{cases} \left(\frac{\lambda(\rho)}{\lambda(\kappa)}\right)^{\bar{z}} \delta_{\underline{\mu},\rho}\delta_{\nu,\kappa\overline{\mu}}\lambda(\kappa\overline{\mu}) & |\mu| \geq |\rho| \\ \left(\frac{\lambda(\rho)\lambda(\overline{\rho})}{\lambda(\underline{\kappa})\lambda(\overline{\rho})}\right)^{\bar{z}} \delta_{\mu,\underline{\rho}}\delta_{\nu\overline{\rho},\kappa}\lambda(\kappa) & |\mu| \leq |\rho| \end{cases} \\
&= \langle S_\mu S_\nu^*, \sigma_{\bar{z}}(S_\rho S_\kappa^*)\rangle,
\end{aligned}
$$

the last line following (when $|\mu| \leq |\rho|$), since the final segments of ρ and κ must agree if the inner product is non-zero. The final condition to check is that $\langle \sigma_1(a^*), b^* \rangle = \langle b, a \rangle$.

First we compute

$$
\begin{aligned}
\langle \sigma_1(S_\mu S_\nu^*), S_\rho S_\kappa^* \rangle &= \frac{\lambda(\mu)}{\lambda(\nu)} \begin{cases} \delta_{\mu,\rho}\delta_{\nu\overline{\rho},\kappa}\lambda(\kappa)g(r(\kappa)) & |\mu| \leq |\rho| \\ \delta_{\mu,\rho}\delta_{\nu,\kappa\overline{\mu}}\lambda(\kappa\overline{\mu})g(r(\mu)) & |\mu| \geq |\rho| \end{cases} \\
&= \begin{cases} \lambda(\rho)g(r(\kappa))\delta_{\mu,\underline{\rho}}\delta_{\nu\overline{\rho},\kappa} & |\mu| \leq |\rho| \\ \lambda(\mu)g(r(\mu))\delta_{\underline{\mu},\rho}\delta_{\nu,\kappa\overline{\mu}} & |\mu| \geq | \end{cases}.
\end{aligned}
$$

Next we have

$$
\begin{aligned}
\langle S_\kappa S_\rho^*, S_\nu S_\mu^* \rangle &= \begin{cases} \delta_{\underline{\kappa},\nu}\delta_{\rho,\mu\overline{\kappa}}\lambda(\mu\overline{\kappa})g(r(\kappa)) & |\kappa| \geq |\nu| \\ \delta_{\kappa,\underline{\nu}}\delta_{\rho\overline{\nu},\mu}\lambda(\mu)g(r(\mu)) & |\kappa| \leq |\nu| \end{cases} \\
&= \begin{cases} \lambda(\rho)g(r(\kappa))\delta_{\underline{\kappa},\nu}\delta_{\rho,\mu\overline{\kappa}} & |\kappa| \geq |\nu| \\ \lambda(\mu)g(r(\mu))\delta_{\kappa,\underline{\nu}}\delta_{\rho\overline{\nu},\mu} & |\kappa| \leq |\nu| \end{cases}.
\end{aligned}
$$

Now for the inner product to be non-zero, we must have $|\rho| + |\nu| = |\kappa| + |\mu|$, and so $|\mu| \leq |\rho| \Leftrightarrow |\nu| \leq |\kappa|$. Comparing the Kronecker deltas in the corresponding cases then yields the desired equality for monomials, and the general case follows by linearity.

THEOREM 4.5. *Let E be a locally finite directed graph. Then there is a one-to-one correspondence between gauge invariant norm lower semicontinuous faithful semifinite functionals on $C^*(E)$ and faithful graph weights on E.*

PROOF. This is proved similarly to Proposition 3.9 in [**PRen**] where the tracial case is considered.

First suppose that (g, λ) is a faithful graph weight on E. Then $(A_c, \phi_{g,\lambda})$ is a modular Hilbert algebra. Since the left representation of A_c on itself is faithful, each p_v, $v \in E^0$, is represented by a non-zero projection. Let the representation be π.

The gauge invariance of $\phi_{g,\lambda}$ shows that for all $z \in \mathbb{T}$, the map $\gamma_z : A_c \to A_c$ extends to a unitary $U_z : \mathcal{H} \to \mathcal{H}$, where $\mathcal{H}$ is the completion of A_c in the Hilbert space norm. It is easy to show that $U_z \pi(a) U_{\bar{z}}(b) = \pi(\gamma_z(a))(b)$ for a, $b \in A_c$. Hence $U_z \pi(a) U_{\bar{z}} = \pi(\gamma_z(a))$, and so $\alpha_z(\pi(a)) := U_z \pi(a) U_{\bar{z}}$ gives a point norm continuous action of $\mathbb{T}$ on $\pi(A_c)$ implementing the gauge action.

We may thus invoke the gauge invariant uniqueness theorem [**BPRS**] to deduce that the representation extends to a faithful representation of $C^*(E)$.

Now $\pi(C^*(E)) \subset \pi(A_c)'' = \overline{\pi(A_c)}^{u.w.}$, the ultra-weak closure. Then Theorem 2.5 in [**Ta**] shows that the functional $\phi_{g,\lambda}$ extends to a faithful, normal semifinite weight $\psi_{g,\lambda}$ on the left von Neumann algebra of A_c, $\pi(A_c)''$.

Restricting the extension $\psi_{g,\lambda}$ to $C^*(E)$ gives a faithful weight. It is norm semifinite, since it is defined on A_c which is dense in $C^*(E)$. Finally, if $a_j \to a$ in norm, then the a_j converge ultra-weakly as well, so $\liminf \psi_{g,\lambda}(a_j) \geq \psi_{g,\lambda}(a)$, which shows that the restriction of $\psi_{g,\lambda}$ to $C^*(E)$ is norm lower semicontinuous.

To get the gauge invariance of $\psi_{g,\lambda}$ we recall that $T \in \pi(A_c)''$ is in the domain of $\psi_{g,\lambda}$ if and only if $T = \pi(\xi)\pi(\eta)^*$ for left bounded elements ξ, $\eta \in \mathcal{H}$. Then $\psi_{g,\lambda}(T) = \psi_{g,\lambda}(\pi(\xi)\pi(\eta)^*) := \langle \xi, \eta \rangle$. As $U_z\xi$ and $U_z\eta$ are also left bounded we have

$$\psi_{g,\lambda}(U_z T U_{\bar{z}}) = \psi_{g,\lambda}(U_z \pi(\xi)\pi(\eta)^* U_{\bar{z}}) = \psi_{g,\lambda}(U_z \pi(\xi)(U_z \pi(\eta))^*)$$
$$= \psi_{g,\lambda}(\pi(\gamma_z(\xi))\pi(\gamma_z(\eta))^*) = \langle U_z\xi, U_z\eta \rangle$$
$$= \langle \xi, \eta \rangle = \psi_{g,\lambda}(T).$$

So $\psi_{g,\lambda}$ is α_z invariant, and $a \mapsto \psi_{g,\lambda}(\pi(a))$ defines a faithful semifinite norm lower semicontinuous gauge invariant weight on $C^*(E)$.

Conversely, suppose that ϕ is a faithful semifinite norm lower semicontinuous weight on $C^*(E)$ which is gauge invariant. Define

$$g(v) = \phi(p_v), \qquad \lambda(e) = \frac{\phi(S_e S_e^*)}{\phi(S_e^* S_e)}.$$

It is readily checked that (g, λ) is a faithful graph weight.

In order to make contact with the index theory for KMS weights set out in [**CNNR**], we require the action associated to our graph weight to be a circle action satisfying the spectral subspace condition, namely that $A_k A_k^*$ should be complemented in the fixed point algebra F.

A sufficient condition to obtain a circle action is that $\lambda(e) = \lambda^{n_e}$ for every edge $e \in E^1$, where now $n : E^1 \to \mathbb{Z}$, and $\lambda \in (0, 1)$. In fact we will simplify matters further and deal here just with a function of the form $n_e \in \{0, 1\}$ for all $e \in E^1$. While this is rather restrictive, it suffices for the examples we consider here. We call such functions *special graph weights*. In order for our special graph weight to accurately reflect the properties of the zhyvot action on our graph, we will also require that $n_e = 1$ if and only if $e \in M^1$.

Also all the graphs we wish to consider are graphs with *finite* zhyvots, with the rest of the graph being composed of trees. It is easy to construct faithful graph traces, i.e. special graph weights with $n_e \equiv 0$, on (unions of) trees, given just the values of the trace on the root(s) [**PRen**]. Thus, it seems that we need only worry about constructing a graph state on the zhyvot.

However, there is a subtlety: *neglecting the trees can affect the existence of special graph weights*.

In fact, we show in the following genus two example that appending trees to the finite graph is necessary for solutions to the graph weight equations by showing that there are no solutions on the finite graph itself. Later we show that adding trees to this example does indeed allow solutions of the graph weight equations.

EXAMPLE. Graph states on $SU_q(2)$. Recall that for $0 \le q < 1$ the C^*-algebra $SU_q(2)$ is (isomorphic to) the graph C^*-algebra of the graph in Figure 3 [**HS**].

FIGURE 3 Graph of $SU_q(2)$.

We want to solve

$$g(v) = \lambda^{n_1} g(v) + \lambda^{n_2} g(w), \quad g(w) = \lambda^{n_3} g(w).$$

First $\lambda^{n_3} = 1$, so for $\lambda \neq 1$ (which we are not interested in), $n_3 = 0$. Then

$$g(v) = \frac{\lambda^{n_2}}{1 - \lambda^{n_1}} g(w).$$

Imposing the requirement that we have a graph state, $g(v) + g(w) = 1$, we get

$$g(v) = \frac{\lambda^{n_2}}{1 - \lambda^{n_1} + \lambda^{n_2}}, \quad g(w) = \frac{1 - \lambda^{n_1}}{1 - \lambda^{n_1} + \lambda^{n_2}}.$$

Observe that if $n_1 = n_2$ we have

$$g(v) = 1 - \lambda^{n_1}, \quad g(w) = \lambda^{n_1}.$$

In this case we get the Haar state by setting $\lambda = q^{2/n_1}$ [**CRT**].

Observe that for $\lambda = 1$ the only non-zero graph trace vanishes on w, and we get the usual trace on the top circle with the kernel of $\phi_g = C(S^1) \otimes \mathcal{K}$. For $\lambda > 1$, we get the same family as before by replacing (n_1, n_2) by $(-n_1, -n_2)$. For a special graph state we must have $n_1 = 1$ and $n_3 = 0$. For n_2 we may choose either value.

So it seems we cannot obtain a special graph weight with $n_e = 1$ for all edges in the zhyvot. However, if we add trees to the graph, the loop on the vertex w will acquire exits, and then it is easy to construct special graph weights with $n_e = 1$ precisely when e is an edge in the zhyvot.

LEMMA 4.6. *Let M be a finite graph and label the vertices $v_1, \ldots, v_n$ so that the sinks, if any, are $v_{r+1}, \ldots, v_n$. Let $p_{jk} \in \mathbb{N} \cup \{0\}$ be the number of edges from v_j to v_k. Then M has a faithful special graph state (g, λ, n) for $\lambda \in (0,1)$ and $n : E^1 \to \{1\} \subset \mathbb{N}$ if and only if the matrix*

$$\begin{pmatrix} (\lambda p_{jk})_{r \times r} & (\lambda p_{jk})_{r \times n-r} \\ 0_{n-r \times r} & Id_{n-r \times n-r} \end{pmatrix}$$

has an eigenvector $(x_1, \ldots, x_n)^T$ with eigenvalue 1 and $x_j > 0$ for $j = 1, \ldots, n$.

PROOF. The equations defining a special graph weight for $\lambda \in (0,1)$ are

$$g(v_j) = \sum_{k=1}^{n} \lambda p_{jk} g(v_k) \quad j = 1, \ldots, r, \qquad g(v_j) = \sum_{k=1}^{n} \delta_{jk} g(v_k) \quad j = r+1, \ldots, n.$$

This gives the necessary and sufficient condition for the existence of a special graph weight $\tilde{g}$ with $\tilde{g}(v_j) = x_j$. To get a state we normalize the eigenvector.

The lemma can obviously be generalized to deal with general graph states on finite graphs. Moreover we note that work in progress is extending the modular index theory to quasi-periodic actions of $\mathbb{R}$, and a modified version of the above lemma will give existence criteria in the quasi-periodic case also.

COROLLARY 4.7. *Let E be a locally finite directed graph without sources and with finite zhyvot $M \subset E$. Let (g, λ, n) be a special graph weight on E for $\lambda \in (0,1)$, $n|_{M^1} \equiv 1$ and $n|_{E^1 \setminus M^1} \equiv 0$. Then $\phi_{g,\lambda}$ extends to a positive norm lower semicontinuous gauge invariant (usual gauge action) functional on $C^*(E)$. The functional $\phi_{g,\lambda}$ is faithful iff (g, λ) is faithful. We have the formula*

$$\phi_{g,\lambda}(ab) = \phi_{g,\lambda}(\sigma(b)a), \quad a, b \in A_c,$$

where $\sigma(S_\mu S_\nu^*) = \frac{\lambda(\nu)}{\lambda(\mu)} S_\mu S_\nu^*$ is a densely defined regular automorphism of $C^*(E)$. In particular, $\phi_{g,\lambda}$ is a KMS weight on $C^*(E)$ for the (modified) zhyvot action

$$\sigma_t(S_\mu S_\nu^*) = \left(\frac{\lambda(\mu)}{\lambda(\nu)} \right)^{it} S_\mu S_\nu^* = \lambda^{(|\mu|_\sigma - |\nu|_\sigma)it} S_\mu S_\nu^*.$$

PROOF. The formula $\phi_{g,\lambda}(ab) = \phi_{g,\lambda}(\sigma(b)a)$ follows from Proposition 4.4. Together with the norm lower semicontinuity and the gauge invariance coming from Theorem 4.5, we see that $\phi_{g,\lambda}$ is a KMS weight on $C^*(E)$.

4.1. The Effect of Field Extensions

Suppose that we start with the infinite graph $\Delta_{\mathbb{K}}/\Gamma$ and we pass to the graph $\Delta_{\mathbb{L}}/\Gamma$, for $\mathbb{L}$ a finite extension of $\mathbb{K}$, by the procedure described in Section 3. As we have seen, this procedure consists of inserting $e_{\mathbb{L}/\mathbb{K}} - 1$ new vertices along edges and attaching infinite trees to the old and new vertices, so that the resulting valence of all vertices is the desired $q^\zeta + 1$, with $\zeta = [\mathbb{L} : \mathbb{K}]/e_{\mathbb{L}/\mathbb{K}}$.

Here we show that, if we have constructed a special graph weight for $\Delta_{\mathbb{K}}/\Gamma$, then we obtain corresponding special graph weights on all the $\Delta_{\mathbb{L}}/\Gamma$ for finite extensions $\mathbb{L} \supset \mathbb{K}$. The special graph weight for $\Delta_{\mathbb{L}}/\Gamma$ is obtained from that of $\Delta_{\mathbb{K}}/\Gamma$ by solving explicit equations.

PROPOSITION 4.8. *Let E be a locally finite directed graph with no sources and with finite zhyvot M. Suppose that (g, λ, n) is a faithful special graph weight on E with $n|_M \equiv 1$ and $n|_{E \setminus M} \equiv 0$, and $\lambda \in (0, 1)$. Let F be the graph obtained from E by inserting some new vertices along edges of M and attaching any positive number of trees to the new vertices and any number of trees to the vertices of E. Then F has finite zhyvot $\tilde{M}$, with $\tilde{M}^0 = \{v \in F^0 : v \in M^0 \text{ or } v = r(e), e \in F^1, s(e) \in M^0\}$ and $\tilde{M}^1 = \{e \in F^1 : r(e) \in \tilde{M}^0\}$, and a faithful special graph weight $(\tilde{g}, \lambda, \tilde{n})$ with the same value of λ and $\tilde{n}|_{\tilde{M}} \equiv 1$, $\tilde{n}|_{F \setminus \tilde{M}} \equiv 0$.*

PROOF. It is clear that F is a graph and that $\tilde{M}$ is a zhyvot for F, since we cannot introduce sources when vertices are only introduced by splitting an existing edge into two, since one of them has the range of the new vertex. Since extending a faithful graph state on the zhyvot $\tilde{M}$ to any graph obtained by adding trees to vertices is possible, we need only be concerned with building a new special graph state on the zhyvot.

The problem turns out to be local, and we refer to Figure 4 for the notation we shall use.

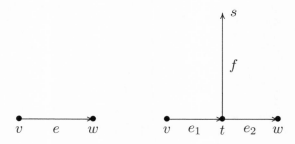

FIGURE 4 Inserting a vertex.

We suppose we are given an edge e with $s(e) = v$ and $r(e) = w$ in a graph with special graph weight (g, λ, n). So we have $g(v) = \lambda(e)g(w) + R$, where $R = \sum_{s(\tilde{e}) = v, \ \tilde{e} \neq e} \lambda(e)g(r(\tilde{e}))$.

We now introduce a new vertex t splitting e into two edges e_1, e_2 with $s(e_1) = v$, $r(e_1) = t$, $s(e_2) = t$, $r(e_2) = w$. We also introduce a new edge f with $s(f) = t$, $r(f) = s$ for some other vertex s. We observe that we could add several edges $f_1, \ldots, f_n$ with source t, and we indicate the modifications required in this case below.

We want to construct a special graph weight $\tilde{g}$ without changing our parameter λ or the values of the graph weight where it is already defined. Thus we would like to solve

$$\tilde{g}(v) = \lambda \tilde{g}(t) + R, \quad \tilde{g}(t) = \lambda \tilde{g}(w) + \lambda \tilde{g}(s), \quad \tilde{g}(v) = g(v), \quad \tilde{g}(w) = g(w).$$

A solution to the above equations is as follows. Define $\tilde{g} = g$ on all previously existing vertices, and on the vertex $s = r(f)$ set $\tilde{g}(s) = \frac{1-\lambda}{\lambda}g(w)$. Then the above equations are satisfied, and we obtain $\tilde{g}(t) = g(w)$. If we have multiple edges $f_1, \ldots, f_n$ then replacing $\tilde{g}(s)$ by $\sum_j \tilde{g}(r(f_j))$ we have a solution provided

$$\sum_j \tilde{g}(r(f_j)) = \frac{1-\lambda}{\lambda}g(w),$$

and thus we may just set the value of $\tilde{g}(r(f_j))$ to be $\frac{1}{n}\frac{1-\lambda}{\lambda}g(w)$.

Finally, define $\tilde{n}$ by making it identically one on edges in $\tilde{M}$ and identically zero on other edges. Observe that f is not an edge in the zhyvot.

5. Modular Index Invariants of Mumford Curves

We have seen that we can associate directed graphs to Mumford curves. These graphs consist of a finite graph along with trees emanating out from some or all of its vertices. Though we do not have a general existence result, we have shown in Lemma 4.6 that generically we can construct "special" graph weights on such graphs. *We will assume in this section that the conditions of Lemma 4.6 hold.* Again we remark that this restriction is likely to prove artificial.

From this we can construct both an equivariant Kasparov module $(X, \mathcal{D})$ and a modular spectral triple $(\mathcal{A}, \mathcal{H}, \mathcal{D})$ as in [**CNNR**]. Here the equivariance is with respect to the modified zhyvot action introduced in Corollary 4.7 To compute the index pairing using the results of [**CNNR**], we need only be able to compute traces of operators of the form $p\Phi_k$, where $p \in F$ is a projection and the Φ_k are spectral projections of the $\mathbb{T}$ action (or of $\mathcal{D}$ or of Δ).

In the specific case of Mumford curves, the modular index pairings we would like to compute are with the modular partial isometries arising from loops in the central graph corresponding to the action on $\Delta_{\mathbb{K}}$ of each one of a chosen set of generators $\{\gamma_1, \dots, \gamma_g\}$ of the Schottky group Γ. These correspond to the fundamental closed geodesics in $\Delta_{\mathbb{K}}/\Gamma$, by analogy with the fundamental closed geodesics in the hyperbolic 3-dimensional handlebody $\mathbb{H}/\Gamma$ considered in [**Man**], [**CM**]. The lengths of these fundamental closed geodesics are the Schottky invariants of $(\Gamma, \{\gamma_1, \dots, \gamma_g\})$ introduced above.

We introduce some notation so that we may effectively describe these projections. The zhyvot of the graph we denote by M. Since outside of M our graph is a union of trees, we may and do suppose that the restriction of our graph weight to the exterior of M is a graph trace. That is, for all $v \notin M$ we have

$$g(v) = \sum_{s(e)=v} g(r(e)),$$

and so for $e \notin M$, $\sigma_t(S_e) = S_e$.

In [**CNNR**] we showed how to construct a Kasparov module for $A = C^*(E)$ and $F = A^\sigma$. We let $\Phi : A \to F$ be the expectation given by averaging over the circle action and define an inner product on A with values in F by setting

$$(a|b) := \Phi(a^*b).$$

We also use the notation $\Theta_{x,y}$ for the rank one endomorphisms $\Theta_{x,y} = x(y|\cdot)$.

We denote the C^*-module completion by X and note that it is a full right F-module. There is an obvious action of A by left multiplication, and this action is adjointable.

On the dense subspace $A_c \subset X$ we define an unbounded operator $\mathcal{D}$ by defining it on generators and extending by linearity. We set

$$\mathcal{D}S_\mu S_\nu^* := (|\mu|_\sigma - |\nu|_\sigma) S_\mu S_\nu^*,$$

so that up to a factor of $\log(\lambda)$, $\mathcal{D}$ is the generator of the zhyvot action. Observe that for a path μ contained in the exterior of M we have $|\mu|_\sigma = 0$. The closure of $\mathcal{D}$ is self-adjoint, regular, and for all $a \in A_c$ the endomorphism of X given by $a(1 + \mathcal{D}^2)^{-1}$ is a compact endomorphism.

It is proved in [**CNNR**] that $(_A X_F, \mathcal{D})$ is an equivariant Kasparov module for A-F (with respect to the zhyvot action), and so it defines a class in $KK^{1,\mathbb{T}}(A, F)$.

Similarly, if we set $\mathcal{H} := \mathcal{H}_{\phi_{g,\lambda}}$ to be the GNS space of A associated to the weight $\phi_{g,\lambda}$, we obtain an unbounded operator $\mathcal{D}$ (with the same definition on $A_c \subset \mathcal{H}$). The triple $(\mathcal{A}, \mathcal{H}, \mathcal{D})$ is not quite a spectral triple.

The compact endomorphisms of the C^*-module X, $End_F^0(X)$, act on $\mathcal{H}$ in a natural fashion [**CNNR**], and we define a von Neumann algebra by $\mathcal{N} = (End_F^0(X))''$. There is a natural trace $\mathrm{Tr}_{\phi_{g,\lambda}}$ on $\mathcal{N}$ satisfying

(5.1) $\mathrm{Tr}_{\phi_{g,\lambda}}(\Theta_{x,y}) = \phi_{g,\lambda}((y|x))$

for all x, $y \in X$. We define a weight $\phi_{\mathcal{D}}$ on $\mathcal{N}$ by

$$\phi_{\mathcal{D}}(T) := \mathrm{Tr}_{\phi_{g,\lambda}}(\lambda^{\mathcal{D}} T), \quad T \in \mathcal{N}.$$

Then the modular group of $\phi_{\mathcal{D}}$ is inner, and we let $\mathcal{M} \subset \mathcal{N}$ denote the fixed point algebra of the modular action. Then $\phi_{\mathcal{D}}$ restricts to a trace on $\mathcal{M}$, and it is shown in [**CNNR**] that

$$f(1 + \mathcal{D}^2)^{-1/2} \in \mathcal{L}^{(1,\infty)}(\mathcal{M}, \phi_{\mathcal{D}}), \quad f \in F.$$

Using this information it is shown in [**CNNR**] that there is a pairing between $(A_c, \mathcal{H}, \mathcal{D})$ and homogeneous (for the zhyvot action) partial isometries $v \in A_c$ with source and range projections in F. The pairing is given by the spectral flow

$$sf_{\phi_{\mathcal{D}}}(vv^*\mathcal{D}, v\mathcal{D}v^*) \in \mathbb{R},$$

this being well defined, since $v\mathcal{D}v^* \in \mathcal{M}$. The numerical spectral flow pairing and the equivariant KK pairing are compatible.

In order to compute the spectral flow, we need explicit formulae for the spectral projections of $\mathcal{D}$ both as an operator on X and as an operator on $\mathcal{H}$.

To this end, if $v \in E^0$ and $m > 0$ we set $|v|_m =$ the number of paths μ with $|\mu|_\sigma = m$ and $r(\mu) = v$. It is important that our graph is locally finite and has no sources, so that $0 < |v|_m < \infty$ for all $v \in E^0$ and $m > 1$.

PROPOSITION 5.1. *The spectral projections of $\mathcal{D}$ can be represented as follows.*
(1) *For $m > 0$*

$$\Phi_m = \sum_{\substack{|\mu|_\sigma = m \\ s(\mu) \in M \\ r(\mu) \in M}} \Theta_{S_\mu, S_\mu}.$$

(2) *For $m = 0$*

$$\Phi_0 = \sum_{v \in E^0} \Theta_{p_v, p_v}.$$

(3) *For $m < 0$, $v \in E^0$*

$$p_v \Phi_m = \frac{1}{|v|_{|m|}} \sum_{\substack{|\mu|_\sigma = |m| \\ r(\mu) = v}} \Theta_{S_\mu^*, S_\mu^*}.$$

In all cases, for (a subprojection of) a vertex projection p_v, the operator $p_v \Phi_m$ is a finite rank endomorphism of the Kasparov module X and in the domain of ϕ_D as an operator in $\mathcal{M} \subset \mathcal{N}$.

PROOF. We first recall that the C^*-module inner product is given by

$$(x|y)_R = \Phi(x^* y).$$

Now let $S_\rho S_\gamma^* \in X$ and with $|\mu|_\sigma > 0$ consider

$$\begin{aligned}
\Theta_{S_\mu, S_\mu} S_\rho S_\gamma^* &= S_\mu (S_\mu | S_\rho S_\gamma^*)_R \\
&= \delta_{|\mu|_\sigma, |\rho|_\sigma - |\gamma|_\sigma} S_\mu S_\mu^* S_\rho S_\gamma^* \\
&= \delta_{|\mu|_\sigma, |\rho|_\sigma - |\gamma|_\sigma} \delta_{\mu, \rho} S_\rho S_\gamma^*.
\end{aligned}$$

Hence this is non-zero only when $|\rho|_\sigma = |\gamma|_\sigma + |\mu|_\sigma \geq |\mu|_\sigma$ and $\rho = \mu$. Thus when $|\rho|_\sigma - |\gamma|_\sigma = m > 0$

$$\sum_{\substack{|\mu|_\sigma = m \\ s(\mu) \in M \\ r(\mu) \in M}} \Theta_{S_\mu, S_\mu} S_\rho S_\gamma^* = \Theta_{\rho, \rho} S_\rho S_\gamma^* = S_\rho S_\gamma^*,$$

and $\sum \Theta_{S_\mu, S_\mu}$ is zero on all other elements of X. Hence the claim for the positive spectral projections is proved, since finite sums of generators $S_\rho S_\gamma^*$ are dense in X. A similar argument proves the claim for the zero spectral projection.

For the negative spectral projections, we observe that

$$\begin{aligned}
\Theta_{S_\mu^*, S_\mu^*} S_\rho S_\gamma^* &= S_\mu^* \delta_{r(\mu), s(\rho)} \delta_{|\mu|_\sigma + |\rho|_\sigma, |\gamma|_\sigma} S_\mu S_\rho S_\gamma^* \\
&= \delta_{r(\mu), s(\rho)} \delta_{|\mu|_\sigma + |\rho|_\sigma, |\gamma|_\sigma} S_\rho S_\gamma^*.
\end{aligned}$$

Summing over all paths μ with $|\mu|_\sigma = m > 0$ and $r(\mu) = s(\rho)$ gives

$$\sum_{\substack{|\mu|_\sigma = m \\ r(\mu) = s(\rho)}} \Theta_{S_\mu^*, S_\mu^*} S_\rho S_\gamma^* = \delta_{|\rho|_\sigma - |\gamma|_\sigma, -m} |s(\rho)|_m S_\rho S_\gamma^*.$$

Hence for a vertex $v \in E^0$

$$\frac{1}{|v|_{|m|}} \sum_{\substack{|\mu|_\sigma = |m| \\ r(\mu) = v}} \Theta_{S_\mu^*, S_\mu^*} S_\rho S_\gamma^* = \delta_{|\rho|_\sigma - |\gamma|_\sigma, -m} \delta_{s(\rho), v} S_\rho S_\gamma^*$$

$$= p_v \Phi_{-m} S_\rho S_\gamma^*.$$

In all cases $p_v \Phi_k$ is a finite sum of rank one endomorphisms, and so has finite rank. In particular they are expressed as finite rank endomorphisms of $\mathrm{dom}(\phi)^{1/2} \subset X$, since for a graph weight g, λ all the S_μ and S_μ^* lie in the domain of the associated weight ϕ. This ensures that these endomorphisms extend by continuity to the Hilbert space completion of $\mathrm{dom}(\phi)^{1/2}$, and by the construction of ϕ_D, each $p_v \Phi_k \in \mathcal{M} \subset \mathcal{N}$ has finite *trace*. Similar comments evidently apply to projections of the form $S_\mu S_\mu^*$, since this is a subprojection of $p_{s(\mu)}$.

For large positive k, the computation of $\phi_D(S_\mu S_\mu^* \Phi_k)$ is extremely difficult, and needs to be handled on a "graph-by-graph" basis. However it turns out that we need only compute for $|k| \leq |\mu|_\sigma$, and this is completely tractable.

LEMMA 5.2. *Let γ be a path in E with $s(\gamma)$, $r(\gamma) \in M$ and $|\gamma|_\sigma > 0$. Then for all $k \in \mathbb{Z}$ with $|\gamma|_\sigma \geq |k|$ we have*

$$\phi_D(S_\gamma S_\gamma^* \Phi_k) = \phi_{g,\lambda}(S_\gamma S_\gamma^*) = \lambda^{|\gamma|_\sigma} g(r(\gamma)).$$

For a path of length zero (i.e. a vertex v) in M and $k < 0$ we have

$$\phi_D(p_v \Phi_k) = \phi_{g,\lambda}(p_v) = g(v).$$

PROOF. We begin with $|\gamma|_\sigma \geq k > 0$. In this case the definitions yield

$$\begin{aligned}
\phi_D(S_\gamma S_\gamma^* \Phi_k) &= \sum_{\substack{|\mu|_\sigma = k \\ s(\mu) \in M \\ r(\mu) \in M}} \phi_D(S_\gamma S_\gamma^* \Theta_{S_\mu, S_\mu}) \\
&= \sum_{\substack{|\mu|_\sigma = k \\ s(\mu) \in M \\ r(\mu) \in M}} \lambda^k \phi_{g,\lambda}(S_\mu^* S_\gamma S_\gamma^* S_\mu) \qquad \text{(by (5.1))} \\
&= \lambda^k \phi_{g,\lambda}(S_{\overline{\gamma}} S_{\overline{\gamma}}^*) \\
&= \lambda^{|\gamma|_\sigma} \phi_{g,\lambda}(S_{\overline{\gamma}}^* S_{\overline{\gamma}}) = \lambda^{|\gamma|_\sigma} \phi_{g,\lambda}(p_{r(\gamma)}) \\
&= \phi_{g,\lambda}(S_\gamma S_\gamma^*).
\end{aligned}$$

So now consider $|\gamma|_\sigma > 0$ or $\gamma = v$ for some vertex $v \in M$ and $k > 0$. In the latter case, $S_\gamma S_\gamma^* = p_v p_v = p_v = S_\gamma^* S_\gamma$. Then

$$\begin{aligned}
\phi_D(S_\gamma S_\gamma^* \Phi_{-k}) &= \frac{1}{|s(\gamma)|_k} \sum_{\substack{|\mu|_\sigma = k \\ r(\mu) = s(\gamma)}} \lambda^{-k} \phi_{g,\lambda}(S_\mu S_\gamma S_\gamma^* S_\mu^*) \\
&= \frac{1}{|s(\gamma)|_k} \sum_{\substack{|\mu|_\sigma = k \\ r(\mu) = s(\gamma)}} \lambda^{|\gamma|_\sigma} \phi_{g,\lambda}(p_{r(\gamma)}) \\
&= \lambda^{|\gamma|_\sigma} \phi_{g,\lambda}(p_{r(\gamma)}) \\
&= \phi_{g,\lambda}(S_\gamma S_\gamma^*).
\end{aligned}$$

This completes the proof.

We now have the necessary ingredients to compute the modular index pairing with S_γ where γ here denotes a loop contained in the finite graph $M = \Delta'_\Gamma / \Gamma$ and corresponding to an element in the chosen set of generators of the Schottky group. We suppose that $k = |\gamma|_\sigma$ is non-zero, so that the loop is nontrivial, and denote $\beta := -\log \lambda$.

We then have the following result, which gives Theorem 1.1.

THEOREM 5.3. *Consider the graph C^*-algebra $C^*(\Delta_{\mathbb{K}}/\Gamma)$ with the zhyvot circle action, under the assumption of Lemma 4.6. The modular index pairing with S_γ, for γ a generator of Γ, determines the translation length $\ell(\gamma)$ and is in turn determined by it.*

PROOF. By Lemma 4.10 of [**CNNR**] and Lemma 5.2 we have

$$sf_{\phi_{\mathcal{D}}}(S_\gamma S_\gamma^* \mathcal{D}, S_\gamma \mathcal{D} S_\gamma^*) = -\sum_{j=0}^{k-1} e^{-\beta j}\, \mathrm{Tr}_\phi(S_\gamma S_\gamma^* \Phi_j)$$

$$= -\sum_{j=0}^{k-1} \phi_{\mathcal{D}}(S_\gamma S_\gamma^* \Phi_j)$$

$$= -k\phi_{g,\lambda}(S_\gamma S_\gamma^*)$$

$$= -k\lambda^k g(r(\gamma)) = -ke^{-\beta k} g(r(\gamma)).$$

Since we assume that (g, λ) are given as part of the data of our special graph weight, we can extract the integer k. Moreover, k determines the value of the index pairing.

Thus, we see that the Schottky invariants of the data $(\Gamma, \{\gamma_1, \ldots, \gamma_g\})$ can be recovered from the modular index pairing and in fact determine it, for a given graph weight (g, λ). This confirms the fact that the noncommutative geometry of the graph algebra $C^*(E) = C^*(\Delta_{\mathbb{K}}/\Gamma)$ maintains the geometric information related to the action of the Schottky group on the Bruhat–Tits tree $\Delta_{\mathbb{K}}$. This is still less information than being able to reconstruct the curve, since the Schottky invariants only depend on the valuation. We show explicitly in the next section how the construction of graph weights works in some simple examples of Mumford curves.

6. Low Genus Examples

We consider here the cases of the elliptic curve with Tate uniformization (genus one case) and the three genus two cases considered in [**CM1**]. In each of these examples we give an explicit construction of graph weights and compute the relevant modular index pairings, showing that one recovers from them the Schottky invariants. Notice that, for the genus two cases, the finite zhyvot graphs Δ_Γ/Γ are the same considered in [**CM1**], which we report here, though in the present setting we work with the infinite graphs $\Delta_{\mathbb{K}}/\Gamma$. We discuss here the graph weight equation on the zhyvot graph and on the infinite graph $\Delta_{\mathbb{K}}/\Gamma$.

EXAMPLE: GENUS ONE. As a first application to Mumford curves we consider the simplest case of genus one. In this case, the Schottky uniformization is the Tate uniformization of p-adic elliptic curves. The p-adic Schottky group is just a copy of $\mathbb{Z}$ generated by a single hyperbolic element in $\mathrm{PGL}_2(\mathbb{K})$. In this case the graph $\Delta_{\mathbb{K}}/\Gamma$ will be always of the form illustrated in Figure 5, with a central polygon with n vertices and trees departing from its vertices.

In fact, in this case, the subtree $\Delta'_{\mathbb{K}} \subset \Delta_{\mathbb{K}}$ spanned by the axes of the elements of Γ consists of a single infinite geodesic in $\Delta_{\mathbb{K}}$, which is the axis $L(\gamma)$ of a generator of $\Gamma \simeq \gamma^{\mathbb{Z}}$, with endpoints consisting of the fixed points $z^{\pm}(\gamma) \in \mathbb{P}^1(\mathbb{K})$. The group Γ acts on this axis by translations of lengths multiple of $\ell(\gamma) = n$; hence one obtains in the quotient a closed ring Δ'_Γ/Γ of length $\ell(\gamma)$, from whose vertices infinite trees are attached, which form the remaining part of $\Delta_{\mathbb{K}}/\Gamma$.

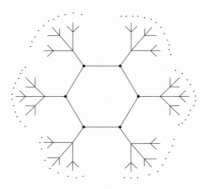

FIGURE 5 The genus one case.

With our convention on the orientations, the edges are oriented in such a way as to go around the central polygon, while the rest of the graph, *i.e.* the trees stemming from the vertices of the polygon, are oriented away from it and towards the boundary $X_\Gamma = \partial\Delta_\mathbb{K}/\Gamma$.

Label the vertices on the polygon by v_i, $i = 1, \ldots, n$. To get a special graph weight we need $0 < \lambda < 1$ and a function g on the vertices such that

$$g(v_i) = \lambda g(v_{i+1}) + B_i, \qquad B_i = \sum_{\substack{v_{i+1} \neq w = r(e), \\ s(e) = v_i}} g(w).$$

To simplify we suppose all the $g(v_i)$ are equal, $\sum g(v_i) = 1$ and all the B_i are equal. Then we obtain a special graph weight for any $\lambda < 1$ by setting

$$g(v_i) = \frac{1}{n}, \qquad B_i = \frac{1 - \lambda}{n}.$$

For each i we can now define the various $g(w)$ appearing in the sum defining B_i by $g(w) = \frac{1}{m_i} g(w)$ where m_i is the number of such $g(w)$. This graph weight can be extended to the rest of the trees as a graph trace, and the associated $\mathbb{T}$ action is nontrivial on each S_e, $e \in M^1$ where M is just the central polygon. Hence choosing γ to be the (directed) path which goes once around the polygon (the choice of $r(\gamma) = s(\gamma)$ is irrelevant) gives

$$\langle [\gamma], \phi_{g,\lambda} \rangle = -\lambda^n.$$

EXAMPLE: GENUS TWO. In the case of genus two, the possible graphs Δ_Γ/Γ and the corresponding special fibers of the algebraic curve are illustrated in Figure 6, which we reproduce from [**CM1**]; see also [**Mar**].

We look in more detail at the various cases. These are the same cases considered in [**CM1**].

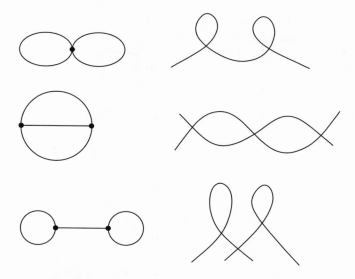

FIGURE 6 The graphs Δ_Γ/Γ for genus $g = 2$ and the corresponding fibers.

CASE 1. In the first case, the tree Δ_Γ is just a copy of the Cayley graph of the free group Γ on two generators as in Figure 7.

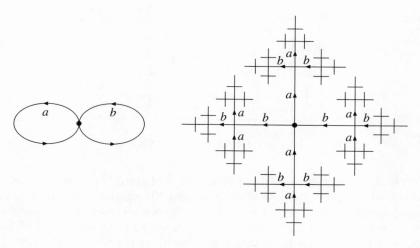

FIGURE 7 Genus two: first case.

The graph algebra of this graph is the Cuntz algebra O_2. The only possible special graph state is $g(v) = 1$ for the single vertex and $\lambda = 1/2$. This corresponds to the usual gauge action and its unique KMS state [**CPR2**]. Once we add trees to this example, many more possibilities for the KMS weights appear.

CASE 2. In the second case, the finite directed graph Δ_Γ/Γ is of the form illustrated in Figure 8. We label by $a = e_1$, $b = e_2$ and $c = e_3$ the oriented edges in the graph Δ_Γ/Γ, so that we have a corresponding set of labels $E = \{a, b, c, \bar{a}, \bar{b}, \bar{c}\}$ for the edges in the covering Δ_Γ. A choice of generators for the group $\Gamma \simeq \mathbb{Z} * \mathbb{Z}$ acting on Δ_Γ is obtained by identifying the generators γ_1 and γ_2 of Γ with the chains of edges ab and $a\bar{c}$, hence the orientation on the tree Δ_Γ and on the quotient graph is as illustrated in the figure.

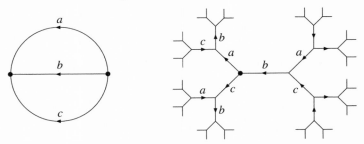

FIGURE 8 Genus two: second case.

There are four special graph states on this graph algebra (up to swapping the roles of the edges a and c). Let $v = s(b)$ and $w = r(b)$. Let $n_1 = n(b)$, $n_2 = n(a)$ and $n_3 = n(c)$ with each $n_j \in \{0, 1\}$. Then the various states are described in Table 1. To fit with our requirement that the zhyvot action "sees" every edge in the zhyvot, we should adopt only the last choice of state. Of course, we are again neglecting the trees, and including them would give us many more options.

n_1	n_2	n_3	λ	$g(v)$	$g(w)$
0	0	0 or 1	—	—	—
0	1	1	$\frac{1}{2}$	$\frac{1}{2}$	$\frac{1}{2}$
1	0	0	$\frac{1}{2}$	$\frac{1}{3}$	$\frac{2}{3}$
1	0	1	$\frac{-1+\sqrt{5}}{2}$	λ^2	λ
1	1	1	$\frac{1}{\sqrt{2}}$	$\frac{1}{1+\sqrt{2}}$	$\frac{\sqrt{2}}{1+\sqrt{2}}$

TABLE 1 Graph states for Case 2.

CASE 3. In the third case the obtained oriented graph is the same as the graph of $SU_q(2)$ of Figure 3. We have already described the graph states for $SU_q(2)$. In this case, the inclusion of trees is necessary to obtain a special graph weight adapted to the zhyvot action. In fact, a choice of generators for the group $\Gamma \simeq \mathbb{Z} * \mathbb{Z}$ acting on Δ_Γ is given by $ab\bar{a}$ and c, so that the obtained orientation is as in Figure 9.

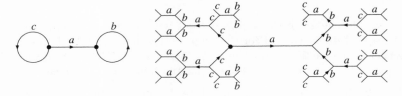

FIGURE 9 Genus two: third case.

When we consider the tree Δ'_Γ instead of Δ_Γ we are typically adding extra vertices. The way the tree Δ'_Γ sits inside the Bruhat–Tits tree $\Delta_{\mathbb{K}}$ and in particular how many extra vertices of $\Delta_{\mathbb{K}}$ are present on the graph Δ'_Γ/Γ with respect to the vertices of Δ_Γ/Γ gives some information on the uniformization, $i.e.$ it depends on where the Schottky group Γ lies in $\mathrm{PGL}_2(\mathbb{K})$, unlike the information on the graph Δ_Γ/Γ which is purely combinatorial. For example, in the genus two case of Figure 6 one can have graphs Δ'_Γ/Γ and $\Delta_{\mathbb{K}}/\Gamma$ of the form as in Figure 10. This case is obtained by embedding a valence three tree, with additional valence two vertices inserted as in the middle panel of Figure 10, inside the valence four Bruhat–Tits tree $\Delta_{\mathbb{Q}_3}$, as in the right hand panel of Figure 10. The group Γ that uniformizes the Mumford curve is generated by the translations of $\Delta_{\mathbb{K}}$ along the two axes of the tree Δ'_Γ in the middle panel of Figure 10, which give as quotient Δ'_Γ/Γ the finite graph in the left panel of Figure 10.

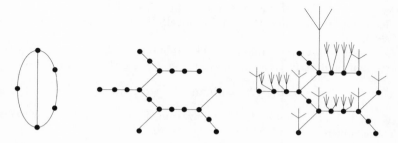

FIGURE 10 An example of a graph Δ'_Γ/Γ, the tree Δ'_Γ and its embedding in $\Delta_{\mathbb{K}}$ for $\mathbb{K} = \mathbb{Q}_3$.

REMARK: HIGHER GENUS. In the higher genus case one knows by ([**Gvp**], p.124) that any stable graph can occur as the graph Δ_Γ/Γ of a Mumford curve. By stable graph we mean a finite graph which is connected and such that each vertex that is not connected to itself by an edge is the source of at least three edges.

Thus, the combinatorial complexity of the graph is pretty much arbitrary. One can also assume, possibly after passing to a finite extension of the field $\mathbb{K}$, that there are infinite homogeneous trees attached to each vertex of the graph Δ_Γ/Γ.

We make the final remark that the restriction to circle actions in this paper is likely an artificial one. In work in progress, KMS index theory is extended to quasi-periodic actions of the reals. As the action associated to *any* graph weight will be quasi-periodic, this will hopefully allow us to prove general existence theorems for (quasi-periodic) graph weights compatible with the zhyvot action.

7. Jacobian and Theta Functions

We recall here briefly the relation between the Jacobian and theta functions of a Mumford curve and the group of currents on the infinite graph $\Delta_{\mathbb{K}}/\Gamma$.

Recall first that a current on a locally finite graph G is an integer-valued function of the oriented edges of G that satisfies the following properties.

(1) Orientation reversal:

(7.1) $$\mu(\bar{e}) = -\mu(e),$$

where $\bar{e}$ is the edge e with the reverse orientation.

(2) Momentum conservation:

$$(7.2) \qquad \sum_{s(e)=v} \mu(e) = 0.$$

One denotes by $\mathcal{C}(G)$ the abelian group of currents on the graph G.

Suppose we are given a tree $\mathcal{T}$. Then the group $\mathcal{C}(\mathcal{T})$ can be equivalently described as the group of finitely additive measures of total mass zero on the space $\partial \mathcal{T}$ of ends of the tree by setting $m(U(e)) = \mu(e)$, where $U(e) \subset \partial \mathcal{T}$ is the clopen set of ends of the infinite half lines starting at the vertex $s(e)$ along the direction e. For $G = \mathcal{T}/\Gamma$, the group of currents $\mathcal{C}(G) = \mathcal{C}(\mathcal{T})^\Gamma$ can be identified with the group of Γ-invariant measures on $\partial \mathcal{T}$, i.e. finitely additive measures of total mass zero on $\partial \mathcal{T}/\Gamma$.

As above, we let $X = X_\Gamma$ be a Mumford curve, uniformized by the p-adic Schottky group Γ. We consider the above applied to the tree $\Delta_{\mathbb{K}}$ with the action of the Schottky group Γ and the infinite, locally finite quotient graph $\Delta_{\mathbb{K}}/\Gamma$ with $\partial \Delta_{\mathbb{K}}/\Gamma = X_\Gamma(\mathbb{K})$.

It is known (see [**vdP**, Lemma 6.3, Theorem 6.4]) that the Jacobian of a Mumford curve can be described, as an analytic variety, via the isomorphism

$$(7.3) \qquad \mathrm{Pic}^0(X) \cong \mathrm{Hom}(\Gamma, \mathbb{K}^*)/c(\Gamma_{ab}),$$

where $\Gamma_{ab} = \Gamma/[\Gamma, \Gamma]$ denotes the abelianization, $\Gamma_{ab} \cong \mathbb{Z}^g$, with g the genus, and the homomorphism

$$(7.4) \qquad c : \Gamma_{ab} \to \mathrm{Hom}(\Gamma_{ab}, \mathbb{K}^*)$$

is defined by the first map in the homology exact sequence

$$(7.5) \qquad 0 \to \mathcal{C}(\Delta_{\mathbb{K}})^\Gamma \xrightarrow{c} \mathrm{Hom}(\Gamma, \mathbb{K}^*) \to H^1(\Gamma, \mathcal{O}(\Omega_\Gamma)^*) \to H^1(\Gamma, \mathcal{C}(\Delta_{\mathbb{K}})) \to 0,$$

associated to the short exact sequence

$$(7.6) \qquad 0 \to \mathbb{K}^* \to \mathcal{O}(\Omega_\Gamma)^* \to \mathcal{C}(\Delta_{\mathbb{K}}) \to 0$$

of Theorem 2.1 of [**vdP**], where $\mathcal{O}(\Omega_\Gamma)^*$ is the group of invertible holomorphic functions on $\Omega_\Gamma \subset \mathbb{P}^1$.

In the sequence (7.5), one uses the fact that $H^i(\Gamma) = 0$ for $i \geq 2$ and the identification

$$(7.7) \qquad \mathcal{C}(\Delta_{\mathbb{K}})^\Gamma = H^0(\Gamma, \mathcal{C}(\Delta_{\mathbb{K}})) = \Gamma_{ab} = \pi_1(\Delta_{\mathbb{K}}/\Gamma)_{ab}$$

(see [**vdP**], Lemma 6.1 and Lemma 6.3). One can then use the short exact sequence

$$(7.8) \qquad 0 \to \mathcal{C}(\Delta_{\mathbb{K}}) \to \mathcal{A}(\Delta_{\mathbb{K}}) \xrightarrow{d} \mathcal{H}(\Delta_{\mathbb{K}}) \to 0,$$

where $\mathcal{C}(\Delta_{\mathbb{K}})$ is the group of currents on the Bruhat–Tits tree $\Delta_{\mathbb{K}}$, $\mathcal{A}(\Delta_{\mathbb{K}})$ is the group of integer-valued functions on the set of edges of $\Delta_{\mathbb{K}}$ satisfying $h(\bar{e}) = -h(e)$ under orientation reversal $e \mapsto \bar{e}$ and $\mathcal{H}(\Delta_{\mathbb{K}})$ is the group of integer-valued functions on the set of vertices of $\Delta_{\mathbb{K}}$. The map d in (7.8) is given by

$$(7.9) \qquad d : \mathcal{A}(\Delta_{\mathbb{K}}) \to \mathcal{H}(\Delta_{\mathbb{K}}), \qquad d(h)(v) = \sum_{s(e)=v} h(e).$$

The long exact homology sequence associated to (7.8) is given by

$$(7.10) \qquad 0 \to \mathcal{C}(\Delta_{\mathbb{K}}/\Gamma) \to \mathcal{A}(\Delta_{\mathbb{K}}/\Gamma) \xrightarrow{d} \mathcal{H}(\Delta_{\mathbb{K}}/\Gamma) \xrightarrow{\Phi} H^1(\Gamma, \mathcal{C}(\Delta_{\mathbb{K}})) \to 0,$$

where one has $H^1(\Gamma, \mathcal{C}(\Delta_{\mathbb{K}})) \cong \mathbb{Z}$ and, under this identification, the last map in the exact sequence is given by

$$(7.11) \qquad \Phi : \mathcal{H}(\Delta_{\mathbb{K}}/\Gamma) \to H^1(\Gamma, \mathcal{C}(\Delta_{\mathbb{K}})) = \mathbb{Z}, \quad \Phi(f) = \sum_{v \in (\Delta_{\mathbb{K}}/\Gamma)^0} f(v).$$

Moreover, one has an identification

$$H^1(\Gamma, \mathcal{O}(\Omega_\Gamma)^*) = H^1(X, \mathcal{O}_X^*) = \mathrm{Pic}(X),$$

the group of equivalence classes of holomorphic (hence by GAGA algebraic) line bundles on the curve X, and the last map in the exact sequence (7.5) is then given by the degree map $\deg : \mathrm{Pic}(X) \to \mathbb{Z}$, whose kernel is the Jacobian $J(X) = \mathrm{Pic}^0(X)$ (see [**vdP**, Lemma 6.3]).

A theta function for the Mumford curve $X = X_\Gamma$ is an invertible holomorphic function $f \in \mathcal{O}(\Omega_\Gamma)^*$ such that

$$\gamma^* f = c(\gamma) f, \quad \forall \gamma \in \Gamma,$$

with $c \in \mathrm{Hom}(\Gamma, \mathbb{K}^*)$ the automorphic factor. The group $\Theta(\Gamma)$ of theta functions of the curve X is then obtained from the exact sequences (7.6) and (7.5) as ([**vdP**])

$$(7.12) \qquad 0 \to \mathbb{K}^* \to \Theta(\Gamma) \to \mathcal{C}(\Delta_{\mathbb{K}})^\Gamma \to 0.$$

More precisely, let $\mathbb{H}_{\mathbb{K}} = \mathbb{P}^1_{\mathbb{K}} \setminus \mathbb{P}^1(\mathbb{K})$ be Drinfeld's p-adic upper half plane. It is well known (see for instance the detailed discussion given in [**BouCar**, §I.1, §I.2]) that $\mathbb{H}_{\mathbb{K}}$ is a rigid analytic space endowed with a surjective map

$$(7.13) \qquad \Lambda : \mathbb{H}_{\mathbb{K}} \to \Delta_{\mathbb{K}}$$

to the Bruhat–Tits tree $\Delta_{\mathbb{K}}$ such that, for vertices $v, w \in \Delta_{\mathbb{K}}^0$ with $v = s(e)$ and $w = r(e)$, for an edge $e \in \Delta_{\mathbb{K}}^1$, the preimages $\Lambda^{-1}(v)$ and $\Lambda^{-1}(w)$ are open subsets of $\Lambda^{-1}(e)$. The picture of the relation between $\mathbb{H}_{\mathbb{K}}$ and $\Delta_{\mathbb{K}}$ through the map Λ is given in Figure 11.

Given a theta function $f \in \Theta(\Gamma)$, the associated current $\mu_f \in \mathcal{C}(\Delta_{\mathbb{K}})^\Gamma$ obtained as in (7.12) is given explicitly by the growth of the spectral norm in the Drinfeld upper half plane when moving along an edge in the Bruhat–Tits tree, that is,

$$(7.14) \qquad \mu(e) = \log_q \|f\|_{\Lambda^{-1}(r(e))} - \log_q \|f\|_{\Lambda^{-1}(s(e))},$$

with $q = \#\mathcal{O}/\mathfrak{m}$, and $\|f\|_{\Lambda^{-1}(v)}$ is the spectral norm

$$\|f\|_{\Lambda^{-1}(v)} = \sup_{z \in \Lambda^{-1}(v)} |f(z)|,$$

with $|\cdot|$ the absolute value with $|\pi| = q^{-1}$ and π a uniformizer, that is, $\mathfrak{m} = (\pi)$.

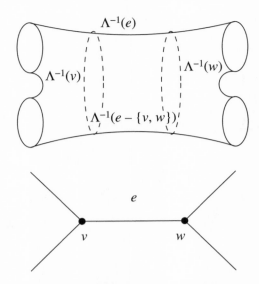

FIGURE 11 The p-adic upper half plane and the Bruhat–Tits tree.

The case of function fields over a finite field $\mathbb{F}_q$ of characteristic p is similar to the p-adic case, with $\mathbb{H}_{\mathbb{K}}$ and $\Delta_{\mathbb{K}}$ the Drinfeld upper half plane and the Bruhat–Tits tree in characteristic p (see for instance [**GekW**]).

7.1 Graph Weights, Currents, and Theta Functions

We now show how to relate theta functions on the Mumford curve to graph weights. The type of graph weights we consider here will in general not be special graph weights such as those we considered in the previous sections. In fact, we will see that, when constructing currents from graph weights, we need to work with functions $\lambda(e)$ of the special form $\lambda(e) = N_e^{-1}$, where N_e is defined as in (7.15) below. Along the outer trees of the graph $\Delta_{\mathbb{K}}/\Gamma$, these are homogeneous trees of valence $q+1$ (or $q^\varsigma + 1$ for field extensions) and the orientation of these trees away from the zhyvot graph Δ'_Γ/Γ gives that $\lambda(e) = \lambda^{n_e}$ for $\lambda = q^{-1} \in (0,1)$ (or $\lambda = q^{-f}$ for field extensions) and $n_e = 1$. Thus, the expression for N_e inside Δ'_Γ/Γ depends on the orientation on Δ'_Γ described in Lemma 2.1. Similarly, as we see below, when we construct (inhomogeneous or signed) graph weights from currents, we use the function $\lambda(e) = N_e^{-1}$, which is again of the form q^{-1} (or q^{-f}) on the outer trees of $\Delta_{\mathbb{K}}/\Gamma$, but which also depends on the given orientation inside Δ'_Γ/Γ. It will be interesting to consider the quasi-periodic actions associated to these types of graph weights.

We first show that the same methods that produce graph weights can be used to construct real-valued currents on the graph $\Delta_{\mathbb{K}}/\Gamma$.

LEMMA 7.1. *Let (g, λ) be a graph weight on the infinite, locally finite graph $\Delta_{\mathbb{K}}/\Gamma$. For an oriented edge $e \in (\Delta_{\mathbb{K}}/\Gamma)^1$ let*

$$(7.15) \qquad N_e := \#\{e' \in (\Delta_{\mathbb{K}}/\Gamma)^1 \,|\, s(e') = s(e)\}.$$

Then the function

$$(7.16) \qquad \mu(e) := \lambda(e)g(r(e)) - \frac{1}{N_e}g(s(e))$$

satisfies the momentum conservation equation (7.2). Moreover, if $g : (\Delta_{\mathbb{K}}/\Gamma)^0 \to [0, \infty)$ is a function on the vertices of the graph such that (g, λ) is a graph weight for $\lambda(e) = N_{\bar{e}}^{-1}$, then the function $\mu : (\Delta_{\mathbb{K}}/\Gamma)^1 \to \mathbb{R}$ given by (7.16) is a real-valued current on $\Delta_{\mathbb{K}}/\Gamma$.

PROOF. The first result is a direct consequence of the equation for graph weights

$$(7.17) \qquad g(v) = \sum_{s(e)=v} \lambda(e) g(r(v)),$$

which gives (7.2) for (7.16). The second statement also follows from (7.17), where in this case the resulting function $\mu : (\Delta_{\mathbb{K}}/\Gamma)^1 \to \mathbb{R}$ given by (7.16) also satisfies the orientation-reversed equation

$$(7.18) \qquad \sum_{r(e)=v} \mu(e) = 0.$$

In particular, it also satisfies

$$\mu(\bar{e}) = \frac{1}{N_e} g(s(e)) - \frac{1}{N_{\bar{e}}} g(r(e)) = -\mu(e),$$

hence it defines a real-valued current on $\Delta_{\mathbb{K}}/\Gamma$,

$$\mu \in \mathcal{C}(\Delta_{\mathbb{K}}/\Gamma) \otimes_{\mathbb{Z}} \mathbb{R}.$$

Conversely, one can use theta functions on the Mumford curve to construct graph weights on the tree, which however do not have the positivity property. We introduce the following notions generalizing that of positive graph weight given in Definition 4.1.

DEFINITION 7.2. Given a graph E, we define an inhomogeneous graph weight to be a triple (g, λ, χ) of non-negative functions $g : E^0 \to \mathbb{R}_+ = [0, \infty)$, $\lambda, \chi : E^1 \to [0, \infty)$ satisfying

$$(7.19) \qquad g(v) + d\chi(v) = \sum_{s(e)=v} \lambda(e) g(r(e)),$$

where, as above, $d\chi(v) = \sum_{s(e)=v} \chi(e)$. A rational virtual graph weight is a pair (g, λ) of functions $g : E^0 \to \mathbb{Q}$ and $\lambda : E^1 \to \mathbb{Q}_+ = \mathbb{Q} \cap [0, \infty)$ such that there exist rational-valued inhomogeneous graph weights $(g^{\pm}, \lambda, \chi)$ with $g(v) = g^+(v) - g^-(v)$ for all $v \in E^1$.

A virtual graph weight satisfies (7.17). The choice of the inhomogeneous weights $(g^{\pm}, \lambda, \chi)$ that give the decomposition $g(v) = g^+(v) - g^-(v)$ is non-unique.

LEMMA 7.3. *Let $f \in \Theta(\Gamma)$ be a theta function for the Mumford curve $X = X_\Gamma$, with $\Theta(\Gamma)$ as in (7.12). Then f defines an associated pair of rational-valued functions (g, λ) on the tree $\Delta_{\mathbb{K}}$, with $\lambda(e) = N_e^{-1}$ and with $g : \Delta_{\mathbb{K}}^0 \to \mathbb{Q}$ satisfying (7.17).*

PROOF. By (7.12) we know that the theta function $f \in \Theta(\Gamma)$ determines an associated integer-valued current μ_f on the graph $\Delta_\mathbb{K}/\Gamma$, that is, an element in $\mathcal{C}(\Delta_\mathbb{K})^\Gamma = \mathcal{C}(\Delta_\mathbb{K}/\Gamma)$. We view μ_f as a current on the tree $\Delta_\mathbb{K}$ that is Γ-invariant. We also know that the current $\mu = \mu_f$ is given by (7.14) in terms of the spectral norm on the Drinfeld p-adic upper half plane.

If we set

$$(7.20) \qquad g(v) := \log_q \|f\|_{\Lambda^{-1}(v)},$$

we see easily that (7.2) for the current $\mu(e) = g(r(e)) - g(s(e))$ implies that the function $g : \Delta_\mathbb{K}^0 \to \mathbb{Z}$ satisfies the weight equation (7.17) with $\lambda(e) = N_e^{-1}$. In fact, we have

$$\sum_{s(e)=v} g(r(e)) = N_v g(v),$$

with $N_v = \#\{e' : s(e') = v\} = N_e$, for all e with $s(e) = v$.

LEMMA 7.4. *The function* $g : \Delta_\mathbb{K}^0 \to \mathbb{Z}$ *associated to a theta function in* $f \in \Theta(\Gamma)$ *is an integer-valued rational virtual graph weight.*

PROOF. The measure $\mu = \mu_f$ can be written (non-uniquely) as a difference

$$(7.21) \qquad \mu(e) = \chi^+(e) - \chi^-(e),$$

with non-negative $\chi^\pm : \Delta_\mathbb{K}^1 \to \mathbb{N} \cup \{0\}$ satisfying

$$(7.22) \qquad \chi^\pm(\bar{e}) = \chi^\mp(e) \quad \text{and} \quad \sum_{s(e)=v} \chi^+(e) = \sum_{s(e)=v} \chi^-(e),$$

for all $e \in \Delta_\mathbb{K}^1$. One then considers the equations

$$(7.23) \qquad g^\pm(r(e)) - g^\pm(s(e)) = \chi^\pm(e).$$

We first see that (7.23) determines unique solutions $g^\pm : \Delta_\mathbb{K}^0 \to \mathbb{Q}_+$ with $g^\pm(v) = 0$ at the basepoint. In fact, suppose we are given a vertex $w \neq v$ in the tree. There is a unique path $P(v, w)$ in $\Delta_\mathbb{K}$ connecting the base vertex v to w. It is given by a sequence $P(v, w) = e_1, \ldots, e_n$ of oriented edges. Let $v = v_0, \ldots, v_n = w$ be the corresponding sequence of vertices. Then (7.23) implies

$$(7.24) \qquad g^\pm(w) = \sum_{j=1}^n \chi^\pm(e_j).$$

This determines uniquely the values of $g^\pm$ at each vertex in $\Delta_\mathbb{K}$. The solutions obtained in this way satisfy $g^+(v) - g^-(v) = g(v)$, where $g(v) = \log_q \|f\|_{\Lambda^{-1}(v)}$ as in Lemma 7.3. The $g^\pm$ satisfy by construction the inhomogeneous weight equation

$$g^\pm(w) + d\chi^\pm(v) = \sum_{s(e)=w} \lambda(e) g^\pm(r(e)),$$

for $\lambda(e) = N_e^{-1}$. Thus, the pair (g, λ) of Lemma 7.3 is a rational virtual weight.

Notice that, even though the current μ_f is Γ-invariant by construction, and the function $\lambda(e) = N_e^{-1}$ is also Γ-invariant, the function $g : \Delta_\mathbb{K}^0 \to \mathbb{Q}$ obtained as above is not in general Γ-invariant, hence it need not descend to a graph weight on $\Delta_\mathbb{K}/\Gamma$.

In fact, for $g(v) = \log_q \|f\|_{\Lambda^{-1}(v)}$, one sees that

$$g(\gamma v) = \log_q \|f\|_{\Lambda^{-1}(\gamma v)} = \log_q \|f \circ \gamma\|_{\Lambda^{-1}(v)}$$
$$= \log_q |c(\gamma)| + \log_q \|f\|_{\Lambda^{-1}(v)} = g(v) + \log_q |c(\gamma)|,$$

where $f(\gamma z) = c(\gamma)f(z)$.

More generally, one has the following result.

LEMMA 7.5. Let (g, λ) be a rational virtual weight on the tree $\Delta_{\mathbb{K}}$, with $g : \Delta_{\mathbb{K}}^0 \to \mathbb{Q}$ and with $\lambda(e) = N_e^{-1}$. Then the function g satisfies

$$(7.25) \qquad g(\gamma v) - g(v) = d\beta_\gamma(v),$$

where $d\beta_\gamma(v) = \sum_{s(e)=v} \beta_\gamma(e)$ and

$$(7.26) \qquad \beta_\gamma(e) = \lambda(e)(g(\gamma r(e)) - g(s(e))).$$

This satisfies $\beta_\gamma(\bar{e}) = -\beta_{\gamma^{-1}}(\gamma e)$ and the 1-cocycle equation

$$(7.27) \qquad d\beta_{\gamma_1 \gamma_2}(v) = d\beta_{\gamma_1}(\gamma_2 v) + d\beta_{\gamma_2}(v).$$

PROOF. First notice that, by (7.17) the function g satisfies

$$(7.27) \qquad g(\gamma v) - g(v) = d\alpha_\gamma(v),$$

where $d\alpha_\gamma(v) = \sum_{s(e)=v} \alpha_\gamma(e)$ with

$$(7.29) \qquad \alpha_\gamma(e) = \lambda(e)\left(g(\gamma r(e)) - g(r(e))\right).$$

Notice moreover that we have

$$\alpha_\gamma(e) = \beta_\gamma(e) - \mu(e),$$

for β_γ as in (7.26) and $\mu = \mu_f$ the Γ-invariant current (7.14). Since $d\mu(v) \equiv 0$ we have $d\alpha_\gamma(v) = d\beta_\gamma(v)$, which gives (7.25). One checks the expression for $\beta_\gamma(\bar{e})$ directly from (7.26), using the Γ-invariance of N_e and $\lambda(e)$. The 1-cocycle equation is also easily verified by

$$g(\gamma_1 \gamma_2 v) - g(v) - g(\gamma_1 \gamma_2 v) + g(\gamma_2 v) - g(\gamma_2 v) + g(v) = 0.$$

In general, the condition for a (virtual) graph weight on the tree $\Delta_{\mathbb{K}}$ to descend to a (virtual) graph weight on the quotient $\Delta_{\mathbb{K}}/\Gamma$ is that the functions (g, λ) satisfy

$$(7.30) \qquad g(v) = \sum_{s(e)=\gamma v} \lambda(e)g(r(e)),$$

for all $\gamma \in \Gamma$. This is clearly equivalent to the vanishing of $d\beta_\gamma(v)$ and to the invariance $g(\gamma v) = g(v)$.

Another possible way of describing (rational) virtual graph weights, instead of using the inhomogeneous equations, is by allowing the function λ to have positive or negative sign, namely we consider $\lambda : E^1 \to \mathbb{Q}$ and look for non-negative solutions $g : E^0 \to \mathbb{Q}_+$ of the original graph weight equation (7.17).

A rational virtual weight (g, λ) defines a solution $(\tilde{g}, \tilde{\lambda})$ as above, with $\tilde{\lambda} : E^1 \to \mathbb{Q}$ and $\tilde{g} : E^0 \to \mathbb{Q}_+$, by setting λ to be $\tilde{\lambda}(e) = \lambda(e)\operatorname{sign}(g(s(e)))\operatorname{sign}(g(r(e)))$ and $\tilde{g}(v) = \operatorname{sign}(g(v))g(v) = |g(v)|$. This definition has an ambiguity when $g(v) = 0$, in which case we can take either $\operatorname{sign}(g(v)) = \pm 1$.

7.2 Theta Functions and K-theory Classes

Another useful observation regarding the relation of theta functions of the Mumford curve X_Γ to properties of the graph algebra of the infinite graph $\Delta_{\mathbb{K}}/\Gamma$ is the fact that one can associate to the theta functions elements in the K-theory of the boundary C^*-algebra $C(\partial\Delta_{\mathbb{K}}) \rtimes \Gamma$.

This is not a new observation: it was described explicitly in [**Rob**] and also used in [**CLM**], though only in the case of finite graphs. The finite graph hypothesis is used in [**Rob**] to obtain the further identification of the Γ-invariant $\mathbb{Z}$-valued currents on the covering tree with the first homology group of the graph.

In our setting, the graph $\Delta_{\mathbb{K}}/\Gamma$ consists of a finite graph Δ'_Γ/Γ together with infinite trees stemming from its vertices. We still have the same result on the identification with the K-theory group $K_0(C(\partial\Delta_{\mathbb{K}}) \rtimes \Gamma)$ of the boundary algebra, as well as with the first homology of the graph $\Delta_{\mathbb{K}}/\Gamma$, which is the same as the first homology of the finite graph Δ'_Γ/Γ.

PROPOSITION 7.6. *There are isomorphisms*

$$(7.31) \qquad \mathcal{C}(\Delta_{\mathbb{K}}, \mathbb{Z})^\Gamma \cong H_1(\Delta_{\mathbb{K}}/\Gamma, \mathbb{Z}) \cong \operatorname{Hom}(K_0(C(\partial\Delta_{\mathbb{K}}) \rtimes \Gamma), \mathbb{Z}).$$

PROOF. The first isomorphism follows directly from (7.7).

To prove the second identification

$$\mathcal{C}(\Delta_{\mathbb{K}}, \mathbb{Z})^\Gamma \cong \operatorname{Hom}(K_0(C(\partial\Delta_{\mathbb{K}}) \rtimes \Gamma), \mathbb{Z}),$$

first notice that $\partial\Delta_{\mathbb{K}}$ is a totally disconnected compact Hausdorff space, hence $K_1(C(\partial\Delta_{\mathbb{K}})) = 0$ and in the exact sequence of [**PV**] for the K-theory of the crossed product by the free group Γ one obtains an isomorphism of $K_0(C(\partial\Delta_{\mathbb{K}}) \rtimes \Gamma)$ with the coinvariants

$$C(\partial\Delta_{\mathbb{K}}, \mathbb{Z})_\Gamma = C(\partial\Delta_{\mathbb{K}}, \mathbb{Z})/\{f \circ \gamma - f \mid f \in C(\partial\Delta_{\mathbb{K}}, \mathbb{Z})\},$$

where $C(\partial\Delta_{\mathbb{K}}, \mathbb{Z})$ is the abelian group of locally constant $\mathbb{Z}$-valued functions on $\partial\Delta_{\mathbb{K}}$, *i.e.* finite linear combinations with integer coefficients of characteristic functions of clopen subsets. We then show that the abelian group $\mathcal{C}(\Delta_{\mathbb{K}}, \mathbb{Z})^\Gamma$ of Γ-invariant currents on the tree $\Delta_{\mathbb{K}}$ can be identified with

$$(7.32) \qquad \mathcal{C}(\Delta_{\mathbb{K}}, \mathbb{Z})^\Gamma \cong \operatorname{Hom}(K_0(C(\partial\Delta_{\mathbb{K}}) \rtimes \Gamma), \mathbb{Z}) = H_1(\Delta_{\mathbb{K}}/\Gamma, \mathbb{Z}).$$

To see that a current $\mu \in \mathcal{C}(\Delta_{\mathbb{K}}, \mathbb{Z})^\Gamma$ defines a homomorphism $\phi : C(\partial\Delta_{\mathbb{K}}, \mathbb{Z})_\Gamma \to \mathbb{Z}$, we use the fact that we can view the current μ on the tree as a measure m of total mass zero on the boundary $\partial\Delta_{\mathbb{K}}$ by setting $m(V(e)) = \mu(e)$, where $V(e)$ is the subset of the boundary determined by all infinite paths starting with the oriented edge e. We then define the functional

$$(7.33) \qquad \phi(f) = \int_{\partial\Delta_{\mathbb{K}}} f \, dm,$$

where the integration is defined by $\phi(\sum_i a_i \chi_{V(e_i)}) = \sum_i a_i \mu(e_i)$ on characteristic functions. To see that ϕ is defined on the coinvariants it suffices to check that it vanishes on elements of the form $f \circ \gamma - f$, for some $\gamma \in \Gamma$. This follows by change of variables and the invariance of the current μ,

$$\int f \circ \gamma \, dm = \int f \, dm \circ \gamma^{-1} = \int f \, dm.$$

Conversely, suppose we are given a homomorphism $\phi : C(\partial\Delta_{\mathbb{K}}, \mathbb{Z})_\Gamma \to \mathbb{Z}$. We define a map $\mu : \Delta_{\mathbb{K}}^1 \to \mathbb{Z}$ by setting $\mu(e) = \phi(\chi_{V(e)})$, where $\chi_{V(e)}$ is the characteristic function of the set $V(e) \subset \partial\Delta_{\mathbb{K}}$. We need to show that this defines a Γ-invariant current on the tree. We need to check that the equation

$$\sum_{s(e)=v} \mu(e) = 0$$

and the orientation reversal condition $\mu(\bar{e}) = -\mu(e)$ are satisfied.

Notice that we have, for any given vertex $v \in \Delta_{\mathbb{K}}^0$, $\cup_{s(e)=v} V(e) = \partial\Delta_{\mathbb{K}}$. If we set

$$h(v) := \sum_{s(e)=v} \phi(\chi_{V(e)}),$$

we obtain a Γ-invariant $\mathbb{Z}$-valued function on the set of vertices $\Delta_{\mathbb{K}}^0$, *i.e.* a $\mathbb{Z}$-valued function on the vertices $(\Delta_{\mathbb{K}}/\Gamma)^0$. In fact, we have

$$\phi(f \circ \gamma) = \phi(f)$$

by the assumption that ϕ is defined on the coinvariants $C(\partial\Delta_{\mathbb{K}}, \mathbb{Z})_\Gamma$; hence

$$h(\gamma v) = \sum_{s(e)=\gamma v} \phi(\chi_{V(e)}) = \sum_{s(e)=v} \phi(\chi_{V(e)} \circ \gamma) = h(v).$$

Since by construction $h = d\mu$, with $\mu(e) = \phi(\chi_{V(e)})$ and $d : \mathcal{A}(\Delta_{\mathbb{K}}^1/\Gamma) \to \mathcal{H}(\Delta_{\mathbb{K}}^0/\Gamma)$ as in (7.10), it is in the kernel of the map Φ of (7.11). This means that

$$\Phi(h) = \sum_{v \in \Delta_{\mathbb{K}}^0/\Gamma} h(v) = 0,$$

but we know that

$$h(v) = \sum_{s(e)=v} \phi(\chi_{V(e)}) = \phi\Big(\sum_{s(e)=v} \chi_{V(e)}\Big) = \phi(\chi_{\partial\Delta_{\mathbb{K}}})$$

so that the condition $\Phi(h) = 0$ implies $h(v) = 0$ for all v, *i.e.* $\phi(\chi_{\partial\Delta_{\mathbb{K}}}) = 0$. This gives

$$\sum_{s(e)=v} \phi(\chi_{V(e)}) = 0,$$

which is the momentum conservation condition for the measure μ. Moreover, the fact that the measure on $\partial\Delta_{\mathbb{K}}$ defined by $\mu(e) = \phi(\chi_{V(e)})$ has total mass zero also implies that

$$0 = \phi(\chi_{\partial\Delta_{\mathbb{K}}}) = \phi(\chi_{V(e)}) + \phi(\chi_{V(\bar{e})}),$$

hence $\mu(\bar{e}) = -\mu(e)$, so that μ is a current. The condition $\phi(f \circ \gamma) = \phi(f)$ shows that it is a Γ-invariant current.

The results of this section relate the theta functions of Mumford curves to the K-theory of a C^*-algebra which is not directly the graph algebra $C^*(\Delta_{\mathbb{K}}/\Gamma)$ we worked with so far, but the "boundary algebra" $C(\partial\Delta_{\mathbb{K}}) \rtimes \Gamma$. However, it is known by the result of Theorem 1.2 of [**KuPa**] that the crossed product algebra $C(\partial\Delta_{\mathbb{K}}) \rtimes \Gamma$ is strongly Morita equivalent to the algebra $C^*(\Delta_{\mathbb{K}}/\Gamma)$. In fact, we use the fact that $\Delta_{\mathbb{K}}$ is a tree and that the p-adic Schottky group Γ acts freely on $\Delta_{\mathbb{K}}$, so that $C^*(\Delta_{\mathbb{K}}) \rtimes \Gamma \simeq C^*(\Delta_{\mathbb{K}}/\Gamma) \otimes \mathcal{K}(\ell^2(\Gamma))$. Thus $C^*(\Delta_{\mathbb{K}}/\Gamma)$ is strongly Morita equivalent to $C^*(\Delta_{\mathbb{K}}) \rtimes \Gamma$. Moreover, $\Delta_{\mathbb{K}}$ is the universal covering tree of

$\Delta_\mathbb{K}/\Gamma$, and Γ can be identified with the fundamental group so that the argument of Theorem 4.13 of [**KuPa**] holds in this case and the result of Theorem 1.2 of [**KuPa**] applies. Thus the K-theory considered here can be also thought of as the K-theory of the latter algebra, and we obtain the following result.

COROLLARY 7.7. *A theta function $f \in \Theta(\Gamma)$ defines a functional*

$$\phi_f \in \mathrm{Hom}(K_0(C^*(\Delta_\mathbb{K}/\Gamma)), \mathbb{Z}).$$

Two theta functions $f, f' \in \Theta(\Gamma)$ define the same $\phi_f = \phi_{f'}$ if and only if they differ by the action of $\mathbb{K}^*$.

PROOF. The first statement follows from Proposition 7.6 and the identification $K_0(C^*(\Delta_\mathbb{K}/\Gamma)) = K_0(C(\partial\Delta_\mathbb{K}) \rtimes \Gamma)$ which follows from the strong Morita equivalence discussed above. The second statement is then a direct consequence of Proposition 7.6 and (7.12).

Corollary 7.7 shows that there is a close relationship between the K-homology of $C^*(\Delta_\mathbb{K}/\Gamma)$ and theta functions. In the next section we make a first step towards constructing theta functions from graphical data and spectral flows.

8. Inhomogeneous Graph Weight Equation and the Spectral Flow

Let E be a graph with zhyvot M with a choice of (not necessarily special) graph weight (g, λ) adapted to the zhyvot action. We would like to construct an inhomogeneous graph weight (G, λ, χ) from these data.

The motivation for the construction is as follows. In the case of a special graph weight, the spectral flow only sees edges and paths in the zhyvot M of E. Consequently

$$-\sum_{s(e)=v} sf_{\phi_\mathcal{D}}(S_e S_e^* \mathcal{D}, S_e \mathcal{D} S_e^*) = \sum_{\substack{s(e)=v \\ e \in M}} \lambda(e)g(r(e)) \le g(v).$$

Thus in some sense the spectral flow is trying to reproduce the graph weight, but it misses information from edges not in M. Alternatively, one may think of restricting (g, λ) to the zhyvot and asking whether it is still a (special) graph weight. This is usually not the case, for exactly the same reason.

So to obtain (G, λ, χ) we begin with the ansatz that λ is the function given to us with our graph weight and that

$$G(v) = g(v) - \alpha(v),$$

so that we require $\alpha(v) \le g(v)$ for all $v \in E$. We now compute

$$\sum_{s(e)=v} \lambda(e)G(r(e)) = g(v) - \sum_{s(e)=v} \lambda(e)\alpha(r(e))$$

$$= g(v) - \alpha(v) + \alpha(v) - \sum_{s(e)=v} \lambda(e)\alpha(r(e))$$

$$= G(v) + \sum_{s(e)=v} \left(\frac{1}{N_e}\alpha(v) - \lambda(e)\alpha(r(e)) \right).$$

Hence to obtain an inhomogeneous graph weight, we must have

$$\chi(e) = \frac{1}{N_e}\alpha(s(e)) - \lambda(e)\alpha(r(e)).$$

Here are some possible choices of α.

(1) $\alpha = g$. This forces $G = 0$, and so we obtain an inhomogeneous graph weight only when $\frac{1}{N_e} = \lambda(e)$.

(2) $\alpha = cg$, $0 < c < 1$. Now $G \neq 0$ but we still need $\frac{1}{N_e} = \lambda(e)$ in order to obtain an inhomogeneous graph weight.

(3) $\alpha(v) = \sum_{\substack{s(e)=v \\ e \in M}} \lambda(e)g(r(e)) = -\sum_{s(e)=v} s f_{\phi_D}(S_e S_e^* D, S_e D S_e^*)$. In this case we obtain

$$\chi(e) = \frac{1}{N_e} \sum_{\substack{s(f)=s(e) \\ f \in M}} \lambda(f)g(r(f)) - \lambda(e) \sum_{\substack{s(h)=r(e) \\ h \in M}} \lambda(h)g(r(h)).$$

This **may** be non-negative for certain values of λ.

(4) $\alpha(v) = \begin{cases} g(v) & v \in M \\ 0 & v \notin M \end{cases}$.

This gives

$$\chi(e) = \begin{cases} \frac{1}{N_e}g(s(e)) - \lambda(e)g(r(e)) & s(e), r(e) \in M \\ \frac{1}{N_e}g(s(e)) & s(e) \in M, r((e) \notin M \\ 0 & s(e), r(e) \notin M \end{cases}.$$

Thus χ is non-negative provided

$$\frac{1}{N_e}g(s(e)) \geq \lambda(e)g(r(e)).$$

The choices 3 and 4 both yield triples (G, λ, χ) satisfying (7.19), and all that remains to understand is the positivity of the function χ.

In fact it is easy to construct examples where choice 4 fails to give a non-negative function χ. However, choice 3 is more subtle. At present we have no way of deciding whether we can always find a g so that the function α in choice 3 is non-negative for λ associated to the zhyvot action. It would seem that passing to field extensions allows us to construct graph weights adapted to the (new) zhyvot so that both choices 3 and 4 fail. The reason is we may increase N_e while keeping $\lambda(e)$ constant.

We describe here another construction of inhomogeneous graph weights adapted to the zhyvot action. This uses special graph weights, with $\lambda(e) \neq 1$ only on the edges inside the zhyvot, so it does not apply to the construction of theta functions, where one needs $\lambda(e) = N_e^{-1}$ (which is $q^{-1} \neq 1$ outside of the zhyvot), but we include it here for its independent interest.

LEMMA 8.1. *Let (g, λ) be a graph weight on the graph E with zhyvot M such that $\lambda(e) \neq 1$ iff $e \in M^1$. With the notation of Section 5, define $\alpha_k : E^0 \to [0, \infty)$ by*

$$\alpha_k(v) = \phi_D(p_v \Phi_k), \quad k = 0, 1, 2, \ldots.$$

Then $\alpha_{k-1}(v) \geq \alpha_k(v)$ for all $v \in E^0$.

PROOF. We first observe that for $k \geq 1$,

$$\Phi_k = \sum_{|\mu|_\sigma = k} \Theta_{S_\mu, S_\mu}.$$

This follows easily by induction from $\Phi_0 = \sum_{v \in E^0} \Theta_{p_v, p_v}$ and the definition of the zhyvot action. Then

$$
\begin{aligned}
\phi_{\mathcal{D}}(p_v \Phi_k) &= \sum_{e \in M^1, |\mu|_\sigma = k-1, s(\mu) = v} \lambda(\mu)\lambda(e) Trace_\phi(\Theta_{S_\mu S_e, S_\mu S_e}) \\
&= \sum_{e \in M^1, |\mu|_\sigma = k-1, s(\mu) = v} \lambda(\mu)\lambda(e)\phi(S_e^* S_\mu^* S_\mu S_e) \\
&\leq \sum_{e \in E^1, |\mu|_\sigma = k-1, s(\mu) = v} \lambda(\mu)\lambda(e)\phi(S_e^* S_\mu^* S_\mu S_e) \\
&= \sum_{e \in E^1, |\mu|_\sigma = k-1, s(\mu) = v} \lambda(\mu)\phi(S_e S_e^* S_\mu^* S_\mu) \\
&= \sum_{|\mu|_\sigma = k-1, s(\mu) = v} \lambda(\mu)\phi(p_{r(\mu)} S_\mu^* S_\mu) \\
&= \phi_{\mathcal{D}}(p_v \Phi_{k-1}).
\end{aligned}
$$

THEOREM 8.2. Let (g, λ) be a graph weight on the graph E with zhyvot M such that $\lambda(e) \neq 1$ iff $e \in M^1$. Then for all $k \geq 1$ the triple $(g - \alpha_k, \lambda, \chi_k)$ is an inhomogeneous graph trace, where $\chi_k(e) = \lambda(e)(\alpha_{k-1}(r(e)) - \alpha_k(r(e)))$.

PROOF. There are a couple of simple observations here. If $v \in E^0 \setminus M^0$, then $\alpha_k(v) = 0$ when $k > 0$. This follows from Lemma 8.1. This means that for $k \geq 1$

$$
\begin{aligned}
\sum_{s(e) = v} \lambda(e)\alpha_k(r(e)) &= \sum_{s(e) = v} \lambda(e)\phi_{\mathcal{D}}(p_{r(e)} \Phi_k) = \sum_{s(e) = v} \lambda(e)\phi_{\mathcal{D}}(S_e^* S_e \Phi_k) \\
&= \sum_{s(e) = v} \phi_{\mathcal{D}}(S_e \Phi_k S_e^*) \\
&= \sum_{s(e) = v} \phi_{\mathcal{D}}(S_e S_e^* \Phi_{k+1}) \\
&= \phi_{\mathcal{D}}(p_v \Phi_{k+1}) = \alpha_{k+1}(v),
\end{aligned}
$$

the last line following from the Cuntz–Krieger relations. Then the fnhomogeneous graph weight equation is simple:

$$
\begin{aligned}
\sum_{s(e) = v} \lambda(e)(g(r(e)) - \alpha_k(r(e))) \\
&= \sum_{s(e) = v} \lambda(e)(g(r(e)) - \alpha_{k-1}(r(e))) + \sum_{s(e) = v} \lambda(e)(\alpha_{k-1}(r(e)) - \alpha_k(r(e))) \\
&= (g(v) - \alpha_k(v)) + \sum_{s(e) = v} \lambda(e)(\alpha_{k-1}(r(e)) - \alpha_k(r(e))).
\end{aligned}
$$

So the inhomogeneous graph weight equation is satisfied if we set

$$\chi_k(e) = \lambda(e)(\alpha_{k-1}(r(e)) - \alpha_k(r(e))),$$

and both $g - \alpha_k$, $\chi_k \geq 0$ by Lemma 8.1.

These inhomogeneous weights are canonically associated to the decompositions $F = F_k \oplus G_k$ of the fixed point algebra arising from the zhyvot action.

8.1. Constructing Theta Functions from Spectral Flows

We show how some of the methods described above that produce inhomogeneous graph weights with $\lambda(e) = N_e^{-1}$ can be adapted to construct theta functions on the Mumford curve.

First notice that, given such a construction of a solution of the inhomogeneous graph weights as above, we can produce a rational virtual graph weight in the following way.

LEMMA 8.3. *Suppose we are given two graph weights g_1, g_2 on the infinite graph $\Delta_{\mathbb{K}}/\Gamma$, both with the same $\lambda(e) = N_e^{-1}$. Suppose we are also given inhomogeneous graph weights of the form $G_i(v) = g_i(v) - \alpha_i(v)$ as above, with $0 \leq \alpha_i(v) \leq g_i(v)$ at all vertices, and with $\chi_i(e) = \frac{1}{N_e}(\alpha_i(s(e)) - \alpha_i(r(e)))$. Then setting $\hat{G}_i(v) = g_i(v) - \alpha(v)$ with $\alpha(v) = \min\{\alpha_1(v), \alpha_2(v)\}$ gives two solutions of the inhomogeneous graph weight equation with the same $\chi(e) = \frac{1}{N_e}(\alpha(s(e)) - \alpha(r(e)))$ and $\lambda(e) = N_e^{-1}$.*

PROOF. We have

$$\hat{G}_i(v) = g_i(v) - \alpha(v) = \sum_{s(e)=v} \frac{1}{N_e} g_i(r(e)) - \alpha(v)$$

$$= \sum_{s(e)=v} \frac{1}{N_e} \hat{G}_i(r(e)) + \sum_{s(e)=v} \frac{1}{N_e}(\alpha(r(e)) - \alpha(s(e)),$$

which shows that both $\hat{G}_i(v)$ are solutions of the inhomogeneous weight equation

$$\hat{G}_i(v) + d\chi(v) = \sum_{s(e)=v} \frac{1}{N_e} \hat{G}_i(r(e)),$$

where $\chi(e) = N_e^{-1}(\alpha(s(e)) - \alpha(r(e)))$.

Thus, whenever we have multiple solutions for the graph weights on $\Delta_{\mathbb{K}}/\Gamma$ we can construct associated rational virtual graph weights by setting

$$(8.1) \qquad\qquad \hat{G}(v) = \hat{G}_1(v) - \hat{G}_2(v).$$

If the graph weights g_i are rational valued, $g_i : E^0 \to \mathbb{Q}_+$, then the virtual graph weight $\hat{G}$ is also rational valued, $\hat{G} : E^0 \to \mathbb{Q}$.

Moreover, since the graph $\Delta_{\mathbb{K}}/\Gamma$ has a finite zhyvot with infinite trees coming out of its vertices, one can obtain a rational virtual graph weight that is integer valued. In fact, along the trees outside the zhyvot Δ_Γ'/Γ of $\Delta_{\mathbb{K}}/\Gamma$, the condition

$$\hat{G}(v) = \sum_{s(e)=v} \frac{1}{N_e} \hat{G}(r(e))$$

is satisfied by extending $\hat{G}(v)$ from the zhyvot by $\hat{G}(r(e)) = \hat{G}(s(e))$ along the trees. In this case, what remains is the finite graph, the zhyvot, which only involves finitely many denominators for a rational-valued $\hat{G}(v)$, which means that one can obtain an integer-valued solution. We will therefore assume that the rational virtual graph weights constructed as in (8.1) and Lemma 8.3 are integer valued, $\hat{G} : E^0 \to \mathbb{Z}$. This in particular includes the cases constructed using the spectral flow.

According to the results of Section 7.1, we then have the following result.

PROPOSITION 8.4. *Suppose we have a virtual graph weight $\hat{G} : E^0(\Delta_{\mathbb{K}}/\Gamma) \to \mathbb{Z}$ as above and a homomorphism $c : \Gamma \to \mathbb{K}^*$. Then there is a theta function f on the Mumford curve X_Γ satisfying $\log_q \|f\|_{\Lambda^{-1}(v)} = \tilde{G}(v)$ and $f(\gamma z) = c(\gamma)f(z)$, where $\tilde{G} : E^1(\Delta_{\mathbb{K}}) \to \mathbb{Z}$ is defined as $\tilde{G}(v) = \hat{G}(v)$ on a fundamental domain of the action of Γ on $\Delta_{\mathbb{K}}$ and extended to $\Delta_{\mathbb{K}}$ by $\tilde{G}(\gamma v) = \log_q |c(\gamma)| + \hat{G}(v)$.*

PROOF. This follows from the identification of the group of theta functions $\Theta(\Gamma)$ with the extension (7.12) of the group $\mathcal{C}(\Delta_{\mathbb{K}})^\Gamma$ of currents on $\Delta_{\mathbb{K}}/\Gamma$ by $\mathbb{K}^*$, and by identifying the current $\mu(e) = \hat{G}(r(e)) - \hat{G}(s(e))$ with $\mu_f(e) = \log_q \|f\|_{\Lambda^{-1}(r(e))} - \log_q \|f\|_{\Lambda^{-1}(s(e))}$.

The last two results highlight the interest in determining whether non-negative α can be found for the function $\lambda(e) = N_e^{-1}$.

References

[BPRS] T. Bates, D. Pask, I. Raeburn, W. Szymanski, *The C^* algebras of row-finite graphs*, New York J. Maths **6** (2000), 301-324.

[BouCar] J.F. Boutot, H. Carayol, *Uniformization p-adique des courbes de Shimura: les théorèmes de Čerednik et de Drinfeld*, "Courbes modulaires et courbes de Shimura" (Orsay, 1987/1988) **7** (1992), 45–158; Astérisque No. (1991), 196-197.

[CNNR] A. Carey, S. Neshveyev, R. Nest, A. Rennie, *Twisted cyclic theory, equivariant KK theory and KMS states*, to appear in Crelle's J.

[CPS2] A. Carey, J. Phillips, F. Sukochev, *Spectral flow and Dixmier traces*, Advances in Math. **173** (2003), 68–113.

[CPR2] A. Carey, J. Phillips, A. Rennie, *Twisted cyclic theory and an index theory for the gauge invariant KMS state on the Cuntz algebra O_n*, to appear in Journal of K-Theory.

[CRT] A. Carey, A. Rennie, K. Tong, *Spectral flow invariants and twisted cyclic theory from the Haar state on $SU_q(2)$*, J. of Geometry and Physics **59** (2009), 1431-1452.

[CoMo] A. Connes, H. Moscovici, *Type III and spectral triples*, Traces in Number Theory, Geometry, and Quantum Fields, Vieweg, 2007, pp. 57–72.

[Cons] C. Consani, *Double complexes and Euler L-factors*, Compositio Math. **111** (1998), 323–358.

[CM] C. Consani, M. Marcolli, *Noncommutative geometry, dynamics and ∞-adic Arakelov geometry*, Selecta Math. (N.S.) **10** (2004 no. 2), 167–251.

[CM1] C. Consani, M. Marcolli, *Spectral triples from Mumford curves*, International Math. Research Notices **36** (2003), 1945–1972.

[CM2] C. Consani, M. Marcolli, *New perspectives in Arakelov geometry*, Number theory, CRM Proc. Lecture Notes 36, American Mathematical Society, Providence, RI, 2004, pp. 81–102.

[CLM] G. Cornelissen, O. Lorscheid, M. Marcolli, *On the K-theory of graph C^*-algebras*, Acta Applicandae Mathematicae, **102** (2008), 57–69.

[CMRV] G. Cornelissen, M. Marcolli, K. Reihani, A. Vdovina, *Noncommutative geometry on trees and buildings*, Traces in Geometry, Number Theory, and Quantum Fields, Vieweg, 2007, pp. 73–98.

[CorMa] G. Cornelissen, M. Marcolli, *Zeta functions that hear the shape of a Riemann surface*, J. of Geometry and Physics **58** (2008), 619–632.

[deJong] J.W. de Jong, *Graphs, spectral triples and Dirac zeta functions*, arXiv:0904.1291.

[DriMan] V.G. Drinfeld, Yu.I. Manin, *Periods of p-adic Schottky groups*, J. Reine Angew. Math. **262/263** (1973), 239–247.

[GekW] E.U. Gekeler, *Analytical construction of Weil curves over function fields*, Journal de théorie des nombres de Bordeaux **7** (1995), 27–49.

[GvP] L. Gerritzen, M. van der Put, *Lecture Notes in Mathematics*, Springer-Verlag, Berlin, 1980, p. 817.

[HS] J.H. Hong, W. Szymanski, *Quantum spheres and projective spaces as graph algebras*, Comm. Math. Phys. **232** (2002), 157-188.

MODULAR INDEX INVARIANTS OF MUMFORD CURVES 73

[KR]   R.V. Kadison, J.R. Ringrose, *Fundamentals of the theory of operator algebras. Vol II: Advanced theory*, Academic Press, 1986.

[kpr]   A. Kumjian, D. Pask, I. Raeburn, *Cuntz–Krieger algebras of directed graphs*, Pacific J. Math. **184** (1998), 161–174.

[KuPa]   A. Kumjian, D. Pask, *C*-algebras of directed graphs and group actions*, Ergod. Th. Dyn. Sys. **19** (1999), 1503–1519.

[Lubo]   A. Lubotsky, *Lattices in rank one Lie groups over local fields*, Geometric and Functional Analysis **1** (1991), 405–431.

[Ma]   Yu.I. Manin, *p-adic automorphic functions*, J. Soviet Math. **5** (1976), 279-333.

[Man]   Yu.I. Manin, *Three-dimensional hyperbolic geometry as ∞-adic Arakelov geometry*, Invent. Math. **104** (1991), 223–244.

[Mar]   M. Marcolli, University Lecture Series **36**, American Mathematical Society, Providence, RI, 2005.

[Mosc]   H. Moscovici, *Local index formula and twisted spectral triples*, arXiv:0902.0835.

[Mum]   D. Mumford, *An analytic construction of degenerating curves over complete local rings*, Compositio Math. **24** (1972), 129–174.

[PRen]   D. Pask, A. Rennie, *The noncommutative geometry of graph C*-algebras I: The index theorem*, J. Funct. Anal. **233** (2006), 92–134.

[PV]   M. Pimsner, D. Voiculescu, *K-groups of reduced crossed products by free groups*, J. Operator Theory, **8** (1982), 131–156.

[R]   I. Raeburn, *Graph algebras: C*-algebras we can see*, CBMS Lecture Notes 103, American Mathematical Society, Providence, RI, 2005.

[Rob]   G. Robertson, *Invariant boundary distributions associated with finite graphs*, preprint.

[Ta]   M. Takesaki *Tomita's theory of modular Hilbert algebras and its applications.* (Lecture Notes in Mathematics 128, Springer-Verlag, 1970).

[T]   M. Tomforde, *Real rank zero and tracial states of C*-algebras associated to graphs*, math.OA/0204095v2.

[vdP]   M. van der Put, *Discrete groups, Mumford curves and theta functions*, Ann. Fac. Sci. Toulouse Math (6) **1** (1992), 399–438.

DIRECTOR, MATHEMATICAL SCIENCES INSTITUTE, JOHN DEDMAN BUILDING, AUSTRALIAN NATIONAL UNIVERSITY, CANBERRA, ACT 0200, AUSTRALIA.
   E-mail address: alan.carey@anu.edu.au

MATHEMATICS DEPARTMENT, MAIL CODE 253-37, CALIFORNIA INSTITUTE OF TECHNOLOGY, 1200 E. CALIFORNIA BLVD., PASADENA, CA 91125, USA.
   E-mail address: matilde@caltech.edu

MATHEMATICAL SCIENCES INSTITUTE, JOHN DEDMAN BUILDING, AUSTRALIAN NATIONAL UNIVERSITY, CANBERRA, ACT 0200, AUSTRALIA.
   E-mail address: adam.rennie@anu.edu.au

Characteristic 1, Entropy and the Absolute Point

Alain Connes and Caterina Consani

ABSTRACT. We show that the mathematical meaning of working in character-istic 1 is directly connected to the fields of idempotent analysis and tropical algebraic geometry and we relate this idea to the notion of the absolute point Spec $\mathbb{F}_1$. After introducing the notion of "perfect" semi-ring of characteris-tic 1, we explain how to adapt the construction of the Witt ring in character-istic $p > 1$ to the limit case of characteristic 1. This construction also unveils an interesting connection with entropy and thermodynamics, while shedding a new light on the classical Witt construction itself. We simplify our earlier construction of the geometric realization of an $\mathbb{F}_1$-scheme and extend our ear-lier computations of the zeta function to cover the case of $\mathbb{F}_1$-schemes with torsion. Then, we show that the study of the additive structures on monoids provides a natural map $M \mapsto A(M)$ from monoids to sets which comes close to fulfilling the requirements for the hypothetical curve $\overline{\mathrm{Spec}\,\mathbb{Z}}$ over the absolute point Spec $\mathbb{F}_1$. Finally, we test the computation of the zeta function on elliptic curves over the rational numbers.

1. Introduction

The main goal of this paper is to explore the mathematical meaning of working in characteristic 1 and to relate this idea to the notion of the absolute point Spec $\mathbb{F}_1$. In the first part of the paper, we explain that an already well-established branch of mathematics supplies a satisfactory answer to the question of the meaning of mathematics in characteristic 1. Our starting point is the investigation of the structure of fields of positive characteristic $p > 1$, in the degenerate case $p = 1$. The outcome is that this limit case is directly connected to the fields of idempotent analysis (see [**KM**]) and tropical algebraic geometry (see [**IMS**], [**Ga**]). In parallel with the development of classical mathematics over fields, its "shadow" idempotent version appears in characteristic 1.

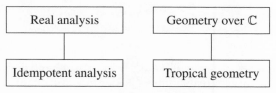

To perform idempotent or tropical mathematics means to perform analysis and/or geometry after replacing the archimedean local field $\mathbb{R}$ with the locally compact *semi-field* $\mathbb{R}_+^{\mathrm{max}}$. This latter structure is defined as the locally compact space $\mathbb{R}_+ = [0, \infty)$ endowed with the ordinary product and a new idempotent addition $x \uplus y = \max\{x, y\}$ that appears as the limit case of the conjugates of ordinary addition by the map $x \mapsto x^h$, when $h \to 0$. The semi-field $\mathbb{R}_+^{\mathrm{max}}$ is trivially isomorphic to the "schedule algebra" (or "max-plus algebra") $(\mathbb{R} \cup \{-\infty\}, \max, +)$, by means of the map $x \mapsto \log x$. The reason we prefer to work with $\mathbb{R}_+^{\mathrm{max}}$ in this paper (rather than with the "max-plus" algebra) is that the structure of $\mathbb{R}_+^{\mathrm{max}}$ makes the role of the "Frobenius" more transparent. On a field of characteristic $p > 1$, the arithmetic Frobenius is given by the "additive" map $x \mapsto x^p$. On $\mathbb{R}_+^{\mathrm{max}}$, the Frobenius flow has the analogous description given by the action $x \mapsto x^\lambda$ ($\lambda \in \mathbb{R}_+^*$) which is also "additive" since it is monotonic and hence compatible with $\max\{x, y\}$.

It is well known that the classification of local fields of positive characteristic reduces to the classification of finite fields of positive characteristic. A local field of characteristic $p > 1$ is isomorphic to the field $\mathbb{F}_q((T))$ of formal power series with finite order pole at 0, over a finite extension $\mathbb{F}_q$ of $\mathbb{F}_p$. There is a strong relation between the p-adic field $\mathbb{Q}_p$ of characteristic zero and the field $\mathbb{F}_p((T))$ of characteristic p. The connection is described by the Ax-Kochen theorem [**AK**] (that depends upon the continuum hypothesis) which states the following isomorphism of ultraproducts

$$(1) \qquad \prod_\omega \mathbb{Q}_p \sim \prod_\omega \mathbb{F}_p((T))$$

for any non-trivial ultrafilter ω on the integers.

The local field $\mathbb{F}_p((T))$ is the unique local field of characteristic p with residue field $\mathbb{F}_p$. Then the Ax-Kochen theorem essentially states that, for sufficiently large p, the field $\mathbb{Q}_p$ resembles its simplification $\mathbb{F}_p((T))$, in which addition forgets the carry over rule in the description of a p-adic number as a power series in p^n. The process that allows one to view $\mathbb{F}_p((T))$ as a limit of fields of characteristic zero obtained by adjoining roots of p to the p-adic field $\mathbb{Q}_p$ was described in [**Kr**]. The role of the Witt vectors is to define a process going backwards from the simplified rule of $\mathbb{F}_p((T))$ to the original algebraic law holding for p-adic numbers.

In characteristic 1, the role of the local field $\mathbb{F}_p((T))$ is played by the local semi-field $\mathbb{R}_+^{\mathrm{max}}$.

Characteristic 0	$\mathbb{Q}_p$, $p > 1$	Archimedean $\mathbb{R}$
Characteristic $\neq 0$	$\mathbb{F}_p((T))$	$p = 1$, $\mathbb{R}_+^{\mathrm{max}}$

The process that allows one to view $\mathbb{R}_+^{\mathrm{max}}$ as a result of a deformation of the usual algebraic structure on real numbers is known as "dequantization" and can be described as a semi-classical limit in the following way. First of all note that, for $\mathbb{R} \ni h > 0$ one has

$$h \ln(e^{w_1/h} + e^{w_2/h}) \to \max\{w_1, w_2\} \quad \text{as } h \to 0.$$

Thus, the natural map

$$D_h : \mathbb{R}_+ \to \mathbb{R}_+ \qquad D_h(u) = u^h$$

satisfies the equation

$$\lim_{h \to 0} D_h(D_h^{-1}(u_1) + D_h^{-1}(u_2)) = \max\{u_1, u_2\}.$$

In the limit $h \to 0$, the usual algebraic rules on $\mathbb{R}_+$ deform to become those of $\mathbb{R}_+^{\max}$. Moreover, in the limit one also obtains a one-parameter group of automorphisms of $\mathbb{R}_+^{\max}$

$$\vartheta_\lambda(u) = u^\lambda, \qquad \lambda \in \mathbb{R}_+^{\max}$$

which corresponds to the arithmetic Frobenius in characteristic $p > 1$.

After introducing the notion of "perfect" semi-ring of characteristic 1 (§2.3.3), our first main result (§2.4) states that one can adapt the construction of the Witt ring to this limit case of characteristic 1. This fact also unveils an interesting deep connection with entropy and thermodynamics, while shedding a new light on the classical Witt construction itself.

Our starting point is the basic physics formula expressing the free energy in terms of entropy from a variational principle. In its simplest mathematical form, it states that

$$(2) \qquad \log(e^a + e^b) = \sup_{x \in [0,1]} S(x) + xa + (1 - x)b$$

where $S(x) = -x \log x - (1 - x) \log(1 - x)$ is the entropy of the partition of 1 as $1 = x + (1 - x)$.

In Theorem 2.20, we explain how (2) allows one to reconstruct addition from its characteristic 1 degenerate limit. This process has an analogue in the construction of the Witt ring which reconstructs the p-adic numbers from their degenerate characteristic p limit $\mathbb{F}_p((t))$. Interestingly, this result also leads to a reformulation of the Witt construction in complete analogy with the case of characteristic 1 (see Proposition 2.23). These developments, jointly with the existing body of results in idempotent analysis and tropical algebraic geometry, supply a strong and motivated expectation on the existence of a meaningful notion of "local" mathematics in characteristic 1. In fact, they also suggest the existence of a notion of "global" mathematics in characteristic 1, pointing to the possibility that the discrete, co-compact embedding of a global field into a locally compact semi-simple non-discrete ring may have an extended version for semi-fields in semi-rings of characteristic 1.

The second part of the paper reconsiders the study of the "absolute point" initiated in [KS]. It is important to explain the distinction between Spec $\mathbb{B}$, where $\mathbb{B}$ is the initial object in the category of semi-rings of characteristic 1 (cf. [Go]) and the sought-for "absolute point" which has been traditionally denoted by Spec $\mathbb{F}_1$. It is generally expected that Spec $\mathbb{F}_1$ sits under Spec $\mathbb{Z}$ in the sense of [TV]:

(3)

In order to reasonably fulfill the property to be "absolute," Spec $\mathbb{F}_1$ should also sit under Spec $\mathbb{B}$. Here, it is important to stress the fact that Spec $\mathbb{B}$ does not qualify to be the "absolute point," since there is no morphism from $\mathbb{B}$ to $\mathbb{Z}$. We

Theorem 2.1 singles out the conditions on a bijection s of the set $\mathbb{K} = H \cup \{0\}$ onto itself (H = abelian group), so that s becomes the addition of 1 in a field whose multiplicative structure is that of the monoid $\mathbb{K}$. These conditions are:

• s commutes with its conjugates for the action of H by multiplication on the monoid $\mathbb{K}$.

• $s(0) \neq 0$.

The degenerate case of characteristic 1 appears when one drops the requirement that s is a bijection and replaces it with the idempotent condition $s \circ s = s$, so that s becomes a retraction onto its range. The degenerate structure of $\mathbb{F}_1$ arises when one drops the condition $s(0) \neq 0$. The trivial solution given by the identity map $s = \mathrm{id}$ then yields such degenerate structure in which, except for the addition of 0, the addition is indeterminate as $0/0$. This fact simply means that, except for 0, one forgets completely the additive structure. This outcome agrees with the point of view developed in the earlier papers [**D1**], [**D2**] and [**TV**].

In [**CC2**], we have introduced a geometric theory of algebraic schemes (of finite type) over $\mathbb{F}_1$ which unifies an earlier construction developed in [**So**] with the approach initiated in our paper [**CC1**]. This theory is reconsidered and fully developed in the present paper. The geometric objects covered by our construction are rational varieties which admit a suitable cell decomposition in toric varieties. Typical examples include Chevalley groups schemes, projective spaces etc. Our initial interest for an algebraic geometry over $\mathbb{F}_1$ originated with the study of Chevalley schemes as (affine) algebraic varieties over $\mathbb{F}_1$. Incidentally, we note that these structures are not toric varieties in general and this shows that the point of view of [**D1**] and [**D2**] is a bit too restrictive. Theorem 2.2 of [**LL**] gives a precise characterization of the schemes studied by our theory in terms of general torifications. This result also points to a deep relation with tropical geometry that we believe is worth exploring in detail.

On the other hand, one should not forget the fact that the class of varieties covered even by this extended definition of schemes over $\mathbb{F}_1$ is still extremely restrictive and, due to the rationality requirement, excludes curves of positive genus. Thus, at first sight, one still seems quite far from the real objective which is to describe the "curve" (of infinite genus) $C = \overline{\mathrm{Spec}\,\mathbb{Z}}$ over $\mathbb{F}_1$, whose zeta function $\zeta_C(s)$ is the complete Riemann zeta function. However, a thorough development of the theory of schemes over $\mathbb{F}_1$ as initiated in [**CC2**] shows that any such scheme X is described (among other data) by a covariant functor $\underline{X}$ from the category $\mathfrak{M}\mathfrak{o}$ of commutative monoids to sets. The functor $\underline{X}$ associates to an object M of $\mathfrak{M}\mathfrak{o}$ the set of points of X which are "defined over M." The adèle class space of a global field $\mathbb{K}$ is a particularly important example of a monoid and this relevant description allows one to evaluate any scheme over $\mathbb{F}_1$ on such monoid. In [**CC2**], we have shown that using the simplest example of a scheme over $\mathbb{F}_1$, namely the curve $\mathbb{P}^1_{\mathbb{F}_1}$, one constructs a perfect set-up to understand conceptually the spectral realization of the zeros of the Riemann zeta function (and more generally of Hecke L-functions over $\mathbb{K}$). The spectral realization appears naturally from the (sheaf) cohomology $H^1(\mathbb{P}^1_{\mathbb{F}_1}, \Omega)$ of very simply defined sheaves Ω on the geometric realization of the scheme $\mathbb{P}^1_{\mathbb{F}_1}$. Such sheaves are obtained from the functions on the points of $\mathbb{P}^1_{\mathbb{F}_1}$ defined over the monoid M of adèle classes. Moreover, the computation of the cohomology $H^0(\mathbb{P}^1_{\mathbb{F}_1}, \Omega)$ provides a description of the graph of the Fourier transform which thus appears conceptually in this picture.

The above construction makes use of a peculiar property of the geometric realization of an $\mathfrak{Mo}$-functor which has no analogue for $\mathbb{Z}$-schemes (or more generally for $\mathbb{Z}$-functors). Indeed, the geometric realization of an $\mathfrak{Mo}$-functor $\underline{X}$ coincides, as a set, with the set of points which are defined over $\mathbb{F}_1$, $i.e.$ with the value $\underline{X}(\mathbb{F}_1)$ of the functor on the most trivial monoid. After a thorough development of the general theory of $\mathfrak{Mo}$-schemes (an essential ingredient in the finer notion of an $\mathbb{F}_1$-scheme), we investigate in full details their geometric realization, which turn out to be schemes in the sense of [**D1**] and [**D2**]. Then, we show that both the topology and the structure sheaf of an $\mathbb{F}_1$-scheme can be obtained in a natural manner on the set $\underline{X}(\mathbb{F}_1)$ of its $\mathbb{F}_1$-points.

At the beginning of §4, we briefly recall our construction of an $\mathbb{F}_1$-scheme and the description of the associated zeta function, under the (local) condition of no-torsion on the scheme ($cf.$ [**CC2**]). We then remove this hypothesis and compute, in particular, the zeta function of the extensions $\mathbb{F}_{1^n}$ of $\mathbb{F}_1$. The general result is stated in Theorem 4.13 and in Corollary 4.14 which present a description of the zeta function of a Noetherian $\mathbb{F}_1$-scheme X as the product

$$\zeta(X, s) = e^{h(s)} \prod_{j=0}^{n} (s - j)^{\alpha_j}$$

of the exponential of an entire function by a finite product of fractional powers of simple monomials. The exponents α_j are rational numbers defined explicitly, in terms of the structure sheaf $\mathcal{O}_X$ in monoids, as follows

$$\alpha_j = (-1)^{j+1} \sum_{x \in X} (-1)^{n(x)} \binom{n(x)}{j} \sum_d \frac{1}{d} \nu(d, \mathcal{O}^\times_{X,x})$$

where $\nu(d, \mathcal{O}^\times_{X,x})$ is the number of injective homomorphisms from $\mathbb{Z}/d\mathbb{Z}$ to the group $\mathcal{O}^\times_{X,x}$ of invertible elements of the monoid $\mathcal{O}_{X,x}$. In order to establish this result, we need to study the case of a counting function

(4) $$\# X(\mathbb{F}_{1^n}) = N(n + 1)$$

no longer polynomial in the integer $n \in \mathbb{N}$. In [**CC2**] we showed that the limit formula that was used in [**So**] to define the zeta function of an algebraic variety over $\mathbb{F}_1$ can be replaced by an equivalent integral formula which determines the equation

(5) $$\frac{\partial_s \zeta_N(s)}{\zeta_N(s)} = -\int_1^\infty N(u) u^{-s-1} du$$

describing the logarithmic derivative of the zeta function $\zeta_N(s)$ associated to the counting function $N(q)$. We use this result to treat the case of the counting function of a Noetherian $\mathbb{F}_1$-scheme and Nevanlinna theory to uniquely extend the counting function $N(n)$ to arbitrary complex arguments $z \in \mathbb{C}$ and finally we compute the corresponding integrals.

In §5 we show that the replacement, in (5), of the integral by the discrete sum

$$\frac{\partial_s \zeta_N^{\mathrm{disc}}(s)}{\zeta_N^{\mathrm{disc}}(s)} = -\sum_1^\infty N(u)u^{-s-1}$$

only modifies the zeta function of a Noetherian $\mathbb{F}_1$-scheme by an exponential factor of an entire function. This observation leads to Definition 5.2 of a modified zeta function over $\mathbb{F}_1$ whose main advantage is that of being applicable to the case of an arbitrary counting function with polynomial growth.

In [**CC2**], we determined the counting distribution $N(q)$, defined for $q \in [1, \infty)$, such that the zeta function (5) gives the complete Riemann zeta function $\zeta_{\mathbb{Q}}(s) = \pi^{-s/2}\Gamma(s/2)\zeta(s)$; *i.e.* such that the following equation holds

$$(6) \qquad \frac{\partial_s \zeta_{\mathbb{Q}}(s)}{\zeta_{\mathbb{Q}}(s)} = -\int_1^\infty N(u)\, u^{-s} d^* u.$$

In §5.2, we use the modified version of zeta function (as described above) to study a simplified form of (5), *i.e.*

$$\frac{\partial_s \zeta(s)}{\zeta(s)} = -\sum_1^\infty N(n)\, n^{-s-1}.$$

This gives, for the counting function $N(n)$, the formula $N(n) = n\Lambda(n)$, where $\Lambda(n)$ is the von-Mangoldt function.[1] Using (4), this shows that, for a suitably extended notion of "scheme over $\mathbb{F}_1$," the hypothetical "curve" $X = \overline{\mathrm{Spec}\,\mathbb{Z}}$ should fulfill the following requirement

$$(7) \qquad \#(X(\mathbb{F}_{1^n})) = \begin{cases} 0, & \text{if } n+1 \text{ is not a prime power} \\ (n+1)\log p, & \text{if } n = p^\ell - 1, \ p \text{ prime.} \end{cases}$$

This expression is neither functorial nor integer valued, but by reconsidering the theory of additive structures previously developed in §2.1, we show in Corollary 5.4 that the natural construction which assigns to an object M of $\mathfrak{Mo}$ the set $A(M)$ of maps $s : M \to M$ such that

• s commutes with its conjugates by multiplication by elements of $M^\times$

• $s(0) = 1$

comes close to solving the requirements of (7) since it gives (for $n > 1$, φ the Euler totient function)

$$\#(A(\mathbb{F}_{1^n})) = \begin{cases} 0, & \text{if } n+1 \text{ is not a prime power} \\ \frac{\varphi(n)}{\log(n+1)} \log p, & \text{if } n = p^\ell - 1, \ p \text{ prime.} \end{cases}$$

In §5.3 we experiment with elliptic curves E over $\mathbb{Q}$, by computing the zeta function associated to a *specific* counting function on E. The specific function $N(q, E)$ is uniquely determined by the following two conditions:

• For any prime power $q = p^\ell$, the value of $N(q, E)$ is the number[2] of points of the reduction of E modulo p in the finite field $\mathbb{F}_q$.

• The function $t(n)$ occurring in the equation $N(n, E) = n + 1 - t(n)$ is weakly multiplicative.

---

[1] With value $\log p$ for powers p^ℓ of primes and zero otherwise.

[2] Including the singular point.

Then, we prove that the obtained zeta function $\zeta_N^{\mathrm{disc}}(s)$ of E fulfills the equation

$$\frac{\partial_s \zeta_N^{\mathrm{disc}}(s)}{\zeta_N^{\mathrm{disc}}(s)} = -\zeta(s+1) - \zeta(s) + \frac{L(s+1, E)}{\zeta(2s+1)M(s+1)}$$

where $L(s, E)$ is the L-function of the elliptic curve, $\zeta(s)$ is the Riemann zeta function and

$$M(s) = \prod_{p \in S}(1 - p^{1-2s})$$

is a finite product of local factors indexed on the set S of primes at which E has bad reduction. In Example 5.8 we exhibit the singularities of $\zeta_E(s)$ in a concrete case.

Acknowledgments. The authors are partially supported by NSF grant DMS-FRG-0652164. The first author thanks C. Breuil for pointing out the reference to M. Krasner [**Kr**]. Both authors acknowledge the paper [**Le**] of P. Lescot, that brings up the role of the idempotent semi-field $\mathbb{B}$.

2. Working in Characteristic 1: Entropy and Witt Construction

In this section we investigate the degeneration of the structure of fields of prime positive characteristic p, in the limit case $p = 1$. A thorough investigation of this process and the algebraic structures of characteristic 1, points to a very interesting link between the fields of idempotent analysis [**KM**] and tropical geometry [**IMS**], with the algebraic structure of a degenerate geometry (of characteristic 1) obtained as the limit case of geometries over finite fields $\mathbb{F}_p$. Our main result is that of adapting the construction of the Witt ring to this limit case of characteristic $p = 1$ while also unraveling a deep connection of the Witt construction with entropy and thermodynamics. Remarkably, we find that this process also sheds new light on the classical Witt construction itself.

2.1. Additive Structure

The multiplicative structure of a field $\mathbb{K}$ is obtained by adjoining an absorbing element 0 to its multiplicative group $H = \mathbb{K}^{\times}$. Our goal here is to understand the additional structure on the multiplicative monoid $H \cup \{0\}$ (the multiplicative monoid underlying $\mathbb{K}$) corresponding to the addition. For simplicity, we shall still denote by $\mathbb{K}$ such a monoid.

We first notice that to define an additive structure on $\mathbb{K}$ it is enough to know how to define the operation

$$(8) \qquad\qquad s_o : \mathbb{K} \to \mathbb{K} \qquad s_o(x) = x + 1 \qquad \forall\, x \in \mathbb{K}$$

since then the definition of the addition follows by setting

$$(9) \qquad + : \mathbb{K} \times \mathbb{K} \to \mathbb{K} \qquad x + y = x\, s_o(yx^{-1}) \qquad \forall\, x, y \in \mathbb{K},\ x \neq 0.$$

Moreover, with this definition given, one has (using the commutativity of the monoid $\mathbb{K}$)

$$x(y + z) = x\, y\, s_o(zy^{-1}) = xy\, s_o(xzy^{-1}x^{-1}) = xy\, s_o(xz(xy)^{-1}).$$

Thus, the distributivity follows automatically

$$x(y + z) = xy + xz \qquad \forall\, x, y, z \in \mathbb{K}.$$

The following result characterizes the bijections s of the monoid $\mathbb{K}$ onto itself such that, when enriched by the addition law (9), the monoid $\mathbb{K}$ becomes a field.

THEOREM 2.1. *Let H be an abelian group. Let s be a bijection of the set $\mathbb{K} = H \cup \{0\}$ onto itself that commutes with its conjugates for the action of H by multiplication on the monoid $\mathbb{K}$. Then, if $s(0) \neq 0$, the operation*

$$(10) \qquad x + y := \begin{cases} y, & \text{if } x = 0 \\ s(0)^{-1}xs(s(0)yx^{-1}), & \text{if } x \neq 0 \end{cases}$$

defines a commutative group law on $\mathbb{K}$. With this law as addition, the monoid $\mathbb{K}$ becomes a commutative field. Moreover the field $\mathbb{K}$ is of characteristic p if and only if

$$s^p = s \circ s \circ \ldots \circ s = id.$$

PROOF. By replacing s with its conjugate, one can assume, since $s(0) \neq 0$, that $s(0) = 1$. Let $A(x,y) = x + y$ be given as in (10). By definition, one has $A(0,y) = y$. For $x \neq 0$, $A(x,0) = xs(0) = x$. Thus 0 is a neutral element for $\mathbb{K}$. The commutation of s with its conjugate for the action of H on $\mathbb{K}$ by multiplication with the element $x \neq 0$ means that

$$(11) \qquad s(x\, s(yx^{-1})) = x\, s(s(y)x^{-1}) \qquad \forall y \in \mathbb{K}.$$

Taking $y = 0$ (and using $s(0) = 1$), the above equation gives

$$(12) \qquad s(x) = x\, s(x^{-1}) \qquad \forall x \neq 0.$$

Assume now that $x \neq 0$ and $y \neq 0$, then from (12) one gets

$$A(x,y) = xs(yx^{-1}) = x(yx^{-1})s(xy^{-1}) = y\, s(xy^{-1}) = A(y,x).$$

Thus A is a commutative law of composition. The associativity of A follows from the commutation of left and right addition which is a consequence of the commutation of the conjugates of s. More precisely, assuming first that all elements involved are non-zero, one has

$$A(A(x,y),z) = A(xs(yx^{-1}),z) = A(z,xs(yx^{-1})) = zs(xs(yx^{-1})z^{-1})$$
$$= zs(xz^{-1}\, s(yx^{-1})).$$

Using (11) one obtains

$$zs(xz^{-1}\, s(yx^{-1})) = z\, xz^{-1}\, s(s(yz^{-1})zx^{-1}) = x\, s(s(yz^{-1})zx^{-1})$$

$$A(x, A(y,z)) = A(x, s(yz^{-1})z) = x\, s(s(yz^{-1})zx^{-1})$$

which yields the required equality. It remains to verify a few special cases. If x or y or z is zero, the equality follows since 0 is a neutral element. Since we never had to divide by $s(a)$, the above argument applies without restriction. Finally let $\theta = s^{-1}(0)$, then for any $x \neq 0$ one has $A(x, \theta x) = xs(\theta) = 0$. This shows that θx is the inverse of x for the law A. We have thus proven that this law defines an abelian group structure on $\mathbb{K}$.

Finally, we claim that the distributive law holds. This means that for any $a \in \mathbb{K}$ one has

$$A(ax, ay) = aA(x,y), \ \forall x,y \in \mathbb{K}.$$

We can assume that all elements involved are $\neq 0$. Then one obtains

$$A(ax, ay) = ax\, s(ay(ax)^{-1}) = ax\, s(yx^{-1}) = a\, A(x, y).$$

This suffices to show that $\mathbb{K}$ is a field.   □

As a corollary, one obtains the following general uniqueness result

COROLLARY 2.2. *Let H be a finite commutative group and let s_j $(j = 1, 2)$ be two bijections of the monoid $\mathbb{K} = H \cup \{0\}$ such that $s_j(0) \neq 0$ and s_j commutes with its conjugates as in Theorem 2.1. Then H is a cyclic group of order $m = p^\ell - 1$ for some prime p and there exists a group automorphism $\alpha \in \mathrm{Aut}(H)$ and an element $g \in H$ such that*

$$s_2 = T \circ s_1 \circ T^{-1}, \ T(x) = g\alpha(x), \ \forall x \in H, \ T(0) = 0.$$

PROOF. By replacing s_j with its conjugate by $g_j = s_j(0)$, one can assume that $s_j(0) = 1$. Each s_j defines on the monoid $\mathbb{K} = H \cup \{0\}$ an additive structure which, in view of Theorem 2.1, turns this set into a finite field $\mathbb{F}_q$, for some prime power $q = p^\ell$. If m is the order of H, then $m = p^\ell - 1 = q - 1$ and H is the cyclic group of order m. The field $\mathbb{F}_q$ with $q = p^\ell$ elements is unique up to isomorphism. Thus there exists a bijection α from $\mathbb{K}$ to itself, which is an isomorphism with respect to the multiplicative and the additive structures given by s_j. This map sends 0 to 0 and 1 to 1, and thus transforms the addition of 1 for the first structure into the addition of 1 for the second one. Since this map also respects the multiplication, it is necessarily given by a group automorphism $\alpha \in \mathrm{Aut}(H)$.   □

2.2. Characteristic $p = 2$

We apply the above discussion to a concrete case. In this subsection we work in characteristic two, thus we consider an algebraic closure $\bar{\mathbb{F}}_2$. It is well known that the multiplicative group $(\bar{\mathbb{F}}_2)^\times$ is *non-canonically* isomorphic to the group $\mathbb{G}^{\mathrm{odd}}$ of roots of unity in $\mathbb{C}$ of odd order. We consider the monoid $\mathbb{K} = \mathbb{G}^{\mathrm{odd}} \cup \{0\}$. The product between non-zero elements in $\mathbb{K}$ is the same as in $\mathbb{G}^{\mathrm{odd}}$ and 0 is an absorbing element (*i.e.* $0 \cdot x = x \cdot 0 = 0, \forall\, x \in \mathbb{K}$). For each positive integer ℓ, there is a unique subgroup $\mu_{(2^\ell - 1)} < \mathbb{G}^{\mathrm{odd}}$ made by the roots of 1 of order $2^\ell - 1$; this is also the subgroup of the invertible elements of the finite subfield $\mathbb{F}_{2^\ell} \subset \bar{\mathbb{F}}_2$. We know how to multiply two elements in $\mathbb{K}$ but we do not know how to add them.

Since we work in characteristic two, the transformation s_o of (8) given by the addition of 1 fulfills $s_o \circ s_o = id$, *i.e.* s_o is an involution.

Let Σ be the set of *involutions* s of $\mathbb{K}$ which commute with all their conjugates by rotations with elements of $\mathbb{G}^{\mathrm{odd}}$ and which fulfill the condition $s(0) = 1$. Thus an element $s \in \Sigma$ is an involution such that $s(0) = 1$ and it satisfies the identity $s \circ (R \circ s \circ R^{-1}) = (R \circ s \circ R^{-1}) \circ s$, for any rotation $R : \mathbb{K} \to \mathbb{K}$.

PROPOSITION 2.3. *For any choice of a group isomorphism*

$$(13) \qquad\qquad j : (\bar{\mathbb{F}}_2)^\times \xrightarrow{\ \sim\ } \mathbb{G}^{\mathrm{odd}},$$

the following map defines an element $s \in \Sigma$

$$s : \mathbb{K} \to \mathbb{K} \qquad s(x) = \begin{cases} j(j^{-1}(x) + 1), & \text{if } x \neq 1 \\ 0, & \text{if } x = 1. \end{cases}$$

Moreover, two pairs $(\bar{\mathbb{F}}_2, j)$ *are isomorphic if and only if the associated symmetries* $s \in \Sigma$ *are the same. Each element of* Σ *corresponds to an uniquely associated pair* $(\bar{\mathbb{F}}_2, j)$.

PROOF. We first show that $s \in \Sigma$. Since $\mathrm{char}(\bar{\mathbb{F}}_2) = 2$, s is an involution. To show that s commutes with its conjugates by rotations with elements of $\mathbb{G}^{\mathrm{odd}}$, it is enough to check that s commutes with its conjugate by $j(y)$, $y \in (\bar{\mathbb{F}}_2)^{\times}$. One first uses the distributivity of the addition to see that

$$j(y)s(xj(y)^{-1}) = j(y)j(j^{-1}(xj(y)^{-1})+1) = j(y(j^{-1}(xj(y)^{-1})+1)) = j(j^{-1}(x)+y).$$

The commutativity of the additive structure on $\bar{\mathbb{F}}_2$ gives the required commutation.

We now prove the second statement. Given two pairs $(\mathbb{K}_i, j_i)$ $(i = 1, 2)$ as in (13), and an isomorphism $\theta : \mathbb{K}_1 \to \mathbb{K}_2$ such that $j_2 \circ \theta = j_1$, one has

$$s_2(x) = j_2(j_2^{-1}(x) + 1) = j_1 \circ \theta^{-1}(\theta(j_1^{-1}(x)) + 1) = s_1(x)$$

since $\theta(1) = 1$. $\qquad\qquad\qquad\qquad\qquad\qquad\qquad\qquad\qquad\qquad\qquad\qquad\quad \square$

Let $s \in \Sigma$ be an involution of $\mathbb{K}$ satisfying the required properties. We introduce the following operation in $\mathbb{K}$

(14)
$$x + y = \begin{cases} y, & \text{if } x = 0 \\ xs(yx^{-1}), & \text{if } x \neq 0. \end{cases}$$

Then it follows as a corollary of Theorem 2.1 that (14) defines a commutative group law on $\mathbb{K} = \mathbb{G}^{\mathrm{odd}} \cup \{0\} \subset \mathbb{C}$. This law and the induced multiplication turn $\mathbb{K}$ into a field of characteristic 2.

REMARK 2.4. We avoid referring to $\bar{\mathbb{F}}_p$ as to "the" algebraic closure of $\mathbb{F}_p$, since for $q = p^n$ a prime power, the finite field $\mathbb{F}_q$ with q elements is well defined only up to an isomorphism. In computations, such as the construction of tables of modular characters, one uses an explicit construction of $\mathbb{F}_q$ as the quotient ring $\mathbb{F}_p[T]/(C_n(T))$, where $C_n(T)$ is a specific irreducible polynomial, called Conway's polynomial. This polynomial is of degree n and fulfills simple algebraic conditions and a normalization property involving the lexicographic ordering (*cf. e.g.* [**Lu**]). In the particular case of characteristic 2, J. Conway was able to produce *canonically* an algebraic closure $\bar{\mathbb{F}}_2$ using inductively defined operations on the ordinals[3] less than $\omega^{\omega^{\omega}}$ (see [**Co**]). His construction also provides one with a natural choice of the isomorphism

$$j : (\bar{\mathbb{F}}_2)^{\times} \overset{\sim}{\longrightarrow} \mathbb{G}^{\mathrm{odd}}.$$

In fact, one can use the well ordering to choose the smallest solution as a root of unity of order $2^{\ell} - 1$, fulfilling compatibility conditions with the previous choices for divisors of ℓ. One obtains in this way a well-defined corresponding symmetry s on $\mathbb{G}^{\mathrm{odd}} \cup \{0\}$. Figure 1 shows the restriction of this symmetry to $\mathbb{G}^{15} \cup \{0\}$, while Figure 2 gives the simple geometric meaning of the commutation of s with its conjugates by rotations.

---

[3]The first author is grateful to Javier Lopez for pointing out this construction of J. Conway.

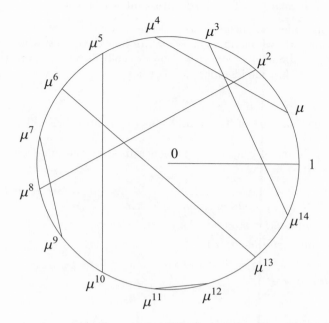

FIGURE 1 The symmetry s for $\mathbb{F}_{16}$ represented by the graph G_s with vertices the 15th roots of 1 and edges $(x, s(x))$.

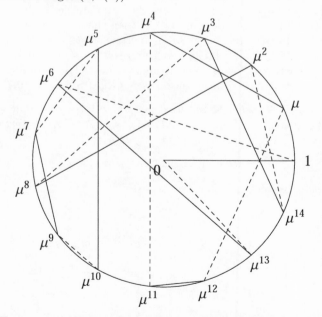

FIGURE 2 The commutativity of s with its conjugates by rotations $R \circ s \circ R^{-1}$ means that the union of the graph G_s with a rotated graph $R(G_s)$ (dashed) is a union of quadrilaterals.

2.3. Characteristic $p = 1$ and Idempotent Analysis

To explain how idempotent analysis and the semi-field $\mathbb{R}_+^{\max}$ appear naturally in the above framework, we shall no longer require that the map s, which represents the "addition of 1," be a bijection of $\mathbb{K}$, but we shall instead require that s is a *retraction, i.e.* it satisfies the idempotent condition

$$s \circ s = s.$$

Before stating the analogue of Theorem 2.1 in this context we will review a few definitions.

The classical theory of rings has been generalized by the more general theory of semi-rings (see [**Go**]).

DEFINITION 2.5. *A semi-ring $(R, +, \cdot)$ is a non-empty set R endowed with operations of addition and multiplication such that*

(a) $(R, +)$ *is a commutative monoid with neutral element* 0

(b) $(R, \cdot)$ *is a monoid with identity* 1

(c) $\forall r, s, t \in R$: $r(s + t) = rs + rt$ *and* $(s + t)r = sr + tr$

(d) $\forall r \in R$: $r \cdot 0 = 0 \cdot r = 0$

(e) $0 \neq 1$.

To any semi-ring R one associates a *characteristic*. The set $\mathbb{N}1_R = \{n1_R | n \in \mathbb{N}\}$ is a commutative subsemi-ring of R called the prime semi-ring. The prime semi-ring is the smallest semi-ring contained in R. If $n \neq m \Rightarrow n1_R \neq m1_R$, then $\mathbb{N}1 \simeq \mathbb{N}$ naturally and the semi-ring R is said to have characteristic zero. On the other hand, if it happens that $k1_R = 0_R$, for some $k > 1$, then there is a least positive integer n with $n1_R = 0_R$. In this case, $\mathbb{N}1 \simeq \mathbb{Z}/n\mathbb{Z}$ and one shows that R itself is a ring of positive characteristic. Finally, if $\ell1_R = m1_R$ for some $\ell \neq m$, $\ell, m \geq 1$ but $k1_R \neq 0$ for all $k \geq 1$, one writes j for $j1_R$ and it follows that $\mathbb{N}1_R = \{0, 1, \dots, n-1\}$. In this case one has the following proposition (see [**Go**], Proposition 9.7).

PROPOSITION 2.6. *Let n be the least positive integer such that $n1_R \in \{i1_R \mid 1 \leq i \leq n-1\}$ and let $i \in \{1, n-1\}$ with $n1_R = i1_R$. Then the prime semi-ring $\mathbb{N}1_R = \{n1_R | n \in \mathbb{N}\}$ is the following semi-ring $B(n, i)$*

$$B(n, i) := \{0, 1, \dots, n-1\}$$

with the following operations, where $m = n - i$

$$x +' y := \begin{cases} x + y, & \text{if } x + y \leq n - 1 \\ \ell, i \leq \ell \leq n-1, \ell = (x + y) \mod m, & \text{if } x + y \geq n \end{cases}$$

$$x \cdot y := \begin{cases} xy, & \text{if } xy \leq n - 1 \\ \ell, i \leq \ell \leq n-1, \ell = (xy) \mod m, & \text{if } xy \geq n. \end{cases}$$

The semi-ring $B(n, i)$ is the homomorphic image of $(\mathbb{N}, +, \cdot)$ by the map π : $\mathbb{N} \to R$, $\pi(k) = k$ for $0 \leq k \leq n - 1$ and for $k \geq n$, $\pi(k)$ is the unique natural number congruent to $k \mod m = n - i$ with $i \leq \pi(k) \leq n - 1$. In this case we say that R has characteristic (n, i).

A semi-ring R is called a *semi-field* when every non-zero element in R has a multiplicative inverse, or equivalently when the set of non-zero elements in R is a (commutative) group for the multiplicative law. Note that if R is a semi-field the only possibility for the semi-ring $B(n, i)$ is when $n = 2$ and $i = 1$. Indeed the subset $\{\ell, i \leq \ell \leq n-1\}$ is stable under product and hence should contain 1 since a finite submonoid of an abelian group is a subgroup. Thus $i = 1$, but in that case one gets $(n-1) \cdot (n-1) = (n-1)$ which is possible only for $n = 2$. The smallest finite (prime) semi-ring structure arises when $n = 2$ (and $i = 1$). We shall denote this structure by $\mathbb{B} = B(2, 1)$. By construction, $\mathbb{B} = \{0, 1\}$ with the usual multiplication law and an addition requiring the idempotent rule $1 + 1 = 1$.

DEFINITION 2.7. *A semi-ring R is said to have characteristic 1 when*

$$1 + 1 = 1$$

in R, i.e. when R contains $\mathbb{B}$ as the prime subsemi-ring.

When R is a semi-ring of characteristic 1, we denote the addition in R, by the symbol $\uplus$. Then it follows from distributivity that

$$a \uplus a = a, \ \forall a \in R.$$

This justifies the term "additively idempotent" frequently used in semi-ring theory as a synonym for "characteristic 1."

THEOREM 2.8. *Let H be an abelian group. Let s be a retraction ($s \circ s = s$) of the set $\mathbb{K} = H \cup \{0\}$ that commutes with its conjugates for the action of H by multiplication on the monoid $\mathbb{K}$. Then, if $s(0) \neq 0$, the operation*

$$(15) \qquad x + y = \begin{cases} y, & \text{if } x = 0 \\ s(0)^{-1} x s(s(0) y x^{-1}), & \text{if } x \neq 0 \end{cases}$$

defines a commutative monoid law on $\mathbb{K}$. With this law as addition, the monoid $\mathbb{K}$ becomes a commutative semi-field of characteristic 1.

PROOF. The proof of Theorem 2.1 applies without modification. Notice that that proof did not use the hypothesis that s is a bijection, except to get the element $\theta = s^{-1}(0)$ which was used to define the additive inverse. The fact that s is a retraction shows that K is of characteristic 1. $\qquad\square$

EXAMPLE 2.9. Let $H = \mathbb{R}_+^*$ be the multiplicative group of the positive real numbers. Let s be the retraction of $\mathbb{K} = H \cup \{0\} = \mathbb{R}_+$ on $[1, \infty) \subset H$ given by

$$s(x) = 1, \ \forall x \leq 1, \ s(x) = x, \ \forall x \geq 1.$$

The conjugate s^λ under multiplication (by λ) is the retraction on $[\lambda, \infty)$ and one easily checks that s commutes with s^λ. The resulting commutative idempotent semi-field is denoted by $\mathbb{R}_+^{\max}$. Thus $\mathbb{R}_+^{\max}$ is the set $\mathbb{R}_+$ endowed with the following two operations:
• addition

$$x \uplus y := \max(x, y), \ \forall x, y \in \mathbb{R}_+.$$

• multiplication is unchanged.

EXAMPLE 2.10. Let H be a group and let 2^H be the set of subsets of H endowed with the following two operations:

• addition

$$(16) \qquad X \uplus Y := X \cup Y, \qquad \forall X, Y \in 2^H$$

• multiplication

$$(17) \qquad X.Y = \{ab \mid a \in X, b \in Y\}, \ \forall X, Y \in 2^H.$$

Addition is commutative and associative and admits the empty set as a neutral element. Multiplication is associative and it admits the empty set as an absorbing element (which we denote by 0 since it is also the neutral element for the additive structure). The multiplication is also distributive with respect to addition.

EXAMPLE 2.11. Let $(M, \uplus)$ be an idempotent semigroup. Then one endows the set $\mathrm{End}(M)$ of endomorphisms $h : M \to M$ such that

$$h(a \uplus b) = h(a) \uplus h(b), \ \forall a, b \in M$$

with the following operations (see [**Go**], I Example 1.14)
• addition

$$(h \uplus g)(a) = h(a) \uplus g(a), \ \forall a \in M, \ \forall h, g \in \mathrm{End}(M)$$

• multiplication

$$(h \cdot g)(a) = h(g(a)), \qquad \forall a \in M, \ \forall h, g \in \mathrm{End}(M).$$

For instance, if one lets $(M, \uplus)$ be the idempotent semi-group given by $(\mathbb{R}, \max)$, the set $\mathrm{End}(M)$ of endomorphisms of M is the set of monotonic maps $\mathbb{R} \to \mathbb{R}$. $\mathrm{End}(M)$ becomes a semi-ring of characteristic 1 with the above operations.

EXAMPLE 2.12. Let $\mathbb{C}_{\mathrm{star}}$ be the set of finitely generated star shaped subsets of the complex numbers $\mathbb{C}$. Thus an element of $\mathbb{C}_{\mathrm{star}}$ is of the form

$$S = \{0\} \bigcup_{z \in F} \{\lambda z \mid z \in F, \ \lambda \in [0, 1]\}$$

for some finite subset $F \subset \mathbb{C}$. One has a natural injection $\mathbb{C}_{\mathrm{star}} \to 2^{\mathbb{C}^*}$ given by

$$S \mapsto S \cap \mathbb{C}^*.$$

The image of $\mathbb{C}_{\mathrm{star}}$ through this injection is stable under the semi-ring operations (16) and (17) in the semi-ring $2^{\mathbb{C}^*}$, where we view $\mathbb{C}^*$ as a multiplicative group. This shows that $\mathbb{C}_{\mathrm{star}}$ is an idempotent semi-ring under the following operations:
• addition

$$X \uplus Y := X \cup Y, \ \forall X, Y \in \mathbb{C}_{\mathrm{star}}$$

• multiplication

$$X.Y = \{ab \mid a \in X, b \in Y\}, \ \forall X, Y \in \mathbb{C}_{\mathrm{star}}.$$

2.3.1. Finite Semi-field of Characteristic 1

We quote the following result from [**Go**] (Example 4.28, Chapter 4).

PROPOSITION 2.13. *The semi-field* $\mathbb{B} = B(2,1)$ *is the only finite semi-field of characteristic* 1.

The semi-field $\mathbb{B} = B(2,1)$ is called the Boolean semi-field (see [**Go**], I Example 1.5).

2.3.2. Lattices

The idempotency of addition in semi-rings of characteristic 1 gives rise to a natural *partial order* which differentiates this theory from that of the more general semi-rings. We recall the following well-known fact

PROPOSITION 2.14. *Let* $(A, \uplus)$ *be a commutative semi-group with an idempotent addition. Define*

$$(18) \qquad\qquad a \preccurlyeq b \quad \Leftrightarrow \quad a \uplus b = b.$$

Then $(A, \preccurlyeq)$ *is a sup-semi-lattice (i.e. a semilattice in which any two elements have a supremum). Furthermore*

$$(19) \qquad\qquad max\{a, b\} = a \uplus b.$$

Conversely, if $(A, \preccurlyeq)$ *is a sup-semi-lattice and* $\uplus$ *is defined to satisfy (19), then* $(A, \uplus)$ *is an idempotent semi-group. These two constructions are inverse to each other.*

PROOF. We check that (18) defines a partial order $\preccurlyeq$ on A. Let $a, b, c \in A$. The reflexive property ($a \leq a$) follows from the idempotency of the addition in A, i.e. $a \uplus a = a$. The antisymmetric property ($a \leq b$, $b \leq a \Rightarrow a = b$) follows from the commutativity of the addition in A, i.e. $b = a \uplus b = b \uplus a = a$. Finally, the transitivity property ($a \leq b$, $b \leq c \Rightarrow a \leq c$) follows from the associativity of the binary operation $\uplus$, i.e. $a \uplus c = a \uplus (b \uplus c) = (a \uplus b) \uplus c = b \uplus c = c$. Thus $(A, \preccurlyeq)$ is a poset and due to the idempotency of the addition, $(A, \preccurlyeq)$ is also a semi-lattice. One defines the join of two elements in A as $a \vee b := a \uplus b$ and due to the closure property of the law $\uplus$ in A (i.e. $a \uplus b \in A$, $\forall a, b \in A$), one concludes that $(A, \preccurlyeq)$ is a sup-semi-lattice (the supremum = maximum of two elements in A being their join).

The converse statement follows too since the above statements on the idempotency and associativity of the operation $\uplus$ hold also in reverse and the closure property derives from (19). $\qquad\qquad\qquad\qquad\qquad\qquad\qquad\qquad\qquad\quad\square$

It follows that any semi-ring of characteristic 1 has a natural partial ordering, denoted $\preccurlyeq$. The addition in the semi-ring is a monotonic operation and 0 is the least element. Distributivity implies that left and right multiplication are semi-lattice homomorphisms and in particular they are monotonic. Thus, any semi-ring of characteristic 1 may be thought of as a semi-lattice ordered semi-group.

2.3.3. Perfect Semi-rings of Characteristic 1

The following notion is the natural extension to semi-rings of the absence of zero divisors in a ring.

DEFINITION 2.15. *A commutative semi-ring is called multiplicatively cancellative when the multiplication by any non-zero element is injective.*

We recall, from ([**Go**] Propositions 4.43 and 4.44) the following result which describes the analogue of the Frobenius endomorphism in characteristic p.

PROPOSITION 2.16. *Let R be a multiplicatively cancellative commutative semi-ring of characteristic 1, then for any integer $n \in \mathbb{N}$, the map $x \mapsto x^n$ is an injective endomorphism of R.*

Recall now that in characteristic $p > 1$ a ring is called *perfect* if the map $x \mapsto x^p$ is surjective.

DEFINITION 2.17. *A multiplicatively cancellative commutative semi-ring of characteristic 1 is called perfect if the endomorphism $R \to R$, $x \mapsto x^n$ is surjective for all n.*

It then follows from Proposition 2.16 that the endomorphism ϑ_n, $x \mapsto x^n$ is bijective and one can invert it and construct the fractional powers

$$\vartheta_\alpha : R \to R, \qquad \vartheta_\alpha(x) = x^\alpha, \ \forall \alpha \in \mathbb{Q}_+^*.$$

Then, by construction, the ϑ_α's are automorphisms $\vartheta_\alpha \in \mathrm{Aut}(R)$ for $\alpha \in \mathbb{Q}_+^*$ and they fulfill the following properties

- $\vartheta_n(x) = x^n$ for all $n \in \mathbb{N}$ and $x \in R$.
- $\vartheta_\lambda \circ \vartheta_\mu = \vartheta_{\lambda\mu}$ for all $\lambda, \mu \in \mathbb{Q}_+^*$.
- $\vartheta_\lambda(x)\vartheta_\mu(x) = \vartheta_{\lambda+\mu}(x)$ for all $\lambda, \mu \in \mathbb{Q}_+^*$ and $x \in R$.

2.3.4. Completion

One can complete a commutative semi-ring of characteristic 1 with respect to any *compatible* metric, *i.e.* a metric d which fulfills the following analogue of the ultrametric inequality

$$d(x \uplus y, z \uplus t) \leq \sup d(x,z), d(y,t), \ \forall x,y,z,t \in R$$

and the following compatibility with the product

$$d(ax, ay) \leq d(0,a)d(x,y), \ \forall a,x,y \in R.$$

Let $\rho \in R$, $1 \preccurlyeq \rho$, be an invertible element, we denote by R^ρ the union

$$R^\rho = \{0\} \cup_{n \in \mathbb{N}} [\rho^{-n}, \rho^n].$$

The following result will be used below.

PROPOSITION 2.18. *Let R be a perfect commutative semi-ring of characteristic 1 with a compatible metric d. Let $\rho \in R$, $1 \preccurlyeq \rho$, be invertible and such that the intervals $[\rho^{-n}, \rho^n]$ are complete and $d(\rho^\alpha, 1) \to 0$ when $\alpha \to 0$. Then for any non-zero $x \in R^\rho$, $d(x^\alpha, 1) \to 0$ when $\alpha \to 0+$.*

PROOF. Let $\epsilon > 0$ and $\delta > 0$ such that $d(\rho^\alpha, 1) < \epsilon$, $\forall \alpha, |\alpha| \leq \delta$. For $x \in [\rho^{-n}, \rho^n]$ and $0 < \alpha < \delta/n$ one has $x^\alpha \in [\rho^{-n\alpha}, \rho^{n\alpha}]$ since θ_α is an automorphism. Thus $\rho^{-n\alpha} \uplus x^\alpha = x^\alpha$, $\rho^{n\alpha} \uplus x^\alpha = \rho^{n\alpha}$ and

$$d(x^\alpha, \rho^{n\alpha}) = d(\rho^{-n\alpha} \uplus x^\alpha, \rho^{n\alpha} \uplus x^\alpha) \leq d(\rho^{-n\alpha}, \rho^{n\alpha}) \leq 2\epsilon$$

so that $d(x^\alpha, 1) \leq 3\epsilon$. $\qquad \square$

2.4. Witt Ring in Characteristic $p = 1$ and Entropy

The places of the global field $\mathbb{Q}$ of the rational numbers fall in two classes: the infinite archimedean place ∞ and the finite places which are labeled by the prime integer numbers. The p-adic completion of $\mathbb{Q}$ at a finite place p determines the corresponding global field $\mathbb{Q}_p$ of p-adic numbers. These local fields have close relatives with simpler structure: the local fields $\mathbb{F}_p((T))$ of formal power series with coefficients in the finite fields $\mathbb{F}_p$ and with finite order pole at 0. We already explained in the introduction that the similarity between the structures of $\mathbb{Q}_p$ and of $\mathbb{F}_p((T))$ is embodied in the Ax-Kochen theorem (see [**AK**]) which states the isomorphism of arbitrary ultraproducts as in (1). By means of the natural construction of the ring of Witt vectors, one recovers the ring $\mathbb{Z}_p$ of p-adic integers from the finite field $\mathbb{F}_p$. This construction (see also §2.5) can be interpreted as a deformation of the ring of formal power series $\mathbb{F}_p[[T]]$ to $\mathbb{Z}_p$.

It is then natural to wonder about the existence of a similar phenomenon at the infinite archimedean place of $\mathbb{Q}$. We have already introduced the semi-field of characteristic 1 $\mathbb{R}_+^{\max}$ as the degenerate version of the field of real numbers. In this subsection we shall describe how the basic physics formula for the free energy involving entropy allows one to move canonically, not only from $\mathbb{R}_+^{\max}$ to $\mathbb{R}$ but in even greater generality from a perfect semi-ring of characteristic 1 to an ordinary algebra over $\mathbb{R}$. This construction is analogous to the construction of the Witt ring.

Let K be a perfect semi-ring of characteristic 1, and let us first assume that it contains $\mathbb{R}_+^{\max}$ as a sub-semi-ring. Thus the operation of raising to a power $s \in \mathbb{Q}_+$ is well defined in K and determines an automorphism $\vartheta_s : K \to K$, $\vartheta_s(x) = x^s$. We first concentrate on the algebraic aspects and use formally the following infinite sum (for the operation $\uplus$) to define a new addition

$$x +' y := \overset{\uplus}{\underset{s \in \mathbb{Q} \cap [0,1]}{\sum}} c(s)\, x^s\, y^{1-s}\,, \ \forall x, y \in K$$

where the function $c(s) \in \mathbb{R}_+$ is defined by

$$c : [0,1] \to \mathbb{R}_+, \quad c(s) = e^{S(s)} = s^{-s}(1-s)^{-(1-s)}.$$

The property of the function $c(s)$ (see Figure 3) which ensures the associativity of the addition law $+'$ is the following: for any $s, t \in [0,1]$ the product $c(s)c(t)^s$ only depends upon the partition of 1 as $st + s(1-t) + (1-s)$. This fact also implies the functional equation

$$(20) \qquad c(u)c(v)^u = c(uv)c(w)^{(1-uv)}\,, \quad w = \frac{u(1-v)}{1-uv}.$$

The function c fulfils the symmetry $c(1-u) = c(u)$ and, by taking it into account, (20) means then that the function on the simplex $\Sigma_2 = \{(s_j) \mid s_j \geq 0, \sum_0^2 s_j = 1\}$ defined by

$$(21) \qquad c_2(s_0, s_1, s_2) = c(s_2)c(\frac{s_0}{s_0 + s_1})^{s_0+s_1}$$

is symmetric under all permutations of the s_j. More generally, for any integer n one may define the function on the n-simplex

$$c_n(s_0, s_1, \ldots, s_n) = \prod \gamma(s_j)^{-1}\,, \quad \gamma(x) = x^x.$$

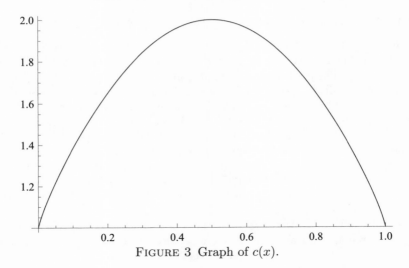

FIGURE 3 Graph of $c(x)$.

LEMMA 2.19. *Let* $x, y \in \mathbb{R}_+$, *then one has*

$$\sup_{s \in [0,1]} c(s)\, x^s\, y^{1-s} = x + y.$$

PROOF. Let $x = e^a$ and $y = e^b$. The function

$$f(s) = \log(c(s)\, x^s\, y^{1-s}) = S(s) + sa + (1-s)b$$

is strictly concave on the interval $[0,1]$ and reaches its unique maximum for $s = \frac{e^a}{e^a + e^b} \in [0,1]$. Its value at the maximum is $\log(e^a + e^b)$.   □

THEOREM 2.20. *Let* K *be a perfect semi-ring of characteristic* 1. *Let* $\rho \in K$, $1 \preccurlyeq \rho$ *be invertible and such that for a compatible metric* d *the intervals* $[\rho^{-n}, \rho^n]$ *are complete and* $d(\rho^\alpha, 1) \to 0$ *when* $\alpha \to 0$. *Let* $K^\rho = \{0\} \cup_{n \in \mathbb{N}} [\rho^{-n}, \rho^n]$ *(see* §2.3.4*). Then the formula*

$$(22) \qquad x +_\rho y := \sum_{s \in \mathbb{Q} \cap [0,1]}^{\uplus} \rho^{S(s)}\, x^s\, y^{1-s}\,, \ \forall x, y \in K^\rho$$

defines an associative law on K^ρ *with* 0 *as a neutral element. The multiplication is distributive with respect to this law. These operations turn the Grothendieck group of the additive monoid* $(K^\rho, +_\rho)$ *into an algebra* $W(K, \rho)$ *over* $\mathbb{R}$ *which depends functorially upon* (K, ρ).

PROOF. Since $1 \preccurlyeq \rho$ one has, using the automorphisms ϑ_λ, that

$$1 \preccurlyeq \vartheta_\lambda(\rho) = \rho^\lambda\,, \ \forall \lambda \in \mathbb{Q}_+.$$

Then it follows that

$$\rho^\lambda \preccurlyeq \rho^{\lambda'}\,, \ \forall \lambda \leq \lambda'.$$

Thus, using completeness to extend the definition of ρ^λ for real values of λ, one obtains that the following map defining a homomorphism of semi-rings

$$\alpha : (\mathbb{R}_+, \max, \cdot) = \mathbb{R}_+^{\max} \to K^\rho, \quad \alpha(\lambda) = \rho^{\log \lambda}, \quad \alpha(0) = 0$$

where we have used the invertibility of ρ in K^ρ to make sense of the negative powers of ρ. Let $w(t) = \rho^{S(t)}$ and $I(n) = \frac{1}{n}\mathbb{Z} \cap [0, 1]$. Let x, y be non-zero elements of K^ρ. Then the partial sums

$$s(n) = \sum_{I(n)} w(\alpha) x^\alpha y^{1-\alpha}$$

form a filtered increasing family for the partial order given by divisibility $n|m$. Moreover one has $s(n) \preccurlyeq \rho^m$ for a fixed m, independent of n as follows using $w(\alpha) \preccurlyeq \rho$ for all $\alpha \in [0, 1]$. Proposition 2.18 shows that the maps

$$\alpha \mapsto w(\alpha), \ \alpha \mapsto x^\alpha, \ \alpha \mapsto y^{1-\alpha}$$

are uniformly continuous for the compatible metric d. It follows using the ultra-metric inequality

$$d(\sum x_i, \sum y_i) \le \mathrm{Sup}(d(x_i, y_i))$$

that the sequence $s(n)$ is a Cauchy sequence. This shows the existence of its limit

$$x +_\rho y := \sum_{s \in \mathbb{Q} \cap [0,1]}^{\uplus} \rho^{S(s)} x^s y^{1-s}.$$

The associativity of the operation $+_\rho$ follows, with $I = \mathbb{Q} \cap (0, 1)$, from

$$(x +_\rho y) +_\rho z = \sum_{s \in I}^{\uplus} w(s) \, \vartheta_s \left(\sum_{t \in I}^{\uplus} w(t) x^t y^{1-t} \right) z^{1-s} =$$

$$= \sum_{s \in I}^{\uplus} \sum_{t \in I}^{\uplus} w(s) w(t)^s x^{ts} y^{(1-t)s} z^{1-s} = \sum_{\Sigma_2}^{\uplus} w_2(s_0, s_1, s_2) x^{s_0} y^{s_1} z^{s_2}$$

which is symmetric in (x, y, z) by making use of (21) and (20). Moreover, one has by homogeneity the distributivity

$$(x +_\rho y)z = xz +_\rho yz, \ \forall x, y, z \in K.$$

We let K^ρ be the semi-ring with the operations $+_\rho$ and the multiplication unchanged. If we endow $\mathbb{R}_+$ with its ordinary addition, we have, by applying Lemma 2.19, that

$$\alpha(\lambda_1) +_\rho \alpha(\lambda_2) = \alpha(\lambda_1 + \lambda_2).$$

Thus α defines a homomorphism

$$\tilde{\alpha} : \mathbb{R}_+ \to (K^\rho, +_\rho, \cdot), \quad \tilde{\alpha}(\lambda) = \rho^{\log \lambda}, \quad \alpha(0) = 0$$

of the semi-ring $\mathbb{R}_+$ (with ordinary addition) into the semi-ring $(K^\rho, +_\rho, \cdot)$.

Let G be the functor from semi-rings to rings which associates to a semi-ring R the Grothendieck group R^Δ of the additive monoid R with the natural extension of the product (see [**Go**]). Let $R[\mathbb{Z}/2\mathbb{Z}]$ be the semi-group ring of $\mathbb{Z}/2\mathbb{Z}$. Elements of $R[\mathbb{Z}/2\mathbb{Z}]$ are given by pairs (m_1, m_2) of elements of R. The addition is given coordinate-wise and the product is defined by

$$(m_1, m_2) \cdot (n_1, n_2) = (m_1 n_1 + m_2 n_2, m_1 n_2 + m_2 n_1).$$

One can view the ring $G(R)$ as the quotient of the semi-ring $R[\mathbb{Z}/2\mathbb{Z}]$ by the ideal of diagonal elements $J = \{(x, x) \mid x \in R\}$ or equivalently by the equivalence relation

$$(m_1, m_2) \sim (n_1, n_2) \Leftrightarrow \exists k \in R, \; m_1 + n_2 + k = m_2 + n_1 + k.$$

This equivalence relation is compatible with the product which turns the quotient $G(R)$ into a ring. We define

$$W(K, \rho) = G((K^\rho, +_\rho, \cdot)).$$

One has $G(\mathbb{R}_+) = \mathbb{R}$. By functoriality of G one thus gets a morphism $G(\tilde{\alpha})$ from $\mathbb{R}$ to $W(K, \rho)$. As long as $1 \neq 0$ in $G((K^\rho, +_\rho, \cdot))$ this morphism endows $W(K, \rho)$ with the structure of an algebra over $\mathbb{R}$. When $1 = 0$ in $G((K^\rho, +_\rho, \cdot))$ one gets the degenerate case $W(K, \rho) = \{0\}$.   □

We did not discuss conditions on ρ which ensure that $(K^\rho, +_\rho, \cdot)$ injects in $W(K, \rho)$, let alone that $W(K, \rho) \neq \{0\}$. The following example shows that the problem comes from how strict the inequality $\rho \preccurlyeq 1$ is assumed to be, *i.e.* where the function $T(x)$ playing the role of absolute temperature actually vanishes.

EXAMPLE 2.21. Let X be a compact space and let $R = C(X, \mathbb{R}_+^{\max})$ be the space of continuous maps from X to the semi-field $\mathbb{R}_+^{\max}$. We endow R with the operations max for addition and the ordinary pointwise product for multiplication. The associated semi-ring is a perfect semi-ring of characteristic 1. Then letting $\rho \in R$, $1 \preccurlyeq \rho$, it is of the form

$$\rho(x) = e^{T(x)}, \; \forall x \in X.$$

Let $\beta(x) = 1/T(x)$ if $T(x) > 0$ and $\beta(x) = \infty$ for $T(x) = 0$. Then the addition $+_\rho$ in R is given by

$$(f +_\rho g)(x) = \left(f(x)^{\beta(x)} + g(x)^{\beta(x)} \right)^{T(x)}, \; \forall x \in X.$$

This follows from Lemma 2.19 which implies more generally that

$$\sup_{s \in [0,1]} e^{TS(s)} x^s y^{1-s} = (x^\beta + y^\beta)^T, \; \beta = \frac{1}{T}.$$

Then, provided that $\rho(x) > 1$ for all $x \in X$, the algebra $W(R, \rho)$ is isomorphic, as an algebra over $\mathbb{R}$, with the real C^*-algebra $C(X, \mathbb{R})$ of continuous functions on X.

2.5. Witt Ring in Characteristic $p > 1$ Revisited

In this subsection we explain in which sense we interpret (22) as the analogue of the construction of the Witt ring in characteristic 1. Let K be a perfect ring of characteristic p. We start by reformulating the construction of the Witt ring in characteristic $p > 1$. One knows (see [Se], Chapter II) that there is a strict p-ring R, unique up to canonical isomorphism, such that its residue ring R/pR is equal to K. One also knows that there exists a unique multiplicative section $\tau : K \to R$ of the residue map. For $x \in K$, $\tau(x) \in R$ is called the Teichmüller representative of x. Every element $z \in R$ can be written uniquely as

$$(23) \qquad\qquad z = \sum_0^\infty \tau(x_n) p^n.$$

The Witt construction of R is functorial and an easy corollary of its properties is the following lemma.

LEMMA 2.22. *For any prime number p, there exists a universal sequence*

$$w(p^n, k) \in \mathbb{Z}/p\mathbb{Z}, \quad 0 < k < p^n$$

such that

$$(24) \qquad \tau(x) + \tau(y) = \tau(x + y) + \sum_{n=1}^{\infty} \tau \left(\sum w(p^n, k) \, x^{\frac{k}{p^n}} y^{1 - \frac{k}{p^n}} \right) p^n.$$

Note incidentally that the fractional powers such as $x^{\frac{k}{p^n}}$ make sense in a perfect ring such as K. The main point here is that (24) suffices to reconstruct the whole ring structure on R and allows one to add and multiply series of the form (23).

PROOF. Formula (24) is a special case of the basic formula

$$s_n = S_n(x_0^{1/p^n}, y_0^{1/p^n}, x_1^{1/p^{n-1}}, y_1^{1/p^{n-1}}, \dots, x_n, y_n)$$

(see [**Se**], Chapter II, §6) which computes the sum

$$\sum_0^{\infty} \tau(x_n) p^n + \sum_0^{\infty} \tau(y_n) p^n = \sum_0^{\infty} \tau(s_n) p^n.$$

$\square$

We can now introduce a natural map from the set I_p of rational numbers in $[0, 1]$ whose denominator is a power of p, to the maximal compact subring of the local field $\mathbb{F}_p((T))$ as follows

$$w_p : I_p \to \mathbb{F}_p((T)), \quad w_p(\alpha) = \sum_{\frac{a}{p^n} = \alpha} w(p^n, a) T^n \in \mathbb{F}_p((T)).$$

(See Table 1 for sample values of $w_p(\alpha)$.) We can then rewrite (24) in a more suggestive form as a deformation of the addition $+$ in $K[[T]]$ to a new addition $+'$. To do this, one first introduces the map

$$\tilde{\tau} : K[[T]] \to R, \quad \tilde{\tau} \left(\sum_0^{\infty} z_n T^n \right) = \sum_0^{\infty} \tau(z_n) p^n.$$

This map is a homomorphism and is multiplicative on monomials, *i.e.*

$$\tilde{\tau}(aT^n Z) = \tilde{\tau}(a)\tilde{\tau}(T)^n \tilde{\tau}(Z), \quad \forall a \in K, \ Z \in K[[T]].$$

Since $\tilde{\tau}$ is not additive, one defines a new addition on $K[[T]]$ by setting

$$X +' Y := \tilde{\tau}^{-1}(\tilde{\tau}(X) + \tilde{\tau}(Y)), \quad \forall X, Y \in K[[T]].$$

PROPOSITION 2.23. *Let us view the ring of formal series $K[[T]]$ as a module over $\mathbb{F}_p[[T]]$. Then one has*

$$x +' y := \sum_{\alpha \in I_p} w_p(\alpha) \, x^\alpha y^{1-\alpha}, \quad \forall x, y \in K.$$

$\frac{1}{125}$	$4T^3$	$\frac{2}{125}$	$3T^3$	$\frac{3}{125}$	$3T^3$	$\frac{4}{125}$	$4T^3$	$\frac{1}{25}$	$4T^2$
$\frac{6}{125}$	$3T^3$	$\frac{7}{125}$	$2T^3$	$\frac{8}{125}$	0	$\frac{9}{125}$	T^3	$\frac{2}{25}$	$3T^2$
$\frac{11}{125}$	$4T^3$	$\frac{12}{125}$	$2T^3$	$\frac{13}{125}$	$2T^3$	$\frac{14}{125}$	$4T^3$	$\frac{3}{25}$	$3T^2+2T^3$
$\frac{16}{125}$	0	$\frac{17}{125}$	0	$\frac{18}{125}$	$3T^3$	$\frac{19}{125}$	$3T^3$	$\frac{4}{25}$	$4T^2+2T^3$
$\frac{21}{125}$	T^3	$\frac{22}{125}$	$3T^3$	$\frac{23}{125}$	0	$\frac{24}{125}$	$3T^3$	$\frac{1}{5}$	$4T$
$\frac{26}{125}$	$3T^3$	$\frac{27}{125}$	0	$\frac{28}{125}$	$2T^3$	$\frac{29}{125}$	T^3	$\frac{6}{25}$	$3T^2+3T^3$
$\frac{31}{125}$	T^3	$\frac{32}{125}$	$2T^3$	$\frac{33}{125}$	$4T^3$	$\frac{34}{125}$	0	$\frac{7}{25}$	$2T^2+T^3$
$\frac{36}{125}$	$3T^3$	$\frac{37}{125}$	T^3	$\frac{38}{125}$	0	$\frac{39}{125}$	T^3	$\frac{8}{25}$	T^3
$\frac{41}{125}$	$2T^3$	$\frac{42}{125}$	$2T^3$	$\frac{43}{125}$	0	$\frac{44}{125}$	$3T^3$	$\frac{9}{25}$	T^2+T^3
$\frac{46}{125}$	$3T^3$	$\frac{47}{125}$	$2T^3$	$\frac{48}{125}$	$4T^3$	$\frac{49}{125}$	$2T^3$	$\frac{2}{5}$	$3T$
$\frac{51}{125}$	T^3	$\frac{52}{125}$	$3T^3$	$\frac{53}{125}$	$4T^3$	$\frac{54}{125}$	T^3	$\frac{11}{25}$	$4T^3$
$\frac{56}{125}$	$2T^3$	$\frac{57}{125}$	T^3	$\frac{58}{125}$	T^3	$\frac{59}{125}$	T^3	$\frac{12}{25}$	$4T^2+T^3$
$\frac{61}{125}$	$4T^3$	$\frac{62}{125}$	T^3	$\frac{63}{125}$	T^3	$\frac{64}{125}$	$4T^3$	$\frac{13}{25}$	$4T^2+T^3$
$\frac{66}{125}$	T^3	$\frac{67}{125}$	T^3	$\frac{68}{125}$	T^3	$\frac{69}{125}$	$2T^3$	$\frac{14}{25}$	$4T^3$
$\frac{71}{125}$	T^3	$\frac{72}{125}$	$4T^3$	$\frac{73}{125}$	$3T^3$	$\frac{74}{125}$	T^3	$\frac{3}{5}$	$3T$
$\frac{76}{125}$	$2T^3$	$\frac{77}{125}$	$4T^3$	$\frac{78}{125}$	$2T^3$	$\frac{79}{125}$	$3T^3$	$\frac{16}{25}$	T^2+T^3
$\frac{81}{125}$	$3T^3$	$\frac{82}{125}$	0	$\frac{83}{125}$	$2T^3$	$\frac{84}{125}$	$2T^3$	$\frac{17}{25}$	T^3
$\frac{86}{125}$	T^3	$\frac{87}{125}$	0	$\frac{88}{125}$	T^3	$\frac{89}{125}$	$3T^3$	$\frac{18}{25}$	$2T^2+T^3$
$\frac{91}{125}$	0	$\frac{92}{125}$	$4T^3$	$\frac{93}{125}$	$2T^3$	$\frac{94}{125}$	T^3	$\frac{19}{25}$	$3T^2+3T^3$
$\frac{96}{125}$	T^3	$\frac{97}{125}$	$2T^3$	$\frac{98}{125}$	0	$\frac{99}{125}$	$3T^3$	$\frac{4}{5}$	$4T$
$\frac{101}{125}$	$3T^3$	$\frac{102}{125}$	0	$\frac{103}{125}$	$3T^3$	$\frac{104}{125}$	T^3	$\frac{21}{25}$	$4T^2+2T^3$
$\frac{106}{125}$	$3T^3$	$\frac{107}{125}$	$3T^3$	$\frac{108}{125}$	0	$\frac{109}{125}$	0	$\frac{22}{25}$	$3T^2+2T^3$
$\frac{111}{125}$	$4T^3$	$\frac{112}{125}$	$2T^3$	$\frac{113}{125}$	$2T^3$	$\frac{114}{125}$	$4T^3$	$\frac{23}{25}$	$3T^2$
$\frac{116}{125}$	T^3	$\frac{117}{125}$	0	$\frac{118}{125}$	$2T^3$	$\frac{119}{125}$	$3T^3$	$\frac{24}{25}$	$4T^2$
$\frac{121}{125}$	$4T^3$	$\frac{122}{125}$	$3T^3$	$\frac{123}{125}$	$3T^3$	$\frac{124}{125}$	$4T^3$	1	1

TABLE 1 Values of $w_5(\alpha)$ modulo T^4.

PROOF. Using the definition of $x +' y$, it is enough to show that

$$\tau(x) + \tau(y) = \tilde{\tau}\left(\sum_{\alpha \in I_p} w_p(\alpha)\, x^\alpha\, y^{1-\alpha}\right)$$

which follows directly from (24) which can be written as

$$\tau(x) + \tau(y) = \tilde{\tau}\left(x + y + \sum_{n=1}^{\infty} \sum_{k=0}^{p^n} w(p^n, k)T^n\, x^{\frac{k}{p^n}}\, y^{1-\frac{k}{p^n}}\right).$$

$\square$

This formula is analogous to (22) of Theorem 2.20 and suggests developing in a deeper way the properties of the $K[[T]]$-valued function $w_p(\alpha)$ in analogy with the classical properties of the entropy function $w_1(\alpha) = e^{S(\alpha)}$ taking its values in $\mathbb{R}_+^{\max}$.

2.6. $\mathbb{B}$, $\mathbb{F}_1$ and the "Absolute Point"

The semi-rings of characteristic 1 provide a natural framework for a mathematics of finite characteristic $p = 1$. Moreover, the semi-field $\mathbb{B}$ is the initial object among semi-rings of characteristic 1. However, Spec $\mathbb{B}$ does not fulfill the requirements of the "absolute point" Spec $\mathbb{F}_1$, as defined in [**KS**]. In particular, one expects that Spec $\mathbb{F}_1$ sits under Spec $\mathbb{Z}$. This property does not hold for Spec $\mathbb{B}$ since there is no unital homomorphism of semi-rings from $\mathbb{B}$ to $\mathbb{Z}$.

We conclude this section by explaining how the "naive" approach to $\mathbb{F}_1$ emerges in the framework of §2.1 and §2.2. First, notice that Proposition 2.3 generalizes to characteristic $p > 2$ by implementing the following modifications:

• The group $\mathbb{G}^{\mathrm{odd}}$ is replaced by the group $\mathbb{G}^{(p)}$ of roots of 1 in $\mathbb{C}$ of order prime to p.

• The involution $s^2 = id$ is replaced by a bijection $s : \mathbb{K} \to \mathbb{K}$ of $\mathbb{K} = \mathbb{G}^{(p)} \cup \{0\}$, satisfying the condition $s^p = id$ and commuting with its conjugates by rotations.

To treat the degenerate case $p = 1$ in §2.3 we dropped the condition that s is a bijection and we replaced it by the idempotency condition $s \circ s = s$. There is however another trivial possibility which is to leave unaltered the condition $s^p = id$ and simply take $s = id$ for $p = 1$. The limit case is then obtained by implementing the following procedure:

• The group $\mathbb{G}^{(p)}$ is replaced by the group $\mathbb{G}$ of all roots of 1 in $\mathbb{C}$.

• The map s is the identity map on $\mathbb{G} \cup \{0\}$.

The additive structure on

$$(25) \qquad \mathbb{F}_{1^\infty} = \mathbb{G} \cup \{0\}$$

degenerates since this case corresponds to setting $s(0) = 0$ in Theorem 2.1, with the bijection s given by the identity map, so that (15) becomes the indeterminate expression $0/0$.

The multiplicative structure on the monoid (25) is the same as the multiplicative structure on the group $\mathbb{G}$, where we consider $\{0\}$ as an absorbing element. By construction, for each integer n the group $\mathbb{G}$ contains a unique cyclic subgroup $\mathbb{G}_n$ of order n. The tower (inductive limit) of the finite subfields $\mathbb{F}_{p^n} \subset \bar{\mathbb{F}}_p$ is replaced in the limit case $p = 1$ by the inductive limit of commutative monoids

$$\mathbb{F}_{1^n} = \mathbb{G}_n \cup \{0\} \subset \mathbb{G} \cup \{0\}$$

where the inductive structure is partially ordered by divisibility of the index n. Notice that on $\mathbb{F}_{1^\infty}$ one can still define

$$0 + x = x + 0 = x, \qquad \forall x \in \mathbb{F}_{1^\infty}$$

and this simple rule suffices to multiply matrices with coefficients in $\mathbb{F}_{1^\infty}$ whose rows have at most one non-zero element. In fact, the multiplicative formula $C_{ik} = \sum A_{ij} B_{jk}$ for the product $C = AB$ of two matrices only makes sense if one can add $0 + x = x + 0 = x$. Hence, with the exception of the special case where 0 is one of the two summands, one considers the addition $x + y$ in $\mathbb{F}_{1^\infty}$ to be indeterminate.

By construction, $\mathbb{F}_1 = \mathbb{G}_1 \cup \{0\} = \{1\} \cup \{0\}$ is the monoid to which $\mathbb{F}_q$ degenerates when $q = 1$, *i.e.* by considering the addition $1 + 1$ on $\mathbb{F}_1$ to be indeterminate. From this simple point of view the category of commutative algebras over $\mathbb{F}_1$ is simply the category $\mathfrak{Mo}$ of commutative monoids with a unit 1 and an absorbing element 0. In particular we see that a (commutative) semi-ring of characteristic 1 is in particular a (commutative) algebra over $\mathbb{F}_1$ in the above sense which is compatible with the absolute property of $\operatorname{Spec} \mathbb{F}_1$ of (3).

REMARK 2.24. By construction, each finite field $\mathbb{F}_{p^n}$ has the same multiplicative structure as the monoid $\mathbb{F}_{1^{p^n-1}}$. However, there exist infinitely many monoidal structures $\mathbb{F}_{1^n}$ which do not correspond to any degeneration of a finite field: $\mathbb{F}_{1^5} = \mathbb{G}_5 \cup \{0\}$ and $\mathbb{F}_{1^9} = \mathbb{G}_9 \cup \{0\}$ are the first two cases. For the sake of clarity, we make it clear that even when $n = p^\ell - 1$ for p a prime number, we shall always consider the algebraic structures $\mathbb{F}_{1^n} = \mathbb{G}_n \cup \{0\}$ only as multiplicative monoids.

3. The Functorial Approach

In this section we give an overview on the geometric theory of algebraic schemes over $\mathbb{F}_1$ that we have introduced in our paper [**CC2**]. The second part of the section contains a new result on the geometric realization of an $\mathfrak{Mo}$-functor. In fact, our latest development of the study of the algebraic geometry of $\mathbb{F}_1$-schemes shows that, unlike the case of a $\mathbb{Z}$-scheme, the topology and the structure sheaf of an $\mathbb{F}_1$-scheme can be obtained naturally on the set of its rational $\mathbb{F}_1$-points.

Our original viewpoint in the development of this theory of schemes over $\mathbb{F}_1$ has been an attempt at unifying the theories developed on the one side by C. Soulé in [**So**] and in our earlier paper [**CC1**] and on the other side by A. Deitmar in ([**D1**], [**D2**]) (following N. Kurokawa, H. Ochiai and A. Wakayama [**KOW**]), by K. Kato in [**Ka**] (with the geometry of logarithmic structures) and by B. Töen and M. Vaquié in [**TV**]. In §2.6, we have described how to obtain a naive version of $\mathbb{F}_1$ leading naturally to the point of view developed by A. Deitmar, where $\mathbb{F}_1$-algebras are commutative monoids (with the slight difference that in our set-up one also requires the existence of an absorbing element 0). It is the analysis performed by C. Soulé in [**So**] of the extension of scalars from $\mathbb{F}_1$ to $\mathbb{Z}$ that lead us to

• Reformulate the notion of a scheme in the sense of K. Kato and A. Deitmar, in functorial terms, *i.e.* as a covariant functor from the category of (pointed) monoids to the category of sets.

• Prove a new result on the geometric realization of functors satisfying a suitable local representability condition.

• Refine the notion of an $\mathbb{F}_1$-scheme by allowing more freedom on the choice of the $\mathbb{Z}$-scheme obtained by base change.

In this section we shall explain this viewpoint in some detail, focusing in particular on the description and the properties of the geometric realization of an $\mathfrak{Mo}$-functor that was only briefly sketched in [**CC2**].

Everywhere in this section we denote by $\mathfrak{Sets}$, $\mathfrak{Mo}$, $\mathfrak{Ab}$ and $\mathfrak{Ring}$ respectively the categories of sets, commutative monoids,[4] abelian groups and commutative (unital) rings.

---

[4]With a unit and a zero element.

3.1. Schemes as Locally Representable $\mathbb{Z}$-functors: A Review

In the following three subsections we will shortly review the basic notions of the theory of $\mathbb{Z}$-functors and $\mathbb{Z}$-schemes: we refer to ([**DG**], Chapters I, III) for a detailed exposition.

DEFINITION 3.1. *A $\mathbb{Z}$-functor is a covariant functor* $\mathcal{F} : \mathfrak{Ring} \to \mathfrak{Sets}$.

Morphisms in the category of $\mathbb{Z}$-functors are natural transformations (of functors).

Schemes over $\mathbb{Z}$ determine a full subcategory $\mathfrak{Sch}$ of the category of $\mathbb{Z}$-functors. In fact, a scheme X over $\operatorname{Spec}\mathbb{Z}$ is *entirely* characterized by the $\mathbb{Z}$-functor

$$(26) \qquad \underline{X} : \mathfrak{Ring} \to \mathfrak{Sets}, \qquad \underline{X}(R) = \operatorname{Hom}_{\mathbb{Z}}(\operatorname{Spec}R, X).$$

To a ring homomorphism $\rho : R_1 \to R_2$ one associates the morphism of (affine) schemes $\rho^* : \operatorname{Spec}(R_2) \to \operatorname{Spec}(R_1)$, $\rho^*(\mathfrak{p}) = \rho^{-1}(\mathfrak{p})$ and the map of sets

$$\rho : \underline{X}(R_1) \to \underline{X}(R_2), \qquad \varphi \to \varphi \circ \rho^*.$$

If $\psi : X \to Y$ is a morphism of schemes then one gets for every ring R a map of sets

$$\underline{X}(R) \to \underline{Y}(R), \qquad \varphi \to \psi \circ \varphi.$$

The functors of the form $\underline{X}$, for X a scheme over $\operatorname{Spec}\mathbb{Z}$, are *local* $\mathbb{Z}$-functors (in the sense that we shall recall in §3.2, Definition 3.2). These functors are also *locally representable* by commutative rings, *i.e.* they have an open cover by representable $\mathbb{Z}$-subfunctors (in the sense explained in §3.2.1 and §3.2.2).

3.2. Local $\mathbb{Z}$-functors

For any commutative (unital) ring R, the geometric space $\operatorname{Spec}(R)$ is the set of prime ideals $\mathfrak{p} \subset R$. The topology on $\operatorname{Spec}(R)$ is the Jacobson topology, *i.e.* the closed subsets are the sets $V(J) = \{\mathfrak{p} \in \operatorname{Spec}(R)|\mathfrak{p} \supset J\}$, where $J \subset R$ runs through the collection of all the ideals of R. The open subsets of $\operatorname{Spec}(R)$ are the complements of the $V(J)$'s, *i.e.* they are the sets

$$D(J) = V(J)^c = \{\mathfrak{p} \in \operatorname{Spec}(R)|\exists f \in J, f \notin \mathfrak{p}\}.$$

It is well known that the open sets $D(f) = V(fR)^c$, for $f \in R$, form a basis of the topology of $\operatorname{Spec}(R)$. For any $f \in R$ one lets $R_f = S^{-1}R$, where $S = \{f^n|n \in \mathbb{Z}_{\geq 0}\}$ denotes the multiplicative system of the non-negative powers of f. One has a natural ring homomorphism $R \to R_f$. Then, for any scheme X over $\operatorname{Spec}\mathbb{Z}$, the associated functor $\underline{X}$ as in (26) fulfills the following *locality* property. For any finite cover of $\operatorname{Spec}(R)$ by open sets $D(f_i)$, with $f_i \in R$ ($i \in I$ = finite index set) the following sequence of maps of sets is exact

$$(27) \qquad \underline{X}(R) \xrightarrow{u} \prod_{i \in I} \underline{X}(R_{f_i}) \underset{w}{\overset{v}{\rightrightarrows}} \prod_{(i,j) \in I \times I} \underline{X}(R_{f_i f_j}).$$

This means that u is injective and the range of u is characterized as the set $\{z \in \prod_i \underline{X}(R_{f_i})|v(z) = w(z)\}$. The exactness of (27) is a consequence of the fact that a morphism of schemes is defined by local conditions. For $\mathbb{Z}$-functors we have the following definition.

DEFINITION 3.2. *A* $\mathbb{Z}$-*functor* $\mathcal{F}$ *is local if for any object* R *of* $\mathfrak{Ring}$ *and a partition of unity* $\sum_{i \in I} h_i f_i = 1$ *in* R *(I = finite index set), the following sequence of sets is exact*

$$(28) \qquad \mathcal{F}(R) \xrightarrow{u} \prod_i \mathcal{F}(R_{f_i}) \underset{w}{\overset{v}{\rightrightarrows}} \prod_{ij} \mathcal{F}(R_{f_i f_j}).$$

EXAMPLE 3.3. This example shows that locality is not automatically verified by a general $\mathbb{Z}$-functor.

The Grassmannian $\mathrm{Gr}(k, n)$ of the k-dimensional linear spaces in an n-dimensional linear space is defined by the functor which associates to a ring R the set of all complemented submodules of *rank* k of the free (right) module R^n. Since any such complemented submodule is projective, by construction we have

$$\mathrm{Gr}(k, n) : \mathfrak{Ring} \to \mathfrak{Sets}, \qquad \mathrm{Gr}(k, n)(R) = \{E \subset R^n | \ E \text{ projective}, \ \mathrm{rk}(E) = k\}.$$

Let $\rho : R_1 \to R_2$ be a homomorphism of rings; the corresponding map of sets is given as follows: for $E_1 \in \mathrm{Gr}(k, n)(R_1)$, one lets $E_2 = E_1 \otimes_{R_1} R_2 \in \mathrm{Gr}(k, n)(R_2)$. If one takes a naive definition of the rank, *i.e.* by just requiring that $E \cong R^k$ as an R-module, one does not obtain a local $\mathbb{Z}$-functor. In fact, let us consider the case $k = 1$ and $n = 2$ which defines the projective line $\mathbb{P}^1 = \mathrm{Gr}(1, 2)$. To show that locality fails in this case, one takes the algebra $R = C(S^2)$ of continuous functions on the sphere S^2 and the partition of unity $f_1 + f_2 = 1$ subordinate to a covering of S^2 by two disks D_j $(j = 1, 2)$, so that $\mathrm{Supp}(f_j) \subset D_j$. One then considers the non-trivial line bundle L on S^2 arising from the identification $S^2 \simeq \mathbb{P}^1(\mathbb{C}) \subset M_2(\mathbb{C})$ which determines an idempotent $e \in M_2(C(S^2))$. The range of e defines a finite projective submodule $E \subset R^2$. The localized algebra R_{f_j} is the same as $C(\bar{D}_j)_{f_j}$ and thus the induced module E_{f_j} is free (of rank one). The modules $E_i = E \otimes_R R_{f_i}$ are free submodules of $R_{f_i}^2$ and the induced modules on $R_{f_i f_j}$ are the same. But since E is not free they do not belong to the image of u and the sequence (28) is not exact in this case.

To obtain a local $\mathbb{Z}$-functor one has to implement a more refined definition of the rank which requires that for any prime ideal $\mathfrak{p}$ of R the induced module on the residue field of R at $\mathfrak{p}$ is a vector space of dimension k.

3.2.1. *Open* $\mathbb{Z}$-*subfunctors*

The $\mathbb{Z}$-functor associated to an affine scheme $\mathrm{Spec} A$ $(A \in \mathrm{obj}(\mathfrak{Ring}))$ is defined by

$$\mathfrak{spec}\, A : \mathfrak{Ring} \to \mathfrak{Sets}, \quad \mathfrak{spec}\, A(R) = \mathrm{Hom}_\mathbb{Z}(\mathrm{Spec} R, \mathrm{Spec} A) \simeq \mathrm{Hom}_\mathbb{Z}(A, R).$$

The open sets of an affine scheme are in general *not* affine and they provide interesting examples of schemes. The subfunctor of $\mathfrak{spec}\, A$

$$\mathrm{Hom}_\mathbb{Z}(\mathrm{Spec} R, D(J)) \subset \mathrm{Hom}_\mathbb{Z}(\mathrm{Spec} R, \mathrm{Spec} A)$$

associated to the open set $D(J) \subset \mathrm{Spec}(A)$ (for a given ideal $J \subset A$) has the following explicit description

$$\underline{D}(J) : \mathfrak{Ring} \to \mathfrak{Sets}, \qquad \underline{D}(J)(R) = \{\rho \in \mathrm{Hom}_\mathbb{Z}(A, R) | \rho(J)R = R\}.$$

This follows from the fact that $\mathrm{Spec}\,(\rho)^{-1}(D(J)) = D(\rho(J)R)$. In general, we say that $\mathcal{U}$ is a subfunctor of a functor $\mathcal{F} : \mathfrak{Ring} \to \mathfrak{Sets}$ if for each ring R, $\mathcal{U}(R)$ is a subset of $\mathcal{F}(R)$ (with the natural compatibility for the maps).

DEFINITION 3.4. *Let $\mathcal{U}$ be a subfunctor of $\mathcal{F} : \mathfrak{Ring} \to \mathfrak{Sets}$. One says that $\mathcal{U}$ is open if for any ring A and any natural transformation $\phi : \mathrm{spec}\,A \to \mathcal{F}$, the subfunctor of $\mathrm{spec}\,A$ inverse image of $\mathcal{U}$ by ϕ*

$$\phi^{-1}(\mathcal{U}) : \mathfrak{Ring} \to \mathfrak{Sets}, \quad \phi^{-1}(\mathcal{U})(R) = \{x \in \mathrm{spec}\,A(R)|\phi(x) \in \mathcal{U}(R) \subset \mathcal{F}(R)\}$$

is of the form $\underline{D}(J)$, for some open set $D(J) \subset Spec(A)$.

Equivalently, using Yoneda's lemma, the above definition can be expressed by saying that, given any ring A and an element $z \in \mathcal{F}(A)$, there exists an ideal $J \subset A$ such that, for any $\rho \in \mathrm{Hom}(A, R)$,

$$(29) \qquad \mathcal{F}(\rho)(z) \in \mathcal{U}(R) \iff \rho(J)R = R.$$

For any open subset $Y \subset X$ of a scheme X the subfunctor

$$\mathrm{Hom}_{\mathbb{Z}}(\mathrm{Spec}R, Y) \subset \mathrm{Hom}_{\mathbb{Z}}(\mathrm{Spec}R, X)$$

is open, and all open subfunctors of $\underline{X}$ arise in this way.

EXAMPLE 3.5. We consider the projective line $\mathbb{P}^1$ identified with $X = \mathrm{Gr}(1, 2)$. Let $\mathcal{U} \subset X$ be the subfunctor described, on a ring R, by the collection of all submodules of rank one of R^2 which are supplements of the submodule $P = \{(0, y)|y \in R\} \subset R^2$. Let p_1 be the projection on the first copy of R, then

$$\mathcal{U}(R) = \{E \in \mathrm{Gr}(1, 2)| \, p_{1|E} \text{ isomorphism}\}.$$

Proving that $\mathcal{U}$ is open is equivalent, using (29), to finding for any ring A and $E \in X(A)$ an ideal $J \subset A$ such that for any ring R and $\rho : A \to R$ one has

$$(30) \qquad p_{1|E \otimes_A R} \text{ isomorphism} \iff R = \rho(J)R.$$

It is easy to see that the ideal J given by the annihilator of the cokernel of $p_{1|E}$ satisfies (30) (see [**DG**], Chapter I, Example 3.9).

3.2.2. Covering by $\mathbb{Z}$-subfunctors

To motivate the definition of a covering of a $\mathbb{Z}$-functor, we start by describing the case of an affine scheme. Let $\underline{X}$ be the $\mathbb{Z}$-functor

$$(31) \qquad \underline{X} : \mathfrak{Ring} \to \mathfrak{Sets}, \quad \underline{X}(R) = \mathrm{Hom}(A, R)$$

that is associated to the affine scheme $\mathrm{Spec}(A)$. We have seen that the open subfunctors $\underline{D}(I) \subset \underline{X}$ correspond to ideals $I \subset A$ with

$$\underline{D}(I)(R) = \{\rho \in \mathrm{Hom}(A, R)|\rho(I)R = R\}.$$

The condition that the open sets $D(I_\alpha)$ ($\alpha \in I$ = index set) form a covering of $\mathrm{Spec}(A)$ is expressed algebraically by the equality $\sum_{\alpha \in I} I_\alpha = A$. We want to describe this condition in terms of the open subfunctors $\underline{D}(I_\alpha)$.

LEMMA 3.6. *Let $\underline{X}$ be as in (31). Then $\sum_\alpha I_{\alpha \in I} = A$ if and only if for any field K one has*

$$\underline{X}(K) = \bigcup_{\alpha \in I} \underline{D}(I_\alpha)(K).$$

PROOF. Assume first that $\sum I_{\alpha \in I} = A$ (a finite sum), *i.e.* $1 = \sum_\alpha a_\alpha$, with $a_\alpha \in I_\alpha$. Let K be a field, then for $\rho \in \mathrm{Hom}(A, K)$, one has $\rho(a_\alpha) \neq 0$ for some $\alpha \in I$. Then $\rho(a_\alpha)K = K$, *i.e.* $\rho \in \underline{D}(I_\alpha)(K)$ so that the union of all $\underline{D}(I_\alpha)(K)$ is $\underline{X}(K)$.

Conversely, assume that $\sum_\alpha I_\alpha \neq A$. Then there exists a prime ideal $\mathfrak{p} \subset A$ containing all I_α's. Let K be the field of fractions of $A/\mathfrak{p}$ and let $\rho : A \to K$ be the natural homomorphism. One has $I_\alpha \subset \mathrm{Ker}\rho$, thus $\rho \notin \bigcup_\alpha \underline{D}(I_\alpha)(K)$. □

Notice that when R is neither a field nor a local ring, the equality

$$\underline{X}(R) = \bigcup_\alpha \underline{D}(I_\alpha)(R)$$

cannot be expected. In fact the range of a morphism $\rho \in \mathrm{Hom}(\mathrm{Spec}(R), \mathrm{Spec}\, A) = \underline{X}(R)$ may not be contained in a single open set of the covering of $\mathrm{Spec}\,(A)$ by the $D(I_\alpha)$ so that ρ belongs to none of the $\underline{D}(I_\alpha)(R) = \mathrm{Hom}(\mathrm{Spec}(R), D(I_\alpha))$.

DEFINITION 3.7. *Let $\underline{X}$ be a $\mathbb{Z}$-functor. Let $\{\underline{X}_\alpha\}_{\alpha \in I}$ be a family of open subfunctors of $\underline{X}$. Then we say that the set $\{\underline{X}_\alpha\}_{\alpha \in I}$ forms a covering of $\underline{X}$ if for any field K one has*

$$\underline{X}(K) = \bigcup_{\alpha \in I} \underline{X}_\alpha(K).$$

For affine schemes, one recovers the usual notion of an open cover. In fact, any open cover of an affine scheme admits a finite subcover. Indeed, the condition is $\sum_\alpha I_\alpha = A$ and if it holds one gets $1 = \sum_{\alpha \in F} a_\alpha$ for some *finite* subset of indices $F \subset I$. For an arbitrary scheme this finiteness condition may *not* hold. However, since a scheme is always "locally affine," one can say, calling the above finiteness condition "quasi-compact," that any scheme is locally quasi-compact.

To conclude this short review of the basic properties of schemes viewed as $\mathbb{Z}$-functors, we quote the main theorem which allows one to consider a scheme as a local and locally representable $\mathbb{Z}$-functor.

THEOREM 3.8. *The $\mathbb{Z}$-functors of the form $\underline{X}$, for X a scheme over $\mathrm{Spec}\,\mathbb{Z}$ are the $\mathbb{Z}$-functors which are local and admit an open cover by representable subfunctors.*

PROOF. See [**DG**], Chapter I, §1, §4.4. □

3.3. Monoids: The Category $\mathfrak{Mo}$

We recall that $\mathfrak{Mo}$ denotes the category of commutative "pointed" monoids $M \cup \{0\}$, *i.e.* M is a semi-group with a commutative multiplicative operation "$\cdot$" (for simplicity we shall use the notation xy to denote the product $x \cdot y$ in $M \cup \{0\}$) and an identity element 1. Moreover, 0 is an absorbing element in M, *i.e.* $0 \cdot x = x \cdot 0 = 0$, $\forall x \in M$.

The morphisms in $\mathfrak{Mo}$ are unital homomorphisms of monoids $\varphi : M \to N$ $(\varphi(1_M) = 1_N)$ satisfying the condition $\varphi(0) = 0$.

We also recall (see [**Gi**], p. 3) that an ideal I of a monoid M is a non-empty subset $I \subset M$ such that $I \supseteq xI = \{xi | i \in I\}$ for each $x \in M$. An ideal $I \subset M$ is *prime* if it is a proper ideal $I \subsetneq M$ and $xy \in I \implies x \in I$ or $y \in I \ \forall x, y \in M$. Equivalently, a proper ideal $\mathfrak{p} \subsetneq M$ is prime if and only if $\mathfrak{p}^c := M \setminus \mathfrak{p}$ is a multiplicative subset of M, i.e. $x \notin \mathfrak{p}$, $y \notin \mathfrak{p} \implies xy \notin \mathfrak{p}$.

It is a standard fact that the pre-image of a prime ideal by a morphism of monoids is a prime ideal. Moreover, it is also straightforward to verify that the complement $\mathfrak{p}_M = (M^\times)^c$ of the set of invertible elements in a monoid M is a prime ideal in M which contains all other prime ideals of the monoid. This interesting fact points to one of the main properties that characterize monoids, namely monoids are local algebraic structures.

We recall that the smallest ideal containing a collection of ideals $\{I_\alpha\}$ of a monoid M is the union $I = \cup_\alpha I_\alpha$.

If $I \subset M$ is an ideal of a monoid M, the relation $\sim$ on M defined by

$$x \sim y \ \Leftrightarrow \ x = y \text{ or } x, y \in I$$

is an example of a *congruence* on M, i.e. an equivalence relation on M which is compatible with the product (see [**Gr**], §1, Proposition 4.6). The quotient monoid $M/I := M/\sim$ (Rees quotient) is identifiable with the pointed monoid $(M \setminus I) \cup \{0\}$, with the operation defined as follows

$$x \star y = \begin{cases} xy, & \text{if } xy \notin I \\ 0, & \text{if } xy \in I. \end{cases}$$

Another interesting example of congruence in a monoid is provided by the operation of *localization* at a multiplicative subset $S \subset M$. One considers the congruence $\sim$ on the submonoid $M \times S \subset M \times M$ generated by the relation

$$(m, s) \sim (m', s') \ \Leftrightarrow \ \exists u \in S \ \ ums' = um's.$$

By introducing the symbol $m/s = (m, s)$, one can easily check that the product $(m/s).(m'/s') = mm'/ss'$ is well defined on the quotient monoid $S^{-1}M = (M \times S)/\sim$. Assuming $1 \in S$ and $0 \notin S$ one has $0/1 \neq 1/1$ in $S^{-1}M$. A particular case of this construction is when, for $f \in M$, one considers the multiplicative set $S = \{f^n; n \in \mathbb{Z}_{\geq 0}\}$: in analogy with rings, the corresponding quotient monoid $S^{-1}M$ is usually denoted by M_f. It is non-zero except when f is nilpotent. When $f^n \neq 0$ for all $n > 0$, the equality

$$\mathfrak{p}(f) = \{x \in M \mid xM \cap \{f^n \mid n > 0\} = \emptyset\}$$

defines a prime ideal $\mathfrak{p}(f)$ of M and the construction of M_f is a special case of the following localization $M \to M_\mathfrak{p}$ of M at any prime ideal $\mathfrak{p}$. By definition $M_\mathfrak{p}$ is $S^{-1}M$ where $S = \mathfrak{p}^c$ is the complement of $\mathfrak{p}$. One has $0 \neq 1$ in $M_\mathfrak{p}$ for all prime ideals $\mathfrak{p}$.

For an ideal $I \subset M$, the set $D(I) = \{\mathfrak{p} \subset M | \mathfrak{p} \text{ prime ideal}, \ \mathfrak{p} \not\supseteq I\}$ determines an open set for the natural topology on the set $\text{Spec}(M) = \{\mathfrak{p} \subset M | \mathfrak{p} \text{ prime ideal}\}$ (cf. [**Ka**]). For $I = \cup_\alpha I_\alpha$ ($\{I_\alpha\}$ a collection of ideals), the corresponding open set $D(I)$ satisfies the property $D(\cup_\alpha I_\alpha) = \cup_\alpha D(I_\alpha)$.

The following equivalent statements characterize the open subsets $D(I) \subset \text{Spec}(M)$.

PROPOSITION 3.9. *Let* $\rho : M \to N$ *be a morphism in the category* $\mathfrak{Mo}$ *and let* $I \subset M$ *be an ideal. Then the following conditions are equivalent:*

(a) $\rho(I)N = N$.

(b) $1 \in \rho(I)N$.

(c) $\rho^{-1}((N^{\times})^c)$ *is a prime ideal belonging to* $D(I)$.

(d) $\rho^{-1}(\mathfrak{p}) \in D(I)$, *for any prime ideal* $\mathfrak{p} \subset N$.

PROOF. One has $(a) \Leftrightarrow (b)$. Moreover, $\rho^{-1}((N^{\times})^c) \not\supset I$ if and only if $\rho(I) \cap N^{\times} \neq \emptyset$ which is equivalent to (a). Thus $(b) \Leftrightarrow (c)$.

If an ideal $J \subset M$ does not contain I, then the same holds obviously for all the subideals of J. Then $(c) \Rightarrow (d)$ since $\mathfrak{p}_N = (N^{\times})^c$ contains all the prime ideals of N. faking $\mathfrak{p} = \mathfrak{p}_N$ one gets $(d) \Rightarrow (c)$. $\square$

The proof of the following lemma is straightforward (see [**Gi**]).

LEMMA 3.10. *Given an ideal* $I \subset M$, *the intersection of the prime ideals* $\mathfrak{p} \subset M$, *such that* $\mathfrak{p} \supset I$ *coincides with the radical of* I

$$\bigcap_{\mathfrak{p} \supset I} \mathfrak{p} = \sqrt{I} := \{x \in M | \exists n \in \mathbb{N}, x^n \in I\}.$$

Given a commutative group H, the following definition determines a pointed monoid in $\mathfrak{Mo}$

$$\mathbb{F}_1[H] := H \cup \{0\} \qquad (0 \cdot h = h \cdot 0 = 0, \quad \forall h \in H).$$

Thus, in $\mathfrak{Mo}$ a monoid of the form $\mathbb{F}_1[H]$ corresponds to a field F ($F = F^{\times} \cup \{0\}$) in the category of commutative rings with unit. It is elementary to verify that the collection of monoids like $\mathbb{F}_1[H]$, for H an abelian group, forms a full subcategory of $\mathfrak{Mo}$ isomorphic to the category $\mathfrak{Ab}$ of abelian groups.

In view of the fact that monoids are local algebraic structures, one can also introduce a notion which corresponds to that of the residue field for local rings and related homomorphism. For a commutative monoid M, one defines the pair $(\mathbb{F}_1[M^{\times}], \epsilon)$ where ϵ is the natural homomorphism

$$(32) \qquad \epsilon : M \to \mathbb{F}_1[M^{\times}] \qquad \epsilon(y) = \begin{cases} 0, & \text{if } y \notin M^{\times} \\ y, & \text{if } y \in M^{\times}. \end{cases}$$

The non-invertible elements of M form a prime ideal; thus ϵ is a multiplicative map. The following lemma describes an application of this idea.

LEMMA 3.11. *Let* M *be a commutative monoid and* $\mathfrak{p} \subset M$ *a prime ideal. Then*

(a) $\mathfrak{p}M_{\mathfrak{p}} \cap (M_{\mathfrak{p}})^{\times} = \emptyset$.

(b) *There exists a unique homomorphism* $\epsilon_{\mathfrak{p}} : M_{\mathfrak{p}} \to \mathbb{F}_1[(M_{\mathfrak{p}})^{\times}]$ *such that*

• $\epsilon_{\mathfrak{p}}(y) = 0, \forall y \in \mathfrak{p}$

• $\epsilon_{\mathfrak{p}}(y) = y, \forall y \in (M_{\mathfrak{p}})^{\times}$.

PROOF. (a) If $j \in \mathfrak{p}$ then the image of j in $M_\mathfrak{p}$ cannot be invertible, since this would imply an equality of the form $sja = sf$ for some $s \notin \mathfrak{p}$, $f \notin \mathfrak{p}$ and hence a contradiction.

(b) The first statement (a) shows that the corresponding (multiplicative) map $\epsilon_\mathfrak{p}$ fulfills the two conditions. To check uniqueness, note that the two conditions suffice to determine $\epsilon_\mathfrak{p}(y)$ for any $y = a/f$ with $a \in M$ and $f \notin \mathfrak{p}$. $\square$

3.4. Geometric Monoidal Spaces

In this subsection we review the construction of the geometric spaces which generalize, in the categorical context of the commutative monoids, the classical theory of the geometric $\mathbb{Z}$-schemes that we have reviewed in §3.1. We refer to §9 in [**Ka**], [**D1**] and [**D2**] for further details.

A geometric monoidal space $(X, \mathcal{O}_X)$ is a topological space X endowed with a sheaf of monoids $\mathcal{O}_X$ (the structural sheaf). Unlike the case of the geometric $\mathbb{Z}$-schemes (see [**DG**], Chapter I, §1, Definition 1.1), there is no need to impose the condition that the stalks of the structural sheaf of a geometric monoidal space are local algebraic structures, since *by construction* any monoid is endowed with this property.

A morphism $\varphi : X \to Y$ of geometric monoidal spaces is a pair $(\varphi, \varphi^\sharp)$ of a continuous map $\varphi : X \to Y$ of topological spaces and a homomorphism of sheaves of monoids $\varphi^\sharp : \mathcal{O}_Y \to \varphi_* \mathcal{O}_X$ which satisfies the property of being *local*, *i.e.* $\forall x \in X$ the homomorphisms connecting the stalks $\varphi_x^\sharp : \mathcal{O}_{Y,\varphi(x)} \to \mathcal{O}_{X,x}$ are local, *i.e.* they fulfill the following definition (see [**D1**]).

DEFINITION 3.12. *A homomorphism* $\rho : M_1 \to M_2$ *of monoids is local if the following property holds*

$$(33) \qquad \rho^{-1}(M_2^\times) = M_1^\times.$$

This locality condition can be equivalently phrased by either one of the following statements

(a) $\rho^{-1}((M_2^\times)^c) = (M_1^\times)^c$.

(b) $\rho((M_1^\times)^c) \subset (M_2^\times)^c$.

The equivalence of (33) and (a) is clear since $\rho^{-1}(S^c) = \rho^{-1}(S)^c$, for any subset $S \subset M_2$. Clearly (a) implies (b). Conversely, if (b) holds, one has $\rho^{-1}((M_2^\times)^c) \supset (M_1^\times)^c$ and using $\rho^{-1}(M_2^\times) \supset M_1^\times$ one gets (a).

We shall denote by $\mathfrak{G}\mathcal{S}$ the category of the geometric monoidal spaces.

Notice that by construction the morphism $\epsilon_\mathfrak{p}$ of Lemma 3.11 (b) is *local*. Thus it is natural to consider, for any geometric monoidal space $(X, \mathcal{O}_X)$, the analogue of the residue field at a point $x \in X$ to be $\kappa(x) = \mathbb{F}_1[\mathcal{O}_{X,x}^\times]$. Then, the associated evaluation map

$$(34) \qquad \epsilon_x : \mathcal{O}_{X,x} \to \kappa(x) = \mathbb{F}_1[\mathcal{O}_{X,x}^\times]$$

satisfies the properties as in Lemma 3.11.(b).

For a pointed monoid M in $\mathfrak{Mo}$, the set $\mathrm{Spec}(M)$ of the prime ideals $\mathfrak{p} \subset M$ is called the *prime spectrum* of M and is endowed with the topology whose closed subsets are

$$V(I) = \{\mathfrak{p} \in \mathrm{Spec}(M) | \mathfrak{p} \supset I\}$$

as I varies among the collection of all ideals of M. Likewise for rings, the subset $V(I) \subset \mathrm{Spec}(M)$ depends only upon the radical of I (see [**B**], II, Chapter 2, §3). Equivalently, one can characterize the topology on $\mathrm{Spec}(M)$ in terms of a basis of open sets of the form $D(fM) = \{\mathfrak{p} \in \mathrm{Spec}(M) | f \notin \mathfrak{p}\}$, as f varies in M.

The sheaf $\mathcal{O}$ of monoids associated to $\mathrm{Spec}\,(M)$ is determined by the following properties:

• The stalk of $\mathcal{O}$ at $\mathfrak{p} \in \mathrm{Spec}(M)$ is $\mathcal{O}_{\mathfrak{p}} = S^{-1}M$, with $S = \mathfrak{p}^c$.

• Let U be an open set of $\mathrm{Spec}(M)$. Then a section $s \in \Gamma(U, \mathcal{O})$ is an element of $\prod_{\mathfrak{p} \in U} \mathcal{O}_{\mathfrak{p}}$ such that its restriction $s_{|D(fM)}$ to any open $D(fM) \subset U$ agrees with an element of M_f.

• The homomorphism of monoids $\varphi : M_f \to \Gamma(D(fM), \mathcal{O})$

$$\varphi(x)(\mathfrak{p}) = a/f^n \in \mathcal{O}_{\mathfrak{p}}, \quad \forall \mathfrak{p} \in D(fM), \ \forall x = a/f^n \in M_f$$

is an isomorphism.

Note that $D(fM)$ is non-empty when f is not nilpotent, and in that case it contains the prime ideal $\mathfrak{p}(f)$ defined above. Moreover, the stalk of $\mathcal{O}$ at $\mathfrak{p}(f)$ is isomorphic to M_f.

For any monoidal geometric space $(X, \mathcal{O}_X)$ one has a canonical morphism $\psi_X : X \to \mathrm{Spec}\,(\Gamma(X, \mathcal{O}_X))$, $\psi_X(x) = \mathfrak{p}_x$, such that $s(x) = 0$ in $\mathcal{O}_{X,x}$, $\forall s \in \mathfrak{p}_x$. It is easy to verify that a monoidal geometric space $(X, \mathcal{O}_X)$ is a *prime spectrum* if and only if the morphism ψ_X is an isomorphism.

DEFINITION 3.13. *A monoidal geometric space $(X, \mathcal{O}_X)$ that admits an open cover by prime spectra $\mathrm{Spec}\,(M)$ is called a geometric $\mathfrak{Mo}$-scheme.*

Prime spectra fulfill the following locality property (see [**D1**]) that will be considered again in §3.5 in the functorial definition of an $\mathfrak{Mo}$-scheme.

LEMMA 3.14. *Let M be an object in $\mathfrak{Mo}$ and let $\{W_\alpha\}_{\alpha \in I}$ be an open cover of the topological space $X = \mathrm{Spec}(M)$. Then $W_\alpha = \mathrm{Spec}(M)$, for some index $\alpha \in I$.*

PROOF. The point $\mathfrak{p}_M = (M^\times)^c \in \mathrm{Spec}(M)$ must belong to some open set W_α, hence $\mathfrak{p}_M \in D(I_\alpha)$ for some index $\alpha \in I$ and this is equivalent to $I_\alpha \cap M^\times \neq \emptyset$, i.e. $I_\alpha = M$. $\qquad\square$

3.5. $\mathfrak{Mo}$-schemes

In analogy to and generalizing the theory of $\mathbb{Z}$-schemes, one develops the theory of $\mathfrak{Mo}$-schemes following both the functorial and the geometrical viewpoint.

3.5.1. $\mathfrak{Mo}$-functors

DEFINITION 3.15. *An $\mathfrak{Mo}$-functor is a covariant functor $\mathcal{F} : \mathfrak{Mo} \to \mathfrak{Sets}$ from commutative (pointed) monoids to sets.*

To a (pointed) monoid M in $\mathfrak{Mo}$ one associates the $\mathfrak{Mo}$-functor

$$\mathfrak{spec}\,M : \mathfrak{Mo} \to \mathfrak{Sets}, \qquad \mathfrak{spec}\,M(N) = \mathrm{Hom}_{\mathfrak{Mo}}\,(M, N).$$

Notice that by applying Yoneda's lemma, a morphism of $\mathfrak{Mo}$-functors (natural transformation) such as $\varphi : \mathfrak{spec}\,M \to \mathcal{F}$ is completely determined by the element $\varphi(id_M) \in \mathcal{F}(M)$ and moreover, any such element gives rise to a morphism $\mathfrak{spec}\,M \to \mathcal{F}$. By applying this fact to the functor $\mathcal{F} = \mathfrak{spec}\,N$, for $N \in \mathrm{obj}(\mathfrak{Mo})$, one obtains an *inclusion* of $\mathfrak{Mo}$ as a full subcategory of the category of $\mathfrak{Mo}$-functors (where morphisms are natural transformations).

An ideal $I \subset M$ defines the sub-$\mathfrak{Mo}$-functor $\underline{D}(I) \subset \mathfrak{spec}\, M$

$$\underline{D}(I) : \mathfrak{Mo} \to \mathfrak{Sets}, \qquad \underline{D}(I)(N) = \{\rho \in \mathfrak{spec}\,(M)(N) | \rho(I)N = N\}.$$

Automatic Locality

We recall (see §3.3) that for $\mathbb{Z}$-functors the "locality" property is defined by requiring, on coverings of prime spectra $\mathrm{Spec}(R)$ by open sets $D(f_i)$ of a basis, the exactness of sequences such as (27) and (28). The next lemma shows that locality is automatically verified for any $\mathfrak{Mo}$-functor.

First of all, notice that for any $\mathfrak{Mo}$-functor $\mathcal{F}$ and any monoid M one has a sequence of maps of sets

$$(35) \qquad \mathcal{F}(M) \xrightarrow{u} \prod_{i \in I} \mathcal{F}(M_{f_i}) \underset{w}{\overset{v}{\rightrightarrows}} \prod_{i,j \in I} \mathcal{F}(M_{f_i f_j})$$

that is obtained by using the open covering of $\mathrm{Spec}(M)$ by the open sets $D(f_i M)$ of a basis (I = finite index set) and the natural homomorphisms $M \to M_{f_i}$, $M_{f_i} \to M_{f_i f_j}$.

LEMMA 3.16. *For any $\mathfrak{Mo}$-functor $\mathcal{F}$ the sequence (35) is exact.*

PROOF. By Proposition 2.18, there exists an index $i \in I$ such that $f_i \in M^\times$. We may also assume that $i = 1$. Then the homomorphism $M \to M_{f_1}$ is invertible, thus u is injective. Let $(x_i) \in \prod_i \mathcal{F}(M_{f_i})$ be a family, with $x_i \in \mathcal{F}(M_{f_i})$ and such that $v(x_i) = (x_i)_{f_j} = (x_j)_{f_i} = w(x_i)$, for all $i, j \in I$. This gives in particular the equality between the image of $x_i \in \mathcal{F}(M_{f_i})$ under the isomorphism $\mathcal{F}(\rho_{i1})$: $\mathcal{F}(M_{f_i}) \to \mathcal{F}(M_{f_i f_1})$ and $\mathcal{F}(\rho_{i1})(x_1) \in \mathcal{F}(M_{f_i f_1})$. By writing $x_1 = \rho_1(x)$ one finds that $u(x)$ is equal to the family (x_i). $\qquad\square$

3.5.2. Open $\mathfrak{Mo}$-subfunctors

In exact analogy with the theory of $\mathbb{Z}$-schemes (see §3.2.1), we introduce the notion of open subfunctors of $\mathfrak{Mo}$-functors.

DEFINITION 3.17. *We say that a subfunctor $\mathcal{G} \subset \mathcal{F}$ of an $\mathfrak{Mo}$-functor $\mathcal{F}$ is open if for any object M of $\mathfrak{Mo}$ and any morphism of $\mathfrak{Mo}$-functors $\varphi : \mathfrak{spec}\, M \to \mathcal{F}$ there exists an ideal $I \subset M$ satisfying the following property. For any object N of $\mathfrak{Mo}$ and for any $\rho \in \mathfrak{spec}\, M(N) = \mathrm{Hom}_{\mathfrak{Mo}}(M, N)$ one has*

$$(36) \qquad \varphi(\rho) \in \mathcal{G}(N) \subset \mathcal{F}(N) \;\Leftrightarrow\; \rho(I)N = N.$$

To clarify the meaning of the above definition we develop a few examples.

EXAMPLE 3.18. The functor

$$\mathcal{G} : \mathfrak{Mo} \to \mathfrak{Sets}, \quad \mathcal{G}(N) = N^\times$$

is an open subfunctor of the (identity) functor $\mathcal{D}^1$

$$\mathcal{D}^1 : \mathfrak{Mo} \to \mathfrak{Sets}, \quad \mathcal{D}^1(N) = N.$$

In fact, let M be a monoid, then by Yoneda's lemma a morphism of functors $\varphi : \mathfrak{spec}\, M \to \mathcal{D}^1$ is determined by an element $z \in \mathcal{D}^1(M) = M$. For any monoid N and $\rho \in \mathrm{Hom}(M, N)$, one has $\varphi(\rho) = \rho(z) \in \mathcal{D}^1(N) = N$, thus the condition $\varphi(\rho) \in \mathcal{G}(N) = N^\times$ means that $\rho(z) \in N^\times$. One takes for I the ideal generated by z in M: $I = zM$. Then it is straightforward to check that (36) is fulfilled.

EXAMPLE 3.19. We start with a monoid M and an ideal $I \subset M$ and define the following sub-$\mathfrak{Mo}$-functor of $\mathfrak{spec}\,(M)$

$$\underline{D}(I) : \mathfrak{Mo} \to \mathfrak{Sets}, \quad \underline{D}(I)(N) = \{\rho \in \mathrm{Hom}(M, N) | \rho(I)N = N\}.$$

This means that for all prime ideals $\mathfrak{p} \subset N$, one has $\rho^{-1}(\mathfrak{p}) \not\supset I$ (see Proposition 3.9). Next, we show that $\underline{D}(I)$ defines an open subfunctor of $\mathfrak{spec}\, M$. Indeed, for any object A of $\mathfrak{Mo}$ and any natural transformation $\varphi : \mathfrak{spec}\, A \to \mathfrak{spec}\, M$ one has $\varphi(id_A) = \eta \in \mathfrak{spec}\, M(A) = \mathrm{Hom}_{\mathfrak{Mo}}(M, A)$; we can take in A the ideal $J = \eta(I)A$. This ideal fulfills the condition (36) for any object N of $\mathfrak{Mo}$ and any $\rho \in \mathrm{Hom}_{\mathfrak{Mo}}(A, N)$. In fact, one has $\varphi(\rho) = \rho \circ \eta \in \mathrm{Hom}_{\mathfrak{Mo}}(M, N)$, and $\varphi(\rho) \in \underline{D}(I)(N)$ means that $\rho(\eta(I))N = N$. This holds if and only if $\rho(J)N = N$.

3.5.3. Open Covering by $\mathfrak{Mo}$-subfunctors

There is a natural generalization of the notion of open cover for $\mathfrak{Mo}$-functors. We recall that the category $\mathfrak{Ab}$ of abelian groups embeds as a full subcategory of $\mathfrak{Mo}$ by means of the functor $H \to \mathbb{F}_1[H]$.

DEFINITION 3.20. Let $\mathcal{F}$ be an $\mathfrak{Mo}$-functor and let $\{\mathcal{F}_\alpha\}_{\alpha \in I}$ (I = an index set) be a family of open subfunctors of $\mathcal{F}$. One says that $\{\mathcal{F}_\alpha\}_{\alpha \in I}$ is an open cover of $\mathcal{F}$ if

$$(37) \qquad \mathcal{F}(\mathbb{F}_1[H]) = \bigcup_{\alpha \in I} \mathcal{F}_\alpha(\mathbb{F}_1[H]), \quad \forall H \in Obj(\mathfrak{Ab}).$$

Since commutative groups (with a zero element) replace fields in $\mathfrak{Mo}$, the above definition is the natural generalization of the definition of open covers for $\mathbb{Z}$-functors (see [**DG**]) within the category of $\mathfrak{Mo}$-functors. The following proposition gives a precise characterization of the properties of the open covers of an $\mathfrak{Mo}$-functor.

PROPOSITION 3.21. Let $\mathcal{F}$ be an $\mathfrak{Mo}$-functor and let $\{\mathcal{F}_\alpha\}_{\alpha \in I}$ be a family of open subfunctors of $\mathcal{F}$. Then the family $\{\mathcal{F}_\alpha\}_{\alpha \in I}$ forms an open cover of $\mathcal{F}$ if and only if

$$\mathcal{F}(M) = \bigcup_{\alpha \in I} \mathcal{F}_\alpha(M), \quad \forall\, M \in obj(\mathfrak{Mo}).$$

PROOF. The condition is obviously sufficient. To show the converse, we assume (37). Let M be a pointed monoid and let $\xi \in \mathcal{F}(M)$; one needs to show that $\xi \in \mathcal{F}_\alpha(M)$ for some $\alpha \in I$. Let $\varphi : \mathfrak{spec}\, M \to \mathcal{F}$ be the morphism of functors with $\varphi(id_M) = \xi$. Since each $\mathcal{F}_\alpha$ is an open subfunctor of $\mathcal{F}$, one can find ideals $I_\alpha \subset M$ such that for any object N of $\mathfrak{Mo}$ and for any $\rho \in \mathfrak{spec}\, M(N) = \mathrm{Hom}_{\mathfrak{Mo}}(M, N)$ one has

$$(38) \qquad \varphi(\rho) \in \mathcal{F}_\alpha(N) \subset \mathcal{F}(N) \Leftrightarrow \rho(I_\alpha)N = N.$$

One applies this to the morphism $\epsilon_M : M \to \mathbb{F}_1[M^\times] = \kappa$ as in (32). One has $\epsilon_M \in \mathfrak{spec}\,(M)(\kappa)$ and $\varphi(\epsilon_M) \in \mathcal{F}(\kappa) = \bigcup_{\alpha \in I} \mathcal{F}_\alpha(\kappa)$. Thus, $\exists \alpha \in I$ such that

$\varphi(\epsilon_M) \in \mathcal{F}_\alpha(\kappa)$. By (38) one concludes that $\epsilon_M(I_\alpha)\kappa = \kappa$ and $I_\alpha \cap M^\times \neq \emptyset$; hence $I_\alpha = M$. Then applying (38) to $\rho = id_M$ one obtains $\xi \in \mathcal{F}_\alpha(M)$ as required.   □

For $\mathcal{F} = \mathfrak{spec}\, M$ and the $\mathcal{F}_\alpha = \underline{D}(I_\alpha)$ open subfunctors corresponding to ideals $I_\alpha \subset M$, the covering condition in Definition 3.20 is equivalent to stating that $\exists \alpha \in I$ such that $I_\alpha = M$. In fact, one takes $H = M^\times$ and $\epsilon : M \to \mathbb{F}_1[M^\times] = \kappa$. Then $\epsilon \in \mathfrak{spec}\, M(\mathbb{F}_1[H])$ and $\exists \alpha,\ \epsilon \in \underline{D}(I_\alpha)(\mathbb{F}_1[H])$ and thus $I_\alpha \cap M^\times \neq \emptyset$; hence $I_\alpha = M$.

Let X be a geometric monoidal space and let $U_\alpha \subset X$ be (a family of) open subsets ($\alpha \in I$). One introduces the following $\mathfrak{Mo}$-functors

(39) $\underline{X}_\alpha : \mathfrak{Mo} \to \mathfrak{Sets},\qquad \underline{X}_\alpha(M) := \operatorname{Hom}(\operatorname{Spec}(M), U_\alpha).$

Notice that $\underline{X}_\alpha(M) \subset \underline{X}(M) := \operatorname{Hom}(\operatorname{Spec}(M), X)$.

PROPOSITION 3.22. *The following conditions are equivalent:*
(a) $\underline{X} = \bigcup_{\alpha \in I} \underline{X}_\alpha$.
(b) $X = \bigcup_{\alpha \in I} U_\alpha$.
(c) $\underline{X}(M) = \bigcup_{\alpha \in I} \underline{X}_\alpha(M)\quad \forall M \in obj(\mathfrak{Mo})$.

PROOF. (a) $\implies$ (b). Assume that (b) fails and let $x \notin \bigcup_\alpha U_\alpha$. Then the local evaluation map $\epsilon_x : \mathcal{O}_{X,x} \to \kappa(x)$ of (34) determines a morphism of geometric monoidal spaces $\operatorname{Spec}(\kappa(x)) \to X$ and for $M = \kappa(x)$ a corresponding element $\epsilon \in \underline{X}(M) = \operatorname{Hom}(\operatorname{Spec}(M), X)$. By applying Definition 3.20, there exists an index α such that $\epsilon \in \underline{X}_\alpha(M)$ and this shows that $x \in U_\alpha$ so that the open sets U_α cover X.

(b) $\implies$ (c). Let $\phi \in \underline{X}(M) = \operatorname{Hom}(\operatorname{Spec}(M), X)$. Then, with $\mathfrak{p}_M$ the maximal ideal of M, one has $\phi(\mathfrak{p}_M) \in X = \bigcup_\alpha U_\alpha$; hence there exists an index α such that $\phi(\mathfrak{p}_M) \in U_\alpha$. It follows that $\phi^{-1}(U_\alpha) \ni \mathfrak{p}_M$ is $\operatorname{Spec}(M)$, and one gets $\phi \in \underline{X}_\alpha(M) = \operatorname{Hom}(\operatorname{Spec}(M), U_\alpha)$.
The implication (c) $\implies$ (a) is straightforward.   □

3.5.4. $\mathfrak{Mo}$-schemes

In view of the fact that an $\mathfrak{Mo}$-functor is local, the definition of an $\mathfrak{Mo}$-scheme simply involves the local representability.

DEFINITION 3.23. *An $\mathfrak{Mo}$-scheme is an $\mathfrak{Mo}$-functor which admits an open cover by representable subfunctors.*

We shall consider several elementary examples of $\mathfrak{Mo}$-schemes

EXAMPLE 3.24. The affine space $\mathcal{D}^n$. For a fixed $n \in \mathbb{N}$, we consider the following $\mathfrak{Mo}$-functor

$$\mathcal{D}^n : \mathfrak{Mo} \to \mathfrak{Sets},\qquad \mathcal{D}^n(M) = M^n.$$

This functor is representable since it is described by

$$\mathcal{D}^n(M) = \operatorname{Hom}_{\mathfrak{Mo}}(\mathbb{F}_1[T_1, \ldots, T_n], M)$$

where

$$\mathbb{F}_1[T_1, \ldots, T_n] := \{0\} \cup \{T_1^{a_1} \cdots T_n^{a_n} | a_j \in \mathbb{Z}_{\geq 0}\}$$

is the pointed monoid associated to the semi-group generated by the variables T_j.

EXAMPLE 3.25. The projective line $\mathbb{P}^1$. We consider the $\mathfrak{Mo}$-functor $\underline{\mathbb{P}}^1$ which associates to an object M of $\mathfrak{Mo}$ the set $\underline{\mathbb{P}}^1(M)$ of complemented submodules $E \subset M^2$ of rank one, where the rank is defined locally. By definition, a complemented submodule is the range of an idempotent matrix $e \in M_2(M)$ (i.e. $e^2 = e$) whose rows have at most[5] one non-zero entry. To a morphism $\rho : M \to N$ in $\mathfrak{Mo}$, one associates the following map

$$\underline{\mathbb{P}}^1(\rho)(E) = N \otimes_M E \subset N^2.$$

In terms of projectors $\underline{\mathbb{P}}^1(\rho)(e) = \rho(e) \in M_2(N)$. The condition of rank one means that for any prime ideal $\mathfrak{p} \in \operatorname{Spec} M$ one has $\epsilon_\mathfrak{p}(e) \notin \{0,1\}$ where $\epsilon_\mathfrak{p} : M \to \mathbb{F}_1[(M_\mathfrak{p})^\times]$ is the morphism of Lemma 3.11 (b) (where $M_\mathfrak{p} = S^{-1}M$, with $S = \mathfrak{p}^c$). Now we compare $\underline{\mathbb{P}}^1$ with the $\mathfrak{Mo}$-functor

$$\mathcal{P}(M) = M \cup_{M^\times} M$$

where the gluing map is given by $x \to x^{-1}$. In other words, we define on the disjoint union $M \cup M$ an equivalence relation given by (using the identification $M \times \{1,2\} = M \cup M$)

$$(x,1) \sim (x^{-1},2) \quad \forall x \in M^\times.$$

We define a natural transformation $e : \mathcal{P} \to \underline{\mathbb{P}}^1$ by observing that the matrices

$$e_1(a) = \begin{pmatrix} 1 & 0 \\ a & 0 \end{pmatrix}, \quad e_2(b) = \begin{pmatrix} 0 & b \\ 0 & 1 \end{pmatrix}, \quad a, b \in M$$

are idempotent ($e^2 = e$) and their ranges also fulfill the following property

$$\operatorname{Im} e_1(a) = \operatorname{Im} e_2(b) \iff ab = 1.$$

LEMMA 3.26. *The natural transformation e is bijective on the objects, i.e.*

$$\mathcal{P}(M) = M \cup_{M^\times} M \cong \underline{\mathbb{P}}^1(M).$$

Moreover, $\underline{\mathbb{P}}^1$ is covered by two copies of representable subfunctors $\mathcal{D}^1$.

PROOF. We refer to [**CC2**], Lemma 3.13.   □

EXAMPLE 3.27. Let M be a monoid and let $I \subset M$ be an ideal. Consider the $\mathfrak{Mo}$-functor $\underline{D}(I)$ of Example 3.19. The next proposition states that this functor is an $\mathfrak{Mo}$-scheme.

PROPOSITION 3.28. *(a) Let $f \in M$ and $I = fM$. Then the subfunctor $\underline{D}(I) \subset$ $\mathfrak{spec}\, M$ is represented by M_f.*
(b) For any ideal $I \subset M$, the $\mathfrak{Mo}$-functor $\underline{D}(I)$ is an $\mathfrak{Mo}$-scheme.

PROOF. (a) For any monoid N and for $I = fM$, one has

$$\underline{D}(I)(N) = \{\rho \in \operatorname{Hom}_{\mathfrak{Mo}}(M,N) | \rho(f) \in N^\times\}.$$

The condition $\rho(f) \in N^\times$ means that ρ extends to a morphism $\tilde{\rho} \in \operatorname{Hom}(M_f, N)$, by setting

$$\tilde{\rho}\left(\frac{m}{f^n}\right) = \rho(m)\rho(f)^{-n} \in N.$$

---

[5]Note that we need the 0 element to state this condition.

Thus one has a canonical and functorial isomorphism $\underline{D}(I)(N) \simeq \mathrm{Hom}_{\mathfrak{M}\mathfrak{o}}(M_f, N)$ which proves the representability of $\underline{D}(I)$ by M_f.

(b) For any $f \in I$, the ideal $fM \subset I$ defines a subfunctor of $\underline{D}(fM) \subset \underline{D}(I)$. This subfunctor is open because it is already open in $\mathfrak{spec}\, M$ and because there are fewer morphisms of type $\mathrm{Spec}(A) \to D(I)$ than those of type $\mathrm{Spec}(A) \to \mathrm{Spec}(M)$, as $\underline{D}(I)(A) \subset \mathfrak{spec}\, M(A)$. Moreover, by (a), this subfunctor is representable. $\square$

3.5.5. Geometric Realization

In this final subsection we describe the construction of the geometric realization of an $\mathfrak{M}\mathfrak{o}$-functor (and scheme) following, and generalizing for commutative monoids, the exposition presented in ([**DG**], §1, n. 4) of the geometric realization of a $\mathbb{Z}$-scheme for rings. Although the general development of this construction presents clear analogies with the case of rings, new features also arise which are specifically inherent to the discussion about monoids. The most important one is that the full subcategory $\mathfrak{Ab}$ of $\mathfrak{M}\mathfrak{o}$ which plays the role of "fields" within monoids, admits a final object. This fact greatly simplifies the description of the geometric realization of an $\mathfrak{M}\mathfrak{o}$-functor as we show in Proposition 3.32 and in Theorem 3.34.

An $\mathfrak{M}\mathfrak{o}$-functor $\mathcal{F}$ can be reconstructed by an inductive limit $\varinjlim \mathfrak{spec}\,(M)$ of representable functors, and the geometric realization is then defined by the inductive limit $\varinjlim \mathrm{Spec}(M)$, $i.e.$ by trading the $\mathfrak{M}\mathfrak{o}$-functors $\mathfrak{spec}\,(M)$ for the geometric monoidal spaces $(\mathrm{Spec}(M), \mathcal{O})$. However, some set-theoretic precautions are also needed since the inductive limits which are taken over pairs (M, ρ), where $M \in \mathrm{obj}(\mathfrak{M}\mathfrak{o})$ and $\rho \in \mathcal{F}(M)$, are indexed over families rather than sets. Another issue arising in the construction is that of seeking that natural transformations of $\mathfrak{M}\mathfrak{o}$-functors form a set. We bypass this problem by adopting the same caution as in [**DG**] (see Conventions générales). Thus, throughout this subsection we shall work with *models* of commutative monoids in a fixed suitable *universe* $\mathbb{U}$. This set-up is accomplished by introducing the full subcategory $\mathfrak{M}\mathfrak{o}' \subset \mathfrak{M}\mathfrak{o}$ of models of commutative monoids in $\mathfrak{M}\mathfrak{o}$, $i.e.$ of commutative (pointed) monoids whose underlying set is an element of the fixed universe $\mathbb{U}$. To simplify the notation and the statements we adopt the convention that all the geometric monoidal spaces considered here only involve monoids in $\mathfrak{M}\mathfrak{o}'$; thus we shall not keep the distinction between the category $\mathfrak{M}\mathfrak{o}\mathfrak{S}$ of $\mathfrak{M}\mathfrak{o}$-functors and the category $\mathfrak{M}\mathfrak{o}'\mathfrak{S}$ of (covariant) functors from $\mathfrak{M}\mathfrak{o}'$ to $\mathfrak{Sets}$.

Given an $\mathfrak{M}\mathfrak{o}$-functor $\mathcal{F} : \mathfrak{M}\mathfrak{o} \to \mathfrak{Sets}$, we denote by $\mathcal{M}_{\mathcal{F}}$ the category of $\mathcal{F}$-*models*, that is, the category whose objects are pairs (M, ρ), where $M \in \mathrm{obj}(\mathfrak{M}\mathfrak{o}')$ and $\rho \in \mathcal{F}(M)$. A morphism $(M_1, \rho_1) \to (M_2, \rho_2)$ in $\mathcal{M}_{\mathcal{F}}$ is given by assigning a morphism $\varphi : M_1 \to M_2$ in $\mathfrak{M}\mathfrak{o}'$ such that $\varphi(\rho_1) = \rho_2 \in \mathcal{F}(M_2)$.

For a geometric monoidal space X, $i.e.$ an object of the category $\mathfrak{G}\mathcal{S}$, we recall the definition of the *functor defined by X* (see (39)). This is the $\mathfrak{M}\mathfrak{o}$-functor

$$\underline{X} : \mathfrak{M}\mathfrak{o} \to \mathfrak{Sets}, \qquad \underline{X}(M) = \mathrm{Hom}_{\mathfrak{G}\mathcal{S}}(\mathrm{Spec}(M), X).$$

This definition can of course be restricted to the subcategory $\mathfrak{M}\mathfrak{o}' \subset \mathfrak{M}\mathfrak{o}$ of models. One also introduces the functor

$$d_{\mathcal{F}} : \mathcal{M}_{\mathcal{F}}^{op} \to \mathfrak{G}\mathcal{S}, \qquad d_{\mathcal{F}}(M, \rho) = \mathrm{Spec}(M).$$

DEFINITION 3.29. *The geometric realization of an $\mathfrak{Mo}$-functor $\mathcal{F}$ is the geometric monoidal space defined by the inductive limit*

$$|\mathcal{F}| = \varinjlim_{(M,\rho) \in \mathcal{M}_\mathcal{F}} d_\mathcal{F}(M, \rho).$$

The functor $|\cdot| : \mathcal{F} \to |\mathcal{F}|$ is called the functor geometric realization.

For each $(M, \rho) \in \mathrm{obj}(\mathcal{M}_\mathcal{F})$, one has a canonical morphism

(40) $$i(\rho) : d_\mathcal{F}(M, \rho) = \mathrm{Spec}(M) \to \varinjlim d_\mathcal{F} = |\mathcal{F}|.$$

The following result shows that the functor geometric realization $|\cdot|$ is left adjoint to the functor $\underline{\cdot}$.

PROPOSITION 3.30. *For every geometric monoidal space X there is a bijection of sets which is functorial in X*

$$\varphi(\mathcal{F}, X) : \mathrm{Hom}_{\mathfrak{GS}}(|\mathcal{F}|, X) \xrightarrow{\sim} \mathrm{Hom}_{\mathfrak{Mo}\,\mathfrak{S}}(\mathcal{F}, \underline{X}), \qquad \varphi(\mathcal{F}, X)(f) = g$$

where $\forall (M, \rho) \in \mathrm{obj}(\mathcal{M}_\mathcal{F})$

$$g(M) : \mathcal{F}(M) \to \underline{X}(M), \qquad g(M)(\rho) = f \circ i(\rho)$$

and $i(\rho) : \mathrm{Spec}(M) \to |\mathcal{F}|$ is the canonical morphism of (40). The functor geometric realization $|\cdot|$ is left adjoint to the functor $\underline{\cdot}$.

PROOF. The proof is similar to the proof given in [**DG**] (see §1, Proposition 4.1). $\qquad\square$

One obtains, in particular, the following morphisms

$$\varphi(\mathcal{F}, |\mathcal{F}|) : \mathrm{Hom}_{\mathfrak{GS}}(|\mathcal{F}|, |\mathcal{F}|) \xrightarrow{\sim} \mathrm{Hom}_{\mathfrak{Mo}\,\mathfrak{S}}(\mathcal{F}, |\mathcal{F}|), \quad \varphi(\mathcal{F}, |\mathcal{F}|)(id_{|\mathcal{F}|}) =: \Psi(\mathcal{F})$$

$$\varphi(\underline{X}, X)^{-1} : \mathrm{Hom}_{\mathfrak{Mo}\,\mathfrak{S}}(\underline{X}, \underline{X}) \xrightarrow{\sim} \mathrm{Hom}_{\mathfrak{GS}}(|\underline{X}|, X), \quad \varphi(\underline{X}, X)^{-1}(id_{\underline{X}}) =: \Phi(X).$$

Notice that as a consequence of the fact that the functor $\mathrm{Spec}(\cdot)$ is fully faithful, $\Psi(\mathfrak{spec}\,(M))$ is invertible; thus $\mathrm{Spec}(M) \simeq |\mathfrak{spec}\,(M)|$, *i.e.* $\Phi(\mathrm{Spec}(M))$ is invertible, for any model M in $\mathfrak{Mo}'$. See also Example 3.35.

We have already said that the covariant functor

$$\mathbb{F}_1[\,\cdot\,] : \mathfrak{Ab} \to \mathfrak{Mo}, \qquad H \mapsto \mathbb{F}_1[H]$$

embeds the category of abelian groups as a full subcategory of $\mathfrak{Mo}$. We shall identify $\mathfrak{Ab}$ to this full subcategory of $\mathfrak{Mo}$. One has a pair of adjoint functors: $H \mapsto \mathbb{F}_1[H]$ from $\mathfrak{Ab}$ to $\mathfrak{Mo}$ and $M \mapsto M^\times$ from $\mathfrak{Mo}$ to $\mathfrak{Ab}$, *i.e.* one has the following natural isomorphism

$$\mathrm{Hom}_{\mathfrak{Mo}}(\mathbb{F}_1[H], M) \cong \mathrm{Hom}_{\mathfrak{Ab}}(H, M^\times).$$

Moreover, any $\mathfrak{Mo}$-functor $\mathcal{F} : \mathfrak{Mo} \to \mathfrak{Sets}$ restricts to $\mathfrak{Ab}$ and gives rise to a functor: its *Weil restriction*

$$\mathcal{F}_{|\mathfrak{Ab}} : \mathfrak{Ab} \to \mathfrak{Sets}$$

taking values into sets.

Let X be a geometric monoidal space. Then the Weil restriction $\underline{X}_{|\mathfrak{Ab}}$ of the $\mathfrak{Mo}$-functor $\underline{X}$ to the full subcategory $\mathfrak{Ab} \subset \mathfrak{Mo}$ is a direct sum of *indecomposable* functors, *i.e.* a direct sum of functors that cannot be decomposed further into a disjoint sum of non-empty subfunctors.

PROPOSITION 3.31. *Let X be a geometric monoidal space. Then the functor*

$$\underline{X}_{|\mathfrak{Ab}} : \mathfrak{Ab} \to \mathfrak{Sets}, \qquad \underline{X}_{|\mathfrak{Ab}}(H) = Hom_{\mathfrak{GS}}(Spec\,\mathbb{F}_1[H], X)$$

is the disjoint union of representable functors

$$\underline{X}_{|\mathfrak{Ab}}(\mathbb{F}_1[H]) = \coprod_{x\in X} (\underline{X}_{|\mathfrak{Ab}})_x(\mathbb{F}_1[H]) = \coprod_{x\in X} Hom_{\mathfrak{Ab}}(\mathcal{O}^\times_{X,x}, H)$$

where x runs through the points of X.

PROOF. Let $\varphi \in \text{Hom}(\text{Spec}\,\mathbb{F}_1[H], X)$. The unique point $\mathfrak{p} \in \text{Spec}\,\mathbb{F}_1[H]$ corresponds to the ideal $\{0\}$. Let $\varphi(\mathfrak{p}) = x \in X$ be its image; there is a corresponding map of the stalks

$$\varphi^\# : \mathcal{O}_{X,\varphi(p)} \to \mathcal{O}_\mathfrak{p} = \mathbb{F}_1[H].$$

This homomorphism is local by hypothesis: this means that the inverse image of $\{0\}$ by $\varphi^\#$ is the maximal ideal of $\mathcal{O}_{X,\varphi(p)} = \mathcal{O}_{X,x}$. Therefore, the map $\varphi^\#$ is entirely determined by the group homomorphism $\rho : \mathcal{O}^\times_{X,x} \to H$ obtained as the restriction of $\varphi^\#$. Thus $\varphi \in \text{Hom}(\text{Spec}\,\mathbb{F}_1[H], X)$ is entirely specified by a point $x \in X$ and a group homomorphism $\rho \in \text{Hom}_{\mathfrak{Ab}}(\mathcal{O}^\times_{X,x}, H)$. □

We recall that the set underlying the geometric realization of a $\mathbb{Z}$-scheme can be obtained by restricting the $\mathbb{Z}$-functor to the full subcategory of fields and passing to a suitable inductive limit (see [**DG**], §4.5). For $\mathfrak{Mo}$-schemes this construction simplifies considerably since the full subcategory $\mathfrak{Ab}$ admits the *final object*

$$\mathbb{F}_1 = \mathbb{F}_1[\{1\}] = \{0, 1\}.$$

Notice though that while $\mathbb{F}_1$ is a final object for the subcategory $\mathfrak{Ab} \subset \mathfrak{Mo}$, it is *not* a final object in $\mathfrak{Mo}$; in fact the following proposition shows that for any monoid $M \in \text{obj}(\mathfrak{Mo})$, the set $\text{Hom}_{\mathfrak{Mo}}(M, \mathbb{F}_1)$ is in canonical correspondence with the points of $\text{Spec}\,M$.

For any geometric monoidal scheme X, we denote by $X^\mathfrak{e}$ its underlying set.

PROPOSITION 3.32. *(a) For any object M of $\mathfrak{Mo}$, the map*

$$\varphi : Spec(M)^\mathfrak{e} \to Hom_{\mathfrak{Mo}}(M, \mathbb{F}_1), \qquad \mathfrak{p} \mapsto \varphi_\mathfrak{p}$$

(41)
$$\varphi_\mathfrak{p}(x) = 0, \ \forall x \in \mathfrak{p}, \quad \varphi_\mathfrak{p}(x) = 1, \ \forall x \notin \mathfrak{p}$$

determines a natural bijection of sets.

(b) For any $\mathfrak{Mo}$-functor $\mathcal{F}$, there is a canonical isomorphism $|\mathcal{F}|^\mathfrak{e} \simeq \mathcal{F}(\mathbb{F}_1)$.

(c) Let X be a geometric monoidal space, and $\mathcal{F} = \underline{X}$. Then there is a canonical bijection $\underline{X}(\mathbb{F}_1) \simeq X$.

PROOF. (a) The map φ is well defined since the complement of a prime ideal $\mathfrak{p}$ in a monoid is a multiplicative set. To define the inverse of φ, one assigns to $\rho \in \text{Hom}_{\mathfrak{Mo}}(M, \mathbb{F}_1)$ its kernel which is a prime ideal of M that uniquely determines ρ. (b) The set $|\mathcal{F}|^\mathfrak{e}$, *i.e.* the set underlying $\varinjlim d_{\mathcal{F}}$ is canonically in bijection with

$$Z = \varinjlim_{(M,\rho)\in\mathcal{M}_{\mathcal{F}}} (\text{Spec}(M))^\mathfrak{e}.$$

By (1), one has a canonical bijection

$$\mathrm{Spec}(M)^{\mathfrak{e}} \simeq \mathrm{Hom}_{\mathfrak{Mo}}(M, \mathbb{F}_1)$$

and using the identification

$$\mathcal{F} = \varinjlim_{(M,\rho) \in \mathcal{M}_{\mathcal{F}}} \mathfrak{spec}(M)$$

we get the canonical bijection

$$Z = \varinjlim_{(M,\rho) \in \mathcal{M}_{\mathcal{F}}} \mathrm{Hom}_{\mathfrak{Mo}}(M, \mathbb{F}_1) = \varinjlim_{(M,\rho) \in \mathcal{M}_{\mathcal{F}}} \mathfrak{spec}(M)(\mathbb{F}_1) = \mathcal{F}(\mathbb{F}_1).$$

(c) An element $\rho \in \mathcal{F}(\mathbb{F}_1) = \underline{X}(\mathbb{F}_1)$ is a morphism of geometric spaces $\mathrm{Spec}\,\mathbb{F}_1 \to X$ and the image of the unique (closed) point of $\mathrm{Spec}\,\mathbb{F}_1$ is a point $x \in X$ which uniquely determines ρ. Conversely any point $x \in X$ determines a morphism $(\varphi, \varphi^{\sharp})$ of geometric spaces $\mathrm{Spec}\,\mathbb{F}_1 \to X$. The homomorphism of sheaves of monoids $\varphi^{\sharp} : \varphi^{-1}\mathcal{O}_X \to \mathcal{O}_{\mathbb{F}_1}$ is uniquely determined by its locality property and sends $\mathcal{O}_{X,x}^{\times}$ to 1 and the complement of $\mathcal{O}_{X,x}^{\times}$ (i.e. the maximal ideal $\mathfrak{m}_x$) to 0. $\qquad\square$

Proposition 3.32 shows how to describe concretely the set underlying the geometric realization $|\mathcal{F}|$ of an $\mathfrak{Mo}$-functor $\mathcal{F}$ in terms of the set $\mathcal{F}(\mathbb{F}_1)$. To describe the topology of $|\mathcal{F}|$ directly on $\mathcal{F}(\mathbb{F}_1)$, we use the following construction of subfunctors of $\mathcal{F}$ defined in terms of arbitrary subsets $P \subset \mathcal{F}(\mathbb{F}_1)$. It corresponds to the construction of ([DG], I, §1, 4.10).

PROPOSITION 3.33. *Let $\mathcal{F}$ be an $\mathfrak{Mo}$-functor and $P \subset \mathcal{F}(\mathbb{F}_1)$ be a subset of the set $\mathcal{F}(\mathbb{F}_1)$. For $M \in obj(\mathfrak{Mo})$ and $\rho \in \mathcal{F}(M)$ the following two conditions are equivalent:*
(a) For every homomorphism $\varphi \in \mathrm{Hom}_{\mathfrak{Mo}}(M, \mathbb{F}_1)$ one has

$$\mathcal{F}(\varphi)(\rho) \in P \subset \mathcal{F}(\mathbb{F}_1).$$

(b) For every $H \in obj(\mathfrak{Ab})$ and every homomorphism $\varphi \in \mathrm{Hom}(M, \mathbb{F}_1[H])$ one has

$$\mathcal{F}(\varphi)(\rho) \in \bigcup_{x \in P} (\mathcal{F}_{|\mathfrak{Ab}})_x(\mathbb{F}_1[H])$$

where P is viewed as a subset of $\varinjlim \mathcal{F}_{|\mathfrak{Ab}} \simeq |\mathcal{F}|$.
The above equivalent conditions define a subfunctor $\mathcal{F}_P \subset \mathcal{F}$.

It is clear that (a) defines a subfunctor $\mathcal{F}_P \subset \mathcal{F}$. We shall omit the proof of the equivalence between conditions (a) and (b) which is only needed to carry on the analogy with the construction in ([DG], I, §1, 4.10).

Thus, *we adopt (a) as the definition of the subfunctor $\mathcal{F}_P \subset \mathcal{F}$.*

Notice that one can reconstruct P from $\mathcal{F}_P$ using the equality

$$P = \mathcal{F}_P(\mathbb{F}_1) \subset \mathcal{F}(\mathbb{F}_1).$$

Let X be a geometric monoidal space. If one endows the subset $P \subset X$ with the induced topology from X and with the structural sheaf $\mathcal{O}_{X|P}$, the functor $\underline{X}_P$ gets identified with $\underline{P} \subset \underline{X}$. Moreover, if P is an open subset in X, then $\underline{X}_P$ is an open subfunctor of $\underline{X}$. In particular, if $\mathcal{F} = \mathfrak{spec}(M)$ then $P \subset \mathrm{Spec}(M) = |\mathcal{F}|$ is open if and only if $\mathcal{F}_P$ is an open subfunctor of $\mathfrak{spec}(M)$.

More generally, the following results hold for arbitrary $\mathfrak{Mo}$-functors.

THEOREM 3.34. *Let $\mathcal{F}$ be an $\mathfrak{M}$o-functor. The following facts hold:*

(a) A subset $P \subset |\mathcal{F}| = \mathcal{F}(\mathbb{F}_1)$ is open if and only if for every $(M, \rho) \in obj(\mathcal{M}_\mathcal{F})$ the set of prime ideals $\mathfrak{p} \in Spec\, M$ such that $\varphi_\mathfrak{p}(\rho) \in P$ is open in $Spec\, M$.

(b) The map $P \to \mathcal{F}_P$ induces a bijection of sets between the collection of open subsets of $|\mathcal{F}| = \mathcal{F}(\mathbb{F}_1)$ and the open subfunctors of $\mathcal{F}$.

(c) The structure sheaf of $|\mathcal{F}| = \mathcal{F}(\mathbb{F}_1)$ is given by

$$\mathcal{O}(U) = Hom_{\mathfrak{M}o\,\mathfrak{S}}(\mathcal{F}_U, \mathcal{D}), \ \forall U \subset |\mathcal{F}| \ open,$$

where $\mathcal{D} : \mathfrak{M}o \to \mathfrak{S}ets$, $\mathcal{D}(M) = M$ is the functor affine line.

PROOF. (a) By definition of the topology on the inductive limit

$$|\mathcal{F}| = \varinjlim_{(M,\rho)\in\mathcal{M}_\mathcal{F}} d_\mathcal{F},$$

a subset P of $|\mathcal{F}|$ is open if and only if its inverse image under the canonical maps $i(\rho)$ of (40) is open. Using the identification $|\mathcal{F}| \simeq \mathcal{F}(\mathbb{F}_1)$ one obtains the required statement.

(b) We give the simple direct argument (see also [**DG**], I, §1, Proposition 4.12). Let $\mathcal{F}_1$ be an open subfunctor of $\mathcal{F}$. By Definition 3.17, given an object M of $\mathfrak{M}o$ and an element $\xi \in \mathcal{F}(M)$, there exists an ideal $I \subset M$, $I = I(M, \xi)$, such that for any object N of $\mathfrak{M}o$ and for any $\rho \in \mathfrak{spec}\, M(N) = Hom_{\mathfrak{M}o}(M, N)$ one has

$$\mathcal{F}(\rho)\xi \in \mathcal{F}_1(N) \subset \mathcal{F}(N) \ \Leftrightarrow \ \rho(I)N = N.$$

Take $N = \mathbb{F}_1$ and $\rho = \varphi_\mathfrak{p}$ for $\mathfrak{p} \in Spec\, M$, then one gets

$$\mathcal{F}(\varphi_\mathfrak{p})(\xi) \in \mathcal{F}_1(\mathbb{F}_1) \ \Leftrightarrow \ 1 \in \varphi_\mathfrak{p}(I) \ \Leftrightarrow \ \mathfrak{p} \in D(I)$$

which shows that $D(I)$ and hence the radical of $I = I(M, \xi)$ is determined by the subset $P = \mathcal{F}_1(\mathbb{F}_1) \subset \mathcal{F}(\mathbb{F}_1)$. Now take $N = M$ and $\rho = id_M$, then one has

$$\xi \in \mathcal{F}_1(M) \subset \mathcal{F}(M) \ \Leftrightarrow \ I = M$$

and this holds if and only if $D(I) = Spec\, M$ or equivalently $\varphi_\mathfrak{p}(\xi) \in \mathcal{F}_1(\mathbb{F}_1)$ for all $\mathfrak{p} \in Spec\, M$ which is the definition of the subfunctor associated to the open subset $P = \mathcal{F}_1(\mathbb{F}_1) \subset \mathcal{F}(\mathbb{F}_1)$.

(c) By using the above identifications, the proof is the same as in ([**DG**], I, §1, Proposition 4.14). The structure of the monoid on $\mathcal{O}(U) = Hom_{\mathfrak{M}o\,\mathfrak{S}}(\mathcal{F}_U, \mathcal{D})$ is given by

$$(\alpha\beta)(M)(x) = \alpha(M)(x)\beta(M)(x) \in M, \ \forall x \in \mathcal{F}_U(M), \ \alpha, \beta \in \mathcal{O}(U).$$

$\square$

EXAMPLE 3.35. Let $M \in obj(\mathfrak{M}o)$. The map $\mathfrak{p} \mapsto \varphi_\mathfrak{p}$ as in (41) defines a natural bijection $Spec\, M \simeq |\mathfrak{spec}\, M| = Hom(M, \mathbb{F}_1)$ and using the topology defined by (a) of Theorem 3.34, one can check directly that this map is a homomorphism. One can also verify directly that the structure sheaf of $|\mathfrak{spec}\, M|$ as defined by (c) of Theorem 3.34 coincides with the structure sheaf of $Spec\, M$. Thus $Spec(M) \simeq |\mathfrak{spec}\,(M)|$, *i.e.* $\Phi(Spec(M))$ and $\Psi(\mathfrak{spec}\,(M))$ are invertible, for any object M in $\mathfrak{M}o$.

COROLLARY 3.36. *Let X be a geometric monoidal space. If $\mathcal{F} \subset \underline{X}$ is an open subfunctor of $\underline{X}$, then $|\mathcal{F}| \subset |X|$ is an open subset of X.*

PROOF. For the proof we refer to ([**DG**], I, §1, Corollary 4.15).   □

The following result describes a property which *does not hold* in general for $\mathbb{Z}$-schemes and provides a natural map $\mathcal{F}(M) \to |\mathcal{F}| \simeq \mathcal{F}(\mathbb{F}_1)$, $\forall M \in \mathrm{obj}(\mathfrak{Mo})$ and any $\mathfrak{Mo}$-functor $\mathcal{F}$.

PROPOSITION 3.37. *Let $\mathcal{F}$ be an $\mathfrak{Mo}$-functor. Then the following facts hold:*
(a) For any monoid $M \in \mathrm{obj}(\mathfrak{Mo})$ there exists a canonical map

$$\pi_M : \mathcal{F}(M) \to |\mathcal{F}| \simeq \mathcal{F}(\mathbb{F}_1)$$

such that

$$\pi_M(\rho) = \varphi_{\mathfrak{p}_M}(\rho), \ \forall \rho \in \mathcal{F}(M).$$

(b) Let U be an open subset of $|\mathcal{F}| = \mathcal{F}(\mathbb{F}_1)$ and $\mathcal{F}_U$ the associated subfunctor of $\mathcal{F}$. Then

$$\mathcal{F}_U(M) = \pi_M^{-1}(U) \subset \mathcal{F}(M).$$

PROOF. (a) follows from the definition of the map $\varphi_{\mathfrak{p}_M}$.

(b) By Theorem 3.34, the set $W = \{\mathfrak{p} \in \mathrm{Spec}\, M \,|\, \varphi_{\mathfrak{p}}(\rho) \in U\}$ is open in $\mathrm{Spec}\, M$. Thus by Proposition 2.18, one has $W = \mathrm{Spec}(M)$ if and only if $\mathfrak{p}_M \in W$ which gives the required statement using the definition of the subfunctor $\mathcal{F}_U$ as in Proposition 3.33.   □

The next result establishes the equivalence between the category of $\mathfrak{Mo}$-schemes and the category of geometric $\mathfrak{Mo}$-schemes, using sufficient conditions for the morphisms

$$\Psi(\mathcal{F}) : \mathcal{F} \to |\mathcal{F}|, \qquad \Phi(X) : |\underline{X}| \to X$$

to be invertible. This result generalizes Example 3.35 where this invertibility has been proven in the case $\mathcal{F} = \mathfrak{spec}\, M$, $X = \mathrm{Spec}\, M$ and for objects M of $\mathfrak{Mo}$.

THEOREM 3.38. *(a) Let X be a geometric $\mathfrak{Mo}$-scheme. Then $\underline{X}$ is an $\mathfrak{Mo}$-scheme and the morphism $\Phi(X) : |\underline{X}| \to X$ is invertible.*
(b) Let $\mathcal{F}$ be an $\mathfrak{Mo}$-scheme. Then the geometric monoidal space $|\mathcal{F}|$ is a geometric $\mathfrak{Mo}$-scheme and $\Psi(\mathcal{F}) : \mathcal{F} \to |\mathcal{F}|$ is invertible.
Thus, the two functors $|\cdot|$ and $\underline{\ \cdot\ }$ induce quasi-inverse equivalences between the category of $\mathfrak{Mo}$-schemes and the category of geometric $\mathfrak{Mo}$-schemes.

PROOF. (a) In view of Proposition 3.32, for any geometric scheme X the map $\Phi(X) : |\underline{X}| \to X$ induces a bijection of the underlying sets. If $\{U_i\}_{i \in \mathbb{U}}$ is a covering of X by prime spectra, it follows that $\Phi(U_i) : |\underline{U_i}| \to U_i$ is an isomorphism. Moreover, since U_i is an open subset of X, it follows from Corollary 3.36 that $|\underline{U_i}|$ is an open subspace of $|\underline{X}|$. Thus, the topologies on the spaces $|\underline{X}|$ and X as well as the related structural sheaves get identified by means of $\Phi(X)$. Hence $\Phi(X)$ is invertible.

(b) Let $\mathcal{F}$ be an $\mathfrak{Mo}$-scheme; we shall show that $|\mathcal{F}|$ is a geometric $\mathfrak{Mo}$-scheme. We let $\{\mathfrak{U}_i\}_{i \in \mathbb{U}}$ be an open covering of $\mathcal{F}$ by (functors) prime spectra such that $\{\mathfrak{U}_{ij\alpha}\}$ is an open covering of $\mathfrak{U}_i \cap \mathfrak{U}_j$ by (functors) prime spectra. Then $\{|\mathfrak{U}_i|\}$ is an open covering of $|\mathcal{F}|$ by prime spectra; thus $|\mathcal{F}|$ verifies the condition of Definition 3.13 and is a geometric $\mathfrak{Mo}$-scheme. In view of the fact that $\Psi(\mathfrak{U}_i) : \mathfrak{U}_i \to |\underline{\mathfrak{U}_i}|$ is invertible, the composite of $\Psi(\mathfrak{U}_i)^{-1}$ and the inclusion $\mathfrak{U}_i \hookrightarrow \mathcal{F}$ gives a collection of morphisms $\Psi_i^{-1}(\mathcal{F}) : |\mathfrak{U}_i| \to \mathcal{F}$. Moreover, one has $\Psi_i^{-1}(\mathcal{F})_{||\mathfrak{U}_{ij\alpha}|} = \Psi_j^{-1}(\mathcal{F})_{||\mathfrak{U}_{ij\alpha}|}$, for all i, j, α. The automatic locality of $\mathcal{F}$ then shows that the Ψ_i^{-1} assemble to a single morphism $|\mathcal{F}| \to \mathcal{F}$ which is the inverse of $\Psi(\mathcal{F}) : \mathcal{F} \to |\mathcal{F}|$. In fact Proposition 3.21 shows that this inverse is surjective.   □

4. $\mathbb{F}_1$-schemes and Their Zeta Functions

A general formalism of category theory allows one to glue together two categories connected by a pair of adjoint functors (see [**CC2**], §4). This construction is particularly interesting when applied to the functorial formalism that describes the most common examples of schemes (of finite type) over $\mathbb{F}_1$ that we have reviewed so far. In all these cases in fact, the two categories naturally involved, *i.e.* $\mathfrak{Mo}$ and $\mathfrak{Ring}$, are linked by a pair of natural adjoint functors. These are

$$\beta : \mathfrak{Mo} \to \mathfrak{Ring}, \qquad \beta(M) = \mathbb{Z}[M]$$

which associates to a monoid the convolution ring $\mathbb{Z}[M]$ and the forgetful functor

$$\beta^* : \mathfrak{Ring} \to \mathfrak{Mo}, \qquad \beta^*(R) = R$$

which associates to a ring its underlying structure of multiplicative monoid. The resulting glued category $\mathfrak{MR} = \mathfrak{Ring} \cup_{\beta,\beta^*} \mathfrak{Mo}$ defines the most natural categorical framework on which one introduces the notions of an $\mathbb{F}_1$-functor and of a scheme over $\mathbb{F}_1$.

DEFINITION 4.1. *An $\mathbb{F}_1$-functor is a covariant functor from the category $\mathfrak{MR} = \mathfrak{Ring} \cup_{\beta,\beta^*} \mathfrak{Mo}$ to the category of sets.*

The conditions imposed in the original definition of a variety over $\mathbb{F}_1$ (see [**So**]) are now applied to a covariant functor $\mathcal{X} : \mathfrak{MR} \to \mathfrak{Sets}$. Such a functor determines a scheme (of finite type) over $\mathbb{F}_1$ if it also fulfills the following properties.

DEFINITION 4.2. *An $\mathbb{F}_1$-scheme is an $\mathbb{F}_1$-functor $\mathcal{X} : \mathfrak{MR} \to \mathfrak{Sets}$, such that:*

- *The restriction $X_{\mathbb{Z}}$ of $\mathcal{X}$ to $\mathfrak{Ring}$ is a $\mathbb{Z}$-scheme.*
- *The restriction $\underline{X}$ of $\mathcal{X}$ to $\mathfrak{Mo}$ is an $\mathfrak{Mo}$-scheme.*
- *The natural transformation $e : \underline{X} \circ \beta^* \to X_{\mathbb{Z}}$ associated to a field is a bijection (of sets).*

Morphisms of $\mathbb{F}_1$-schemes are natural transformations of the corresponding functors.

4.1. Torsion Free Noetherian $\mathbb{F}_1$-schemes

The first theorem stated in this subsection shows that for any Noetherian, torsion-free $\mathbb{F}_1$-scheme the function counting the number of rational points (for the associated $\mathbb{Z}$-scheme) over a finite field $\mathbb{F}_q$ is *automatically* polynomial. The same theorem also provides a complete description of its zeta function over $\mathbb{F}_1$ (originally introduced upside-down in [**So**]).

By following ([**So**], §6, Lemma 1) and implementing the correction concerning the inversion in the formula of the zeta function, the definition of the zeta function of an algebraic variety $X = (\underline{X}, X_{\mathbb{C}}, e_X)$ over $\mathbb{F}_1$ such that $\#X_{\mathbb{Z}}(\mathbb{F}_q) = N(q)$, with $N(x)$ a polynomial function and $\#\underline{X}(\mathbb{F}_{1^n}) = N(q)$ if $n = q - 1$, is given as the limit

$$(42) \qquad \zeta_X(s) := \lim_{q \to 1} Z(X, q^{-s})(q-1)^{N(1)}.$$

Here, the function $Z(X, q^{-s})$ (*i.e.* the Hasse-Weil exponential series) is defined by

$$Z(X, T) := \exp\left(\sum_{r \geq 1} N(q^r)\frac{T^r}{r}\right).$$

One obtains

$$N(x) = \sum_{k=0}^{d} a_k x^k \implies \zeta_X(s) = \prod_{k=0}^{d} (s-k)^{-a_k}.$$

For instance, in the case of the projective line $\mathbb{P}^1$ one has $\zeta_{\mathbb{P}^1}(s) = \dfrac{1}{s(s-1)}$.

We say that an $\mathbb{F}_1$-scheme is Noetherian if the associated $\mathfrak{Mo}$- and $\mathbb{Z}$-schemes are Noetherian (see [**CC2**], Definition 4.12). An $\mathbb{F}_1$-scheme is said to be torsion free if the groups $\mathcal{O}_{X,x}^{\times}$ of the invertible elements of the monoids $\mathcal{O}_{X,x}$ (X is the associated geometric $\mathfrak{Mo}$-scheme) are torsion free $\forall x \in X$. The following result is related to Theorem 1 of [**D3**], but applies also to non-toric varieties.

THEOREM 4.3. *Let $\mathcal{X}$ be a torsion free, Noetherian $\mathbb{F}_1$-scheme and let X be the geometric realization of its restriction $\underline{X}$ to $\mathfrak{Mo}$. Then*
(a) There exists a polynomial $N(x + 1)$ with positive integral coefficients such that

$$\# X(\mathbb{F}_{1^n}) = N(n+1), \; \forall n \in \mathbb{N}.$$

(b) For each finite field $\mathbb{F}_q$, the cardinality of the set of points of the $\mathbb{Z}$-scheme $X_{\mathbb{Z}}$ which are rational over $\mathbb{F}_q$ is equal to $N(q)$.
(c) The zeta function of $\mathcal{X}$ has the following description

$$(43) \qquad \zeta_{\mathcal{X}}(s) = \prod_{x \in X} \frac{1}{\left(1 - \frac{1}{s}\right)^{\otimes n(x)}}$$

where $\otimes$ denotes Kurokawa's tensor product and $n(x)$ is the local dimension of X at the point x.

In (43), we use the convention that when $n(x) = 0$ the expression

$$\left(1 - \frac{1}{s}\right)^{\otimes n(x)} = s.$$

We refer to [**Ku**] and [**M**] for the details of the definition of Kurokawa's tensor products and zeta functions and to [**CC2**] for the proof of the theorem.

4.2. Extension of the Counting Functions in the Torsion Case

The remaining part of this section is dedicated to the computation of the zeta function of an arbitrary Noetherian $\mathbb{F}_1$-scheme $\mathcal{X}$. For simplicity, we will always denote by X the geometric realization (*i.e.* the geometric monoidal scheme) of the restriction $\underline{X}$ of $\mathcal{X}$ to $\mathfrak{Mo}$ and we will shortly refer to X as to the $\mathbb{F}_1$-scheme. In the general case, we shall prove that the counting function

$$\# X(\mathbb{F}_{1^n}) = N_X(n+1), \; \forall n \in \mathbb{N}$$

of an $\mathbb{F}_1$-scheme is no longer a polynomial function of n and that its description involves periodic functions. First, we will show that there is a *canonical* extension of the counting function $N_X(n)$ to the complex plane, as an entire function $N_X(z)$, whose growth is well controlled. Then we will explain how to compute the zeta function (42) using $N_X(q)$ for arbitrary real values of $q \geq 1$.

The simplest example of a Noetherian $\mathbb{F}_1$-scheme is $X = \mathrm{Spec}\,(\mathbb{F}_{1^m})$. The number of rational points of X over $\mathbb{F}_{1^n}$ is the cardinality of the set of group homomorphisms $\mathrm{Hom}(\mathbb{Z}/m\mathbb{Z}, \mathbb{Z}/n\mathbb{Z})$, *i.e.*

$$(44) \qquad \# X(\mathbb{F}_{1^n}) = \# \mathrm{Hom}(\mathbb{Z}/m\mathbb{Z}, \mathbb{Z}/n\mathbb{Z}) = \gcd(n, m).$$

This is a *periodic* function of n. More generally the following statement holds.

LEMMA 4.4. *Let X be a Noetherian $\mathbb{F}_1$-scheme. The counting function is a finite sum of monomials of the form*

$$(45) \qquad \# X(\mathbb{F}_{1^n}) = \sum_{x \in X} n^{n(x)} \prod_j \gcd(n, m_j(x))$$

where $n(x)$ denotes the local dimension at $x \in X$ and the natural numbers $m_j(x)$ are the orders of the finite cyclic groups which compose the residue "field" $\kappa(x) = \mathbb{F}_1[\mathcal{O}_{X,x}^\times]$, i.e.

$$\mathcal{O}_{X,x}^\times = \mathbb{Z}^{n(x)} \prod_j \mathbb{Z}/m_j(x)\mathbb{Z}.$$

PROOF. It follows from the proof of the first part of Theorem 4.3 that

$$\# X(\mathbb{F}_{1^n}) = \sum_{x \in X} \mathrm{Hom}_{\mathfrak{Ab}}\,(\mathcal{O}_{X,x}^\times, \mathbb{Z}/n\mathbb{Z})$$

and the result follows from (44). $\qquad\qquad\qquad\qquad\qquad\qquad\qquad\qquad\qquad$ □

We continue by explaining how to extend canonically a function such as (45) to an entire function on the complex plane.

4.2.1. Extension of Functions from $\mathbb{Z}$ to $\mathbb{C}$

Let $f(z)$ be an entire function of $z \in \mathbb{C}$ and let $T(R, f)$ be its *characteristic function*

$$T(R, f) = \frac{1}{2\pi} \int_0^{2\pi} \log^+ |f(Re^{i\theta})|\, d\theta.$$

This function may be interpreted as the average magnitude of $\log^+ |f(z)|$. One sets

$$\log^+ x = \begin{cases} \log x, & \text{if } x \geq 1 \\ 0, & \text{if } 0 < x < 1 \end{cases}$$

and one denotes by

$$N(a, R) = \sum_{\substack{f(z)=a \\ |z|<R}} \log \left|\frac{R}{z}\right|$$

the sum over the zeros of $(f - a)(z)$ inside the disk of radius R. The First Fundamental Theorem of Nevanlinna theory states as follows.

THEOREM 4.5. (*First Fundamental Theorem*) *If* $a \in \mathbb{C}$ *then*

$$N(a, R) \leq T(R, f) - \log |f(0) - a| + \epsilon(a, R)$$

where

$$|\epsilon(a, R)| \leq \log^+ |a| + \log 2 \,.$$

When $a = 0$ and f has a zero of order n at 0, the inequality applied to $F(z) = f(z)z^{-n}$ shows that for $R \geq 1$

$$N'(0, R) = \sum_{\substack{f(z)=0 \\ |z|<R,\ z \neq 0}} \log |\frac{R}{z}| \leq T(R, f) - \log |F(0)| + \log 2$$

since $T(R, f) \geq T(R, F)$ for $R \geq 1$. If an entire function $f(z)$ vanishes on $\mathbb{Z} \subset \mathbb{C}$, one has

$$N'(0, R) \geq \sum_{\substack{z \in \mathbb{Z} \setminus 0 \\ |z|<R}} \log |\frac{R}{z}| \sim 2R$$

and hence the above inequality shows that when $R \to \infty$

$$(46) \qquad\qquad \underline{\lim} \frac{T(R, f)}{R} \geq 2.$$

We shall now use this inequality to prove the following result.

PROPOSITION 4.6. *Let* $N(n)$ *be a real function on* $\mathbb{Z}$ *of the form*

$$(47) \qquad\qquad N(n) = \sum_{j \geq 0} T_j(n) \, n^j$$

where the $T_j(n)$ *are periodic functions with integral periods. Then there exists a unique polynomial* $Q(n)$ *and a unique entire function* $f : \mathbb{C} \to \mathbb{C}$ *such that*

- $\overline{f(\bar{z})} = f(z)$.
- $\overline{\lim} \frac{T(R,f)}{R} < 2$.
- $N(n) - (-1)^n Q(n) = f(n), \quad \forall\, n \in \mathbb{Z}$.

PROOF. Choose $L \in \mathbb{N}$ such that all the $T_j(n)$ fulfill $T_j(m + L) = T_j(m)$, for all $m \in \mathbb{Z}$. For each $j \in \mathbb{N}$, one has a Fourier expansion of the form

$$T_j(m) = \sum_{\alpha \in I_L} t_{j,\alpha} e^{2\pi i \alpha m} + s_j(-1)^m, \qquad I_L = \{\frac{k}{L} \mid |k| < L/2\},$$

where $t_{j,-\alpha} = \bar{t}_{j,\alpha}$ for $\alpha \in I_L$ and $s_j \in \mathbb{R}$. We define $Q(n) = \sum_j s_j n^j$ and the function $f(z)$ by

$$f(z) = \sum_j \sum_{\alpha \in I_L} t_{j,\alpha} e^{2\pi i \alpha z} z^j.$$

One has $\overline{f(\bar{z})} = f(z)$. Since $|\alpha| \leq c < \frac{1}{2}$, $\forall\, \alpha \in I_L$, one gets that

$$\log^+ |f(Re^{i\theta})| \leq 2\pi R c |\sin\theta| + o(R)$$

which in turn implies $\overline{\lim} \frac{T(R,f)}{R} < 2$, using $\int_0^{2\pi} |\sin\theta| d\theta = 4$ and $c < 1/2$. The equality $N(n) - (-1)^n Q(n) = f(n)$ for all $n \in \mathbb{Z}$ holds by construction and it remains to show the uniqueness. This amounts to showing that a non-zero entire function $f(z)$ such that $\overline{f(\bar{z})} = f(z)$ and $f(n) = (-1)^n Q(n)$ for all $n \in \mathbb{Z}$ and for some polynomial $Q(n)$ automatically fulfills (46). For $x \in \mathbb{R}$ one has $f(x) \in \mathbb{R}$. Let $s_k n^k$ be the leading term of $Q(n) = \sum s_j n^j$. Then outside a finite set of indices n the sign of $Q(n)$ is independent of n for $n > 0$ and for $n < 0$. It follows that the function f changes sign in each interval $[n, n+1]$ and hence admits at least one zero in such intervals. Thus,

$$N'(0, R) = \sum_{\substack{f(z)=0 \\ |z|<R, \ z \neq 0}} \log |\frac{R}{z}|$$

fulfills

$$\varliminf \frac{N(0, R)}{R} \geq 2$$

and hence, by Theorem 4.5 one gets

$$\varliminf \frac{T(R, f)}{R} \geq 2.$$

The uniqueness follows. $\qquad \square$

DEFINITION 4.7. *Let $N(n)$ be a real function on $\mathbb{Z}$ of the form (47). Then the canonical extension $N(z)$, $z \in \mathbb{C}$ is given by*

$$N(z) = f(z) + cos(\pi z)Q(z)$$

where f and Q are uniquely determined by Proposition 4.6.

We shall explicitly compute this extension in the case of the function $N_X(q)$ for $X = \mathrm{Spec}\,(\mathbb{F}_1[H])$ (see Figure 4).

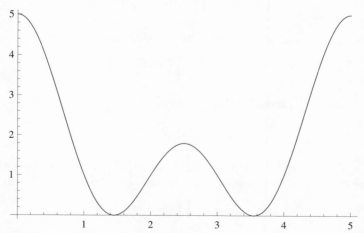

FIGURE 4 Graph of the canonical extension of $N(q)$, for $X = \mathrm{Spec}\,(\mathbb{F}_1[\mathbb{Z}/5\mathbb{Z}])$.

4.2.2. *The Counting Function for $X = Spec\,(\mathbb{F}_1[H])$*

Let H be a finitely generated abelian group. In this subsection we compute the counting function $N_X(n)$ for $X = \mathrm{Spec}\,(\mathbb{F}_1[H])$. We start by considering the cyclic case $H = \mathbb{Z}/m\mathbb{Z}$.

LEMMA 4.8. *Let $X = Spec\,(\mathbb{F}_{1^m})$. Then the canonical extension of the counting function $N_X(n)$ is given by*

$$(48) \qquad N(z) = \sum_{d|m} \frac{\varphi(d)}{d} \left(\sum_{|k|<d/2} e^{2i\pi\frac{(z-1)k}{d}} + \epsilon_d\,cos(\pi(z-1)) \right), \quad z \in \mathbb{C}$$

where $\epsilon_d = 1$ if $d \in \mathbb{Z}$ is even and $\epsilon_d = 0$ otherwise.

PROOF. It follows from (44) that $N(n) = \gcd(n-1, m)$. Moreover, one also knows that

$$\gcd(n,m) = \sum_{\substack{d|n \\ d|m}} \varphi(d)$$

where φ is the Euler totient function. Thus

$$(49) \qquad \gcd(n,m) = \sum_{d|m} \frac{\varphi(d)}{d} \sum e^{2i\pi\frac{nk}{d}}$$

where the sum $\sum e^{2i\pi\frac{nk}{d}}$ is taken over all characters of the additive group $\mathbb{Z}/d\mathbb{Z}$ evaluated on n mod. d. This sum can be written as

$$\sum e^{2i\pi\frac{nk}{d}} = \sum_{|k|<d/2} e^{2i\pi\frac{nk}{d}} + \epsilon_d\cos(\pi n)$$

which gives the canonical extension of $N(n)$ after a shift of one in the variable. $\square$

There is a conceptual explanation for the expansion (48). Let C_d be the dth cyclotomic polynomial

$$C_d(x) = \prod_{r|d} (x^{d/r} - 1)^{\mu(r)}.$$

One has the decomposition into irreducible factors over $\mathbb{Q}$

$$x^m - 1 = \prod_{d|m} C_d(x)$$

which determines the decomposition

$$\mathbb{Q}[\mathbb{Z}/m\mathbb{Z}] = \bigoplus_{d|m} \mathbb{K}_d$$

where $\mathbb{K}_d = \mathbb{Q}(\xi_d)$ is the cyclotomic field extension of $\mathbb{Q}$ by a primitive dth root of 1. This decomposition corresponds to a factorization of the zeta function

$$\zeta_X(s) = \prod_p \zeta_{X/\mathbb{F}_p}(s)$$

as a product

$$\zeta_X(s) = \prod_{d|m} \zeta_{X/\mathbb{K}_d}(s)$$

where each term of the product is in turn a product of L-functions

$$\zeta_{X/\mathbb{K}_d}(s) = \prod_{\chi \in \hat{G}(d)} L(s, \chi), \qquad G(d) = (\mathbb{Z}/d\mathbb{Z})^*$$

and the order of $G(d)$ is $\varphi(d)$. The decomposition

$$\zeta_X(s) = \prod_{d|m} \prod_{\chi \in \hat{G}(d)} L(s,\chi)$$

corresponds precisely to the decomposition (49).

Now let $H = \prod_j \mathbb{Z}/m_j\mathbb{Z}$ be a finite abelian group. Then the extension of the group ring $\mathbb{Z}[H]$ over $\mathbb{Q}$ can be equivalently interpreted as the tensor product

$$\mathbb{Q}[H] = \bigotimes_j (\bigoplus_{d|m_j} \mathbb{K}_d).$$

In fact, by applying the Galois correspondence, the decomposition of H into finite cyclic groups corresponds to the disjoint union of orbits for the action of the Galois group $\hat{\mathbb{Z}}^*$ on the cyclotomic extension $\mathbb{Q}^{\mathrm{ab}}/\mathbb{Q}$ (*i.e.* the maximal abelian extension of $\mathbb{Q}$). If we denote the group H additively, two elements $x, y \in H$ are on the same Galois orbit if and only if $\{nx \mid n \in \mathbb{Z}\} = \{ny \mid n \in \mathbb{Z}\}$, *i.e.* if and only if they generate the same subgroup of H. Thus, the set Z of orbits is the set of cyclic subgroups of H.

For a cyclic group C of order d we let, as in (48)

$$(50) \qquad N(z,C) = \frac{\varphi(d)}{d} \left(\sum_{|k|<d/2} e^{2i\pi \frac{(z-1)k}{d}} + \epsilon_d \cos(\pi(z-1)) \right).$$

Then we obtain the following lemma.

LEMMA 4.9. *Let* $\Gamma = \mathbb{Z}^k \times H$ *be a finitely generated abelian group (H is a finite abelian torsion group) and* $X = Spec\,(\mathbb{F}_1[\Gamma])$. *Then the canonical extension of the counting function* $N_X(q)$ *is given by*

$$(51) \qquad N(z) = (z-1)^k \sum_{C \subset H} N(z,C), \qquad z \in \mathbb{C}$$

where the sum is over the cyclic subgroups $C \subset H$.

PROOF. It is enough to show that the function $N(n+1)$ as in (51) gives the number of homomorphisms from Γ to $\mathbb{Z}/n\mathbb{Z}$. Indeed, as a function of z it is already in the canonical form of Definition 4.7. Then by duality we just need to compare the number $\iota(n,H)$ of homomorphisms ρ from $\mathbb{Z}/n\mathbb{Z}$ to H with $\sum_{C \subset H} N(n+1,C)$. The cyclic subgroup $C = \mathrm{Im}\,\rho$ is uniquely determined, thus

$$\iota(n,H) = \sum_{C \subset H} \mu(n,C)$$

where $\mu(n,C)$ is the number of surjective homomorphisms from $\mathbb{Z}/n\mathbb{Z}$ to C. If d denotes the order of C, this number is 0 unless $d|n$ and in the latter case it is $\varphi(d)$; thus using (50) it coincides in all cases with $N(n+1,C)$. $\square$

4.3. An Integral Formula for the Logarithmic Derivative of $\zeta_N(s)$

Let $N(q)$ be a real-valued continuous function on $[1, \infty)$ satisfying a polynomial bound $|N(q)| \leq Cq^k$ for some finite positive integer k and a fixed positive integer constant C. Then the associated generating function is

$$Z(q, T) = \exp\left(\sum_{r \geq 1} N(q^r) T^r / r\right)$$

and one knows that the power series $Z(q, q^{-s})$ converges for $\Re(s) > k$. The zeta function over $\mathbb{F}_1$ associated to $N(q)$ is defined as follows

$$\zeta_N(s) := \lim_{q \to 1} (q - 1)^\chi Z(q, q^{-s}), \quad \chi = N(1).$$

This definition requires some care to ensure its convergence. To eliminate the ambiguity in the extraction of the finite part, one works instead with the logarithmic derivative

$$\frac{\partial_s \zeta_N(s)}{\zeta_N(s)} = -\lim_{q \to 1} F(q, s)$$

where

$$F(q, s) = -\partial_s \sum_{r \geq 1} N(q^r) \frac{q^{-rs}}{r}.$$

One then has (see [**CC2**]) the following result.

LEMMA 4.10. *With the above notations and for $\Re(s) > k$*

$$\lim_{q \to 1} F(q, s) = \int_1^\infty N(u) u^{-s} d^* u, \quad d^* u = du/u$$

and

$$(52) \qquad \frac{\partial_s \zeta_N(s)}{\zeta_N(s)} = -\int_1^\infty N(u) u^{-s} d^* u.$$

4.4. Analyticity of the Integrals

The statement of Lemma 4.10 and the explicit form of the function $N(q)$ as in Lemma 4.8 clearly indicate that one needs to analyze basic integrals of the form

$$(53) \qquad f(s, a) = \int_1^\infty e^{iau} u^{-s} d^* u.$$

LEMMA 4.11. *For $a > 0$*

$$(54) \qquad f(s, a) = a^s \int_a^\infty e^{iu} u^{-s} d^* u$$

defines an entire function of $s \in \mathbb{C}$.

PROOF. For $\Re(s) > 0$, the integral in (54) is absolutely convergent. Integrating by parts one obtains

$$\int_a^\infty e^{iu} u^{-s-1} du + \int_a^\infty \frac{1}{i} e^{iu}(-s-1) u^{-s-2} du = [\frac{1}{i} e^{iu} u^{-s-1}]_a^\infty$$

which supplies the equation

(55) $$af(s,a) = -i(s+1)f(s+1,a) + ie^{ia}.$$

This shows that f is an entire function of s, since after iterating (55) the variable s can be moved to the domain of absolute convergence. □

The function $f(s,a)$ as in (53) can be expressed in terms of hypergeometric functions in the following way

$$f(s,a) = e^{-\frac{1}{2}i\pi s} a^s \Gamma(-s) + i\frac{a}{s-1} H\left[\frac{1-s}{2}, (\frac{3}{2}, \frac{3-s}{2}), -\frac{a^2}{4}\right]$$
$$+ \frac{1}{s} H\left[-\frac{s}{2}, (\frac{1}{2}, \frac{2-s}{2}), -\frac{a^2}{4}\right]$$

where the hypergeometric function H is defined by

$$H(\alpha, (\beta, \gamma), z) = \sum_{k=0}^\infty \frac{(\alpha)_k}{(\beta)_k (\gamma)_k} \frac{z^k}{k!}.$$

One sets $(x)_k = \prod_{j=0}^{k-1} (x+j)$. Note that the formulas simplify as follows

$$\frac{1}{s-1} H\left[\frac{1-s}{2}, (\frac{3}{2}, \frac{3-s}{2}), -\frac{a^2}{4}\right] = \sum_{k=0}^\infty \frac{(-a^2)^k}{(2k+1)!(s-(2k+1))}$$

and

$$\frac{1}{s} H\left[-\frac{s}{2}, (\frac{1}{2}, \frac{2-s}{2}), -\frac{a^2}{4}\right] = \sum_{k=0}^\infty \frac{(-a^2)^k}{(2k)!(s-2k)}.$$

By using (53) and expanding the exponential in powers of a in the integral

$$\int_0^1 e^{iau} u^{-s} d^*u = \sum_{n=0}^\infty \int_0^1 \frac{(iau)^n}{n!} u^{-s} d^*u$$

the resulting expression for $f(s,a)$ in terms of hypergeometric functions becomes

(56) $$f(s,a) = e^{-\frac{1}{2}i\pi s} a^s \Gamma(-s) + \sum_{n=0}^\infty \frac{(ia)^n}{n!(s-n)}.$$

Thus one gets the following lemma.

LEMMA 4.12. *For $a > 0$, the following identity holds*

$$f(s,a) = e^{-\frac{1}{2}i\pi s} a^s \Gamma(-s) - \sum_{poles} \frac{residue}{s-n}.$$

PROOF. The last term of (56) is a sum of the form

$$\sum_{\text{poles } g} \frac{\text{residue}(g)}{s - n}$$

where $g(s) = e^{-\frac{1}{2}i\pi s} a^s \Gamma(-s)$. This suffices to conclude the proof since one knows from Lemma 4.11 that $f(s, a)$ is an entire function of s. One can also check directly that the expression for the residues in (56) is correct by using *e.g.* the formula of complements.   □

4.5. Zeta Function of Noetherian $\mathbb{F}_1$-schemes

The above discussion has shown that the elementary constituents of the zeta function of a Noetherian $\mathbb{F}_1$-scheme are of the following type

$$(57) \qquad \xi_d(s) = s^{-\frac{\varphi(d)}{d}} e^{-\frac{\varphi(d)}{d} H_d(s)}$$

where $H_d(s)$ is the primitive, vanishing for $\Re(s) \to \infty$, of the entire function

$$(58) \qquad \partial_s H_d(s) = \sum_{\substack{|k| \leq d/2 \\ k \neq 0}} e^{-i\frac{2\pi k}{d}} f\left(s, \frac{2\pi k}{d}\right).$$

(See Figure 5.) When d is even, one divides by 2 the contribution of $k = \frac{d}{2}$. Note that, by Lemma 4.12, the function $\partial_s H_d(s)$ is obtained by applying the transformation

$$F \mapsto F - \sum_{\text{poles}} \frac{\text{residue}}{s - n}$$

to the function

$$F_d(s) = \Gamma(-s) \sum_{\substack{|k| \leq d/2 \\ k \neq 0}} \cos\left(\frac{2\pi k}{d} + \frac{\pi s}{2}\right)\left(2\pi \frac{k}{d}\right)^s.$$

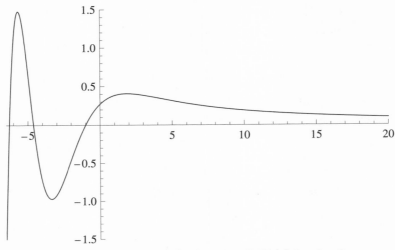

FIGURE 5 Graph of the function $\partial_s H_d(s)$ for $d = 3$.

THEOREM 4.13. *Let X be a Noetherian $\mathbb{F}_1$-scheme. Then the zeta function of X over $\mathbb{F}_1$ is a finite product*

$$\zeta_X(s) = \prod_{x \in X} \zeta_{X_x}(s) \tag{59}$$

where X_x denotes the constituent of X over the residue field $\kappa(x) = \mathbb{F}_1[\mathcal{O}_{X,x}^*]$ and $\mathcal{O}_{X,x}^* = \Gamma$ is a finitely generated abelian group such as $\Gamma = H \times \mathbb{Z}^n$ (H = finite torsion group). The zeta function of each constituent X_x takes the form

$$\zeta_{X_x}(s) = \prod_{k \geq 0} \zeta(\mathbb{F}_1[H], s - n + k)^{(-1)^k \binom{n}{k}} \tag{60}$$

where $\zeta(\mathbb{F}_1[H], s)$ is the product

$$\zeta(\mathbb{F}_1[H], s) = \prod_{d \mid |H|} \xi_d(s)^{\gamma(H,d)}$$

where $\gamma(H, d)$ denotes the number of cyclic subgroups $C \subset H$ of order d.

PROOF. The proof of (59) follows the same lines as that given in Theorem 4.3. The expression (60) for $\zeta_{X_x}(s)$ in terms of $\zeta(\mathbb{F}_1[H], s)$ follows from (51) and the simple translation in s that the multiplication of the counting function by a power of q generates using Lemma 4.10. By (51) and (50) it remains to show that the zeta function $Z_d(s)$ associated to the counting function (50) coincides with $\xi_d(s)$ defined in (57). By Lemma 4.10 and (52) the logarithmic derivative of $Z_d(s)$ is

$$\frac{\partial_s Z_d(s)}{Z_d(s)} = -\int_1^\infty \frac{\varphi(d)}{d} \left(\sum_{|k| < d/2} e^{2i\pi \frac{(u-1)k}{d}} + \epsilon_d \cos(\pi(u-1)) \right) u^{-s} d^* u.$$

The term coming from $k = 0$ contributes $-\frac{\varphi(d)}{d} \frac{1}{s}$ and thus corresponds to the fractional power $s^{-\frac{\varphi(d)}{d}}$ in (57). The other terms contribute integrals of the form

$$I(s, \alpha) = \int_1^\infty e^{2\pi i \alpha (u-1)} u^{-s} d^* u.$$

One has

$$I(s, \alpha) = e^{-2\pi i \alpha} \int_1^\infty e^{2\pi i \alpha u} u^{-s} d^* u = e^{-2\pi i \alpha} (2\pi \alpha)^s \int_{2\pi \alpha}^\infty e^{iv} v^{-s} d^* v.$$

Thus with the notation (54) one obtains

$$I(s, \alpha) = e^{-ia} f(s, a), \quad a = 2\pi \alpha.$$

The equality $Z_d(s) = \xi_d(s)$ follows using (58). □

COROLLARY 4.14. *Let X be a Noetherian $\mathbb{F}_1$-scheme. Then the zeta function of X is the product*

$$\zeta_X(s) = e^{h(s)} \prod_{j=0}^n (s - j)^{\alpha_j}$$

of an exponential of an entire function $h(s)$ by a finite product of fractional powers of simple monomials. The exponents α_j are rational numbers given explicitly by

$$(61) \qquad \alpha_j = (-1)^{j+1} \sum_{x \in X} (-1)^{n(x)} \binom{n(x)}{j} \sum_d \frac{1}{d} \nu(d, \mathcal{O}_{X,x}^\times)$$

where $n(x)$ is the local dimension (i.e. the rank of $\mathcal{O}_{X,x}^\times$), and $\nu(d, \mathcal{O}_{X,x}^\times)$ is the number of injective homomorphisms from $\mathbb{Z}/d\mathbb{Z}$ to $\mathcal{O}_{X,x}^\times$.

PROOF. By (57) one has

$$\zeta(\mathbb{F}_1[H], s) = \prod_{d || H|} \xi_d(s)^{\gamma(H,d)} = s^{-\epsilon_H} e^{-\sum_{d || H|} \frac{\varphi(d)\gamma(H,d)}{d}} H_d(s)$$

where

$$\epsilon_H = \sum_{d || H|} \frac{\varphi(d)\gamma(H,d)}{d}.$$

Next, by (60) one has

$$\zeta(\mathbb{F}_1[\Gamma], s) = \prod_{k \geq 0} \zeta(\mathbb{F}_1[H], s - n + k)^{(-1)^k \binom{n}{k}}$$

which is a product of the form

$$\zeta(\mathbb{F}_1[\Gamma], s) = \prod_{j \geq 0} (s - j)^{-(-1)^{n-j}\binom{n}{j}\epsilon_H} e^{h(s)}$$

where $h(s)$ is an entire function. Thus one gets that the exponent α_j of $(s - j)$ is

$$\alpha_j = (-1)^{j+1} \sum_{x \in X} (-1)^{n(x)} \binom{n(x)}{j} \sum_{d || H_x|} \frac{\varphi(d)\gamma(H_x, d)}{d}$$

where $n(x)$ is the local dimension and H_x is the finite group in the decomposition of $\mathcal{O}_{X,x}^\times$. The above expression gives (61) since $\varphi(d)\gamma(H_x, d)$ is the number of injective homomorphisms from $\mathbb{Z}/d\mathbb{Z}$ to $\mathcal{O}_{X,x}^\times$. $\qquad \square$

5. Beyond $\mathbb{F}_1$-schemes

One may naturally wonder about the adeptness of the method of the computation of the zeta function over $\mathbb{F}_1$ that we have described in the above section, for a possible generalization beyond the case of $\mathbb{Z}$-schemes associated to Noetherian $\mathbb{F}_1$-schemes. In [**CC2**], we have already given evidence on the effectiveness of the integral formula of Lemma 4.10 of §4.3, by showing that there exists a *uniquely* defined distribution $N(u)$ on $[1, \infty)$ which describes the counting function for the hypothetical curve $C = \overline{\operatorname{Spec} \mathbb{Z}}$ over $\mathbb{F}_1$ whose zeta function $\zeta_C(s)$ (over $\mathbb{F}_1$) coincides with the completed Riemann zeta function $\zeta_{\mathbb{Q}}(s) = \pi^{-s/2}\Gamma(s/2)\zeta(s)$ over $\mathbb{C}$.

The goal of this section is to show that as long as one is only interested in the singularities of the zeta function, one may, in the case of Noetherian $\mathbb{F}_1$-schemes, replace the integral which appears in (52) by the corresponding infinite sum, *i.e.*

$$(62) \qquad -\int_1^\infty N(u) u^{-s} d^* u \rightsquigarrow -\sum_{n \geq 1} N(n) n^{-s-1}$$

which is now defined without any need for interpolation. This substitution provides one with a natural definition of a modified zeta function that is defined by the formula

$$\frac{\partial_s \zeta_N^{\text{disc}}(s)}{\zeta_N^{\text{disc}}(s)} = -\sum_{n=1}^{\infty} N(n)\, n^{-s-1}.$$

In §5.3 we compute this function for a *particular* choice of the extension of the counting function $N(q)$ in the case of elliptic curves over $\mathbb{Q}$.

5.1. Modified Zeta Function

Within the category of the Noetherian $\mathbb{F}_1$-schemes, one has the following result.

PROPOSITION 5.1. *The replacement (62) in the definition of the logarithmic derivative $\frac{\partial_s \zeta_N(s)}{\zeta_N(s)}$ in (52) does not alter the singularities of the zeta function associated to a Noetherian $\mathbb{F}_1$-scheme. In fact, the ratio of these two zeta functions $\zeta_N(s)$ and $\zeta_N^{\text{disc}}(s)$ is the exponential of an entire function.*

PROOF. It is enough to treat the case of a counting function $N(u)$, of the real variable u, of the form

$$N(u) = \frac{\varphi(d)}{d}\, u^{\ell} \left(\sum_{|k|<d/2} e^{2i\pi \frac{(u-1)k}{d}} + \epsilon_d \cos(\pi(u-1)) \right)$$

and show that the following is an entire function of $s \in \mathbb{C}$

$$h(s) = \int_1^{\infty} N(u)\, u^{-s} d^* u - \sum_{n=1}^{\infty} N(n)\, n^{-s-1}.$$

Notice that the term u^{ℓ} generates a shift in s and thus we can assume that $\ell = 0$. Let $G(d) = (\mathbb{Z}/d\mathbb{Z})^*$ be the multiplicative group of residues modulo d which are prime to d. By extending the characters $\chi \in \hat{G}(d)$ by 0 on residues modulo d which are not prime to d, one obtains the equality

$$\sum_{\chi \in \hat{G}(d)} \chi(n) = \begin{cases} \varphi(d), & \text{if } n = 1 \\ 0, & \text{if } n \neq 1. \end{cases}$$

Thus the restriction of $N(u)$ to the integers agrees with the following sum

$$N(n) = \sum_{\chi \in \hat{G}(d)} \chi(n)$$

and one gets

$$\sum_{n \geq 1} N(n)\, n^{-s-1} = \sum_{\chi \in \hat{G}(d)} L(\chi, s+1).$$

When $\chi \neq 1$, the function $L(\chi, s+1)$ is entire, thus the only singularity arises from the function $L(1, s+1)$ which is known to have a unique pole at $s = 0$ with residue $\prod_{p|d}(1 - \frac{1}{p})$. The only singularity of the integral $\int_1^{\infty} N(u)\, u^{-s} d^* u$ is due to the contribution of the constant term, *i.e.*

$$\int_1^{\infty} \frac{\varphi(d)}{d}\, u^{-s} d^* u = \frac{\varphi(d)}{d} \frac{1}{s}.$$

Thus the equality $\prod_{p|d}(1 - \frac{1}{p}) = \frac{\varphi(d)}{d}$ shows that the function $h(s)$ is entire.   □

We can therefore adopt the following definition.

DEFINITION 5.2. *Let X be an $\mathbb{F}_1$-scheme. The modified zeta function $\zeta_X^{\mathrm{disc}}(s)$ is defined by*

(63)
$$\frac{\partial_s \zeta_X^{\mathrm{disc}}(s)}{\zeta_X^{\mathrm{disc}}(s)} = - \sum_{n \geq 1} N(n)\, n^{-s-1}$$

where $N(n+1) = \#X(\mathbb{F}_{1^n})$.

By Proposition 5.1, the singularities of this modified zeta function are the same as the singularities of $\zeta_X(s)$: see Corollary 4.14. The advantage of this more general definition is that it requires no choice of an interpolating function. This means that one can define $\zeta_X^{\mathrm{disc}}(s)$, up to a multiplicative normalization factor, by just requiring some polynomial control on the size of growth of the finite set $X(\mathbb{F}_{1^n})$.

REMARK 5.3. There is a superficial resemblance between the right hand side of (63) that defines the logarithmic derivative of $\zeta_X^{\mathrm{disc}}(s)$ and the "absolute" Igusa zeta function associated in [**DKK**] to an $\mathbb{F}_1$-scheme. However, there is a shift of 1 in the argument n which is not easy to eliminate and that introduces a fundamental difference. Thus in terms of points defined over $\mathbb{F}_{1^n}$ we obtain

$$\sum_{n \geq 1} N(n)\, n^{-s-1} = \sum_{n \geq 1} X(\mathbb{F}_{1^{n-1}})\, n^{-s-1}.$$

On the other hand, the zeta function associated in [**DKK**] is given by

$$\zeta^I(s, X) = \sum_{n \geq 1} X(\mathbb{F}_{1^n})\, n^{-s}.$$

The shift of 1 in s is easily accounted for, but not the replacement of $X(\mathbb{F}_{1^{n-1}})$ by $X(\mathbb{F}_{1^n})$. In the next subsection we will show how this shift manifests itself in a concrete example.

5.2. Modified Zeta Function and $\overline{\mathrm{Spec}\,\mathbb{Z}}$

In [**CC2**], we have shown how to determine the counting distribution $N(q)$, defined for $q \in [1, \infty)$ in such a way that the zeta function in (52) gives the complete Riemann zeta function, *i.e.* so that the following equation holds

(64)
$$\frac{\partial_s \zeta_{\mathbb{Q}}(s)}{\zeta_{\mathbb{Q}}(s)} = - \int_1^\infty N(u)\, u^{-s} d^* u$$

where

$$\zeta_{\mathbb{Q}}(s) = \pi^{-s/2} \Gamma(s/2) \zeta(s).$$

Moreover, using convergence in the sense of distributions one obtains

$$N(q) = q - \sum_{\rho \in Z} \mathrm{order}(\rho)\, q^\rho + 1.$$

This is in perfect analogy with the theory for function fields. Indeed, for a smooth curve X over a finite field $\mathbb{F}_{q_0}$ and for any power $q = q_0^\ell$ of the order q_0 of the finite field of constants one has

$$\#X(\mathbb{F}_q) = N(q) = q - \sum_{\rho \in Z} \operatorname{order}(\rho)\, q^\rho + 1.$$

In the above formula, we have written the complex eigenvalues of the characteristic polynomial of the Frobenius operator acting on the cohomology of the curve in the form $\alpha = q_0^\rho$.

While these results supply a strong indication on the coherence of the quest for an arithmetic theory over $\mathbb{F}_1$, they also indicate that the similarities with the geometry over finite fields are elusive and that the $\mathbb{F}_1$-counterpart of the classical theory will be enriched with new ingredients. In this section, we will study what type of additional information one can gather about the hypothetical curve $C = \overline{\operatorname{Spec}\,\mathbb{Z}}$ by using the modified zeta function, *i.e.* the simplified formula (72). Since this simplified formula clearly neglects the contribution of the archimedean place, it seems natural to replace the complete zeta function $\zeta_{\mathbb{Q}}(s)$ by the Riemann zeta function $\zeta(s)$. Thus, the simplified form of (64) takes the form

$$\frac{\partial_s \zeta(s)}{\zeta(s)} = -\sum_1^\infty N(n)\, n^{-s-1}.$$

This supplies for the counting function $N(n)$ the formula

$$N(n) = n\Lambda(n)$$

where $\Lambda(n)$ is the von-Mangoldt function with value $\log p$ for powers p^ℓ of primes and zero otherwise. We now use the relation $N(n+1) = \#X(\mathbb{F}_{1^n})$ of Definition 5.2 and conclude that the sought-for "scheme over $\mathbb{F}_1$" should fulfill the following condition

$$(65) \qquad \#(X(\mathbb{F}_{1^n})) = \begin{cases} 0, & \text{if } n+1 \text{ is not a prime power} \\ (n+1)\log p, & \text{if } n = p^\ell - 1, \ p \text{ prime.} \end{cases}$$

The two obvious difficulties are that $(n+1)\log p$ is not an integer, and that multiples $m = nk$ of numbers of the form $n = p^\ell - 1$ are no longer of a similar form, while there are morphisms from $\mathbb{F}_{1^n}$ to $\mathbb{F}_{1^m}$. This shows that functoriality can only hold in a more restrictive manner. Moreover, at first sight, it seems hopeless to find a natural construction starting from a monoid M which singles out among the monoids $M = \mathbb{F}_1[\mathbb{Z}/n\mathbb{Z}] = \mathbb{Z}/n\mathbb{Z} \cup \{0\}$ those for which $n = p^\ell - 1$, for some prime power. However, this is precisely what is achieved by the study of the additive structure in §2.1. Indeed, one has the following result.

COROLLARY 5.4. *Let $M = \mathbb{F}_1[\mathbb{Z}/n\mathbb{Z}] = \mathbb{Z}/n\mathbb{Z} \cup \{0\}$ viewed as a commutative (pointed) monoid. Then $n = p^\ell - 1$ for a prime power p^ℓ if and only if the following set $A(M)$ is non-empty. $A(M)$ is the set of maps $s : M \to M$ such that*

- *s commutes with its conjugates by multiplication by elements of $M^\times$*
- *$s(0) = 1$.*

PROOF. This would follow from Corollary 2.2 if we assumed that s is bijective. So, we need to prove that this assumption is unnecessary. The proof of Theorem 2.1 shows that, for $s \in A(M)$, M becomes a semi-field, where we use (10) to define the addition. Moreover, since M is finite, if it is not a field, M contains a (prime) semi-ring $B(n, i)$ for finite n and i. As explained after Proposition 2.6, this implies $n = 2$ and $i = 1$ so that M is an idempotent semi-field. Finally, by Proposition 2.13 one gets that unless M is a field it is equal to $\mathbb{B}$. Thus we obtain that the cardinality of M is a prime power.   □

The cardinality of the set $A(M)$ can be explicitly computed. By using Corollary 2.2 and the fact that the automorphisms of the finite field $\mathbb{F}_q$ are powers of the Frobenius, one gets for[6] $n > 1$

$$\#(A(\mathbb{F}_{1^n})) = \begin{cases} 0, & \text{if } n+1 \text{ is not a prime power} \\ \varphi(p^\ell - 1)/\ell, & \text{if } n = p^\ell - 1, \ p \text{ prime} \end{cases}$$

where φ is the Euler totient function. This produces a result which, although not exactly equal to the (non-integer) value as in (65), comes rather close. Moreover, one should acknowledge the fact that even though the definition of $A(M)$ is natural it is only functorial for isomorphisms and not for arbitrary morphisms.

5.3. Elliptic Curves

Definition 5.2 applies any time one has a reasonable guess for the definition of an extension of the counting function of points from the case $q = p^\ell$ to the case of an arbitrary positive integer. For elliptic curves E over $\mathbb{Q}$, one may use the associated modular form

$$F_E(\tau) = \sum_{n=1}^{\infty} a(n)q^n, \qquad q = e^{2\pi i \tau}$$

since the sequence $a(n)$ of the coefficients of the above series is defined for all values of n. Moreover, these Fourier coefficients also fulfill equations

(66) $$N(p, E) = \#E(\mathbb{F}_p) = p + 1 - a(p)$$

at each prime number p of good reduction for the curve E. This choice is, however, a bit too naive since (66) does not continue to hold at primes powers.

We say that a function $t(n)$ is weakly multiplicative when

$$t(nm) = t(n)t(m) \text{ if } n \text{ is relatively prime to } m.$$

We define the function $t(n)$ as the only weakly multiplicative function which agrees with $a(p)$ at each prime and for which (66) continues to hold for prime powers. Then we have the following result.

LEMMA 5.5. Let $t(m)$ be the only weakly multiplicative function such that $t(1) = 1$ and

(67) $$N(q, E) = q + 1 - t(q)$$

---

[6]One has $\#(A(\mathbb{F}_1)) = 2$.

for any prime power q. Then the generating function

$$R(s, E) = \sum_{n \geq 0} t(n) \, n^{-s}$$

has the following description

(68) $$R(s, E) = \frac{L(s, E)}{\zeta(2s - 1)M(s)}.$$

$L(s, E)$ *is the L-function of the elliptic curve,* $\zeta(s)$ *is the Riemann zeta function, and*

$$M(s) = \prod_{p \in S} (1 - p^{1-2s})$$

is a finite product of simple factors over the set S of primes of bad reduction for E.

PROOF. One has an Euler product for the L-function $L(s, E) = \sum a(n)n^{-s}$ of the elliptic curve of the form

$$L(s, E) = \prod_{p} L_p(s, E).$$

For almost all primes, *i.e.* for $p \notin S$ with S a finite set

(69) $$L_p(s, E) = \left((1 - \alpha_p p^{-s})(1 - \bar{\alpha}_p p^{-s}) \right)^{-1} = \sum_{n=0}^{\infty} a(p^n) p^{-ns}$$

where $\alpha_p = \sqrt{p} \, e^{i\theta_p}$. Since the function $t(n)$ is weakly multiplicative one has an Euler product

(70) $$R(s, E) = \sum_{n \geq 0} t(n) \, n^{-s} = \prod_{p} \left(\sum_{n \geq 0} t(p^n) p^{-ns} \right) = \prod_{p} R_p(s, E).$$

Let $p \notin S$ be a prime of good reduction. Then for $q = p^\ell$, one has

$$N(q, E) = q + 1 - \alpha_p^\ell - \bar{\alpha}_p^\ell.$$

We now show that

(71) $$(1 - p^{1-2s}) \sum_{n=0}^{\infty} a(p^n) p^{-ns} = \sum_{n=0}^{\infty} t(p^n) p^{-ns}.$$

This relation follows from (69) and the equality

$$\frac{(1 - p x^2)}{(1 - \alpha_p x)(1 - \bar{\alpha}_p x)} = \frac{1}{(1 - \alpha_p x)} + \frac{1}{(1 - \bar{\alpha}_p x)} - 1$$

for $x = p^{-s}$. The above formula appears in the expression of the Poisson kernel (see [**R**], §5.24, and §11.5)

$$P_r(\theta) = \sum_{-\infty}^{\infty} r^{|n|} e^{in\theta} = \frac{1 - r^2}{1 - 2r\cos\theta + r^2}.$$

Using (71) one gets

$$(72) \qquad R_p(s, E) = (1 - p^{1-2s}) \sum_{n=0}^{\infty} a(p^n) p^{-ns}.$$

When $p \in S$, the local factor $L_p(s, E)$ of the L-function takes one of the following descriptions (see [**Si**], Appendix C, §16):

- $L_p(s, E) = \sum_{n=0}^{\infty} p^{-ns}$ when E has split multiplicative reduction at p.

- $L_p(s, E) = \sum_{n=0}^{\infty} (-1)^n p^{-ns}$ when E has non-split multiplicative reduction at p.

- $L_p(s, E) = 1$ when E has additive reduction at p.

We prove now that in any of the above cases one has

$$(73) \qquad R_p(s, E) = L_p(s, E).$$

One knows that in the singular case, the elliptic curve has exactly one singular point. This point is already defined over $\mathbb{F}_p$ since if a cubic equation has a double root this root is rational (see §6). Thus for any power $q = p^\ell$ one has

$$N(q, E) = 1 + N_{\mathrm{ns}}(q, E)$$

where the cardinality $N_{\mathrm{ns}}(q, E)$ of the set of the non-singular points is given, for any power $q = p^\ell$, as follows (see [**Si**], exercise 3.5, p. 104):

- $N_{\mathrm{ns}}(q, E) = q - 1$, if E has split multiplicative reduction over $\mathbb{F}_q$.
- $N_{\mathrm{ns}}(q, E) = q + 1$, if E has non-split multiplicative reduction over $\mathbb{F}_q$.
- $N_{\mathrm{ns}}(q, E) = q$, if E has additive reduction over $\mathbb{F}_q$.

Thus it follows from the definition (67) that the value $t(p^\ell)$ assumes one of the following three descriptions:

- $t(p^\ell) = 1$ for all ℓ, if E has split multiplicative reduction at p.
- $t(p^\ell) = (-1)^\ell$, if E has non-split multiplicative reduction at p.
- $t(p^\ell) = 0$, if E has additive reduction at p.

The second case arises since the split property of E depends upon the parity of ℓ. The equality (73) is checked for all primes of bad reduction $p \in S$. Together with (72) and (70), this implies (68). □

REMARK 5.6. Note that the function $N(n) = n + 1 - t(n)$ is always positive. Indeed, it is enough to show that for all n one has $|t(n)| \leq n$. Since $t(n)$ is weakly multiplicative, it is enough to show the above inequality when $n = p^k$ is a prime power. In the case of good reduction, one has $t(p^k) = \alpha_p^k + \bar{\alpha}_p^{\,k}$ and since the modulus of α_p is $\sqrt{p}$ one gets

$$|t(q)| \leq 2\sqrt{q}, \qquad q = p^k.$$

This proves that $|t(q)| \leq q$ for all q satisfying $2\sqrt{q} \leq q$, *i.e.* when $q \geq 4$. For $q = 2$ one has $2\sqrt{2} \cong 2.828 < 3$ and since $|t(2)|$ is an integer one obtains $|t(2)| \leq 2$. Similarly, for $q = 3$ one has $2\sqrt{3} \cong 3.4641 < 4$ and thus $|t(3)| \leq 3$. In the case of bad reduction one has always $|t(q)| \leq 1$.

Definition 5.2 gives the following equation for the modified zeta function $\zeta_N^{\mathrm{disc}}(s)$ associated to the counting function $N(q, E)$

$$\frac{\partial_s \zeta_N^{\mathrm{disc}}(s)}{\zeta_N^{\mathrm{disc}}(s)} = -\sum_{n=1}^{\infty} (n + 1 - t(n))\, n^{-s-1} = -\zeta(s+1) - \zeta(s) + R(s+1, E).$$

Using Lemma 5.5 we then obtain the following theorem.

THEOREM 5.7. *Let E be an elliptic curve over $\mathbb{Q}$ and let $L(s, E)$ be the associated L-function. Then the modified zeta function $\zeta_N^{\mathrm{disc}}(s)$ of E fulfills the following equality*

$$\frac{\partial_s \zeta_N^{\mathrm{disc}}(s)}{\zeta_N^{\mathrm{disc}}(s)} = -\zeta(s+1) - \zeta(s) + \frac{L(s+1, E)}{\zeta(2s+1) M(s+1)}$$

where $\zeta(s)$ is the Riemann zeta function.

The Riemann zeta function has trivial zeros at the points $-2n$ for $n \geq 1$. They generate singularities of $\zeta_E(s)$ at the points $s = -n - \frac{1}{2}$, *i.e.* at $s = -\frac{3}{2}, -\frac{5}{2}, \ldots$. If RH holds, the non-trivial zeros of the Riemann zeta function generate singularities at points which have real part $-\frac{1}{4}$. Note that the poles of the archimedean local factor $N_E^{s/2} (2\pi)^{-s} \Gamma(s)$ ($N_E = $ conductor) of the L-function of E determine trivial zeros. These zeros occur for $s \in -\mathbb{N}$ and do not cancel the above singularities. Finally, we have all the zeros of $M(s+1)$. Each factor $(1 - p^{-(2s-1)})$ contributes by an arithmetic progression with real part $-\frac{1}{2}$

$$s \in -\frac{1}{2} + \bigcup_{p \in S} \frac{\pi i}{\log p} \mathbb{Z}.$$

Note that the pole at $-\frac{1}{2}$ has order the cardinality of S.

For a better understanding of the role played by the primes of bad reduction we consider the following example.

EXAMPLE 5.8. Let us consider the elliptic curve in $\mathbb{P}^2(\mathbb{Q})$ given by the zeros of the homogeneous equation

$$E: \quad Y^2 Z + Y Z^2 = X^3 - X^2 Z - 10 X Z^2 - 20 Z^3.$$

Then the sequence $a(n)$ is given by the coefficients of the modular form

$$F_E(q) = \sum_{n=1}^{\infty} a(n) q^n = q \prod_{n=1}^{\infty} (1 - q^n)^2 (1 - q^{11n})^2.$$

These coefficients are easy to compute, for small values of n, using the Euler formula

$$\prod_{n=1}^{\infty} (1 - q^n) = 1 + \sum_{n=1}^{\infty} (-1)^n (q^{(3n^2 - n)/2} + q^{(3n^2 + n)/2}).$$

The first values are

$$a(0) = 0, a(1) = 1, a(2) = -2, a(3) = -1, a(4) = 2, a(5) = 1, a(6) = 2, a(7) = -2, \ldots.$$

The function $N(n)$ is given by $N(n) = n + 1 - t(n)$. It is important to check that it takes positive values. Its graph is depicted in Figure 6. In this example, the only prime of bad reduction is $p = 11$ and the singularities of the modified zeta function $\zeta_N^{\mathrm{disc}}(s)$ are described in Figure 7.

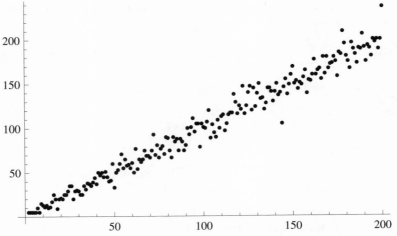

FIGURE 6 Graph of the function $N(n) = n + 1 - t(n)$ for $n \leq 200$, for the elliptic curve in Example 5.8.

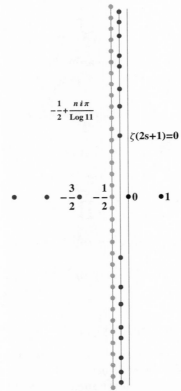

FIGURE 7 Singularities of $\zeta_N^{\mathrm{disc}}(s)$.

The challenge is to define a natural functor $\underline{E}$ to finite sets, such that $\#\underline{E}(\mathbb{F}_{1^n}) = N(n+1)$ for all n.

6. Appendix: Primes of Bad Reduction of an Elliptic Curve

In this appendix we supply more details on the process of counting points used in the proof of Lemma 5.5. In order to reduce modulo p an elliptic curve E over $\mathbb{Q}$, the first step is to take an equation of the curve which is in minimal Weierstrass form over the local field $\mathbb{Q}_p$, *i.e.* of the form

$$y^2 + a_1 xy + a_3 y = x^3 + a_2 x^2 + a_4 x + a_6$$

with $a_j \in \mathbb{Z}_p$ such that the power of p dividing the discriminant is minimal. Such a minimal Weierstrass equation can always be found (see [**Si**], Proposition VII, §1.3) and is unique up to a change of variables of the form

$$x = u^2 x' + r, \quad y = u^3 y' + u^2 s x' + t$$

for $u \in \mathbb{Z}_p^*$ and $r, s, t \in \mathbb{Z}_p$. One then considers the reduced equation

(74)
$$y^2 + \bar{a}_1 xy + \bar{a}_3 y = x^3 + \bar{a}_2 x^2 + \bar{a}_4 x + \bar{a}_6$$

where $\bar{a}_j$ is the residue of a_j modulo p. This equation is unique up to the standard changes of coordinates on $\mathbb{F}_p$ and defines a curve over $\mathbb{F}_p$. One then looks for points of this curve, *i.e.* solutions of (74) (together with the point at ∞), in the extensions $\mathbb{F}_q$, $q = p^\ell$. When $p > 2$, one can always write the reduced equation in the form

$$y^2 = x^3 + \bar{a}_2 x^2 + \bar{a}_4 x + \bar{a}_6.$$

Thus assuming $p \neq 2$, a singular point has second coordinate $y = 0$ while x is the common root of $x^3 + \bar{a}_2 x^2 + \bar{a}_4 x + \bar{a}_6 = 0$, with $3x^2 + 2\bar{a}_2 x + \bar{a}_4 = 0$. This proves that the elliptic curve has a unique singular point and that both its coordinates belong to $\mathbb{F}_p$. After a translation of x one can then write the equation of the curve in the form $y^2 = x^2(x - \beta)$ with $\beta \in \mathbb{F}_p$. There are three cases:

- $\beta = 0$ is the case of additive reduction: one has a cusp at $(0,0)$.
- $\beta \neq 0$, $-\beta \in \mathbb{F}_p^2$ means split multiplicative reduction.
- $\beta \neq 0$, $-\beta \notin \mathbb{F}_p^2$ is the remaining case of non-split multiplicative reduction.

Now let $\mathbb{K}$ be a perfect field of characteristic 2. One starts with a Weierstrass equation in general form

$$y^2 + a_1 xy + a_3 y = x^3 + a_2 x^2 + a_4 x + a_6.$$

Then when the curve is singular it can be written, using affine transformations of (x, y) with coefficients in $\mathbb{K}$, in the form

(75)
$$y^2 = x^3 + a_4 x + a_6.$$

We refer to [**Si**] (Appendix A, Propositions 1.2 and 1.1) for details. Indeed, the curve is singular if and only if its discriminant vanishes (see Proposition 1.2 in [**Si**]). Then one can use the description of the curve as in Proposition 1.1c of [**Si**] to deduce that $a_3 = 0$, since $\Delta = 0$. The point at ∞ is always non-singular. A point (x, y) is singular if and only if (75) holds together with $3x^2 + a_4 = 0$. This equation is equivalent to $x^2 = a_4$ and uniquely determines x since the field is perfect. Then one gets $y^2 = a_6$ since $x^3 + a_4 x = x(x^2 + a_4) = 0$. This in turns uniquely determines y

since $\mathbb{K}$ is perfect. Thus for any finite field of characteristic two, one concludes that, if the curve is singular, it has exactly one singular point. In fact, for elliptic curves over $\mathbb{Q}$, there are only 4 cases to look at in characteristic two. They correspond to the values $a_4 \in \{0,1\}$ and $a_6 \in \{0,1\}$.

One can always assume that $a_6 = 0$ by replacing y with $y + 1$, if needed. For $a_4 = a_6 = 0$ one has a cusp at $(0,0)$ and the points over $\mathbb{F}_q$ are labeled by the x coordinate. Thus there are $q + 1$ points over $\mathbb{F}_q$ when $q = 2^\ell$. For $a_4 = 1$, $a_6 = 0$, the singular point is $P = (1,0)$. After shifting x by 1 the equation becomes $y^2 = x^2(x + 1)$ and thus the singular point is also a cusp. The points over $\mathbb{F}_q$ are labeled as $(1 + t^2, t(1 + t^2))$ for $t \in \mathbb{F}_q$ (together with the point at ∞). Again, there are $q + 1$ points over $\mathbb{F}_q$ when $q = 2^\ell$.

References

[AK]   J. Ax and S. Kochen, *Diophantine problems over local fields* I, American Journal of Mathematics **87** (1965), 605–630.

[B]   N. Bourbaki, *Algèbre commutative*, Eléments de Math. Hermann, Paris (1961–65).

[CC1]   A. Connes and C. Consani, *On the notion of geometry over* $\mathbb{F}_1$, Journal of Algebraic Geometry arXiv08092926v2 [mathAG] (to appear).

[CC2]   A. Connes and C. Consani, *Schemes over* $\mathbb{F}_1$ *and zeta functions*, Compositio Mathematica **146** 6 (2010), 1383–1415.

[Co]   J. Conway, *On Numbers and Games*, Academic Press, London (1976).

[D1]   A. Deitmar, *Schemes over* $\mathbb{F}_1$, in Number Fields and Function Fields, Two Parallel Worlds (Ed. by G. van der Geer, B. Moonen and R. Schoof), Progr. in Math **239** (2005).

[D2]   A. Deitmar, *F1-schemes and toric varieties*, Contributions to Algebra and Geometry **49** 2 (2008), 517–525.

[D3]   A. Deitmar, *Remarks on zeta functions and K-theory over F1*, Proc. Japan Acad. Ser. A Math. Sci. **82** 8 (2006), 141–146.

[DG]   M. Demazure and P. Gabriel, *Groupes algébriques*, Masson & CIE, Éditeur Paris (1970).

[DKK]   A. Deitmar, S. Koyama and N. Kurokawa, *Absolute zeta functions*, Proc. Japan Acad. Ser. A Math. Sci. **84** 8 (2008), 138–142.

[Ga]   A. Gathmann, *Tropical algebraic geometry*, Jahresber. Deutsch. Math.-Verein **108** 1 (2006), 3–32.

[Gi]   R. Gilmer, *Commutative semigroup rings*, University of Chicago Press, Chicago (1980).

[Go]   J. Golan, *Semi-rings and their applications (updated and expanded version of The theory of semi-rings, with applications to mathematics and theoretical computer science)*, Kluwer Academic Publishers, Dordrecht (1999).

[Gr]   P. A. Grillet, *Semigroups. An introduction to the structure theory*, Mercel Dekker, New York, Basel, Hong Kong (1995).

[IMS]   I. Itenberg, G. Mikhalkin and E. Shustin *Tropical algebraic geometry*, second edition, Oberwolfach Seminars 35, Birkhäuser Verlag, Basel (2009).

[KM]   V. Kolokoltsov and V. Maslov, *Idempotent analysis and its applications. Translation of Idempotent analysis and its application in optimal control (Russian) in "Nauka" Moscow, 1994. Translated by V. E. Nazaikinskii. With an appendix by Pierre Del Moral.*, Mathematics and Its Applications 401, Kluwer Academic Publishers, Dordrecht (1997).

[KS]   M. Kapranov and A. Smirnov *Cohomology determinants and reciprocity laws* (prepublication).

[KOW]   N. Kurokawa, H. Ochiai and A. Wakayama, *Absolute derivations and zeta functions*, Documenta Math. extra volume: Kazuya Kato Fiftieth Birthday (2003), 565–584.

[Ka]   K. Kato, *Toric singularities*, American Journal of Mathematics **116** 5 (1994), 1073–1099.

[Kr]   M. Krasner, *Approximation des corps valués complets de caractéristique* $p \neq 0$ *par ceux de caractéristique* 0, Colloque d'algèbre supérieure, tenu à Bruxelles du 19 au 22 décembre 1956, Centre Belge de Recherches Mathématiques Établissements Ceuterick, Louvain; Librairie Gauthier-Villars, Paris, 129–206.

[Ku]   N. Kurokawa, *Multiple zeta functions: An example in Zeta functions in geometry (Tokyo, 1990)*, Adv. Stud. Pure Math. **21** (1992), 219–226.

[LL]   J. López Peña and O. Lorscheid, *Mapping $\mathbb{F}_1$-land: An overview of geometries over the field with one element* (this volume).

[Le]   P. Lescot, *Algèbre absolue, arXiv:0911.1989*.

[Lu]   F. Luebeck, *www.math.rwth-aachen.de/Frank.Luebeck/data/ConwayPol/index.html*.

[M]   Y. I. Manin, *Lectures on zeta functions and motives (according to Deninger and Kurokawa)*, Astérisque **228** (4) (1995), 121–163.

[R]   W. Rudin, *Functional analysis*, International Series in Pure and Applied Mathematics. McGraw-Hill, New York (1991).

[Se]   J. P. Serre, *Local fields*, Graduate Texts in Mathematics 67, Springer-Verlag, New York (1979).

[Si]   J. Silverman, *The arithmetic of elliptic curves*, Graduate Texts in Mathematics 106, Springer-Verlag, New York (1986).

[So]   C. Soulé, *Les variétés sur le corps à un élément*, Mosc. Math. J. **4** 1 (2004), 217–244.

[TV]   B. Töen and M. Vaquié, *Au dessous de Spec($\mathbb{Z}$)*, J. K-Theory **3** (2009), 437–500.

COLLÈGE DE FRANCE, 3 RUE D'ULM, 75231, PARIS CEDEX 05, FRANCE, AND DEPARTMENT OF MATHEMATICS, 1326 STEVENSON CENTER, VANDERBILT UNIVERSITY, NASHVILLE, TN 37240, USA.

E-mail address: `alain@connes.org`

DEPARTMENT OF MATHEMATICS, JOHNS HOPKINS UNIVERSITY, 3400 N CHARLES ST., BALTIMORE, MD 21218, USA.

E-mail address: `kc@math.jhu.edu`

The Gauss-Bonnet Theorem for the Noncommutative Two Torus

Alain Connes and Paula Tretkoff

ABSTRACT. In this paper we shall show that the value at the origin, $\zeta(0)$, of the zeta function of the Laplacian on the noncommutative two torus, endowed with its canonical conformal structure, is independent of the choice of the volume element (Weyl factor) given by a (non-unimodular) state. We had obtained, in the late eighties, in an unpublished computation, a general formula for $\zeta(0)$ involving modified logarithms of the modular operator of the state. We give here the detailed computation and prove that the result is independent of the Weyl factor as in the classical case, thus proving the analogue of the Gauss-Bonnet theorem for the noncommutative two torus.

1. Introduction

The main result of this paper, namely the analogue of the Gauss-Bonnet theorem for the noncommutative two torus $\mathbb{T}^2_\theta$, follows from the computation of the value at the origin, $\zeta(0)$, of the zeta function of the analogue of the Laplacian for non-unimodular geometric structures on $\mathbb{T}^2_\theta$. The two authors did this computation at the end of the 1980's. The result was mentioned in [Co], but it was never published except as an MPI preprint ([CoC]). At the time the computation was done, the result was formulated in terms of the modular operator but its significance was unclear. In the mean time the following two theories have been developed:

(1) The spectral action
(2) The non-unimodular spectral triples

The spectral action ([CC1], [MCC]) allows one to interpret gravity coupled to the standard model from spectral invariants of a geometry which encodes a fine structure of space-time. In the two dimensional case the constant term in the spectral action is (up to adding the dimension of the kernel of the operator) given by the value at the origin, $\zeta(0)$, of the zeta function. This value is a topological invariant in classical Riemannian geometry.

The twisted (or non-unimodular) spectral triples appeared very naturally ([CM]) in the study of type III examples of foliation algebras.

Thus, in more modern terminology, our computation was that of the spectral action for spectral triples on the noncommutative two torus $\mathbb{T}^2_\theta$ both in the usual and the twisted cases. Initially, the complexity of the computation and the lack of simplicity of the result made us reticent about publishing it. It is only recently

141

that, by pushing the computation further, we could prove the expected (*cf.* [**CC2**]) conformal invariance.

THEOREM 1.1. *Let θ be an irrational number and consider, on the noncommutative two torus $\mathbb{T}_\theta^2$, a translation invariant conformal structure. Let k be an invertible positive element of $C^\infty(\mathbb{T}_\theta^2)$ considered as a Weyl factor rescaling the metric. Then the value at the origin of the zeta function $\zeta(s)$ of the Laplacian of the rescaled metric is independent of k.*

In fact we must retract a bit after stating this theorem since we have only performed the computation for the simplest translation invariant conformal structure but we do not expect that the general case will be different.

There are two main actors in the computation and they display an interesting interplay between two theories which look a priori quite distinct, but use the same notation for their key ingredient:

(1) The spectral theory of the Laplacian $\triangle$
(2) The modular operator Δ of states on operator algebras

The second is part of the scene because of the non-unimodularity of the spectral triple, or in simpler terms because the state defining the volume form is no longer assumed to be a trace and hence inherits a modular operator Δ. While both the Laplacian and the modular operator will be denoted by the capital letter delta, we hope the distinction[1] between these two operators will remain clear throughout to the alert reader.

Acknowledgments. The second author was supported in part by NSF Grant DMS-0800311.

2. Preliminaries

Recall that for a classical Riemann surface Σ with metric g, to the Laplacian $\triangle_g = d^* d$, where d is the de Rham differential operator acting on functions on the Riemann surface, one associates the zeta function

$$\zeta(s) = \sum \lambda_j^{-s}, \quad \mathrm{Re}(s) > 1,$$

where the summation is over the non-zero eigenvalues λ_j of $\triangle_g$ (see, for example, [**R**]). The meromorphic continuation of $\zeta(s)$ to $s = 0$, where it has no pole, gives the important information

$$\zeta(0) + \mathrm{Card}\{j \mid \lambda_j = 0\} = \frac{1}{12\pi} \int_\Sigma R = \frac{1}{6} \chi(\Sigma),$$

where R is the scalar curvature and $\chi(\Sigma)$ the Euler-Poincaré characteristic. This vanishes when Σ is the classical 2-torus $\mathbb{R}^2/\mathbb{Z}^2$, for example, and is an invariant within the conformal class of the metric, that is under the transformation $g \to e^f g$ for f a smooth real-valued function on Σ.

2.1. The Noncommutative Two Torus $\mathbb{T}_\theta^2$

Fix a real irrational number θ. We consider the irrational rotation C^*-algebra A_θ, with two unitary generators which satisfy

$$VU = e^{2\pi i\theta} UV, \quad U^* = U^{-1}, \ V^* = V^{-1}.$$

---

[1] We use a slightly different typography for the two cases.

We introduce the dynamical system given by the action of $\mathbb{T}^2$, $\mathbb{T} = \mathbb{R}/2\pi\mathbb{Z}$ on A_θ by the 2-parameter group of automorphisms $\{\alpha_s\}$, $s \in \mathbb{R}^2$, determined by

$$\alpha_s(U^n V^m) = e^{is.(n,m)}U^n V^m, \quad (s \in \mathbb{R}^2).$$

We define the sub-algebra A_θ^∞ of smooth elements of A_θ to be those x in A_θ such that the mapping

$$\mathbb{R}^2 \to A_\theta, \quad s \mapsto \alpha_s(x)$$

is smooth. Expressed as a condition on the coefficients of the element $a \in A_\theta$,

$$a = \sum a(n,m)U^n V^m,$$

this is the same as saying that they be of rapid decay, so that $\{|n|^k\,|m|^q\,|a(n,m)|\}$ is bounded for any positive k, q. The derivations associated to the above group of automorphisms are given by the defining relations

$$\delta_1(U) = U, \quad \delta_1(V) = 0,$$

$$\delta_2(U) = 0, \quad \delta_2(V) = V.$$

The derivations δ_1, δ_2 are analogues of the differential operators $\frac{1}{i}\partial/\partial x$, $\frac{1}{i}\partial/\partial y$ on the smooth functions on $\mathbb{R}^2/2\pi\mathbb{Z}^2$.

2.2. Conformal Structure on $\mathbb{T}_\theta^2$

As θ is supposed irrational, there is a unique trace τ on A_θ determined by the orthogonality properties

(1) $\tau(U^n V^m) = 0$ if $(n,m) \neq (0,0)$,   and $\tau(1) = 1$.

We can construct a Hilbert space $\mathcal{H}_0$ from A_θ by completing with respect to the inner product

(2) $$\langle a, b \rangle = \tau(b^* a), \quad a, b \in A_\theta,$$

and, using the derivations δ_1, δ_2, we introduce a complex structure by defining

(3) $$\partial = \delta_1 + i\delta_2, \quad \partial^* = \delta_1 - i\delta_2,$$

where (extending ∂, ∂^* to unbounded operators on $\mathcal{H}_0$) ∂^* is the adjoint of ∂ with respect to the inner product defined by τ. As an appropriate analogue of the space of $(1,0)$-forms on the classical 2-torus, one takes the unitary bimodule $\mathcal{H}^{(1,0)}$ over A_θ^∞ given by the Hilbert space completion of the space of finite sums $\sum a\,\partial b$, $a, b \in A_\theta^\infty$, with respect to the inner product

(4) $$\langle a\partial b, a'\partial b' \rangle = \tau((a')^* a(\partial b)(\partial b')^*), \quad a, a', b, b' \in A_\theta^\infty.$$

The information on the conformal structure is encoded by the positive Hochschild two cocycle (cf. [CCu], [CB]) on A_θ^∞ given by

(5) $$\psi(a, b, c) = -\tau(a\partial b\partial^* c).$$

2.3. Modular Automorphism Δ

In order to vary inside the conformal class of a metric we consider the family of positive linear functionals $\varphi = \varphi_h$, parameterized by $h = h^* \in A_\theta^\infty$, a self-adjoint element of A_θ^∞, and defined on A_θ by

$$\varphi(a) = \tau(ae^{-h}), \quad a \in A_\theta.$$

Note that, whereas for τ we have the trace relation

$$\tau(b^*a) = \tau(ab^*), \quad a, b \in A_\theta,$$

for φ we have

$$\varphi(ab) = \varphi(be^{-h}ae^h) = \varphi(b\sigma_i(a)), \quad a \in A_\theta,$$

which is the KMS condition at $\beta = 1$ for the 1-parameter group σ_t, $t \in \mathbb{R}$, of inner automorphisms

$$\sigma_t(x) = e^{ith}xe^{-ith}.$$

This group is $\sigma_t = \Delta^{-it}$, where the modular operator is

$$\Delta(x) = e^{-h}xe^h,$$

which is a positive operator fulfilling

$$(6) \qquad \langle \Delta^{1/2}x, \Delta^{1/2}x \rangle_\varphi = \langle x^*, x^* \rangle_\varphi, \quad \forall x \in A_\theta,$$

where we define the inner product $\langle \ , \ \rangle_\varphi$ on A_θ by

$$\langle a, b \rangle_\varphi = \varphi(b^*a), \quad a, b \in A_\theta.$$

We let $\mathcal{H}_\varphi$ be the Hilbert space completion of A_θ for the inner product $\langle \ , \ \rangle_\varphi$. It is a unitary left module on A_θ by construction. The 1-parameter group σ_t is generated by the derivation $-\log \Delta$

$$-\log \Delta(x) = [h, x], \quad x \in A_\theta^\infty.$$

2.4. Laplacian and Weyl Factor on $\mathbb{T}_\theta^2$

The Laplacian $\triangle$ on functions on $\mathbb{T}_\theta^2$ is given by

$$(7) \qquad \triangle = \partial^*\partial = \delta_1^2 + \delta_2^2,$$

where ∂ is viewed as an unbounded operator from $\mathcal{H}_0$ to $\mathcal{H}^{(1,0)}$.

When one modifies the volume form on $\mathbb{T}_\theta^2$ by replacing the trace τ by the functional φ, the modified Laplacian $\triangle'$ is given by

$$(8) \qquad \triangle' = \partial_\varphi^*\partial_\varphi,$$

where ∂_φ is the same operator ∂ but viewed as an unbounded operator from $\mathcal{H}_\varphi$ to $\mathcal{H}^{(1,0)}$. By construction $\triangle'$ is a positive unbounded operator in $\mathcal{H}_\varphi$.

LEMMA 2.1. *The operator $\triangle'$ is anti-unitarily equivalent to the positive unbounded operator $k\triangle k$ in the Hilbert space $\mathcal{H}_0$, where $k = e^{h/2} \in A_\theta$ acts in $\mathcal{H}_0$ by left multiplication.*

PROOF. The right multiplication by k extends to an isometry W from $\mathcal{H}_0$ to $\mathcal{H}_\varphi$,

$$(9) \qquad\qquad Wa = ak\,, \ \forall a \in A_\theta,$$

since

$$\langle Wa, Wb \rangle_\varphi = \tau((bk)^*(ak)k^{-2}) = \tau(b^*a)\,, \ \forall a, b \in A_\theta.$$

The operator $\partial_\varphi \circ W$ from $\mathcal{H}_0$ to $\mathcal{H}^{(1,0)}$ is given by

$$\partial_\varphi \circ W(a) = \partial(ak) = \partial \circ R_k,$$

where R_k is the right multiplication by k. Thus $\triangle' = \partial_\varphi^* \partial_\varphi$ is unitarily equivalent to $(\partial \circ R_k)^* \partial \circ R_k = R_k \partial^* \partial R_k$. Now, let J be the anti-unitary involution on $\mathcal{H}_0$ given by the star operation $Ja = a^*$ for all $a \in A_\theta$. The operator J commutes with $\triangle$ and fulfills $JR_kJ = k$ using $k^* = k$. This gives the required equivalence.

It is the dependence on k in the computations of the behavior of the zeta function of $k\triangle k$ at the origin that will feature in what follows.

2.5. Spectral Triples

With the notations of the previous section, consider the Hilbert space and operator

$$(10) \qquad\qquad \mathcal{H} = \mathcal{H}_\varphi \oplus \mathcal{H}^{(1,0)}, \quad D = \begin{pmatrix} 0 & \partial_\varphi^* \\ \partial_\varphi & 0 \end{pmatrix}.$$

We recall that $\mathcal{H}^{(1,0)}$ is naturally a unitary bimodule over A_θ.

LEMMA 2.2. *(1) The left action of A_θ in $\mathcal{H} = \mathcal{H}_\varphi \oplus \mathcal{H}^{(1,0)}$ and the operator D yield an even spectral triple $(A_\theta, \mathcal{H}, D)$.*

(2) Let J_φ be the Tomita antilinear unitary in $\mathcal{H}_\varphi$ and $a \mapsto J_\varphi a^* J_\varphi$ the corresponding unitary right action of A_θ in $\mathcal{H}_\varphi$. Then the right action $a \mapsto a^{\mathrm{op}}$ of A_θ in $\mathcal{H} = \mathcal{H}_\varphi \oplus \mathcal{H}^{(1,0)}$ and the operator D yield an even twisted spectral triple $(A_\theta^{\mathrm{op}}, \mathcal{H}, D)$; i.e., the following operators are bounded

$$(11) \qquad\qquad D\, a^{\mathrm{op}} - (k^{-1}ak)^{\mathrm{op}} D\,, \ \forall a \in A_\theta.$$

(3) The zeta function of the operator D, i.e., $\zeta_D(s) = \mathrm{Trace}(|D|^{-s})$, is equal to $2\, \mathrm{Trace}((k\triangle k)^{-s/2})$.

PROOF. (1) In order to show that $[D, a]$ is bounded, it is enough to check that $[\partial_\varphi, a]$ is bounded which follows from the derivation property of ∂_φ and the equivalence of the norms $\|.\|_\varphi$ and $\|.\|_0$.

(2) Let us first show that the twisted commutator $\partial_\varphi\, a^{\mathrm{op}} - (k^{-1}ak)^{\mathrm{op}}\, \partial_\varphi$ is bounded or equivalently that

$$(12) \qquad\qquad \partial_\varphi\, (kak^{-1})^{\mathrm{op}} - a^{\mathrm{op}}\, \partial_\varphi$$

is bounded as an operator from $\mathcal{H}_\varphi$ to $\mathcal{H}^{(1,0)}$. The isometry W from $\mathcal{H}_0$ to $\mathcal{H}_\varphi$ defined in (9) replaces ∂_φ by $\partial \circ R_k$ and replaces a^{op} by the right multiplication R_a by a for any $a \in A_\theta$. Thus it replaces the operator (12) by

$$\partial \circ R_k \circ R_{kak^{-1}} - R_a \circ \partial \circ R_k = R_{\partial a} \circ R_k,$$

which is a bounded operator from $\mathcal{H}_0$ to $\mathcal{H}^{(1,0)}$. Thus (12) is bounded and so is its adjoint

$$((kak^{-1})^{\mathrm{op}})^*\partial_\varphi^* - \partial_\varphi^*(a^{\mathrm{op}})^*.$$

Thus the boundedness of (11) follows from the equality

$$((kak^{-1})^{\mathrm{op}})^* = ((kak^{-1})^*)^{\mathrm{op}} = (k^{-1}a^*k)^{\mathrm{op}},$$

so that $\partial_\varphi^* a^{\mathrm{op}} - (k^{-1}ak)^{\mathrm{op}} \partial_\varphi^*$ is bounded for all $a \in A_\theta$.

(3) We have seen in Lemma 2.1 that the spectrum of $\triangle' = \partial_\varphi^*\partial_\varphi$ is the same as that of $k\triangle k$. Since the non-zero part of the spectrum of a product A^*A is the same as the non-zero part of the spectrum of AA^*, using the unitary equivalence given by the polar decomposition, one gets the required result.

3. Statement of the Theorem

With the notation of §1, we study the Laplacian zeta function defined for $\mathrm{Re}(s) > 1$ by the Mellin transform

$$\zeta(s) = \frac{1}{\Gamma(s)} \int_0^\infty \mathrm{Trace}^+(e^{-t\triangle'})\, t^{s-1}\, dt = \mathrm{Trace}(\triangle'^{-s}).$$

Here, $\triangle' \sim k\triangle k$, and

$$\mathrm{Trace}^+(e^{-t\triangle'}) = \mathrm{Trace}\,(e^{-t\triangle'}) - \mathrm{Dim}\,\mathrm{Ker}(\triangle'),$$

where by $\mathrm{Trace}(\cdot)$ we understand the ordinary trace of the operator. The definition of $\zeta(s)$ can be extended by meromorphic continuation to all values of s, barring $s = 1$, where the function has a simple pole. We now state the main result.

THEOREM 3.1. *Let θ be an irrational number and k an invertible positive element of A_θ^∞. Then the value at the origin of the zeta function $\zeta(s)$ of the operator $\triangle' \sim k\triangle k$ is independent of k.*

Our proof relies on a long explicit computation whose main ingredient is the following lemma.

3.1. Main Technical Lemma

LEMMA 3.2. *Let θ be an irrational number and k an invertible positive element of A_θ^∞. Then the value at the origin of the zeta function $\zeta(s)$ of the operator $\triangle' \sim k\triangle k$ is given by*[2]

$$(13) \qquad \zeta(0) + 1 = 2\pi\, \varphi(f(\Delta)(\delta_j(k))\delta_j(k)),$$

where φ is the functional $\varphi(x) = \tau(xk^{-2})$, Δ is the modular operator of φ and the function $f(u)$ is given by

$$(14) \qquad f(u) = \frac{1}{6}u^{-1/2} - \frac{1}{3} + \mathcal{L}_1(u) - 2(1+u^{1/2})\mathcal{L}_2(u) + (1+u^{1/2})^2\,\mathcal{L}_3(u),$$

where $\mathcal{L}_m$, m a positive integer, stands for the modified logarithm

$$\mathcal{L}_m(u) = (-1)^m\,(u-1)^{-(m+1)}\left(\log u - \sum_{j=1}^m (-1)^{j+1}\frac{(u-1)^j}{j}\right).$$

---

[2]Summation of repeated indices over $j = 1, 2$ is understood.

The proof of this lemma will occupy the remaining sections of the paper. In the present section, we shall show how it implies Theorem 3.1. The statement of Lemma 3.2 is the same[3] as that of the main theorem in [**CoC**], where the right hand side is written as $\tau(h(\theta, k))$, where

$$(15) \qquad h(\theta, k) = \frac{\pi}{3} k^{-1} \delta_j^2(k) - \frac{2\pi}{3} k^{-1} \delta_j(k) \delta_j(k) k^{-1} + 2\pi \mathcal{D}_1 (k^{-1} \delta_j(k)) \delta_j(k) k^{-1}$$

$$-4\pi \mathcal{D}_2 (1 + \Delta^{1/2}) (k^{-1} \delta_j(k)) \delta_j(k) k^{-1} + 2\pi \mathcal{D}_3 (1 + 2\Delta^{1/2} + \Delta)(k^{-1} \delta_j(k)) \delta_j(k) k^{-1}$$

and $\mathcal{D}_n = \mathcal{L}_n(\Delta)$. One has indeed

$$\tau(k^{-1} \delta_j^2(k)) = -\tau(\delta_j(k^{-1}) \delta_j(k)) =$$
$$\tau(k^{-1} \delta_j(k) k^{-1} \delta_j(k)) = \tau(k^{-2} \Delta^{-1/2} (\delta_j(k)) \delta_j(k)),$$

which accounts for the first term on the right hand side of (14), while the other terms are easily compared since the left multiplication by k^{-1} commutes with any function of Δ.

3.2. Result in Terms of $\log(k)$

The function $f(u)$ is of the form $h(\log(u))$, where h is the entire function

$$(16) \qquad h(x) = -\frac{e^{-x/2} \left(-1 + 3e^{x/2} + 3e^x + 6e^{3x/2}x - 3e^{2x} - 3e^{5x/2} + e^{3x}\right)}{6 \left(-1 + e^{x/2}\right)^4 \left(1 + e^{x/2}\right)^2}$$

(see Figure 1). Thus one can rewrite (13) in the form

$$(17) \qquad \zeta(0) + 1 = 2\pi \, \varphi(h(\log \Delta)(\delta_j(k)) \delta_j(k)).$$

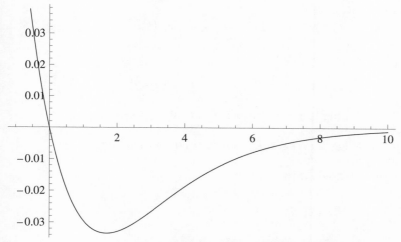

FIGURE 1 Graph of the function h.

As pointed out in §1, in the commutative case the corresponding value for the zeta function at the origin is zero. In fact in that case one has $\log \Delta = 0$. The

---

[3]Up to adding 1 to $\zeta(0)$.

function h vanishes at 0 and its Taylor expansion there is

$$h(x) = -\frac{x}{20} + \frac{x^2}{40} - \frac{x^3}{210} + \frac{x^4}{3360} + \frac{x^5}{201600} + O[x]^6.$$

In fact, one obtains a further simplification in general by expressing the result in terms of the element $\psi = \log k$. We introduce the function

(18) $$K(x) = -\frac{x - \mathrm{sh}\left[\frac{x}{2}\right] - \mathrm{sh}[x] + \frac{1}{3}\mathrm{sh}\left[\frac{3x}{2}\right]}{x^2 \mathrm{sh}\left[\frac{x}{2}\right]^2}.$$

LEMMA 3.3. *With the notations of Theorem 3.1, one has*

(19) $$\zeta(0) + 1 = 2\pi\, \tau(K(\log\Delta)(\delta_j(\log k))\delta_j(\log k)).$$

PROOF. One uses the formula

(20) $$k^{-1}\delta_j(k) = \int_0^1 \Delta^{s/2}(\delta_j(\log k))ds,$$

which gives

(21) $$k^{-1}\delta_j(k) = 2\frac{\Delta^{1/2} - 1}{\log\Delta}(\delta_j(\log k))$$

and similarly

$$\delta_j(k)k^{-1} = -2\frac{\Delta^{-1/2} - 1}{\log\Delta}(\delta_j(\log k)).$$

One then rewrites (13) in the form

$$\zeta(0) + 1 = 2\pi\, \tau(f(\Delta)(k^{-1}\delta_j(k))\delta_j(k)k^{-1})$$

and one uses the following identity, where F is an entire function:

(22) $$\tau(aF(\log\Delta)(b)) = \tau(F(-\log\Delta)(a)b)\,, \quad \forall a, b \in A_\theta^\infty,$$

which shows that (19) holds with

$$K(x) = 4\left(-1 + e^{x/2}\right)^2 x^{-2}\, h(x),$$

which simplifies to (18).

3.3. Proof of Theorem 3.1

By (18) the function $K(x)$ is an odd function. Thus using (22) one has

$$\tau(K(\log\Delta)(\delta_j(\log k))\delta_j(\log k)) = -\tau(K(\log\Delta)(\delta_j(\log k))\delta_j(\log k)) = 0.$$

4. Pseudo-differential Calculus

With the notation of the preceding sections, we introduce in the present one the notion of a pseudo-differential operator given the dynamical system $(A_\theta^\infty, \alpha_s)$ as developed in [**B**] and [**C**]. First of all, for a non-negative integer n, we define the vector space of differential operators of order at most n to be those polynomial expressions in δ_1, δ_2 of the form

$$P(\delta_1, \delta_2) = \sum_{|j| \le n} a_j \, \delta_1^{j_1} \delta_2^{j_2}, \quad a_j \in A_\theta^\infty, \quad j = (j_1, j_2) \in \mathbb{Z}_{\ge 0}^2, \quad |j| = j_1 + j_2.$$

To extend this definition, let $\mathbb{R}^2$ be the group dual to $\mathbb{R}^2$ and introduce the class of operator-valued distributions given by those complex linear functions

$$P : C^\infty(\mathbb{R}^2) \to A_\theta^\infty,$$

which are continuous with respect to the semi-norms p_{i_1, i_2} determined by

$$p_{i_1, i_2}(P(\varphi)) = \|\delta_1^{i_1} \delta_2^{i_2}(P(\varphi))\|, \quad i_1, i_2 \in \mathbb{Z}_{\ge 0}, \quad \varphi \in C^\infty(\mathbb{R}^2).$$

We use the notation $y_1 = e^{2\pi i \xi_1}$, $y_2 = e^{2\pi i \xi_2}$, $\xi = (\xi_1, \xi_2) \in \mathbb{R}^2$ for the canonical coordinates of $\mathbb{R}^2$, and $\partial_1 = \partial/\partial \xi_1$, $\partial_2 = \partial/\partial \xi_2$ for the corresponding derivations. We may now introduce the algebra of pseudo-differential operators via the algebra of operator-valued symbols.

DEFINITION 4.1. *An element $\rho = \rho(\xi) = \rho(\xi_1, \xi_2)$ of $C^\infty(\mathbb{R}^2, A_\theta^\infty)$ is a symbol of order the integer n if and only if for all non-negative integers i_1, i_2, j_1, j_2*

$$p_{i_1, i_2}(\partial_1^{j_1} \partial_2^{j_2} \rho(\xi)) \le c \, (1 + |\xi|)^{n - |j|},$$

where c is a constant depending only on ρ, and if there exists an element $k = k(\xi_1, \xi_2)$ of $C^\infty(\mathbb{R}^2 - \{0, 0\}, A_\theta^\infty)$ such that

$$\lim_{\lambda \to \infty} \lambda^{-n} \rho(\lambda \xi_1, \lambda \xi_2) = k(\xi_1, \xi_2).$$

We denote the space of symbols of order n by S_n, the union $S = \cup_{n \in \mathbb{Z}} S_n$ forming an algebra. Symbols of non-integral order are not required for this paper. An example of a symbol of order n a positive integer is provided by the polynomial $\rho(\xi) = \sum_{|j| \le n} a_j \xi_1^{j_1} \xi_2^{j_2}$, $a_j \in A_\theta^\infty$, and one has $\rho(n, m) = \sum_{|j| \le n} a_j \, n^{j_1} m^{j_2}$, so that $\rho(n, m) \, U^n V^m = \sum_{|j| \le n} a_j \, \delta_1^{j_1} \delta_2^{j_2}(U^n V^m)$. For an element $a = \sum_{n, m} a(n, m) \, U^n V^m$ of A_θ^∞ one therefore has $\sum_{n, m} \rho(n, m) \, a(n, m) \, U^n V^n = \sum_{|j| \le n} a_j \, \delta_1^{j_1} \delta_2^{j_2}(a)$, associating to the symbol ρ the differential operator $P_\rho = P(\delta_1, \delta_2) = \sum_{|j| \le n} a_j \, \delta_1^{j_1} \delta_2^{j_2}$ on A_θ^∞.

For every integer n, a symbol ρ of that order determines an operator on A_θ^∞ via the map $\psi : \rho \mapsto P_\rho$ given by the general formula

(23)
$$P_\rho(a) = (2\pi)^{-2} \int \int e^{-is \cdot \xi} \rho(\xi) \, \alpha_s(a) ds d\xi.$$

In our case this gives, using

$$\alpha_s(U^n V^m) = e^{is \cdot (n, m)} U^n V^m,$$

the simpler formula

$$(24) \qquad P_\rho(a) = \sum_{n,m \in \mathbb{Z}} \rho(n,m) \, a(n,m) \, U^n \, V^m, \quad a = \sum_{n,m} a(n,m) \, U^n \, V^m.$$

For example, the image under ψ of the symbol $(1 + |\xi|^2)^{-k}$, $k \geq 1$, of order $-2k$, acts on A_θ^∞.

DEFINITION 4.2. *The space ψ of pseudo-differential operators is given by the image of the algebra S under the map ψ.*

DEFINITION 4.3. *The equivalence $\rho \sim \rho'$ between two symbols ρ, ρ' in S_k, $k \in \mathbb{Z}$, holds if and only if $\rho - \rho'$ is a symbol of order n for all integers n.*

DEFINITION 4.4. *The class of pseudo-differential operators is the space ψ modulo addition by an element of $\psi(Z)$, where Z is the sub-algebra of S with elements equivalent to the zero symbol.*

It is possible to invert the map ψ to obtain for each element P of ψ a unique symbol $\sigma(P)$ up to equivalence. Recall from §1 that the trace τ on A_θ^∞ enables one to define the adjoint of operators acting on A_θ^∞ via their extension to $\mathcal{H}_0$. By direct analogy with [**G**] (Chapter 1, theorem, p. 16), one may deduce the following result.

PROPOSITION 4.5. *For an element P of ψ with symbol $\sigma(P) = \rho = \rho(\xi)$, the symbol of the adjoint P^* satisfies*

$$\sigma(P^*) \sim \sum_{(\ell_1,\ell_2)\in(\mathbb{Z}_{\geq 0})^2} (1/(\ell_1)! \, (\ell_2)!) \, [\partial_1^{\ell_1} \, \partial_2^{\ell_2} \, \delta_1^{\ell_1} \, \delta_2^{\ell_2} (\rho(\xi))^*].$$

If Q is an element of ψ with symbol $\sigma(Q) = \rho' = \rho'(\xi)$, then the product PQ is also in ψ and has symbol

$$\sigma(PQ) \sim \sum_{(\ell_1,\ell_2)\in(\mathbb{Z}_{\geq 0})^2} (1/(\ell_1)! \, (\ell_2)!) \, [\partial_1^{\ell_1} \, \partial_2^{\ell_2} (\rho(\xi)) \, \delta_1^{\ell_1} \, \delta_2^{\ell_2} (\rho'(\xi))].$$

Notice that in the above proposition as throughout the present paper, given symbols $\{\rho_j\}_{j=0}^\infty$, the relation $\rho \sim \sum_{j=0}^\infty \rho_j$ signifies that for any given k there exists a positive integer h such that for all $n > h$, the difference $\rho - \sum_{j=0}^n \rho_j$ is in S_k. The elliptic pseudo-differential operators are those whose symbols fulfil the criterion which follows.

DEFINITION 4.6. *Let n be an integer and ρ a symbol of order n. Then $\rho = \rho(\xi)$ is elliptic if it is invertible within the algebra $C^\infty(\mathbb{R}^2, A_\theta^\infty)$ and if its inverse satisfies*

$$\|\rho(\xi)^{-1}\| \leq c(1 + |\xi|)^{-n}$$

for a constant c depending only on ρ and for $|\xi| = (\xi_1^2 + \xi_2^2)^{1/2}$ sufficiently large.

An example of an elliptic operator is provided by the Laplacian $\triangle = \delta_1^2 + \delta_2^2$ on A_θ^∞ introduced in §1 which has the corresponding invertible symbol $\sigma(\triangle) = |\xi|^2$.

5. Understanding the Computations

The arguments of this section are kept brief, being direct analogues of standard ones. Bearing in mind the definition of the zeta function given in §2, we observe that by Cauchy's formula we have

$$e^{-t\triangle'} = (1/2\pi i) \int_C e^{-t\lambda}(\triangle' - \lambda\mathbf{1})^{-1}\,d\lambda$$

where λ is a complex number but is not real non-negative, and C encircles the non-negative real axis in the anti-clockwise direction without touching it. One then obtains a workable estimate of $(\triangle' - \lambda\mathbf{1})^{-1}$ by passing to the algebra of symbols with the important nuance that the scalar λ is now treated as a symbol of order two in all calculations. Using the definition of a symbol, one can replace the trace in the formula for the zeta function by an integration in the symbol space (argument along the diagonal), namely,

$$\zeta(s) = (1/\Gamma(s)) \int_0^\infty \int \tau(\sigma(e^{-t\triangle'})(\xi))\, t^{s-1}\,d\xi\,dt.$$

The function $\Gamma(s)$ has a simple pole at $s = 0$ with residue 1, so that

$$\zeta(0) = \text{Res}_{s=0} \int_0^\infty \int \tau(\sigma(e^{-t\triangle'})(\xi))\, t^{s-1}\,d\xi\,dt.$$

Just as in the arguments employed in the derivation of the asymptotic formula (see, for example, [**G**])

$$\int \tau(\sigma(e^{-t\triangle'})(\xi))\,d\xi \sim t^{-1} \sum_{n=0}^\infty B_{2n}(\triangle')\,t^n, \quad t \to 0_+,$$

one may appeal to the Cauchy formula quoted above. In particular, if B_λ denotes (a chosen approximation to) the inverse operator of $(\lambda\mathbf{1} - \triangle')$, its symbol has an expansion of the form

$$\sigma(B_\lambda) = \sigma(B_\lambda)(\xi) = b_0(\xi) + b_1(\xi) + b_2(\xi) + \dots,$$

where j ranges over the non-negative integers and $b_j(\xi) = b_j(\xi, \lambda)$ is a symbol of order $-2 - j$. As we shall explain at more length in §6, these symbols may be calculated inductively using the symbol algebra formulae beginning with $b_0(\xi) = (\lambda - k^2|\xi|^2)^{-1}$ which is the principal (highest homogeneous degree in ξ) symbol of $(\lambda\mathbf{1} - \triangle')^{-1}$. It turns out that $\zeta(0)$ equals the coefficient of λ^{-1} in $\int \tau(b_2(\xi))\,d\xi$ by

(25)
$$\zeta(0) = \int \tau(b_2(\xi))\,d\xi.$$

6. Computational Proof of the Main Lemma

Following on from the arguments of §4, by homogeneity there is no loss of generality in placing $\lambda = -1$ throughout the computation of $\zeta(0)$ and multiplying the final answer by -1. The problem is then to derive in the symbol algebra a recursive solution of the form $\sigma = b_0(\xi) + b_1(\xi) + b_2(\xi) + \dots$ to the equation

$$\sigma \cdot (\sigma(\triangle' - \lambda)) = 1 + 0(|\xi|^{-3}).$$

The accuracy to order -3 in ξ on the right hand side is in practice sufficient as we are only interested in evaluating σ up to $b_2(\xi)$. Throughout this section the convention of summation over repeated indices in the range $i, j = 1, 2$ is observed.

LEMMA 6.1. *The operator* $\triangle'$ *has symbol* $\sigma(\triangle') = a_2(\xi) + a_1(\xi) + a_0(\xi)$, *where, with summation over repeated indices in the range* $i = 1, 2$, *one has*

$$a_2 = a_2(\xi) = k^2 \xi_i \, \xi_i$$
$$a_1 = a_1(\xi) = 2 \, \xi_i (k \, \delta_i(k))$$
$$a_0 = a_0(\xi) = k \, \delta_i \, \delta_i(k).$$

These expressions are derived by applying the product formula within the algebra of symbols given in Proposition §3 to $\sigma_1(\xi) = \xi_i \, \xi_i$ *and* $\sigma_2(\xi) = k$ *and then multiplying on the left by* k.

To begin the inductive calculation of the inverse of the symbol of $\triangle' - \lambda$, set

$$(26) \qquad b_0 = b_0(\xi) = (k^2 \, |\xi|^2 + 1)^{-1}$$

and compute to order -3 in ξ the product $b_0 \cdot ((a_2 + 1) + a_1 + a_0)$. By singling out terms of the appropriate degree -1 in ξ and using the proposition of §3, one obtains

$$(27) \qquad b_1 = -(b_0 \, a_1 \, b_0 + \partial_i(b_0) \, \delta_i(a_2) \, b_0).$$

Note that b_0 appears on the right in this formula, unlike what is given in the formula on p. 52 of [**G**] which is valid for scalar principal symbols only. In a similar fashion, collecting terms of degree -2 in ξ and using (27) one obtains

$$b_2 = -(b_0 \, a_0 \, b_0 + b_1 \, a_1 \, b_0 + \partial_i(b_0) \, \delta_i(a_1) \, b_0$$
$$(28) \qquad + \partial_i(b_1) \, \delta_i(a_2) \, b_0 + (1/2) \, \partial_i \, \partial_j(b_0) \, \delta_i \, \delta_j(a_2) \, b_0).$$

A direct computation of these terms (up to the last multiplication by b_0) gives, ignoring the terms which are odd in ξ,

$$-b_0 k \delta_1^2(k) - b_0 k \delta_2^2(k) + \left(\xi_2^2 + 5\xi_1^2\right)\left(k^2 b_0^2\right) k \delta_1^2(k) + \left(5\xi_2^2 + \xi_1^2\right)\left(k^2 b_0^2\right) k \delta_2^2(k) +$$
$$2\xi_2^2\left(k^2 b_0^2\right) \delta_1(k)\delta_1(k) + 6\xi_1^2\left(k^2 b_0^2\right) \delta_1(k)\delta_1(k) + \xi_2^2\left(k^2 b_0^2\right) \delta_1^2(k)k + \xi_1^2\left(k^2 b_0^2\right) \delta_1^2(k)k +$$
$$6\xi_2^2\left(k^2 b_0^2\right) \delta_2(k)\delta_2(k) + 2\xi_1^2\left(k^2 b_0^2\right) \delta_2(k)\delta_2(k) + \xi_2^2\left(k^2 b_0^2\right) \delta_2^2(k)k + \xi_1^2\left(k^2 b_0^2\right) \delta_2^2(k)k -$$
$$4\xi_2^2\xi_1^2\left(k^4 b_0^3\right) k \delta_1^2(k) - 4\xi_1^4\left(k^4 b_0^3\right) k \delta_1^2(k) - 4\xi_2^4\left(k^4 b_0^3\right) k \delta_2^2(k) - 4\xi_2^2\xi_1^2\left(k^4 b_0^3\right) k \delta_2^2(k) -$$
$$8\xi_2^2\xi_1^2\left(k^4 b_0^3\right) \delta_1(k)\delta_1(k) - 8\xi_1^4\left(k^4 b_0^3\right) \delta_1(k)\delta_1(k) - 4\xi_2^2\xi_1^2\left(k^4 b_0^3\right) \delta_1^2(k)k - 4\xi_1^4\left(k^4 b_0^3\right) \delta_1^2(k)k -$$
$$8\xi_2^4\left(k^4 b_0^3\right) \delta_2(k)\delta_2(k) - 8\xi_2^2\xi_1^2\left(k^4 b_0^3\right) \delta_2(k)\delta_2(k) - 4\xi_2^4\left(k^4 b_0^3\right) \delta_2^2(k)k - 4\xi_2^2\xi_1^2\left(k^4 b_0^3\right) \delta_2^2(k)k +$$
$$2\xi_2^2 b_0 k \delta_1(k) b_0 k \delta_1(k) + 6\xi_1^2 b_0 k \delta_1(k) b_0 k \delta_1(k) + 2\xi_2^2 b_0 k \delta_1(k) b_0 \delta_1(k)k + 2\xi_1^2 b_0 k \delta_1(k) b_0 \delta_1(k)k -$$
$$4\xi_2^2\xi_1^2 b_0 k \delta_1(k)\left(k^2 b_0^2\right) k \delta_1(k) - 4\xi_1^4 b_0 k \delta_1(k)\left(k^2 b_0^2\right) k \delta_1(k) - 4\xi_2^2\xi_1^2 b_0 k \delta_1(k)\left(k^2 b_0^2\right) \delta_1(k)k -$$
$$4\xi_1^4 b_0 k \delta_1(k)\left(k^2 b_0^2\right) \delta_1(k)k + 6\xi_2^2 b_0 k \delta_2(k) b_0 k \delta_2(k) + 2\xi_1^2 b_0 k \delta_2(k) b_0 k \delta_2(k) +$$
$$2\xi_2^2 b_0 k \delta_2(k) b_0 \delta_2(k)k + 2\xi_1^2 b_0 k \delta_2(k) b_0 \delta_2(k)k - 4\xi_2^4 b_0 k \delta_2(k)\left(k^2 b_0^2\right) k \delta_2(k) -$$
$$4\xi_2^2\xi_1^2 b_0 k \delta_2(k)\left(k^2 b_0^2\right) k \delta_2(k) - 4\xi_2^4 b_0 k \delta_2(k)\left(k^2 b_0^2\right) \delta_2(k)k - 4\xi_2^2\xi_1^2 b_0 k \delta_2(k)\left(k^2 b_0^2\right) \delta_2(k)k -$$
$$2\xi_2^4\left(k^2 b_0^2\right) k \delta_1(k) b_0 k \delta_1(k) - 16\xi_2^2\xi_1^2\left(k^2 b_0^2\right) k \delta_1(k) b_0 k \delta_1(k) - 14\xi_1^4\left(k^2 b_0^2\right) k \delta_1(k) b_0 k \delta_1(k) -$$
$$2\xi_2^4\left(k^2 b_0^2\right) k \delta_1(k) b_0 \delta_1(k)k - 12\xi_2^2\xi_1^2\left(k^2 b_0^2\right) k \delta_1(k) b_0 \delta_1(k)k - 10\xi_1^4\left(k^2 b_0^2\right) k \delta_1(k) b_0 \delta_1(k)k +$$
$$4\xi_2^4\xi_1^2\left(k^2 b_0^2\right) k \delta_1(k)\left(k^2 b_0^2\right) k \delta_1(k) + 8\xi_2^2\xi_1^4\left(k^2 b_0^2\right) k \delta_1(k)\left(k^2 b_0^2\right) k \delta_1(k) +$$
$$4\xi_1^6\left(k^2 b_0^2\right) k \delta_1(k)\left(k^2 b_0^2\right) k \delta_1(k) + 4\xi_2^4\xi_1^2\left(k^2 b_0^2\right) k \delta_1(k)\left(k^2 b_0^2\right) \delta_1(k)k +$$

$$8\xi_2^2\xi_1^4\left(k^2b_0^2\right)k\delta_1(k)\left(k^2b_0^2\right)\delta_1(k)k + 4\xi_1^6\left(k^2b_0^2\right)k\delta_1(k)\left(k^2b_0^2\right)\delta_1(k)k-$$
$$14\xi_2^4\left(k^2b_0^2\right)k\delta_2(k)b_0k\delta_2(k) - 16\xi_2^2\xi_1^2\left(k^2b_0^2\right)k\delta_2(k)b_0k\delta_2(k) - 2\xi_1^4\left(k^2b_0^2\right)k\delta_2(k)b_0k\delta_2(k)-$$
$$10\xi_2^4\left(k^2b_0^2\right)k\delta_2(k)b_0\delta_2(k)k - 12\xi_2^2\xi_1^2\left(k^2b_0^2\right)k\delta_2(k)b_0\delta_2(k)k - 2\xi_1^4\left(k^2b_0^2\right)k\delta_2(k)b_0\delta_2(k)k+$$
$$4\xi_2^6\left(k^2b_0^2\right)k\delta_2(k)\left(k^2b_0^2\right)k\delta_2(k) + 8\xi_2^4\xi_1^2\left(k^2b_0^2\right)k\delta_2(k)\left(k^2b_0^2\right)k\delta_2(k)+$$
$$4\xi_2^2\xi_1^4\left(k^2b_0^2\right)k\delta_2(k)\left(k^2b_0^2\right)k\delta_2(k) + 4\xi_2^6\left(k^2b_0^2\right)k\delta_2(k)\left(k^2b_0^2\right)\delta_2(k)k+$$
$$8\xi_2^4\xi_1^2\left(k^2b_0^2\right)k\delta_2(k)\left(k^2b_0^2\right)\delta_2(k)k + 4\xi_2^2\xi_1^4\left(k^2b_0^2\right)k\delta_2(k)\left(k^2b_0^2\right)\delta_2(k)k-$$
$$2\xi_2^4\left(k^2b_0^2\right)\delta_1(k)kb_0k\delta_1(k) - 12\xi_2^2\xi_1^2\left(k^2b_0^2\right)\delta_1(k)kb_0k\delta_1(k) - 10\xi_1^4\left(k^2b_0^2\right)\delta_1(k)kb_0k\delta_1(k)-$$
$$2\xi_2^4\left(k^2b_0^2\right)\delta_1(k)kb_0\delta_1(k)k - 8\xi_2^2\xi_1^2\left(k^2b_0^2\right)\delta_1(k)kb_0\delta_1(k)k - 6\xi_1^4\left(k^2b_0^2\right)\delta_1(k)kb_0\delta_1(k)k+$$
$$4\xi_2^4\xi_1^2\left(k^2b_0^2\right)\delta_1(k)k\left(k^2b_0^2\right)k\delta_1(k) + 8\xi_2^2\xi_1^4\left(k^2b_0^2\right)\delta_1(k)k\left(k^2b_0^2\right)k\delta_1(k)+$$
$$4\xi_1^6\left(k^2b_0^2\right)\delta_1(k)k\left(k^2b_0^2\right)k\delta_1(k) + 4\xi_2^4\xi_1^2\left(k^2b_0^2\right)\delta_1(k)k\left(k^2b_0^2\right)\delta_1(k)k+$$
$$8\xi_2^2\xi_1^4\left(k^2b_0^2\right)\delta_1(k)k\left(k^2b_0^2\right)\delta_1(k)k + 4\xi_1^6\left(k^2b_0^2\right)\delta_1(k)k\left(k^2b_0^2\right)\delta_1(k)k-$$
$$10\xi_2^4\left(k^2b_0^2\right)\delta_2(k)kb_0k\delta_2(k) - 12\xi_2^2\xi_1^2\left(k^2b_0^2\right)\delta_2(k)kb_0k\delta_2(k) - 2\xi_1^4\left(k^2b_0^2\right)\delta_2(k)kb_0k\delta_2(k)-$$
$$6\xi_2^4\left(k^2b_0^2\right)\delta_2(k)kb_0\delta_2(k)k - 8\xi_2^2\xi_1^2\left(k^2b_0^2\right)\delta_2(k)kb_0\delta_2(k)k - 2\xi_1^4\left(k^2b_0^2\right)\delta_2(k)kb_0\delta_2(k)k+$$
$$4\xi_2^6\left(k^2b_0^2\right)\delta_2(k)k\left(k^2b_0^2\right)k\delta_2(k) + 8\xi_2^4\xi_1^2\left(k^2b_0^2\right)\delta_2(k)k\left(k^2b_0^2\right)k\delta_2(k)+$$
$$4\xi_2^2\xi_1^4\left(k^2b_0^2\right)\delta_2(k)k\left(k^2b_0^2\right)k\delta_2(k) + 4\xi_2^6\left(k^2b_0^2\right)\delta_2(k)k\left(k^2b_0^2\right)\delta_2(k)k+$$
$$8\xi_2^4\xi_1^2\left(k^2b_0^2\right)\delta_2(k)k\left(k^2b_0^2\right)\delta_2(k)k + 4\xi_2^2\xi_1^4\left(k^2b_0^2\right)\delta_2(k)k\left(k^2b_0^2\right)\delta_2(k)k+$$
$$8\xi_2^4\xi_1^2\left(k^4b_0^3\right)k\delta_1(k)b_0k\delta_1(k) + 16\xi_2^2\xi_1^4\left(k^4b_0^3\right)k\delta_1(k)b_0k\delta_1(k) + 8\xi_1^6\left(k^4b_0^3\right)k\delta_1(k)b_0k\delta_1(k)+$$
$$8\xi_2^4\xi_1^2\left(k^4b_0^3\right)k\delta_1(k)b_0\delta_1(k)k + 16\xi_2^2\xi_1^4\left(k^4b_0^3\right)k\delta_1(k)b_0\delta_1(k)k + 8\xi_1^6\left(k^4b_0^3\right)k\delta_1(k)b_0\delta_1(k)k+$$
$$8\xi_2^6\left(k^4b_0^3\right)k\delta_2(k)b_0k\delta_2(k) + 16\xi_2^4\xi_1^2\left(k^4b_0^3\right)k\delta_2(k)b_0k\delta_2(k) + 8\xi_2^2\xi_1^4\left(k^4b_0^3\right)k\delta_2(k)b_0k\delta_2(k)+$$
$$8\xi_2^6\left(k^4b_0^3\right)k\delta_2(k)b_0\delta_2(k)k + 16\xi_2^4\xi_1^2\left(k^4b_0^3\right)k\delta_2(k)b_0\delta_2(k)k + 8\xi_2^2\xi_1^4\left(k^4b_0^3\right)k\delta_2(k)b_0\delta_2(k)k+$$
$$8\xi_2^4\xi_1^2\left(k^4b_0^3\right)\delta_1(k)kb_0k\delta_1(k) + 16\xi_2^2\xi_1^4\left(k^4b_0^3\right)\delta_1(k)kb_0k\delta_1(k) + 8\xi_1^6\left(k^4b_0^3\right)\delta_1(k)kb_0k\delta_1(k)+$$
$$8\xi_2^2\xi_1^2\left(k^4b_0^3\right)\delta_1(k)kb_0\delta_1(k)k + 16\xi_2^2\xi_1^4\left(k^4b_0^3\right)\delta_1(k)kb_0\delta_1(k)k + 8\xi_1^6\left(k^4b_0^3\right)\delta_1(k)kb_0\delta_1(k)k+$$
$$8\xi_2^6\left(k^4b_0^3\right)\delta_2(k)kb_0k\delta_2(k) + 16\xi_2^4\xi_1^2\left(k^4b_0^3\right)\delta_2(k)kb_0k\delta_2(k) + 8\xi_2^2\xi_1^4\left(k^4b_0^3\right)\delta_2(k)kb_0k\delta_2(k)+$$
$$8\xi_2^6\left(k^4b_0^3\right)\delta_2(k)kb_0\delta_2(k)k + 16\xi_2^4\xi_1^2\left(k^4b_0^3\right)\delta_2(k)kb_0\delta_2(k)k + 8\xi_2^2\xi_1^4\left(k^4b_0^3\right)\delta_2(k)kb_0\delta_2(k)k.$$

It is extremely useful during the computation to exploit the fact that in the target formula for $\zeta(0)$ given in (25), one invokes the trace, so that members of the factors of the individual summands may be permuted cyclically without loss of generality for the answer. In our case this means that instead of multiplying the above sum by b_0 on the right we can just multiply it on the left, thus simply adding one to the exponent of the first occurrence of b_0. By (25), one then has to sum the integrals of each of these terms over the whole ξ-plane. After multiplying by b_0 on the left, passing in polar coordinates and integrating the angular variable one gets, up to an overall factor of 2π,

$$-b_0^2k\delta_1^2(k) - b_0^2k\delta_2^2(k) + 3r^2\left(k^2b_0^3\right)k\delta_1^2(k) + 3r^2\left(k^2b_0^3\right)k\delta_2^2(k) + 4r^2\left(k^2b_0^3\right)\delta_1(k)\delta_1(k)+$$
$$r^2\left(k^2b_0^3\right)\delta_1^2(k)k + 4r^2\left(k^2b_0^3\right)\delta_2(k)\delta_2(k) + r^2\left(k^2b_0^3\right)\delta_2^2(k)k - 2r^4\left(k^4b_0^4\right)k\delta_1^2(k)-$$
$$2r^4\left(k^4b_0^4\right)k\delta_2^2(k) - 4r^4\left(k^4b_0^4\right)\delta_1(k)\delta_1(k) - 2r^4\left(k^4b_0^4\right)\delta_1^2(k)k - 4r^4\left(k^4b_0^4\right)\delta_2(k)\delta_2(k)-$$
$$2r^4\left(k^4b_0^4\right)\delta_2^2(k)k + 4r^2b_0^2k\delta_1(k)b_0k\delta_1(k) + 2r^2b_0^2k\delta_1(k)b_0\delta_1(k)k-$$
$$2r^4b_0^2k\delta_1(k)\left(k^2b_0^2\right)k\delta_1(k) - 2r^4b_0^2k\delta_1(k)\left(k^2b_0^2\right)\delta_1(k)k + 4r^2b_0^2k\delta_2(k)b_0k\delta_2(k)+$$
$$2r^2b_0^2k\delta_2(k)b_0\delta_2(k)k - 2r^4b_0^2k\delta_2(k)\left(k^2b_0^2\right)k\delta_2(k) - 2r^4b_0^2k\delta_2(k)\left(k^2b_0^2\right)\delta_2(k)k-$$
$$8r^4\left(k^2b_0^3\right)k\delta_1(k)b_0k\delta_1(k) - 6r^4\left(k^2b_0^3\right)k\delta_1(k)b_0\delta_1(k)k + 2r^6\left(k^2b_0^3\right)k\delta_1(k)\left(k^2b_0^2\right)k\delta_1(k)+$$
$$2r^6\left(k^2b_0^3\right)k\delta_1(k)\left(k^2b_0^2\right)\delta_1(k)k - 8r^4\left(k^2b_0^3\right)k\delta_2(k)b_0k\delta_2(k) - 6r^4\left(k^2b_0^3\right)k\delta_2(k)b_0\delta_2(k)k+$$
$$2r^6\left(k^2b_0^3\right)k\delta_2(k)\left(k^2b_0^2\right)k\delta_2(k) + 2r^6\left(k^2b_0^3\right)k\delta_2(k)\left(k^2b_0^2\right)\delta_2(k)k-$$

$$6r^4 \left(k^2 b_0^3\right) \delta_1(k) k b_0 k \delta_1(k) - 4r^4 \left(k^2 b_0^3\right) \delta_1(k) k b_0 \delta_1(k) k + 2r^6 \left(k^2 b_0^3\right) \delta_1(k) k \left(k^2 b_0^2\right) k \delta_1(k) +$$
$$2r^6 \left(k^2 b_0^3\right) \delta_1(k) k \left(k^2 b_0^2\right) \delta_1(k) k - 6r^4 \left(k^2 b_0^3\right) \delta_2(k) k b_0 k \delta_2(k) - 4r^4 \left(k^2 b_0^3\right) \delta_2(k) k b_0 \delta_2(k) k +$$
$$2r^6 \left(k^2 b_0^3\right) \delta_2(k) k \left(k^2 b_0^2\right) k \delta_2(k) + 2r^6 \left(k^2 b_0^3\right) \delta_2(k) k \left(k^2 b_0^2\right) \delta_2(k) k +$$
$$4r^6 \left(k^4 b_0^4\right) k \delta_1(k) b_0 k \delta_1(k) + 4r^6 \left(k^4 b_0^4\right) k \delta_1(k) b_0 \delta_1(k) k + 4r^6 \left(k^4 b_0^4\right) k \delta_2(k) b_0 k \delta_2(k) +$$
$$4r^6 \left(k^4 b_0^4\right) k \delta_2(k) b_0 \delta_2(k) k + 4r^6 \left(k^4 b_0^4\right) \delta_1(k) k b_0 k \delta_1(k) + 4r^6 \left(k^4 b_0^4\right) \delta_1(k) k b_0 \delta_1(k) k +$$
$$4r^6 \left(k^4 b_0^4\right) \delta_2(k) k b_0 k \delta_2(k) + 4r^6 \left(k^4 b_0^4\right) \delta_2(k) k b_0 \delta_2(k) k,$$

where $b_0 = (r^2 k^2 - \lambda)^{-1}$ and where the integration is in rdr and from 0 to ∞.

6.1. Terms with All b_0 on the Left

Using the trace property these terms give the following:

$$-b_0^2 k \delta_1^2(k) - b_0^2 k \delta_2^2(k) + r^2 (3k^2 b_0^3 k \delta_1^2(k) - 2k^4 r^2 b_0^4 k \delta_1^2(k) + 3k^2 b_0^3 k \delta_2^2(k) -$$
$$2k^4 r^2 b_0^4 k \delta_2^2(k) + 4k^2 b_0^3 \delta_1(k) \delta_1(k) - 4k^4 r^2 b_0^4 \delta_1(k) \delta_1(k) + k^2 b_0^3 \delta_1^2(k) k -$$
$$2k^4 r^2 b_0^4 \delta_1^2(k) k + 4k^2 b_0^3 \delta_2(k) \delta_2(k) - 4k^4 r^2 b_0^4 \delta_2(k) \delta_2(k) + k^2 b_0^3 \delta_2^2(k) k - 2k^4 r^2 b_0^4 \delta_2^2(k) k),$$

which gives the same result as

$$\left(4k^2 r^2 b_0^3 - 4k^4 r^4 b_0^4\right) \left(\delta_1(k)^2 + \delta_2(k)^2\right) +$$
$$\left(-k b_0^2 + 4k^3 r^2 b_0^3 - 4k^5 r^4 b_0^4\right) \left(\delta_1^2(k) + \delta_2^2(k)\right)$$

and the coefficient of $1/\lambda$ in the integral $\int \bullet \, rdr$ gives, up to an overall coefficient of 2π,

$$(29) \qquad -\frac{1}{3} k^{-2} \left(\delta_1(k)^2 + \delta_2(k)^2\right) + \frac{1}{6} k^{-1} \left(\delta_1^2(k) + \delta_2^2(k)\right),$$

which corresponds to the first two terms of the formula for $h(\theta, k)$ in the statement of the lemma in the form (15).

6.2. Terms with b_0^2 in the Middle

They are the following:

$$-2r^4 b_0^2 k \delta_1(k) \left(k^2 b_0^2\right) k \delta_1(k) - 2r^4 b_0^2 k \delta_1(k) \left(k^2 b_0^2\right) \delta_1(k) k -$$
$$2r^4 b_0^2 k \delta_2(k) \left(k^2 b_0^2\right) k \delta_2(k) - 2r^4 b_0^2 k \delta_2(k) \left(k^2 b_0^2\right) \delta_2(k) k +$$
$$2r^6 \left(k^2 b_0^3\right) k \delta_1(k) \left(k^2 b_0^2\right) k \delta_1(k) + 2r^6 \left(k^2 b_0^3\right) k \delta_1(k) \left(k^2 b_0^2\right) \delta_1(k) k +$$
$$2r^6 \left(k^2 b_0^3\right) k \delta_2(k) \left(k^2 b_0^2\right) k \delta_2(k) + 2r^6 \left(k^2 b_0^3\right) k \delta_2(k) \left(k^2 b_0^2\right) \delta_2(k) k +$$
$$2r^6 \left(k^2 b_0^3\right) \delta_1(k) k \left(k^2 b_0^2\right) k \delta_1(k) + 2r^6 \left(k^2 b_0^3\right) \delta_1(k) k \left(k^2 b_0^2\right) \delta_1(k) k +$$
$$2r^6 \left(k^2 b_0^3\right) \delta_2(k) k \left(k^2 b_0^2\right) k \delta_2(k) + 2r^6 \left(k^2 b_0^3\right) \delta_2(k) k \left(k^2 b_0^2\right) \delta_2(k) k.$$

One has

$$\partial_r(b_0) = -2k^2 r b_0^2$$

and one can use integration by parts in r to transform terms such as

$$\int_0^\infty r^6 (k^2 b_0^3) \delta_1(k) k (k^2 b_0^3) \delta_1(k) k r dr$$

and replace them by

$$\frac{1}{2} \int_0^\infty \partial_r \left(r^6 (k^2 b_0^3)\right) \delta_1(k) k b_0 \delta_1(k) k dr,$$

where now b_0 only appears at the first power on the second occurrence.

After performing this operation, and combining with those which have b_0 in the middle, namely,

$$4r^2 b_0^2 k\delta_1(k) b_0 k\delta_1(k) + 2r^2 b_0^2 k\delta_1(k) b_0 \delta_1(k)k + 4r^2 b_0^2 k\delta_2(k) b_0 k\delta_2(k)+$$

$$2r^2 b_0^2 k\delta_2(k) b_0 \delta_2(k)k - 8r^4 \left(k^2 b_0^3\right) k\delta_1(k) b_0 k\delta_1(k) - 6r^4 \left(k^2 b_0^3\right) k\delta_1(k) b_0 \delta_1(k)k-$$

$$8r^4 \left(k^2 b_0^3\right) k\delta_2(k) b_0 k\delta_2(k) - 6r^4 \left(k^2 b_0^3\right) k\delta_2(k) b_0 \delta_2(k)k - 6r^4 \left(k^2 b_0^3\right) \delta_1(k) k b_0 k\delta_1(k)-$$

$$4r^4 \left(k^2 b_0^3\right) \delta_1(k) k b_0 \delta_1(k)k - 6r^4 \left(k^2 b_0^3\right) \delta_2(k) k b_0 k\delta_2(k) - 4r^4 \left(k^2 b_0^3\right) \delta_2(k) k b_0 \delta_2(k)k+$$

$$4r^6 \left(k^4 b_0^4\right) k\delta_1(k) b_0 k\delta_1(k) + 4r^6 \left(k^4 b_0^4\right) k\delta_1(k) b_0 \delta_1(k)k + 4r^6 \left(k^4 b_0^4\right) k\delta_2(k) b_0 k\delta_2(k)+$$

$$4r^6 \left(k^4 b_0^4\right) k\delta_2(k) b_0 \delta_2(k)k + 4r^6 \left(k^4 b_0^4\right) \delta_1(k) k b_0 k\delta_1(k)+$$

$$4r^6 \left(k^4 b_0^4\right) \delta_1(k) k b_0 \delta_1(k)k + 4r^6 \left(k^4 b_0^4\right) \delta_2(k) k b_0 k\delta_2(k) + 4r^6 \left(k^4 b_0^4\right) \delta_2(k) k b_0 \delta_2(k)k,$$

one obtains the following terms:

$$T = -2 \left(r^2 b_0^2\right) \left(k\delta_1(k) b_0 \delta_1(k)k + k\delta_2(k) b_0 \delta_2(k)k\right) + 2 \left(k^2 r^4 b_0^3\right) \left(k\delta_1(k) b_0 k\delta_1(k)+\right.$$

$$2k\delta_1(k) b_0 \delta_1(k)k + k\delta_2(k) b_0 k\delta_2(k) + 2k\delta_2(k) b_0 \delta_2(k)k + \delta_1(k) k b_0 \delta_1(k)k+$$

$$\delta_2(k) k b_0 \delta_2(k)k\Big) - 2 \left(k^4 r^6 b_0^4\right) \left(k\delta_1(k) b_0 k\delta_1(k) + k\delta_1(k) b_0 \delta_1(k)k + k\delta_2(k) b_0 k\delta_2(k)+\right.$$

$$k\delta_2(k) b_0 \delta_2(k)k + \delta_1(k) k b_0 k\delta_1(k) + \delta_1(k) k b_0 \delta_1(k)k + \delta_2(k) k b_0 k\delta_2(k) + \delta_2(k) k b_0 \delta_2(k)k\Big),$$

which all have b_0 in the middle.

6.3. Terms with b_0 in the Middle

Since we are in the noncommutative case, when in particular k and $\delta_i(k)$, $i = 1, 2$, do not commute, the computation of such terms requires us to permute k with elements of A_θ^∞. This is achieved using the following lemma.

LEMMA 6.2. *For every element ρ of A_θ^∞ and every non-negative integer m one has*

$$(30) \qquad \int_0^\infty \frac{k^{2m+2} u^m}{(k^2 u + 1)^{m+1}} \rho \frac{1}{(k^2 u + 1)} \, du = \mathcal{D}_m(\rho),$$

where the modified logarithm function is $\mathcal{D}_m = \mathcal{L}_m(\Delta)$, Δ is the operator introduced in §1 and

$$(31) \qquad \mathcal{L}_m(u) = \int_0^\infty \frac{x^m}{(x + 1)^{m+1}} \frac{1}{(xu + 1)} \, dx.$$

PROOF. On effecting the change of variables $u = e^s$ one obtains, with $k = e^{f/2}$,

$$\int_0^\infty \frac{k^{2m+2} u^m}{(k^2 u + 1)^{m+1}} \rho \frac{1}{(k^2 u + 1)} du$$

$$= \int_{-\infty}^\infty \frac{e^{(m+1)f+ms}}{(e^{(s+f)} + 1)^{m+1}} \rho \frac{e^s}{(e^{(s+f)} + 1)} ds$$

$$= \int_{-\infty}^\infty \frac{e^{(m+1/2)(s+f)}}{(e^{(s+f)} + 1)^{m+1}} \Delta^{-1/2}(\rho) \frac{e^{(s+f)/2}}{(e^{(s+f)} + 1)} ds$$

$$= \int_{-\infty}^\infty \frac{e^{(m+1/2)(s+f)}}{(e^{(s+f)} + 1)^{m+1}} \Delta^{-1/2}(\rho) \int_{-\infty}^\infty \frac{e^{it(s+f)}}{e^{\pi t} + e^{-\pi t}} dt \, ds$$

$$= \int_{-\infty}^\infty \frac{e^{(m+1/2)(s+f)}}{(e^{(s+f)} + 1)^{m+1}} \int_{-\infty}^\infty \frac{e^{it(s+f)}}{e^{\pi t} + e^{-\pi t}} \Delta^{-1/2+it}(\rho) \, dt \, ds$$

$$= \int_{-\infty}^\infty \frac{e^{itf}}{e^{\pi t} + e^{-\pi t}} \left(\int_{-\infty}^\infty \frac{e^{(m+1/2)(s+f)} e^{its}}{(e^{(s+f)} + 1)^{m+1}} ds \right) \Delta^{-1/2+it}(\rho) \, dt.$$

Now one has

$$\int_{-\infty}^\infty \frac{e^{(m+1/2)(s+f)} e^{its}}{(e^{(s+f)} + 1)^{m+1}} ds = e^{-itf} F_m(t),$$

where F_m is the Fourier transform of the function

$$h_m(s) = \frac{e^{(m+1/2)s}}{(e^s + 1)^{m+1}}.$$

We thus get

$$\int_0^\infty \frac{k^{2m+2} u^m}{(k^2 u + 1)^{m+1}} \rho \frac{1}{(k^2 u + 1)} du$$

$$= \int_{-\infty}^\infty \frac{F_m(t)}{e^{\pi t} + e^{-\pi t}} \Delta^{-1/2+it}(\rho) \, dt.$$

Moreover one has

$$\int_{-\infty}^\infty \frac{F_m(t)}{e^{\pi t} + e^{-\pi t}} u^{-it} \, dt$$

$$= \int_0^\infty \frac{x^{(m+1/2)}}{(x+1)^{m+1}} \frac{x^{-1/2} u^{1/2}}{x^{-1} u + 1} \frac{dx}{x} = u^{-1/2} \mathcal{L}_m(u^{-1}).$$

Replacing u by Δ^{-1} one gets the required equality (30). One can check the normalization taking $u = 1$. Then the integral $\mathcal{L}_m$ equals I_m,

$$I_m = \int_0^\infty v^m / (v + 1)^{m+2} \, dv = 1/(m+1),$$

for every positive integer m. On the other hand, by inspection one sees that $\mathcal{L}_m$ is of the form

$$\mathcal{L}_m(u) = c_m \left(\log(u) - P(u) \right) / (u - 1)^{m+1},$$

where P is a polynomial of degree at most m. In the neighborhood of $u = 1$ one has

$$\log(u) = \sum_{j=1}^{\infty} \frac{(-1)^{j+1}}{j}(u-1)^j,$$

and from its value at $u = 1$, where $\mathcal{L}_m$ is non-singular, one sees from this last expression that $\mathcal{L}_m$ is the modified logarithm $\mathcal{D}_m$ introduced in §2, where

$$\mathcal{D}_m(\Delta) = ((-1)^m/(\Delta-1)^{m+1})\left\{\log(\Delta) - \sum_{j=1}^{m} \frac{(-1)^{j+1}}{j}(\Delta-1)^j\right\}.$$

This completes the proof of the lemma.

We now split the sum T of §6.2 as a sum of three terms $T = T_1 + T_2 + T_3$ and compute the coefficient of $1/\lambda$ in the integral with respect to rdr in each of them using the above lemma.

6.3.1. Terms Involving $\mathcal{D}_1$

These terms come from

$$T_1 = -2\left(r^2 b_0^2\right)\left(k\delta_1(k)b_0\delta_1(k)k + k\delta_2(k)b_0\delta_2(k)k\right).$$

With $u = r^2$ the integrand is $rdr = \frac{1}{2}du$ and thus (up to the overall factor of 2π) these terms give, by setting $\lambda = -1$ and changing the overall sign,

$$(32) \quad \tau(k^{-2}\int_0^{\infty} \frac{k^4 u}{(k^2 u + 1)^2}\delta_i(k)\frac{1}{(k^2 u + 1)}du\,\delta_i(k)) = \tau(\mathcal{D}_1(\delta_i(k))\,\delta_i(k)\,k^{-2}).$$

6.3.2. Terms Involving $\mathcal{D}_2$

These terms come from

$$T_2 = 2\left(k^2 r^4 b_0^3\right)(k\delta_1(k)b_0 k\delta_1(k) + 2k\delta_1(k)b_0\delta_1(k)k + k\delta_2(k)b_0 k\delta_2(k) + 2k\delta_2(k)b_0\delta_2(k)k + \delta_1(k)kb_0\delta_1(k)k + \delta_2(k)kb_0\delta_2(k)k).$$

Since k commutes with b_0 and one works under the trace, they give the same result as

$$4\left(k^2 r^4 b_0^3\right)(k\delta_j(k)kb_0\delta_j(k) + k^2\delta_j(k)b_0\delta_j(k)).$$

One has $k\delta_j(k)k = k^2\Delta^{1/2}(\delta_j(k))$. Thus after setting $\lambda = -1$ and changing the overall sign, one gets

$$-2\tau(k^{-2}\int_0^{\infty} \frac{k^6 u^2}{(k^2 u + 1)^3}(\Delta^{1/2}(\delta_i(k)) + \delta_i(k))\frac{1}{(k^2 u + 1)}du\,\delta_i(k))$$

$$(33) \qquad = -2\tau((\mathcal{D}_2\,\Delta^{1/2})(\delta_i(k))\,\delta_i(k)\,k^{-2}) - 2\tau(\mathcal{D}_2(\delta_i(k))\,\delta_i(k)\,k^{-2}).$$

6.3.3. Terms Involving $\mathcal{D}_3$

These terms come from

$$T_3 = -2\left(k^4 r^6 b_0^4\right)(k\delta_1(k)b_0 k\delta_1(k) + k\delta_1(k)b_0\delta_1(k)k + k\delta_2(k)b_0 k\delta_2(k) +$$
$$k\delta_2(k)b_0\delta_2(k)k + \delta_1(k)kb_0 k\delta_1(k) + \delta_1(k)kb_0\delta_1(k)k + \delta_2(k)kb_0 k\delta_2(k) + \delta_2(k)kb_0\delta_2(k)k).$$

Since k commutes with b_0 and one works under the trace, they give the same result as

$$-2\left(k^4 r^6 b_0^4\right)\left(k^2\delta_j(k)b_0\delta_j(k) + 2k\delta_j(k)kb_0\delta_j(k) + \delta_j(k)k^2 b_0\delta_j(k)\right).$$

One has $k\delta_j(k)k = k^2\Delta^{1/2}(\delta_j(k))$ and $\delta_j(k)k^2 = k^2\Delta(\delta_j(k))$. Thus after setting $\lambda = -1$ and changing the overall sign, one gets

$$(34) \quad \tau\left(\left(\mathcal{D}_3(\delta_i(k))\,\delta_i(k) + 2(\mathcal{D}_3\,\Delta^{1/2})(\delta_i(k))\,\delta_i(k) + (\mathcal{D}_3\,\Delta)(\delta_i(k))\,\delta_i(k)\right)k^{-2}\right).$$

References

[B]   S. Baaj, *Calcul pseudo-différentiel et produits croisés de C^*-algèbres*. I *et* II, C.R. Acad. Sc. Paris **307, Série I** (1988), 581–586, 663–666.

[CC1]   A. Chamseddine, A. Connes, *The spectral action principle*, Comm. Math. Phys. **186** (1997), 731–750.

[CC2]   A. Chamseddine, A. Connes, *Scale invariance in the spectral action*, J. Math. Phys. **47** 6 (2006), 063504.

[MCC]   A. Chamseddine, A. Connes, M. Marcolli, *Gravity and the standard model with neutrino mixing*, hep-th/0610241.

[Co]   P. B. Cohen, *On the noncommutative torus of real dimension two*, Number Theory and Physics, Springer Proc. in Physics **47**, Springer-Verlag, Berlin, Heidelberg, 1990, pp. 151–156.

[CoC]   P. B. Cohen, A. Connes, *Conformal geometry of the irrational rotation algebra*, Preprint MPI **92-93** (1992).

[C]   A. Connes, *C^*-algèbres et géométrie différentielle*, C.R. Acad. Sc. Paris **290, Série A** (1980), 599–604.

[CB]   A. Connes, *Noncommutative geometry*, Academic Press, 1994.

[CCu]   A. Connes, J. Cuntz, *Quasi homomorphismes, cohomologie cyclique et positivité*, Comm. Math. Phys. **114** (1988), 515–526.

[CM]   A. Connes, H. Moscovici, *Type III and spectral triples*, ArXiv:math/0609703.

[G]   P. Gilkey, *The index theorem and the heat equation*, Publish or Perish, Boston, 1974.

[R]   S. Rosenberg, *The Laplacian in a Riemannian manifold*, LMS Student Texts 31, CUP, 1997.

COLLÈGE DE FRANCE, 3 RUE D'ULM, 75231, PARIS CEDEX 05, FRANCE, AND DEPARTMENT OF MATHEMATICS, 1326 STEVENSON CENTER, VANDERBILT UNIVERSITY, NASHVILLE, TN 37240, USA.

E-mail address: `alain@connes.org`

DEPARTMENT OF MATHEMATICS, TEXAS A&M, TAMU 3368, COLLEGE STATION, TX 77843, USA.

E-mail address: `ptretkoff@math.tamu.edu`

Zeta Phenomenology

David Goss

This paper is dedicated to my wife, Rita.

ABSTRACT. It is well known that Euler experimentally discovered the functional equation of the Riemann zeta function. Indeed he detected the fundamental $s \mapsto 1 - s$ invariance of $\zeta(s)$ by looking only at special values. In particular, via this functional equation, the permutation group on two letters, $S_2 \simeq \mathbb{Z}/(2)$, is realized as a group of symmetries of $\zeta(s)$. In this paper, we use the theory of special values of our characteristic p zeta functions to experimentally detect a natural symmetry group $S_{(q)}$ for these functions of cardinality $\mathfrak{c} = 2^{\aleph_0}$ (where $\mathfrak{c}$ is the cardinality of the continuum); $S_{(q)}$ is a realization of the permutation group on $\{0, 1, 2 \ldots\}$ as homeomorphisms of $\mathbb{Z}_p$ stabilizing both the nonpositive and nonnegative integers. We present a number of distinct instances in which $S_{(q)}$ acts (or appears to act) as symmetries of our functions. In particular, we present a natural, but highly mysterious, action of $S_{(q)}$ on a large subset of the domain of our functions that appears to stabilize zeta zeroes. As of this writing, we do not yet know an overarching formalism that unifies these examples; however, it would seem that this formalism will involve an interplay between the 1-unit group U_1—playing the role of a "gauge group"—and $S_{(q)}$. Furthermore, we show that $S_{(q)}$ may be naturally realized as an automorphism group of the convolution algebras of characteristic p valued measures.

1. Introduction

Euler's work on the zeta function has been an inspiration to us for many years. This work is briefly summarized in Section 2, but we highly recommend [**Ay1**] to the reader. Euler was able to compute the values of the Riemann zeta function at the positive even integers and at the negative integers. By very cleverly comparing them, he found the basic symmetry given by the famous functional equation of $\zeta(s)$. In particular, the lesson Euler taught us was that the special values are a window allowing one to glimpse very deep internal structure of the zeta function.

In the characteristic p theory, we have long had good results on special values in the basic case where the base ring A is $\mathbb{F}_q[T]$. At the positive integers, one had the brilliant analog of Euler's results due to L. Carlitz in the 1930's and 1940's (complete with an analogs of $2\pi i$, Bernoulli numbers, factorials, etc.). At the negative integers, one also had good formulas for the values of the characteristic

159

p zeta function. However, all attempts to put the positive and negative integers together in an "$s \mapsto 1 - s$" fashion failed.

The theory of these characteristic p functions works for any of Drinfeld's rings A (the affine ring of a complete smooth curve over $\mathbb{F}_q$ minus a fixed closed point ∞). It is however substantially harder to do explicit calculations for general A and so there are not yet many specific examples to study.

In the 1990's Dinesh Thakur [**Th1**] looked at the "trivial zeroes" of these zeta functions for certain nonpolynomial rings A. He found the intriguing phenomenon that such trivial zeroes may have a higher order of vanishing than naturally arises from the theory (current theory only gives a very classical looking lower bound on this vanishing in general, not the exact order!). More recently Javier Diaz-Vargas [**DV2**] extended Thakur's calculations. Both Thakur and Diaz-Vargas experimentally found that this general higher vanishing at $-j$ appears to be associated with j of a very curious type: the sum of the q-adic digits of j must be bounded.

We call a trivial zero "regular" if its order is exactly the classical lower bound and "irregular" otherwise. (See Subsection 6.3.1 for a discussion of this terminology.)

We have been trying to come to grips with the implications of Thakur's and Diaz-Vargas's inspired calculations for a few years; see [**Go2**]. Recently we discovered a huge group of symmetries that seems to underlie their calculations. It is our purpose here to describe this group and its relationship to their calculations as well as other instances when it arises. In particular, we shall see how this group conjecturally allows one to establish certain *finiteness* results on trivial zeroes for characteristic p zeta functions. Moreover, calculations in the polynomial ring case also indicate that this group acts on the zeroes of the zeta function (see Subsection 6.2.1).

This paper is written to quickly explain these implications to the reader. We try as much as possible to stay away from general theory and keep the paper as self-contained as possible.

The symmetry group $S_{(q)}$ is introduced in Section 5. It is a group of homeomorphisms of $\mathbb{Z}_p$ obtained by simply rearranging the q-expansion coefficients. In particular, we readily see that $S_{(q)}$ stabilizes *both* the nonpositive and nonnegative integers; there is no mixing as in $s \mapsto 1 - s$. Thus, perhaps, we have the "true" explanation for the failure to somehow put the positive and negative integers together as Euler did. We will see in Subsection 6.1 that the action of $S_{(q)}$ appears to preserve the orders of trivial zeroes coming from Thakur's and Diaz-Vargas' calculations (which is how we actually discovered it).

Moreover, all may not be lost here in terms of relating the positive and negative integers. Indeed, we shall see in Subsection 6.3 that $S_{(q)}$ may also be realized as symmetries of Carlitz's "von Staudt-Clausen" result where he calculated the denominator of his Bernoulli analogs at the positive integers in the basic $\mathbb{F}_q[T]$ case. This is very exciting and highly mysterious to us.

Remarkably, we further establish for $A = \mathbb{F}_q[T]$ that $S_{(q)}$ determines the degree of the "special polynomials" (see Definition 9) that arise in the theory; i.e., these degrees are an *invariant* for the action of $S_{(q)}$. Since the trivial zeroes are zeroes of the special polynomials, knowledge of the degrees of these polynomials would obviously bound the orders of trivial zeroes.

As is well known by now, these characteristic p zeta functions are analytically continued by "summing according to degree." For each fixed degree d, one obtains continuous functions from $\mathbb{Z}_p$ into our finite characteristic fields. We further show that the zero sets of these functions, as subsets of $\mathbb{Z}_p$, are *stable* under the action of $S_{(q)}$, and we present some evidence that the same result holds for the v-adic interpolations of $\zeta(s)$ at the finite primes v.

We shall frequently use the essential paper of J. Sheats [**Sh1**] which, in particular, established rigorously some results mentioned much earlier by L. Carlitz [**Ca4**]. Sheats did this as he proved the "Riemann hypothesis for $\mathbb{F}_q[T]$." In other words, our functions are naturally defined on the space $\mathbb{S}_\infty$ (See Definition 6) which is a product of a characteristic p variable x and p-adic variable y; Sheats then shows that, upon fixing y, all zeroes in x^{-1} are simple and uniquely determined by their absolute values. So Galois invariance immediately implies that the zeroes must lie on the line $\mathbb{F}_q((1/T))$.

Note that the classical functional equation, $s \mapsto 1 - s$ becomes $t \mapsto -t$ upon setting $s = 1/2 + it$ and obviously $|t| = |-t|$.

In the function field situation, our whole theory begins by choosing a notion of "positive" (generalizing the notion of "monic polynomial") and then fixing a positive parameter π. So for instance if we use the standard notion of monic, one could choose $\pi = 1/T$. Once we have chosen π, we are then able to "exponentiate ideals" and define L-series. Passing from one positive parameter π_1 to another π_2 multiplies the x-coordinate of a zeta zero with a certain power of the "phase" $u = \pi_1/\pi_2$ (which is a 1-unit; see Lemma 3). Clearly multiplication by a power of u does not change the absolute value of the x-coordinate of a given zero, but it *does* alter the various expansions of the zeroes in terms of π_1 and π_2 (as described in Subsection 6.2.1).

In Subsection 6.2.1 we shall also present examples, in the polynomial case, where zeta zeroes are acted upon via digit permutations and the group $S_{(q)}$. More specifically one takes the expansion of a zero in a parameter π and then permutes the coefficients of this expansion in a prescribed way to obtain other zeroes. This does NOT work for all the coefficients but only seems to for those coefficients which are *invariant* under the change from one parameter to another! (So in this regard, we are obtaining finer information than simply the absolute value.) This passing to the invariants is also what gives the analogy to the use of gauge groups in physics.

Much more work will need to be done before we are able to place the ideas presented here in their proper context.

Finally in Subsection 6.3.1 we discuss how the notions of extra vanishing, etc., also arise in the classical theory of Bernoulli numbers. This builds on the famous results of von Staudt-Clausen, Adams, and Kummer.

Acknowledgments. This paper grew out of my lecture at the workshop "Noncommutative Geometry and Geometry over the Field with One Element" at Vanderbilt University in May 2008 as well as lectures at subsequent conferences. It is my great pleasure to thank the organizers of these very interesting meetings for their kind hospitality and support. It is also my pleasure to acknowledge very useful input from Warren Sinnott and Bernd Kellner.

2. Euler's Creation of Zeta Phenomenology

We recall here very briefly the fabulous first example of zeta phenomenology: Euler's numerical discovery of the functional equation of the Riemann zeta function

$\zeta(s)$. Our treatment here follows that of [**Ay1**]; we have also covered these ideas in [**Go2**].

DEFINITION 1. The Bernoulli numbers, B_n, are defined by

$$\frac{x}{e^x - 1} = \sum_{n=0}^{\infty} \frac{B_n x^n}{n!}.$$

After many years of work, Euler computed the values $\zeta(2n)$, $n = 1, 2, \cdots$ in terms of Bernoulli numbers and obtained the famous formula

(1) $$B_{2n} = (-1)^{n+1} \frac{2(2n)!}{(2\pi)^{2n}} \zeta(2n).$$

Euler then turned his attention to the values of $\zeta(s)$ at the negative integers where his work on special values becomes divinely inspired! Indeed, Euler did not have the notion of analytic continuation of complex valued functions to work with. Thus he relied on his instincts for beauty while working with divergent series; nevertheless, he obtained the right values.

Euler begins with the very well known expansion

(2) $$\frac{1}{1 - x} = 1 + x + x^2 + x^3 + \cdots + x^n + \cdots.$$

Clearly this expansion is only valid when $|x| < 1$, but that does not stop Euler. Upon putting $x = -1$, he deduces

(3) $$1/2 = 1 - 1 + 1 - 1 + 1 \cdots,$$

where we simply ignore questions of convergence! He then applies the operator $x \frac{d}{dx}$ to Equation 2 and again evaluates at $x = -1$ obtaining

(4) $$1/4 = 1 - 2 + 3 - 4 + 5 \cdots.$$

Applying the operator again, Euler finds the "trivial zero"

(5) $$0 = 1 - 2^2 + 3^2 - \cdots,$$

and so on. Euler recognizes the sum on the right of these equations to be the values at the negative integers of the modified zeta function

(6) $$\zeta^*(s) := (1 - 2^{1-s})\zeta(s) = \sum_{n=1}^{\infty} (-1)^{n-1}/n^s.$$

The wonderful point is, of course, that these values *are* the values rigorously obtained much later by Riemann. (N.B.: in [**Ay1**], our $\zeta^*(s)$ is denoted $\phi(s)$.)

Nine years later, Euler notices, at least for small $n \geq 2$, that his calculations imply

(7) $$\frac{\zeta^*(1 - n)}{\zeta^*(n)} = \begin{cases} \frac{(-1)^{(n/2)+1}(2^n - 1)(n-1)!}{(2^{n-1} - 1)\pi^n} & \text{if } n \text{ is even} \\ 0 & \text{if } n \text{ is odd}. \end{cases}$$

Upon rewriting Equation 7 using his gamma function $\Gamma(s)$ and the cosine, Euler then "hazards" to conjecture

(8) $$\frac{\zeta^*(1 - s)}{\zeta^*(s)} = \frac{-\Gamma(s)(2^s - 1)\cos(\pi s/2)}{(2^{s-1} - 1)\pi^s},$$

which translates easily into the functional equation of $\zeta(s)$!

REMARK 1. Note the important role played by the trivial zeroes in Equation 7 in that they render harmless our inability to calculate explicitly $\zeta^*(n)$, or $\zeta(n)$, at odd integers > 1.

Euler then calculates both sides of Equation 8 at $s = 1$ and obtains the same answer. To Euler, this is "strong justification" for his conjecture. Of course history has proved him to be spectacularly right!

From now on, until Subsection 6.3.1, where we return to classical theory, the symbol "$\zeta(s)$" will be reserved for characteristic p valued functions.

3. The Factorial Ideal

In order to later define Bernoulli elements in characteristic p, and so explain Carlitz's von Staudt-Clausen result, we clearly need a good notion of "factorial."

We begin by reviewing the basic set-up of the characteristic p theory. We let $q = p^{n_0}$, where p is prime and n_0 is a positive integer. Let X be a smooth projective geometrically connected curve over the finite field $\mathbb{F}_q$ with q-elements. Choose ∞ to be a fixed closed point on X of degree d_∞ over $\mathbb{F}_q$. Thus, $X - \infty$ is an affine curve and we let A be the ring of its global functions. Note that A is a Dedekind domain with finite class group and that $A^* = \mathbb{F}_q{}^*$. We let k denote the quotient field of A. The completion of k at ∞ is denoted k_∞ and the completion of a fixed algebraic closure of k_∞ (under the canonical topology) is denoted $\mathbb{C}_\infty$. We let $\mathbb{F}_\infty \subset k_\infty$ be the associated finite field. Set $q_\infty := q^{d_\infty}$ so that $\mathbb{F}_\infty \simeq \mathbb{F}_{q_\infty}$.

Of course the simplest example of such an A is $\mathbb{F}_q[T]$, $k = \mathbb{F}_q(T)$. In general though, A will not be Euclidean or factorial.

Let x be a transcendental element.

DEFINITION 2. 1. For $i = 1, 2, \cdots$, we set $[i](x) := x^{q^i} - x$.

2. We define $L_0(x) \equiv 1$ and for $i = 1, 2, \cdots$, we set $L_i(x) := [i](x)[i - 1](x) \cdots [1](x)$.

3. We define $D_0(x) \equiv 1$ and for $i = 1, 2, \cdots$, we set $D_i(x) := [i](x)[i - 1](x)^q \cdots [1](x)^{q^{i-1}}$.

Elementary considerations of finite fields allow one to show the following proposition (see Proposition 3.1.6 in [**Go1**]). In it, the elements just defined in Definition 2 are considered as members of the polynomial ring $\mathbb{F}_q[x]$.

PROPOSITION 1. 1. $[i](x)$ *is the product of all monic irreducible polynomials in x whose degree divides i.*
2. $L_i(x)$ *is the least common multiple of all polynomials in x of degree i.*
3. $D_i(x)$ *is the product of all monic polynomials in x of degree i.*

As we will readily see later on (Proposition 3) the polynomials $D_i(x)$ and $L_i(x)$ are universal for the exponential and logarithm of general Drinfeld modules.

Our next goal is to use the functions $D_i(x)$ to define a factorial function à la Carlitz. Let j be an integer that we write q-adically as $j = \sum_{e=0}^{w} c_e q^e$, where $0 \le c_e < q$ all e.

DEFINITION 3. We set

$$\Pi_j(x) := \prod_{e=0}^{w} D_e(x)^{c_e}.$$

As a function of the integer j, $\Pi_j(x)$ satisfies many of the same divisibility results as the classical $n!$.

Let A be an arbitrary affine ring as above. We now define the basic ideals of A of interest to us. Let $f(x) \in \mathbb{F}_q[x]$.

DEFINITION 4. We set $\tilde{f} := (f(a))_{a \in A}$; i.e., $\tilde{f}$ is the ideal generated by the values of $f(x)$ on the elements of A.

In general one would expect these ideals to be trivial (i.e., equal to A itself) as the example $f(x) = x + 1$ shows. However, for the functions given in Definition 2, they are highly nontrivial.

EXAMPLE 1. We show here that $[\tilde{i}] = \prod \mathfrak{P}$, where the product ranges over all primes of degree (over $\mathbb{F}_q$) dividing i. Let $\mathfrak{P}$ have degree dividing i; then modulo $\mathfrak{P}$ we have $a^{q^i} = a$ (or $a^{q^i} - a = 0$) for any a. Thus, $\mathfrak{P}$ must divide $[\tilde{i}]$. Now let $a \in A$ be a uniformizer at $\mathfrak{P}$. Then clearly so is $[i](a)$. Therefore $\mathfrak{P}^2$ does not divide $[\tilde{i}]$. Finally, a moment's thought along these lines also shows that the only possible prime divisors of $[\tilde{i}]$ are those whose degree divides i.

Let $\mathfrak{P}$ be a prime of A with additive valuation $v_{\mathfrak{P}}$. Thakur observed that for a function $f(x)$, $v_{\mathfrak{P}}(\tilde{f}) = v_{\mathfrak{P}}(\tilde{f}_{\mathfrak{P}})$, where $\tilde{f}_{\mathfrak{P}}$ is the analog of $\tilde{f}$ constructed locally on the completion $A_{\mathfrak{P}}$ of A at $\mathfrak{P}$. As a consequence, we need only compute these valuations on $A = \mathbb{F}_q[T]$ where it is known that the ideals associated to the functions of Definition 2 and Definition 3 are generated by their values at $x = T$. Thus, using Theorem 9.1.1 of [**Go1**], we have the following basic factorization of $\tilde{\Pi}_j$.

PROPOSITION 2. *Let $\mathfrak{P}$ be a prime of A of degree d. Then*

$$v_{\mathfrak{P}}(\tilde{\Pi}_j) = \sum_{e \geq 1}[j/q^{ed}],$$

where $[w]$ is the greatest integer function, $w \in \mathbb{Q}$.

In the fundamental case $A = \mathbb{F}_q[T]$, Proposition 2 was proved by W. Sinnott; it is clearly a direct analog of the calculation of the p-adic valuation of $n!$.

Finally, we explain the relationship with Drinfeld modules that the reader may skip as it is not needed for the remainder of the paper. As before, let k be the quotient field of A. Let L be a finite extension of k with O_L the ring of A-integers in L. Let ψ be a Drinfeld module of arbitrary rank over L with coefficients in O_L. Let $e(z) = z + \sum_{i \geq 1} e_i z^{q^i}$ and $l(z) = z + \sum_{i \geq 1} l_i z^{q^i}$ be the exponential and logarithm of the Drinfeld module (obtained, say, by embedding L into $\mathbb{C}_\infty$). Let $a \in A$.

PROPOSITION 3. *The elements $D_i(a)e_i$ and $L_i(a)l_i$ lie in O_L.*

PROOF. One has the basic recurrence relations

$$e(az) = \psi_a(e(z))$$

and

$$al(z) = l(\psi_a(z)).$$

The result now follows by induction and the definition of $D_i(x)$ and $L_i(x)$. □

4. Zeta Functions and Integral Zeta Values

4.1. Exponentiation of Ideals

As mentioned in Section 1, we shall define these functions and values here with a minimum of theory and refer the reader to Chapter 8 of [**Go1**] for the elided details. Our goal is to define an analog of n^j, where n is a positive integer and j is an arbitrary integer. However, as general A is *not* factorial, we have to define "$\mathcal{J}^j$" as an element of $\mathbb{C}_\infty^*$ for nonprincipal $\mathcal{J}$. Here we immediately run into a notational issue in that the symbol "$\mathcal{J}^j$" is universally reserved for taking the j-th power of the *ideal* $\mathcal{J}$ in the Dedekind domain A. We do *not* change this; rather we will use "$\mathcal{J}^{(j)}$" for the above element of $\mathbb{C}_\infty^*$, so that there will be no confusion.

Recall that the completion of k at ∞ is denoted k_∞ with $\mathbb{F}_\infty \subset k_\infty$ being the associated finite field; recall also that $q_\infty = q^{d_\infty}$. Fix an element $\pi \in k_\infty^*$ of order 1. Every element $x \in k_\infty^*$ has a unique decomposition:

$$(9) \qquad\qquad x = \zeta_x \pi^{v_\infty(x)} u_x,$$

where $\zeta_x \in \mathbb{F}_\infty^*$, $v_\infty(x) \in \mathbb{Z}$, and $u_x \in k_\infty$ is a 1-unit (i.e., congruent to 1 modulo (π)) and depends on π. We say x is "positive" or "monic" if and only if $\zeta_x = 1$. Clearly the positive elements form a subgroup of finite index of k_∞^*.

DEFINITION 5. We set $\langle x \rangle = \langle x \rangle_\pi := u_x$, where u_x is defined in Equation 9.

As mentioned, the element $\langle x \rangle$ depends on π, but no confusion will result by not making this dependence explicit. We will have more to say about this later.

Note that $x \mapsto \langle x \rangle$ is a homomorphism from k_∞^* to its subgroup $U_1(k_\infty)$ of 1-units.

As above, X is the smooth projective curve associated to k. For any fractional ideal I of A, we let $\deg_k(I)$ be the degree over $\mathbb{F}_q$ of the divisor associated to I on the affine curve $X - \infty$. For $\alpha \in k^*$, one sets $\deg_k(\alpha) = \deg_k((\alpha))$, where (α) is the associated fractional ideal; this clearly agrees with the degree of a polynomial in $\mathbb{F}_q[T]$.

DEFINITION 6. Set $\mathbb{S}_\infty := \mathbb{C}_\infty^* \times \mathbb{Z}_p$.

The space $\mathbb{S}_\infty$ plays the role of the complex numbers in our theory in that it is the domain of "n^s." Indeed, let $s = (x, y) \in \mathbb{S}_\infty$ and let $\alpha \in k$ be positive. The element $v = \langle \alpha \rangle - 1$ has absolute value < 1; thus, $\langle \alpha \rangle^y = (1 + v)^y$ is easily defined and computed via the binomial theorem.

DEFINITION 7. We set

$$(10) \qquad\qquad \alpha^s := x^{\deg_k(\alpha)} \langle \alpha \rangle^y.$$

Clearly $\mathbb{S}_\infty$ is a group whose operation is written additively. Suppose that $j \in \mathbb{Z}$ and α^j is defined in the usual sense of the canonical $\mathbb{Z}$-action on the multiplicative group. Let $\pi_* \in \mathbb{C}_\infty^*$ be a fixed d_∞-th root of π. Set $s_j := (\pi_*^{-j}, j) \in \mathbb{S}_\infty$. One checks easily that Definition 7 gives $\alpha^{s_j} = \alpha^j$. When there is no chance of confusion, we denote s_j simply by "j."

In the basic case $A = \mathbb{F}_q[T]$ one can now proceed to define zeta values. However, in general A has nonprincipal and positively generated ideals. Fortunately there is a canonical and simple procedure to extend Definition 7 to them as follows. Let $\mathcal{I}$ be the group of fractional ideals of the Dedekind domain A and let $\mathcal{P} \subseteq \mathcal{I}$ be the subgroup of principal ideals. Let $\mathcal{P}^+ \subseteq \mathcal{P}$ be the subgroup of principal ideals which have positive generators. It is a standard fact that $\mathcal{I}/\mathcal{P}^+$ is a finite abelian group. The association

$$(11) \qquad\qquad \mathfrak{h} \in \mathcal{P}^+ \mapsto \langle \mathfrak{h} \rangle := \langle \lambda \rangle_\pi,$$

where λ is the unique positive generator of $\mathfrak{h}$, is obviously a homomorphism from $\mathcal{P}^+$ to $U_1(k_\infty) \subset \mathbb{C}_\infty^*$.

Let $U_1(\mathbb{C}_\infty) \subset \mathbb{C}_\infty^*$ be the group of 1-units defined in the obvious fashion. The binomial theorem, again, shows that $U_1(\mathbb{C}_\infty)$ is a $\mathbb{Z}_p$-module. However, it is also closed under the unique operation of taking p-th roots; as such, $U_1(\mathbb{C}_\infty)$ is a $\mathbb{Q}_p$-vector space.

LEMMA 1. *The mapping* $\mathcal{P}^+ \to U_1(\mathbb{C}_\infty)$ *given by* $\mathfrak{h} \mapsto \langle \mathfrak{h} \rangle_\pi$ *has a unique extension to* $\mathfrak{I}$ *(which we also denote by* $\langle ? \rangle_\pi$*).*

PROOF. As $U_1(\mathbb{C}_\infty)$ is a $\mathbb{Q}_p$-vector space, it is a divisible group; thus, the extension follows by general theory. The uniqueness then follows by the finitude of $\mathcal{I}/\mathcal{P}^+$. □

The next lemma explaining the dependence on π will play a fundamental role for us.

LEMMA 2. *Let* π_1 *and* π_2 *be two positive parameters and let* I *be a nonzero ideal of* A. *Then*

$$(12) \qquad\qquad \langle I \rangle_{\pi_1} = (\pi_1/\pi_2)^{\deg_k(I)/d_\infty} \langle I \rangle_{\pi_2}.$$

PROOF. The formula obviously works when I is principal and positively generated and both sides of Equation 12 are homomorphisms as functions of I. Therefore, the uniqueness of the extension from $\mathcal{P}^+$ to all ideals finishes the result. □

If $s \in \mathbb{S}_\infty$ and I are chosen as above, we now set

$$(13) \qquad\qquad I^s := x^{\deg_k(I)} \langle I \rangle^y.$$

Thus, if $\alpha \in k$ is positive one sees that $(\alpha)^s$ agrees with α^s as in Equation 10.

For a fractional ideal $\mathfrak{I}$ and integer j, as promised, we now put $\mathfrak{I}^{(j)} := \mathfrak{I}^{s_j}$. Thus, if $a \in k$ is positive then $(a)^{(j)} = a^j$ by definition. The reader can check that $\mathfrak{I}^{(j)}$ is independent of π and π_* up to possible multiplication of the d_∞-th root of unity.

The values $\mathfrak{I}^{(j)}$ are obviously determined multiplicatively by $\mathfrak{I}^{(1)}$. Furthermore, suppose $\mathfrak{I}^t = (i)$, where i is positive where t is a positive integer (which always exists, as the ideal class group is finite) and put $\mathfrak{i} = \mathfrak{I}^{(1)}$. Then we have the basic formula

$$(14) \qquad\qquad \mathfrak{i}^t = i.$$

From this it is very easy to see that the values $\mathfrak{I}^{(1)}$ generate a finite extension V of k in $\mathbb{C}_\infty$ which is called the *value field*. It is also easy to see that $\mathfrak{I}$ becomes principal in this field and is generated by $\mathfrak{I}^{(1)}$.

4.2 The Zeta Values

Let j be an arbitrary integer.

DEFINITION 8. We formally put

$$\zeta(j) := \sum_{\mathfrak{J}} \mathfrak{J}^{(-j)} = \sum_{\mathfrak{J}} \mathfrak{J}^{s-j}$$

where $\mathfrak{J}$ ranges over the nonzero ideals of A and $s_{-j} \in \mathbb{S}_\infty$ was defined after Definition 7.

Because the analysis is nonarchimedean, the sum $\zeta(j)$ clearly converges to an element of $\mathbb{C}_\infty$ for $j > 0$. At the nonpositive integers we must regroup the sum. More precisely, for $j \geq 0$ we write

$$(15) \qquad \zeta(-j) = \sum_{e=0}^{\infty} \left(\sum_{\deg_k(\mathfrak{J})=e} \mathfrak{J}^{(j)} \right),$$

where, as above, $\deg_k(\mathfrak{J})$ is the degree over $\mathbb{F}_q$ of $\mathfrak{J}$. As is now well known (see, e.g., Chapter 8 of [**Go1**]) for sufficiently large e the sum in parentheses vanishes. Thus, the value is an algebraic integer over A.

It is known that $\zeta(-j) = 0$ for j a positive integer divisible by $q_\infty - 1$. These are the *trivial zeroes*.

DEFINITION 9. For $j \geq 0$, we set

$$(16) \qquad z(x,-j) = z_A(x,-j) := \sum_{e=0}^{\infty} x^{-e} \left(\sum_{\deg_k(\mathfrak{J})=e} \mathfrak{J}^{(j)} \right).$$

As the sum in parentheses in Equation 15 vanishes for sufficiently large e, we see that $z(x,-j)$ is a *polynomial* in x^{-1}. These polynomials themselves also occur as special zeta values, and, as such, are called the *special polynomials*.

We note that the values given above are all special values of the function given in the next definition.

DEFINITION 10. We define

$$(17) \qquad \zeta(s) = \zeta_A(s) = \zeta_{A,\pi}(s) := \sum_{\mathfrak{J}} \mathfrak{J}^{-s}$$

for all $s \in \mathbb{S}_\infty$ and nonzero ideals $\mathfrak{J}$ of A.

In Lemma 3 below we will make explicit the dependence of $\zeta_{A,\pi}(s)$ on π.

Let $s = (x,y)$. One then rewrites $\zeta_{A,\pi}(s)$ as

$$(18) \qquad \zeta_{A,\pi}(s) = \sum_{e=0}^{\infty} x^{-e} \left(\sum_{\deg_k(\mathfrak{J})=e} \langle \mathfrak{J} \rangle^{-y} \right).$$

The analytic continuation of $\zeta_{A,\pi}(s)$ (and all such arithmetic Dirichlet series) is accomplished by showing that these power series are actually entire in x^{-1}, with very strong continuity properties on all of $\mathbb{S}_\infty$.

LEMMA 3. *Let π_1 and π_2 be two positive parameters, $\alpha \in \mathbb{C}_\infty^*$ and $y_0 \in \mathbb{Z}_p$. Then we have*

$$(19) \qquad \zeta_{A,\pi_2}(\alpha, y_0) = \zeta_{A,\pi_1}((\pi_1/\pi_2)^{-y_0/d_\infty} \alpha, y_0).$$

PROOF. This follows from Lemma 2 upon unraveling the definitions.   □

Let $t = (x_0, y_0) \in \mathbb{S}_\infty$. By the "order of zero of $\zeta_{A,\pi}(s)$ at t," one means the order of zero of the entire power series $\zeta_{A,\pi}(x, y_0)$ at $x = x_0$.

5. The Group $S_{(q)}$

In this section we will introduce the automorphism groups of interest to us. These will be subgroups of the group of homeomorphisms of $\mathbb{Z}_p$ and they will stabilizee—and so permute—both the nonpositive and nonnegative integers sitting in $\mathbb{Z}_p$.

Let q continue to be a power of p and let $x \in \mathbb{Z}_p$. Write x q-adically as

$$(20) \qquad x = \sum_{i=0}^{\infty} c_i q^i,$$

where $0 \leq c_i < q$ for all i. If x is a nonnegative integer (so that the sum in Equation 20 is obviously finite), then we set

$$(21) \qquad \ell_q(x) = \sum_i c_i.$$

Let ρ be a permutation of the set $\{0, 1, 2, \dots \}$.

DEFINITION 11. We define $\rho_*(x)$, $x \in \mathbb{Z}_p$, by

$$(22) \qquad \rho_*(x) := \sum_{i=0}^{\infty} c_i q^{\rho(i)}.$$

Clearly $x \mapsto \rho_*(x)$ gives a representation of ρ as a set permutation (in fact, as we will see in Proposition 4, a homeomorphism) of $\mathbb{Z}_p$.

DEFINITION 12. We let $S_{(q)}$ be the group of permutations of $\mathbb{Z}_p$ obtained as ρ varies over all permutations of $\{0, 1, 2, \dots \}$.

REMARK 2. We use the notation "$S_{(q)}$" to avoid confusion with the symmetric group S_q on q-elements.

Note that if q_0 and q_1 are powers of p, and $q_0 \mid q_1$, then $S_{(q_1)}$ is naturally realized as a subgroup of $S_{(q_0)}$.

The next proposition gives the basic properties of the mapping $\rho_*(x)$.

PROPOSITION 4. *Let $\rho_*(x)$ be defined as above.*
1. *The mapping $x \mapsto \rho_*(x)$ is continuous on $\mathbb{Z}_p$.*
2. *("Semi-additivity") Let x, y, z be three p-adic integers with $z = x + y$ and where there is no carry over of q-adic digits. Then $\rho_*(z) = \rho_*(x) + \rho_*(y)$.*
3. *The mapping $x \mapsto \rho_*(x)$ stabilizes the nonnegative integers.*
4. *The mapping $x \mapsto \rho_*(x)$ stabilizes the negative integers.*
5. *Let n be a nonnegative integer. Then $\ell_q(n) = \ell_q(\rho_*(n))$.*
6. *Let n be an integer. Then $n \equiv \rho_*(n) \pmod{q-1}$.*

PROOF. To see Part 1, let j be a positive integer. We want to show that the first q^j expansion coefficients of $\rho_*(x)$ and $\rho_*(y)$ are the same if $x \equiv y \pmod{q^t}$ for some positive integer t. Let ϕ be the inverse permutation to ρ (as functions on the nonnegative integers). Choose t greater than $\phi(e)$ for $e = 0, \dots, j-1$. Parts 2 and 3 are obvious. To see Part 4, let n be a negative integer and let j be a positive integer chosen so that $q^j + n$ is nonnegative. Then q-adically we have

$$(23) \qquad n = (q^j + n) - q^j = (q^j + n) + (q-1)q^j + (q-1)q^{j+1} + \cdots,$$

as $-1 = q - 1 + (q-1)q + (q-1)q^2 \cdots$. On the other hand, $\rho_*(n)$ will now clearly also have almost all of its q-adic coefficients equal to $q-1$ and the result is clear. Part 5 is clear and implies Part 6 for nonnegative n, as then we have $n \equiv \ell_q(n) \pmod{q-1}$. Thus, suppose n is negative. As in Equation 23 write $n = (q^j + n) - q^j$ with $q^j + n$ nonnegative. By Part 2, we have

$$(24) \qquad \rho_*(n) = \rho_*(q^j + n) + \rho_*(-q^j).$$

Clearly $\rho_*(-q^j)$ has almost all coefficients equal to $q-1$ with the rest equaling 0; thus, we can write $\rho_*(-q^j) = m - q^t$ for some t, where m is positive and divisible by $q-1$. Part 6 for nonnegative integers now implies that modulo $q-1$ we have

$$\begin{aligned}
\rho_*(n) &= \rho_*(q^j + n) + m - q^t \\
&\equiv (q^j + n) - q^t \\
&\equiv 1 + n - 1 \\
&\equiv n.
\end{aligned}$$

Thus, by Parts 3 and 4 of Proposition 4, ρ_* permutes both the nonpositive and nonnegative integers. $\qquad \square$

Notice further that the injection $x \mapsto p^e x$ (e a positive integer) is not in $S_{(p)}$, as it is not surjective. However, let n be a positive integer. Then clearly $p^e n = \rho_*(n)$ for infinitely many $\rho_* \in S_{(p)}$ (which may vary with n). Note, however, that multiplication by p will change the set of q-adic digits of an integer if $q > p$, etc. Thus, in this case pn will not equal $\rho_*(n)$ for any $\rho_* \in S_{(q)}$.

The reader can readily see that the cardinality of $S_{(p)}$ is $\mathfrak{c}$ (where $\mathfrak{c}$ is the cardinality of the continuum), as this is the cardinality of the group of permutations of $\{0, 1, 2, \cdots\}$.

DEFINITION 13. Let ρ be as above and let $x \in \mathbb{Z}_p$. We define

$$(25) \qquad \hat{\rho}_*(x) := -\rho_*(-x).$$

It is clear that $\hat{\rho}_*$ also stabilizes the nonnegative and nonpositive integers. Moreover, one can easily find examples of ρ and x such that $\hat{\rho}_*(x)$ is *not* given by digit permutations.

REMARK 3. Let $x \in (0, 1)$ be a real number with decimal expansion

$$(26) \qquad x = .x_0 x_1 \dots.$$

In order to have this expansion be unique, we require that infinitely many of the x_i do not equal 9. We can then play the same permutation game with these decimal coefficients and we thereby obtain homeomorphisms of $(0, 1)$. Similar remarks work for an arbitrary base b.

It is quite remarkable that the groups $S_{(q)}$ have a very natural relationship with binomial coefficients considered modulo p. This is given in our next two results.

PROPOSITION 5. *Let* $\sigma \in S_{(p)}$, $y \in \mathbb{Z}_p$, *and* k *a nonnegative integer. Then we have*

$$(27) \qquad \binom{y}{k} \equiv \binom{\sigma y}{\sigma k} \quad (\text{mod } p).$$

PROOF. This follows immediately from Lucas' formula.   □

Next, we recall the definition of the algebra of divided power series over a field L. These are formal sums of the form

$$\sum_{i=0}^{\infty} c_i \frac{z^i}{i!},$$

where $\{c_i\} \subseteq L$ and one has the obvious multiplication

$$\frac{z^i}{i!} \cdot \frac{z^j}{j!} := \binom{i+j}{i} \frac{z^{i+j}}{(i+j)!}.$$

As the binomial coefficient is an integer, this definition works in all characteristics.

PROPOSITION 6. *Let* i *and* j *be two nonnegative integers. Let* $\sigma \in S_{(p)}$. *Then*

$$(28) \qquad \binom{i+j}{i} \equiv \binom{\sigma i + \sigma j}{\sigma i} \quad (\text{mod } p).$$

PROOF. Lucas' formula shows that if there is any carry over of p-adic digits in the addition of i and j, then $\binom{i+j}{i}$ is 0 modulo p. However, there is carry over of the p-adic digits in the sum of i and j if and only if there is carry over in the sum of σi and σj; in this case both sums are 0 modulo p. If there is no carry over, then the result follows from Part 2 of Proposition 4.   □

COROLLARY 1. *Let* σ *be as in the proposition. Then the mapping* $\frac{z^i}{i!} \mapsto \frac{z^{\sigma i}}{\sigma(i)!}$ *is an algebra automorphism of the divided power series in characteristic* p.

As was explained in Subsection 8.22 of [**Go2**] and [**Co1**], the algebras of measures in characteristic p are isomorphic to divided power series algebras. More precisely, let $R := \mathbb{F}_q[[u]]$. One first picks a basis for the Banach space of continuous $\mathbb{F}_q$-linear functions; then, using the q-adic expansion of an integer t, one obtains an associated basis for the Banach algebra of *all* continuous functions from R to itself. One then sees readily that the algebra of R-valued measures on R (equipped with convolution as usual) is thus isomorphic to the formal divided power series algebra over R. Therefore the next corollary follows immediately.

COROLLARY 2. *The group* $S_{(p)}$ *acts as automorphisms of the convolution algebra of* R-*valued measures on* R.

EXAMPLE 2. As an example of how Corollary 2 may be used, consider the field $k_\infty(\pi_*) \simeq \mathbb{F}_{q_\infty}((\pi_*))$. It contains the ring $\mathbb{F}_{q_\infty}[[\pi_*]]$, which is obviously the completion of $\mathbb{F}_{q_\infty}[\pi_*]$ at the ideal generated by π_*. One can then, e.g., use the Carlitz polynomial basis, as in Subsection 8.22 of [**Go1**], to obtain the isomorphism with the divided power series algebra.

REMARK 4. At first glance, classical theory would indicate that there should be an extension of the action of $S_{(p)}$ to all of $\mathbb{S}_\infty$, which is analytic in the first variable and which takes integer powers to integer powers. However, one knows that the bijective rigid analytic maps from $\mathbb{G}_m$ to itself are of the form $x \mapsto cx^{\pm 1}$ for some nonzero constant c. There simply do not appear to be enough of these functions to extend the action of $S_{(p)}$ on the integers inside $\mathbb{S}_\infty$ to all of $\mathbb{S}_\infty$ *analytically* (in the first variable).

The above remark leads us to look at nonanalytic maps in hopes of extending the action of $S_{(q)}$ to $\mathbb{C}_\infty^*$ and thus all of $\mathbb{S}_\infty$. Calculations, to be given in Subsecton 6.2.1, lead to the following construction.

DEFINITION 14. We let $K_1 := \mathbb{F}_\infty((\pi_*))$.

Obviously K_1 is totally ramified extension of K. Let $x \in K_1^*$, which we write as $x = \sum_{i \gg -\infty} c_i \pi_*^i$.

DEFINITION 15. Let ρ be a permutation of $\{0, 1, 2, \dots\}$ as before and let $x \in K_1^*$ be written as just above. We set

$$(29) \qquad \rho_*(x) := \sum_{i \gg -\infty} c_i \pi_*^{\rho_*(i)} \in K_1^*.$$

This definition precisely works because, by Proposition 4, we know that ρ_* stabilizes both the nonpositive and nonnegative integers. Moreover, one can easily modify the proof of Proposition 4 to deduce the continuity of ρ_* on K_1^*.

DEFINITION 16. We set

$$\mathbb{S}_{\infty, \pi_*} := K_1^* \times \mathbb{Z}_p \subset \mathbb{S}_\infty.$$

For $(x, y) \in \mathbb{S}_{\infty, \pi_*}$, we set

$$(30) \qquad \rho_*(x, y) := (\rho_*(x), \hat{\rho}_*(y)) \in \mathbb{S}_{\infty, \pi_*},$$

with $\hat{\rho}$ as in Definition 13.

One then easily computes that for an integer j

$$(31) \qquad \rho_*(s_{-j}) = s_{-\rho_*(j)},$$

which we would like for any extension of the action of ρ on $\mathbb{Z}_p$ and is very natural in terms of trivial zeroes (see Subsection 6.1). As of this writing, we know of no natural way to extend Definition 16 to all of $\mathbb{S}_\infty$.

REMARK 5. Let v be a prime of A, where A is now arbitrary. We discuss here briefly the v-adic theory associated to $\zeta_{A,\pi}(s)$. Let k_v be the associated completion of k with fixed algebraic closure $\bar{k}_v$ equipped with the canonical absolute value, etc. Let $\mathbb{C}_v$ be the associated completion. As explained in Subsection 8.3 of [**Go1**], the special polynomials (Definition 9) interpolate to two-variable functions on

$$\mathbb{C}_v^* \times \mathbb{S}_v,$$

where $\mathbb{S}_v = \mathbb{Z}_p \times \mathbb{Z}/(q^\delta - 1)$, and δ is determined by the degree of v plus a choice of injection, over k, of the value field V into $\mathbb{C}_v$. (This construction is quite similar to the p-adic interpolation of classical L-values.) The appropriate group of homeomorphisms of $\mathbb{S}_v$ is then $S_{(q^\delta)}$, as its elements preserve congruence classes modulo $q^\delta - 1$.

Finally, we finish this section with the following finiteness result whose easy proof will be left to the reader. Note first that by Part 5 of Proposition 4, for any constant c, the set $X(q,c)$ consisting of all positive integers n with $\ell_q(n) = c$ is stable under $S_{(q)}$.

PROPOSITION 7. *The set $X(q,c)$ consists of finitely many orbits of $S_{(q)}$.*

6. $S_{(q)}$ as Symmetries of $\zeta(s)$

In this last section, we present the evidence showing how $S_{(q)}$ and its subgroups arise as symmetries of the zeta functions and values of Subsection 4.2. The evidence we have is Eulerian by its very nature, as we *only* use the special values. However, unlike Euler, we cannot now guess at the mechanism that exhibits these groups as automorphisms of the full two-variable zeta function.

As we saw in Proposition 4, the group $S_{(q)}$ permutes both the positive and negative integers. Each set of integers separately gives evidence that $S_{(q)}$ acts as symmetries of $\zeta(s)$. However, it may ultimately turn out that both types of evidence are really manifestations of the same underlying symmetries.

We begin with the evidence from the negative integers.

6.1. Evidence from Zeta Values at Negative Integers

In this subsection we present the evidence that $S_{(q)}$ acts as symmetries of $\zeta(s)$ arising from the negative integers. We believe that this evidence has greater impact than the evidence given Subsection 6.3 (at the positive integers) because it represents actual symmetries associated to the zeroes of $\zeta(s)$. We shall see, experimentally at least, that the orders of vanishing of $\zeta(s)$ at negative integers (as recalled in Subsection 4.2) appear to be invariants of the action of $S_{(q)}$.

Here is what is known about such vanishing in general. Recall that $q_\infty := q^{d_\infty}$, where d_∞ is the degree of ∞ over $\mathbb{F}_q$. Let j be a positive integer which is divisible by $q_\infty - 1$. Then it is known that $\zeta(-j) = 0$ (see, e.g., Subsection 8.13 of [**Go1**]). Current theory naturally gives very classical looking *lower bounds* on the order of vanishing of these "trivial zeroes." In the case $d_\infty = 1$ this bound is 1; when $d_\infty > 1$, it may be greater than 1. As our examples here all have $d_\infty = 1$, we refer the interested reader to [**Go1**] for the general case.

The first example is $A = \mathbb{F}_q[T]$. Here it is known [**Sh1**] that *all* zeroes are simple and that $\zeta(-j) \neq 0$ for $j \not\equiv 0 \pmod{q-1}$. By Part 6 of Proposition 4, we have $j \equiv \rho_*(j) \pmod{q-1}$. Thus, the next proposition follows immediately.

PROPOSITION 8. *Let $A = \mathbb{F}_q[T]$. Then the order of vanishing of $\zeta(s)$ at $-j$, j positive, is an invariant of the action of $S_{(q)}$ on the positive integers.*

By itself, Proposition 8 is certainly not overwhelming evidence for realizing $S_{(q)}$ as a symmetry of the full zeta function. However, in the case of some nonpolynomial A with ∞ rational, Dinesh Thakur [**Th1**], Theorem 5.4.9 of [**Th2**], and Javier Diaz-Vargas [**DV2**] have produced some fundamentally important calculations on zeroes at negative integers. They found examples of trivial zeroes where the natural lower bound was *not* the exact order of vanishing; of course, this is something that never happens in the classical analytic theory of L-series (but see Subsection 6.3.1 for analogies in the classical theory of Bernoulli numbers as well as justification for the terminology which follows). Zeroes arising at the negative integers with order of vanishing strictly greater than the predicted lower bound will be called "irregular," otherwise they are said to be "regular." (N.B.: In earlier versions of this work, as well as other papers, we called such zeroes "nonclassical.")

The calculations of Thakur and Diaz-Vargas seem to imply that the irregular trivial zeroes which occur at $-j$, j a nonnegative integer, actually occur where $\ell_q(j)$ is *bounded*. Moreover, their calculations *also* continue to exhibit $S_{(q)}$ invariance of the orders of vanishing of even these irregular zeroes.

As such we expect Proposition 8 to remain true for general A, where q will need to be replaced by q_∞.

REMARK 6. If Proposition 8 is true for general A, and the irregular trivial zeroes are characterized by having bounded sums of q-adic digits, then Proposition 7 immediately implies that there are only finitely many possibilities for the order of zero at irregular trivial zeroes. This result is quite reasonable.

As analytic objects on $\mathbb{S}_\infty$, our zeta functions are naturally 1-parameter families of entire power series, where the parameter is $y \in \mathbb{Z}_p$. Having such a huge group acting on $\mathbb{Z}_p$ may ultimately give us good control of the family. We will see serious evidence for this in our next subsection.

6.2. Evidence from Special Polynomials

We begin by recalling some relevant history as well as results. Let i and e be nonnegative integers. Let $A = \mathbb{F}_q[T]$ and let $A^+(e)$ be the monic polynomials of degree e inside A. Define

$$(32) \qquad S_e(i) := \sum_{f \in A^+(e)} f^i.$$

Let π be a uniformizer in k_∞ as before, with associated 1-unit parts $\langle f \rangle = \pi^e f$. Let $t \in \mathbb{Z}_p$. Define

$$(33) \qquad \tilde{S}_e(t) := \sum_{f \in A^+(e)} \langle f \rangle^t.$$

Clearly $\tilde{S}_e(t)$ is continuous in t. Obviously, for i a nonnegative integer, $\pi^{ei} S_e(i) = \tilde{S}_e(i)$, and so both sides are nonzero for the same i.

In [**Ca4**], L. Carlitz mentions the following necessary and sufficient criterion for $S_e(i)$ to be nonzero: There should be an expression

$$(34) \qquad i = i_0 + i_1 + \cdots + i_e,$$

such that all i_j are nonnegative, and such that
  (1) there is no carryover of p-adic digits in the sum,
  (2) for $0 \le j < e$, we have $i_j > 0$ and divisible by $q - 1$.
We agree to call such an expression for i an "admissible representation for i relative to e." The necessity actually follows easily from expanding $f \in A^+(e)$ by the multinomial theorem.

Dinesh Thakur astutely realized that Carlitz's criterion, along with a corresponding formula for the degree of $S_e(i)$, could be used to compute the Newton polygons associated to the power series arising from ζ_A (which do *not* depend on the choice of π). However, it fell to J. Sheats, in [**Sh1**], to give the first rigorous proof of Carlitz's assertions for general q. Along the way, Sheats also established the following results.

PROPOSITION 9. 1. $\tilde{S}_e(t) = 0$ *implies that* t *is a nonnegative integer.*
2. *Let* i *be a nonnegative integer. Then*

$$S_e(i) = 0 \Rightarrow S_{e+1}(i) = 0,$$

for all $e \geq 0$.

PROPOSITION 10. *Let* i *be a nonnegative integer with* $i = i_0 + i_1 + \cdots + i_e$ *an admissible representation of* i *relative to* e. *Let* $\rho_* \in S_{(q)}$. *Then* $\rho_*(i) = \rho_*(i_0) + \cdots + \rho_*(i_e)$ *is an admissible representation of* $\rho_*(i)$ *relative to* e.

PROOF. This follows immediately from Parts 2 and 6 of Proposition 4. $\square$

COROLLARY 3. *Let* e, i *and* ρ *be as above. Then we have*

(35) $S_e(i) = 0 \Longleftrightarrow S_e(\rho_*(i)) = 0.$

PROOF. This follows immediately from the proposition and Sheats' results. $\square$

COROLLARY 4. *Let* j *be a nonnegative integer with associated special polynomial* $z(x, -j)$ *(Definition 9). Let* ρ_* *be as in the proposition. Then* $z(x, -j)$ *and* $z(x, -\rho_*(j))$ *have the same degree in* x^{-1}.

COROLLARY 5. *Let* $X_e \subset \mathbb{Z}_p$ *be the zero set of* $\tilde{S}_e(t)$. *Then* X_e *is stable under the action of* $S_{(q)}$ *on* t.

PROOF. By Sheats we know that any zeroes of $\tilde{S}_e(t)$ must be nonnegative integers. By the proposition, we therefore see that

$$\tilde{S}_e(t) \neq 0 \Longleftrightarrow \tilde{S}_e(\rho_*(t)) \neq 0$$

for all t and homeomorphisms ρ_*. So the result follows immediately. $\square$

REMARK 7. Calculations suggest that X_e consists of finitely many orbits of $S_{(q)}$.

REMARK 8. Warren Sinnott [**Si1**] has studied a large class of functions on $\mathbb{Z}_p$ containing uniform limits of $\tilde{S}_e(t)$ (for arbitrary e of course). Let $f(t)$ be one nontrivial such function. Then Sinnott shows that the zero set of $f(t)$ cannot contain an open set, unlike general continuous functions on $\mathbb{Z}_p$. The results given above suggest that the zero set of $f(t)$ may in fact be countable.

REMARK 9. Gebhard Böckle has informed me that Sheats' results imply that the degree of $z(x, -j)$ as a polynomial in x^{-1} equals

$$\min_i \{ [\ell_q(jp^i)/(q-1)] \},$$

where $i = 0, 1, \ldots, n_0 - 1$ and $q = p^{n_0}$.

REMARK 10. The results of Jeff Sheats just mentioned were established along the way in Sheats' proof of the Riemann hypothesis for $\zeta_{\mathbb{F}_q[T]}(s)$. That is, as mentioned before, following D. Wan, D. Thakur, J. Diaz-Vargas and B. Poonen, J. Sheats was able to establish that $\zeta_{\mathbb{F}_q[T]}(s)$ has *at most* one (including multiplicity) zero in x^{-1} of a given absolute value for each fixed $y \in \mathbb{Z}_p$; this implies that the the zeroes in x^{-1} must be simple and "lie on the line" $\mathbb{F}_q((1/T))$.

It seems reasonable to expect that generalizations of the above results should hold for arbitrary A, etc.

Let K be a local nonarchimedean field of characteristic 0. A standard argument using Krasner's Lemma implies that K has only finitely many extensions of bounded degree. The same argument works in finite characteristic for degrees less than the characteristic (so one avoids inseparability issues).

Therefore, suppose that $\mathfrak{C}$ is an orbit of the nonnegative integers under $S_{(q_\infty)}$. Suppose that the special polynomials associated to elements in $\mathfrak{C}$ all have the same degree, which we assume is less than p. We can conclude that their zeroes all belong to a finite extension of k_∞.

QUESTION 1. Is it true in general that the zeroes of special polynomials i and j, where i and j are in the same orbit under $S_{(q_\infty)}$, lie in a finite extension of k_∞?

The results of Sheats for $A = \mathbb{F}_q[T]$ imply that all the zeroes of special polynomials lie in k_∞ in this case and so here the answer is obviously yes. Moreover, in Subsection 6.2.1 we will exhibit some evidence that the zeroes of special polynomials in the same orbit are related via $S_{(q_\infty)}$, so it makes sense to wonder whether this relationship carries over to their splitting fields.

Finally, let $A = \mathbb{F}_q[T]$ again for the rest of this subsection. Let $v = (f)$ be a prime of degree d in A with f monic. Recall that in Remark 5 we explained briefly how our functions interpolated v-adically; in the case at hand, it is easy to see that $\delta = d$. Let $A^+(e)_v$ be the monic polynomials of degree e which are prime to v. Define

$$(36) \qquad \tilde{S}_{e,v}(t) := \sum_{f \in A_v^+(e)} f^t,$$

where t now belongs to $\mathbb{S}_v = \mathbb{Z}_p \times \mathbb{Z}/(q^d - 1)$. It is also reasonable to expect our results just given to hold v-adically. Let $X_{e,v} \subseteq \mathbb{S}_v$ be the zero set of $\tilde{S}_{e,v}(t)$. We simply point out that we already can deduce many orbits of $S_{(q^d)}$ on the nonnegative integers which lie in $X_{e,v}$. Indeed, if $e < d$ then $A_v^+(e) = A^+(e)$; so $X_e \subseteq X_{e,v}$ (recall X_e consists of nonnegative integers which also project densely into $\mathbb{S}_v$) and is obviously stable under $S_{(q^d)}$.

Suppose now that $e \geq d$. Let i be a nonnegative integer. Clearly,

$$\tilde{S}_{e,v}(i) = S_e(i) - f^i S_{e-d}(i).$$

Thus, by Part 2 of Proposition 9 we see that $X_{e-d} \subseteq X_{e,v}$ and is obviously stable under $S_{(q^d)}$.

6.2.1. *Associated Computations*

In this subsection A will always be $\mathbb{F}_q[T]$. We shall present here some calculations, as well as a related result, that indicate rather strongly that the zeroes of $\zeta_A(s)$ are "stable" under the action given in Definition 16. Indeed, in Remark 10, we recalled that Sheats has established that all the zeroes of $\zeta_{\mathbb{F}_q[T]}(s)$ have x-coordinate in $\mathbb{F}_q((1/T))$ *and* that there is at most one zero of any given absolute value for fixed $y \in \mathbb{Z}_p$; so, in particular, as elements of $\mathbb{S}_\infty$, the zeroes actually lie in $\mathbb{S}_{\infty,\pi_*} = \mathbb{S}_{\infty,\pi}$.

We put the word "stable," above, in quotes because the situation is far from clear at the moment. Indeed, what actually appears to be stable is that part of the zeroes that is *invariant* upon changing one positive parameter for another, as indicated in Lemma 3. However, we do not yet have a clear general statement of this even for $\mathbb{F}_q[T]$. Moreover, for more general A, one knows that there are zeroes of $\zeta_A(s)$ lying *outside* $\mathbb{S}_{\infty,\pi_*}$.

PROPOSITION 11. *The set of positive parameters is a principal homogeneous space for the group $U_1(K)$ of 1-units.*

PROOF. Let π_1 and π_2 be two positive parameters. Then the quotient π_1/π_2 obviously belongs to $U_1(K)$. The result follows. $\qquad\square$

REMARK 11. Let π be a positive parameter and let $u \in K$ be a unit $\not\equiv 1$ (mod π). Then $u\pi$ will be a positive parameter for a different choice of "positive."

REMARK 12. In quantum field theory one is interested in the wave function ψ. This is a complex-valued function whose absolute value $|\psi|$ gives predictions for results in physical experiments. In other words, only $|\psi|$ has physical meaning. Let $U := \{e^{i\theta}\}$ be the "gauge group" of complex numbers of norm 1; obviously multiplication by an element of U does *not* change the absolute value (and conversely). One is thus free to multiply ψ by functions of the form $e^{ip(x,t)}$ for arbitrary real-valued functions $p(x,t)$ depending on position and time. From this freedom many remarkable things may be deduced. In the function field case, the "gauge group" $U_1(K)$ acts on the parameters and the values in line with Lemma 3. However, as we shall see, we are interested in more invariants than those just given by the absolute value!

Let π_1 and π_2 be two positive uniformizers. Lemma 3 tells us that if (α, y_0) is a zero of $\zeta_{A,\pi_1}(s)$ then $((\pi_1/\pi_2)^{y_0}\alpha, y_0)$ is a zero of $\zeta_{A,\pi_2}(s)$. It is then reasonable to ask what invariants arise in passing from α to $u^y\alpha$ for $u = \pi_1/\pi_2 \in U_1(K)$. For instance, obviously $|\alpha|_\infty = |u^y\alpha|_\infty$, where $|?|_\infty$ is the absolute value associated to the place ∞ (which is in keeping with the analogy to quantum theory just given). Let $t \in \mathbb{Z}$ be chosen so that $|\alpha/\pi_1^t|_\infty = |u^y\alpha/\pi_2^t|_\infty = 1$. Thus, we can write

$$\alpha = a\pi_1^t + \{\text{lower terms}\}$$

and

$$u^y\alpha = b\pi_2^t + \{\text{lower terms}\},$$

where a and b are nonzero elements of $\mathbb{F}_q$; as u is a principal unit, we conclude $a = b$. In other words, the whole monomial of least degree is an invariant of our action. We shall see shortly that there may be other invariant monomials.

We now present some elementary calculations that appear to indicate an action of $S_{(q)}$ on zeta zeroes.

EXAMPLE 3. Let $A = \mathbb{F}_3[T]$, $\pi = 1/T$ and $i = 13 = 1 + 3 + 3^2$. Let $H_{13} \subset S_{(3)}$ be the isotropy subgroup of 13. Let $\sigma_* \in H_{13}$; note that σ_* *must* permute the first three digits of a 3-adic integer (otherwise $\sigma_*(13) \neq 13$); this is simple but is also the key observation. Hand calculations then give

$$(37) \qquad \zeta_{A,\pi}(x, -13) = 1 - (\pi^4 + \pi^{10} + \pi^{12})x^{-1}.$$

Note that

$$\sigma_*(\pi^4 + \pi^{10} + \pi^{12}) = \pi^{\sigma_*(4)} + \pi^{\sigma_*(10)} + \pi^{\sigma_*(12)}$$
$$= \pi^4 + \pi^{10} + \pi^{12},$$

as $\sigma_*(4) = \sigma_*(1+3) = 1+3, 1+9,$ or $3+9$, etc., since the first three digits of 13 are permuted! Thus,

$$(38) \qquad \sigma_*(\pi^4 + \pi^{10} + \pi^{12}, -13) = (\pi^4 + \pi^{10} + \pi^{12}, -13),$$

and the zero is obviously fixed under the action arising from digit permutations as given in Definition 16.

EXAMPLE 4. Let $A = \mathbb{F}_2[T]$, $\pi = 1/T$ and $i = 3 = 1 + 2$ or $i = 5 = 1 + 4 = 1 + 2^2$. For $i = 3$ one finds

$$(39) \qquad \zeta_{A,\pi}(x, -3) = (1 - \alpha x^{-1})(1 - \beta x^{-1}),$$

where $\alpha = \pi^3$ (this is the trivial zero, which will arise for all i, as $q - 1 = 1$ in this case) and $\beta = \pi + \pi^2$. As above, if $\sigma_* \in H_3 =$ the isotropy subgroup of 3, then $\sigma_*(\alpha) = \alpha$ and $\sigma_*(\beta) = \beta$, so that σ_* fixes both $(\alpha, -3)$ and $(\beta, -3)$. A similar calculation holds for $i = 5$, where there is the trivial zero $(\pi^5, -5)$ and another zero $(\pi + \pi^4, -5)$. Note also that there exists permutations ρ such that $\rho_*(3) = 5$. One then finds immediately that

$$(40) \qquad \rho_*(\pi^3, -3) = (\pi^5, -5)$$

and

$$(41) \qquad \rho_*(\pi + \pi^2, -3) = (\pi + \pi^4, -5).$$

REMARK 13. The reader should note that having an action of $S_{(q)}$ on the zeroes as in the above examples appears to put extremely strong constraints on the form that the zeroes may take. For instance, let $q = 2$ and $i = 3$ as above. Note that if the second zero was $(\pi + \pi^4, -3)$ we would run into great difficulty, as π^4 has an infinite orbit under H_3.

In the arithmetic of function fields, the truly canonical ideas do *not* depend on the choice of positive parameter π. For instance, let $q = 2$ and $i = 3$ as in Example 4. The reader can easily find examples of positive parameters π so that when we transform the zero $(\pi + \pi^2, -3)$ using Lemma 3, we obtain a zero with expansion terms involving higher powers of π. As we just remarked, this causes absolute havoc with our hoped-for action of $S_{(2)}$. *However*, the answer to this appears to lie in the use of our "gauge group" $U_1(K)$ as in our next example.

EXAMPLE 5. Let $q = 2$ and now put $\pi_0 = 1/T$. As in Example 4 we see that $\zeta_{A,1/T}(x, -3) = (1 - \alpha x^{-1})(1 - \beta x^{-1})$ with $\alpha = \pi_0^3$ and $\beta = \pi_0 + \pi_0^2$. Let π be *any* other uniformizer and let $\tilde{\alpha}$, $\tilde{\beta}$ be the transformed zeroes under Lemma 3. It follows immediately that $\tilde{\pi}_0^3 = \pi^3$ and

$$\widetilde{\pi_0 + \pi_0^2} = (\pi/\pi_0)^3(\pi_0 + \pi_0^2) = \pi^3(\pi_0^{-1} + \pi_0^{-2})$$
$$= \pi^3(T + 1/T^2).$$

Now $\pi_0 = \pi + a_2\pi^2 + a_3\pi^3 + \cdots$, where $a_i \in \mathbb{F}_2$ and may be completely arbitrary (as π is a completely arbitrary positive uniformizer). Using long division to compute $T = \pi_0^{-1}$, we find that

$$(42) \qquad \widetilde{\pi_0 + \pi_0^2} = \pi + \pi^2 + 0 \cdot \pi^3 + O(\pi^4).$$

In other words, the terms "$\pi + \pi^2 + 0 \cdot \pi^3$" are invariant of the choice of π, *and* these are the ones that transform correctly under our action of $S_{(2)}$!

One can readily find other such examples. Finally we present a rather general result that gives further support for believing that the zeroes of $\zeta_{\mathbb{F}_q[T]}(s)$ are acted upon by $S_{(q)}$ in line with the above examples.

Let $q = p^{n_0}$ as above and let $y \in \mathbb{Z}_p$ be written as

$$(43) \qquad y = \sum_{i=0}^{N} c_i q^{e_i},$$

where $0 \neq c_i < q$ for all i, N may be infinite, and $e_0 < e_1 < \cdots$.

DEFINITION 17. We set

$$(44) \qquad y_c := \sum_{i=0}^{N} c_i q^i,$$

and call it the *q-adic collapse of y*.

For instance the 2-adic collapse of 5 is 3.

Let j now be a positive integer with associated q-adic collapse j_c. Let ρ be a permutation of $\{0, 1, \dots\}$ which has, in the above notation, the special values $\rho(i) = e_i$, for $i = 0, \dots, N$. (Obviously, as j is a positive integer, $N < \infty$.)

By Corollary 4, we see that $\zeta(x, -j_c)$ and $\zeta(x, -j)$ have the same degree d in x^{-1}. From Sheats we know that the zeroes lie in K and that there is at most one zero of each polynomial of a given absolute value. Let $\{\alpha_1, \dots, \alpha_d\}$ be the zeroes of $\zeta(x, -j_c)$, and let $\{\beta_1, \dots, \beta_d\}$ be the zeroes of $\zeta(x, -j)$, where

$$\operatorname{ord}_\pi \alpha_1 < \operatorname{ord}_\pi \alpha_2 < \cdots < \operatorname{ord}_\pi \alpha_d$$

and

$$\operatorname{ord}_\pi \beta_1 < \cdots < \operatorname{ord}_\pi \beta_d.$$

PROPOSITION 12. *Let j be as above, but assume that all q-adic digits of j are $< p$, (for instance, when $q = p$). Then*

$$(45) \qquad \operatorname{ord}_\pi(\beta_i) = \rho_*(\operatorname{ord}_\pi \alpha_i)$$

for $i = 1, \dots, d$.

PROOF. This follows from the calculations of these orders given by Diaz-Vargas in [**DV1**].   □

Calculations seem to indicate that the result is true for arbitrary $y \in \mathbb{Z}_p$. We hope to return to these themes in future work.

6.3. Evidence from Zeta Values at Positive Integers

In this subsection we discuss the evidence from the positive integers and, in particular, the evidence arising from Carlitz's analog, Theorem 2, of the classical von Staudt-Clausen result (which computes the denominators of Bernoulli numbers and is recalled in Subsection 6.3.1 when we return to classical number theory).

The work of David Hayes on "sign-normalized" rank one Drinfeld modules (see [**Hay1**] or Chapter 7 of [**Go1**]) shows the existence of a special Drinfeld module ψ with the following properties: It is defined over the ring of integers in a certain Hilbert class field of k (ramified at ∞) lying in $\mathbb{C}_\infty$ which we denote by H^+. Let $\mathfrak{J}$ be an ideal of A. Then the product of all $\mathfrak{J}$-division values of ψ lies in H^+ and is an explicit generator of $\mathcal{O}^+\mathfrak{J}$, where $\mathcal{O}^+$ is the ring of A-integers. The lattice L associated to ψ may be written $A\xi$ for a transcendental element $\xi \in \mathbb{C}_\infty$.

Let $T := V \cdot H^+$ be the compositum of V and H^+, where V is defined after Equation 14. The following result is shown in [**Go1**] (Theorem 8.18.3).

THEOREM 1. *Let j be a positive integer divisible by $q_\infty - 1$ and let $\zeta(j)$ be defined as in* Definition 8. *Then*

$$(46) \qquad\qquad 0 \neq \zeta(j)/\xi^j \in T.$$

Theorem 1 was established in the basic $\mathbb{F}_q[T]$-case by L. Carlitz in the 1930's; in this case, both V and H^+ equal k.

Let $\mathcal{O}_T$ be the A-integers of T.

DEFINITION 18. *Let j be divisible by $q_\infty - 1$. We define $\widetilde{BC}_j$ to be the $\mathcal{O}_T$ fractional ideal generated by $\tilde{\Pi}_j\zeta(j)/\xi^j$.*

We call the fractional ideal $\widetilde{BC}_j$ the "j-th Bernoulli-Carlitz fractional ideal" as, again, these were originally defined (as elements in $\mathbb{F}_q(T)$) by Carlitz in the 1930's and are clearly analogous to the classical Bernoulli numbers (Definition 1).

DEFINITION 19. *Let $\mathfrak{d}_j := \{a \in A \mid a\widetilde{BC}_j \subseteq \mathcal{O}_T\}$.*

We call $\mathfrak{d}_j$ the "A-denominator of $\widetilde{BC}_j$;" it is obviously an ideal of A.

For $A = \mathbb{F}_q[T]$, Carlitz ([**Ca1**], [**Ca2**], [**Ca3**]) gives an explicit calculation of $\mathfrak{d}_j$ which we now recall. Let $q = p^{n_0} > 2$ for the moment.

THEOREM 2. (Carlitz) *There are two conditions on j:*
1. $h := \ell_p(j)/(p-1)n_0$ *is integral.*
2. $q^h - 1$ *divides j.*
If j satisfies both conditions, then $\mathfrak{d}_j$ is the product of all prime ideals of degree h. If j does not satisfy both conditions, then $\mathfrak{d}_j = (1)$.

Carlitz's result gives us the first (historically) indication of an action of $S_{(q)}$ on zeta values. This is given in the following corollary.

COROLLARY 6. *Let* $\mathfrak{P}$ *be a prime ideal of* A *of degree* h *with additive valuation* $v_{\mathfrak{P}}$. *Then* $v_{\mathfrak{P}}(\mathfrak{d}_j)$ *is an invariant of the action of* $S_{(q^h)}$ *on the positive integers divisible by* $q - 1$.

What about $q = 2$? Here ([**Ca3**]) the same result holds if $h \neq 2$. More precisely Carlitz established the following result.

THEOREM 3. (Carlitz) *Let* $q = 2$ *and consider the system given in* Theorem 2. *If this system is consistent for* $h = \ell_2(j) \neq 2$, *then* $\mathfrak{d}_j$ *is the product of all prime ideals of degree* h. *If it is consistent for* $h = 2$, *then for* j *even we have*

$$(47) \qquad\qquad \mathfrak{d}_j = (T^2 + T + 1),$$

while, for j *odd, we have*

$$(48) \qquad\qquad \mathfrak{d}_j = (T^2 + T \cdot T^2 + T + 1).$$

If the system is inconsistent and j *is of the form* $2^\alpha + 1$ *(so* $\ell_2(j) = 2$*), then*

$$(49) \qquad\qquad \mathfrak{d}_j = (T^2 + T).$$

If it is inconsistent and j *cannot be written as* $2^\alpha + 1$, *then* $\mathfrak{d}_j = (1)$.

COROLLARY 7. *(*$q = 2$*) Let* $\mathfrak{P}$ *be a prime of* A *of degree* h *and suppose first that* $h \neq 1$. *Then* $v_{\mathfrak{P}}(\mathfrak{d}_j)$ *is an invariant of the action of* $S_{(2^h)}$ *on the positive integers. If* $h = 1$, *then* $v_{\mathfrak{P}}(\mathfrak{d}_j)$ *is an invariant of the subgroup* $\tilde{S}_{(4)}$ *of* $S_{(4)}$ *arising from permutations of* $\{0, 1, 2, \ldots\}$ *fixing* 0.

It is reasonable to expect these symmetries to persist when Carlitz's results are generalized to arbitrary A where, again, one will need to replace q with q_∞.

Let $A = \mathbb{F}_q[T]$ and let $\mathfrak{P}$ be a prime of A of degree h. For simplicity assume that $q > 2$. Let i and j be two positive integers which are in the same orbit of $S_{(q^h)}$ and are divisible by $q - 1$. Suppose i satisfies Carlitz's two conditions given in Theorem 2; then by Corollary 6 so does j. In this case we see that $\widetilde{BC}_i$ and $\widetilde{BC}_j$ have the same order, -1, at $\mathfrak{P}$.

QUESTION 2. Let $\mathfrak{P}$ be a prime of $A = \mathbb{F}_q[T]$ of degree h. Suppose i and j are two nonnegative integers which are divisible by $q - 1$ and in the same orbit of $S_{(q^h)}$. Do $\widetilde{BC}_i$ and $\widetilde{BC}_j$ have the same order at $\mathfrak{P}$?

One can formulate variants of Question 2 for arbitrary A. We discuss a classical variant in our next subsection.

6.3.1. Classical Bernoulli Numbers

In this final subsection we formulate an analog of Question 2 for classical Bernoulli numbers (Section 2) as well as present a reasonable solution to it as an exercise in p-adic continuity.

Let p now be an odd prime and let B_n be the classical Bernoulli number for a positive integer n, so that the functional equation for the Riemann zeta function gives

$$(50) \qquad\qquad -B_n/n = \zeta(1 - n),$$

where $\zeta(s)$ now is the Riemann zeta function as in Section 2.

Next we recall some fundamental, and famous, classical results on which we shall build. First, we have the classical *Adams congruences* which state that if $p-1$ does not divide n, then B_n/n is integral at p. The following result now follows immediately.

PROPOSITION 13. *Let n be a positive integer which is even and not divisible by $p-1$. Let v_p denote the additive valuation associated to p. Then*

$$(51) \qquad\qquad v_p(B_n) \geq v_p(n).$$

Let $t := v_p(n)$. If $t > 0$ then we say that "B_n has a trivial zero of order at least t at p." In general, if $t = v_p(B_n)$, then we say that n is "regular" with respect to p; otherwise we say it is "irregular."

Note obviously that n is regular with respect to p if and only if $v_p(B_n/n) = 0$; thus, we see the connection with the usual notion of "regularity" from the theory of cyclotomic fields. This simple analogy is the motivation for our use of this terminology also for zeta zeroes at negative integers (Subsection 6.1).

Next suppose that $p-1$ divides n, n even. As above, let $\mathfrak{d}_n$ be the denominator of B_n. In this case the classical *von Staudt-Clausen Theorem* implies that $v_p(\mathfrak{d}_n) = 1$ and otherwise vanishes (N.B.: $v_2(\mathfrak{d}_n)$ is identically 1). We say that "B_n has a simple pole at p."

Finally, we recall the powerful *Kummer congruences* which state the following: Let $i > 0$ be a positive integer which is not divisible by $p-1$. Let $j > 0$ be another integer and assume that

$$(52) \qquad\qquad i \equiv j \pmod{p^{b-1}(p-1)}.$$

Then

$$(53) \qquad (1-p^{i-1})B_i/i \equiv (1-p^{j-1})B_j/j \pmod{p^b}.$$

Suppose now that $\rho_* \in S_{(p)}$ and let $m := \rho_*(n)$, where n is an even integer. We have seen that $m \equiv n \pmod{p-1}$ and, in particular, is also even. Question 2 leads us to consider whether $v_p(B_n) = v_p(B_m)$ in general. In this formulation, the answer is "no," as one sees with $p = 5$, $n = 2$ and $m = 10 = 2 \cdot 5$. Indeed, the numerator of B_2 is 1 and the numerator of $B_{10} = 5$ (so both 2 and 10 are regular with respect to $p = 5$ and B_{10} has a trivial zero at 5).

Note, however, that the classical von Staudt-Clausen result is simpler than its function field counterpart and only depends upon divisibility of n by $p-1$. As such, we suppress the action of $S_{(p)}$ here and concentrate on the more elementary notion of congruence. Under this assumption, a reasonable analogy to Question 2 may then be given as follows.

QUESTION 3. Let n be a positive even integer. Does there exist an integer $t \geq 0$, which depends on n, such that if m is another positive integer with $m \equiv n \pmod{(p-1)p^t}$, then $v_p(B_n) = v_p(B_m)$?

The following result provides a positive answer.

PROPOSITION 14. *Let p be an odd prime and n a positive even integer. Let*

$$t_0 := \max\{v_p(n), v_p(B_n/n)\}$$

and $t = t_0 + 1$. Suppose that m is another positive integer with

$$m \equiv n \pmod{(p-1)p^t}.$$

Then $v_p(B_n) = v_p(B_m)$.

PROOF. One sees immediately that $v_p(n) = v_p(m)$. Moreover, the Kummer congruences tell us that $v_p(B_n/n) = v_p(B_m/m)$. The result follows. $\square$

It may be interesting to find an upper bound for the integer t given in the proposition which depends only on n. Moreover, I do not know how often an integer m satisfying the hypothesis of Proposition 3 can be obtained from n by permuting its p-adic digits.

References

[Ay1] R. Ayoub, *Euler and the zeta function*, Amer. Math. Monthly **81**, 1067-1086.

[Ca1] L. Carlitz, *An analogue of the von Staudt-Clausen theorem*, Duke Math. J. **3**, 503-517.

[Ca2] L. Carlitz, *An analogue of the Staudt-Clausen theorem*, Duke Math. J. **7**, 62-67.

[Ca3] L. Carlitz, *An analogue of the Bernoulli polynomials*, Duke Math. J. **8**, 405-412.

[Ca4] L. Carlitz, *Finite sums and interpolation formulas over $GF[p^n, x]$*, Duke Math. J. **15**, 1001-1012.

[Co1] K. Conrad, *The digit principle*, J. Number Theory **84**, 230-257.

[DV1] J. Diaz-Vargas, *Riemann hypothesis for $\mathbb{F}_p[T]$*, J. Number Theory **59**, 313-318.

[DV2] J. Diaz-Vargas, *On zeros of characteristic p zeta function*, J. Number Theory **117**, 241-262.

[Go1] D. Goss, *Basic Structures of Function Field Arithmetic*, Springer-Verlag, Berlin.

[Go2] D. Goss, *Zeroes of L-series in characteristic p*, Int. J. Appl. Math. Stat. **11**, 69-80.

[Hay1] D. Hayes, *A brief introduction to Drinfeld modules*, in The Arithmetic of Function Fields (D. Goss et al., eds.), de Gruyter, pp. 1-32.

[Si1] W. Sinnott, *Dirichlet series in function fields*, J. Number Theory **128**, 1893-1899.

[Sh1] J. Sheats, *The Riemann hypothesis for the Goss zeta function for $\mathbb{F}_q[T]$*, J. Number Theory **71**, 121-157.

[Th1] D. Thakur, *On characteristic p zeta functions*, Comp. Mathematica **99**, 231-247.

[Th2] D. Thakur, *Function Field Arithmetic*, World Scientific.

DEPARTMENT OF MATHEMATICS, OHIO STATE UNIVERSITY, 231 W 18TH AVE., COLUMBUS, OHIO 43210, USA.

E-mail address: goss@math.ohio-state.edu

Renormalization by Birkhoff-Hopf Factorization and by Generalized Evaluators: A Case Study

Li Guo, Sylvie Paycha, and Bin Zhang

ABSTRACT. We compare different techniques used to evaluate divergent multiple sums and integrals, either being inspired by the Birkhoff-Hopf approach of Connes and Kreimer in quantum field theory renormalization or arising from generalized evaluators, as well as different procedures within the Birkhoff-Hopf framework. We then apply this study to evaluate divergent Riemann integrals indexed by rooted trees and by rooted trees decorated by symbols.

1. Introduction

Evaluating divergent multiple sums and divergent multiple integrals with constraints is a natural problem that arises both from perturbative quantum field theory and from number theory. From a mathematical point of view a Feynman integral associated with a Feynman diagram with a fixed number of loops[1] can be viewed as a multiple integral in the internal momenta with affine constraints expressed in terms of the external momenta imposed by the conservation of momenta at the vertices of the diagram. Similarly, multiple zeta functions can be seen as multiple sums with conical constraints. Whereas physicists have developed very sophisticated techniques to evaluate divergent Feynman diagrams, some of which have inspired mathematicians to evaluate divergent multiple zeta values, a unified mathematical approach which encompasses both divergent sums and divergent integrals is still not completed in spite of the substantial progress made in recent years. Indeed, the seminal work of Connes and Kreimer [4] on the algebraic approach of renormalization in perturbative quantum field theory brought the physical theory of renormalization once again to the forefront of active mathematical study [3,5,6,7,9,10] and allowed the method of renormalization to be applied to pure mathematics subjects such as multiple zeta values [18,27]. One of the major questions in both directions is the uniqueness of the values obtained by different procedures, or how to compare the results derived by different evaluation techniques. This question, which in physics is formulated in terms of renormalization groups, still remains mysterious in the mathematical context.

In a pure mathematical context considered here, there are no physical constraints usually imposed on a physical model. However a basic principle in physics,

---

[1]We henceforth identify the diagram and its associated integral.

the principle of locality, can be translated into mathematical terms. Locality means that distant objects cannot have direct influence on one another: an object is influenced directly only by its immediate surroundings.[2] The locality principle in physics implies that the evaluation of two concatenated Feynman diagrams corresponds to the product of the evaluated individual diagrams. From a mathematical point of view, it says that evaluation of multiple integrals or multiple sums with constraints should factorize over independent sets of constraints. From this point of view, the locality principle translates to a compatibility of the evaluator with the product at hand, namely the concatenation in the context of Feynman diagrams or stuffle product in the context of multiple zeta functions. Evaluating while preserving compatibility with the underlying product and hence locality is indeed our main concern in the paper. In some cases, Speer's generalized evaluators[3] do the job; in others, a Birkhoff-Hopf factorization[4] is more appropriate.

As an attempt to address these questions, this paper discusses the relationship between different evaluation techniques and, as a case study, applies these techniques to evaluate divergent Riemann integrals indexed by (decorated) rooted trees, with the latter case as a toy model of Feynman integrals initiated in the work of Connes and Kreimer. Although we are aware that renormalization in physics goes beyond just evaluating Feynman integrals with fixed loop number while preserving locality, in view of the interpretation of the locality principle suggested above, we shall abusively use the term "renormalization procedure" for techniques that evaluate divergent quantities while preserving the underlying product structure.[5]

The paper essentially consists of two parts, the first one describing the theoretical background for the applications described in the second part. The first part, consisting of Section 2, is dedicated to the description and comparison of various renormalization procedures, especially the renormalization by Birkhoff-Hopf factorization and by generalized evaluators. Renormalization by Birkhoff-Hopf factorization has become a classical renormalization tool

(1) leading to renormalized Feynman integrals by means of a Hopf algebra of Feynman diagrams [**4,22**], and

(2) leading to multiple zeta values obeying stuffle relations by means of the stuffle Hopf algebra [**18,19,27**].

On the other hand, there is also renormalization by generalized evaluators which dates back to [**31**] and has recently received renewed attention [**17,29,30**]. In contrast with the Birkhoff-Hopf renormalization used in physics based on dimensional regularization which introduces one regularizing parameter from a modification of the ambient dimension, this approach by generalized evaluators requires introducing various regularizing parameters. While it might seem somewhat unnatural in

---

[2]This was stated as follows by Albert Einstein [**11**]: "The following idea characterizes the relative independence of objects far apart in space (A and B): external influence on A has no direct influence on B; this is known as the Principle of Local Action, which is used consistently only in field theory. If this axiom were to be completely abolished, the idea of the existence of quasienclosed systems, and thereby the postulation of laws which can be checked empirically in the accepted sense, would become impossible."

[3]This approach is known in the physics literature under the name of analytic renormalization; see [**31**].

[4]See [**25**] for a mathematical description.

[5]The compatibility one expects of the evaluation procedures for Feynman diagrams in a perturbative expansion with different loop numbers is ensured here by the compatibility with a natural filtration in the number of variables.

the context of dimensional regularization, it is more satisfactory from a mathematical point of view, as it enables a precise description of poles arising from the singularities to be cured by renormalization. Comparing it with Birkhoff-Hopf renormalization therefore involves identifying the regularizing parameters. In Section 2 we describe renormalization via generalized evaluators and compare it with the Birkhoff-Hopf renormalization, and we prove that the latter can be obtained from renormalization by generalized evaluators under some natural conditions. We also compare renormalization procedures within the framework of Connes-Kreimer by algebraic Birkhoff decomposition.

The second part, consisting of the remaining sections of the paper, implements these methods on iterated integrals naturally associated to rooted trees. In Section 3 we recall the Hopf algebra structure on decorated rooted trees and their characterization as free operated algebras. Then in Section 4 we implement various renormalization techniques described in the first part to renormalize iterated integrals associated with rooted trees. These techniques are generalized in Section 5 to renormalize iterated integrals associated with rooted trees decorated by pseudodifferential symbols. We compute Birkhoff-Hopf renormalized iterated integrals associated with rooted trees, which we then compare with the values obtained using generalized evaluators. This extends our previous work [**17**] in which we showed (see Section 6.6.3 in [**17**]) that depth 2 renormalized multiple zeta values at negative integers derived using a generalized (symmetrized) evaluator coincide with those derived using an algebraic Birkhoff factorization.

Comparing different renormalization procedures, as we do here for the toy model given by iterated integrals associated with rooted trees, is helpful to get a better grasp on the renormalization group. This we hope to achieve in a forthcoming paper via a close study of renormalized sums of polynomials on cones using different renormalization procedures, one of which involves a Hopf algebra on cones inspired by [**2**].

Acknowledgments. L. Guo acknowledges the support from NSF grant DMS-0505643 and thanks the Center of Mathematics at Zhejiang University for its hospitality. L. Guo and S. Paycha are grateful to JAMI at Johns Hopkins University and thank K. Consani for inviting them to deliver a series of lectures at Johns Hopkins University on topics related to the contents of this paper. B. Zhang acknowledges the support from NSFC grant 10631050. The authors also thank the referee for helpful comments.

2. Comparison of Renormalization Schemes

We compare the Birkhoff-Hopf renormalization in the algebraic framework of Connes-Kreimer [**5**] with the renormalization by generalized evaluators, as well as comparing different Birkhoff-Hopf renormalization schemes. After providing background definitions and examples, we show that a class of Birkhoff-Hopf renormalizations can be realized by generalized evaluators. We finally give some results on the comparison of renormalization schemes in the Connes-Kreimer framework.

2.1. Tensor Algebras of Meromorphic Functions and Generalized Evaluators

2.1.1. Tensor Algebras of Meromorphic Functions

We first define the kind of functions that we would like to consider.

DEFINITION 2.1.
(1) A **filtered tensor family** is a family $\mathcal{I} := \{I_k\}_{k\geq 1}$, where I_k is a set of
    linear forms (i.e., linear functions) in the variables $z_1, \cdots, z_k$ such that
    (i) $I_k \subset I_{k+1}, k \geq 1$, where a function in the variables $z_1, \cdots, z_k$ is viewed
        as a function in the variables $z_1, \cdots, z_{k+1}$ which is constant in z_{k+1};
    (ii) for every $L_k \in I_k$ and $L_{k'} \in I_{k'}$, we have $L_k \otimes L_{k'} \in I_{k+k'}$.
(2) For a filtered tensor family $\mathcal{I}$, the **filtered algebra with filtered tensor
    family of poles** $\mathcal{I}$, or **type** $\mathcal{I}$ **filtered algebra** in short, is the union
    $\mathcal{F}_\otimes := \mathcal{F}_\mathcal{I} = \cup_{k=1}^\infty \mathcal{F}_k$, where $\mathcal{F}_k$ is the set of functions $f : \mathbb{C}^k \to \mathbb{C}$ whose
    poles are controlled by $\mathcal{I}$. More precisely, for each $f \in \mathcal{F}_k$, there is a map

$$m : I_k \to \mathbb{N}, \quad L_k \mapsto m_{L_k}, \quad L_k \in I_k$$

with finite support such that the map $f(\underline{z}) \prod_{L_k \in I_k} (L_k(\underline{z}))^{m_{L_k}}$ is holomorphic around $\underline{z} = 0$. Here we denote $\underline{z} = (z_1, \cdots, z_k)$.

The term and tensor notation of a type $\mathcal{I}$ filtered algebra $\mathcal{F}_\otimes$ are justified by the following property.

LEMMA 2.2. *If $\mathcal{I}$ is a filtered tensor family of linear forms, then $\mathcal{F}_\mathcal{I}$ is a filtered algebra for the tensor product.*

PROOF. For $k \geq 1$, since $I_k \subseteq I_{k+1}$, we clearly have $\mathcal{F}_k \subseteq \mathcal{F}_{k+1}$. So $\mathcal{F}$ is filtered. Let

$$f = \frac{h}{\prod_{L_k \in I_k} (L_k)^{m_{L_k}}} \in \mathcal{F}_k \quad \text{and} \quad f' = \frac{h'}{\prod_{L_{k'} \in I_{k'}} (L_{k'})^{m_{L_{k'}}}} \in \mathcal{F}_{k'}$$

for some holomorphic functions h on $\mathbb{C}^k$ and h' on $\mathbb{C}^{k'}$. Then

$$f \otimes f' = \frac{h \otimes h'}{\prod_{L_k \in I_k} (L_k)^{m_{L_k}} \otimes \prod_{L_{k'} \in I_{k'}} (L_{k'})^{m_{L_{k'}}}}$$

lies in $\mathcal{F}_{k+k'}$ since $L_k \otimes L_{k'}$ is in $L_{k+k'}$.   □

REMARK 2.3. The set $I_k = \{L_1, \cdots, L_{k_i}\}$ corresponds to an arrangement of hyperplanes $H_i : L_i(z_1, \cdots, z_k) = 0$ in $\mathbb{C}^k$ in the sense of [8] and [12].

EXAMPLE 2.4. If we take I_k to be the set of all linear forms in k variables, then $\mathcal{I} = \{I_k\}_{k\geq 1}$ is a filtered tensor family. We denote the corresponding filtered algebra by $\mathcal{M}_\otimes$ and call it the **algebra of germs of meromorphic functions with linear poles**.

EXAMPLE 2.5. The set $\mathcal{B}_\otimes = \cup_{k\geq 1} \mathcal{B}_k$ with

$$\mathcal{B}_k := \left\{ f : \mathbb{C}^k \mapsto \mathbb{C} \;\middle|\; \begin{array}{c} f(z_1, \cdots, z_k) \prod_{\tau \in \Sigma_k} \left(\prod_{j=1}^k (z_{\tau(1)} + \cdots + z_{\tau(j)})^{m_{\tau,j}} \right) \\ \text{is holomorphic around } \underline{z} = 0 \end{array} \right\},$$

equipped with the tensor product is a type $\mathcal{I}$ filtered algebra for the filtered tensor family $\mathcal{I} = \{I_k\}_{k\geq 1}$ with $I_k = \{z_{\tau(1)} + \cdots + z_{\tau(j)} \mid 1 \leq j \leq k, \tau \in \Sigma_k\}, k \geq 1$. The hyperplanes of poles correspond to the arrangement

$$H_{\tau,j} : z_{\tau(1)} + \cdots + z_{\tau(j)} = 0, \quad \tau \in \Sigma_k, \quad j \in \{1, \cdots, k\}.$$

EXAMPLE 2.6. The set $\mathcal{S}_\otimes = \cup_{k \geq 1} \mathcal{S}_k$ with

$$\mathcal{S}_k := \left\{ f : \mathbb{C}^k \mapsto \mathbb{C} \;\middle|\; \begin{array}{c} \exists m_{ij} \in \mathbb{Z}_{\geq 0} \text{ such that } f(z_1, \cdots, z_k) \prod_{1 \leq i < j \leq k} (z_i - z_j)^{m_{ij}} \\ \text{is holomorphic around } \underline{z} = 0 \end{array} \right\},$$

equipped with the tensor product is a type $\mathcal{I}$ filtered algebra for the tensor filtered family $\mathcal{I} = \{I_k\}_{k \geq 1}$ with $I_k = \{z_i - z_j \mid 1 \leq i < j \leq k\}$, $k \geq 1$. The hyperplanes of poles correspond to braid arrangements (see e.g., [**12**])

$$H_{ij} : x_i = x_j, \quad 1 \leq i < j \leq k.$$

2.1.2. Generalized Evaluators

Let us introduce the concept of generalized evaluators inspired by Speer's analytic renormalization approach.

DEFINITION 2.7. ([**29,30,31**]) A **generalized evaluator**[6] at 0 is a character

$$\mathcal{E} : \mathcal{F}_\otimes \to \mathbb{C}$$

on a type $\mathcal{I}$ filtered algebra $\mathcal{F} = \bigcup_{k=1}^\infty \mathcal{F}_k$ which coincides with the evaluation at zero on holomorphic functions around zero. More precisely, denoting $\mathcal{E}_k = \mathcal{E}|_{\mathcal{F}_k}$, we have

(1) $\mathcal{E}_k$ is linear,
(2) $\mathcal{E}_k$ is compatible with the filtration on $\mathcal{F}$: $\mathcal{E}_{k+1}\big|_{\mathcal{F}_k} = \mathcal{E}_k$,
(3) $\mathcal{E}_k$ coincides with the evaluation at 0 on holomorphic functions around 0, and
(4) $\mathcal{E}_k$ is multiplicative on tensor products, i.e.,

$$(1) \qquad \mathcal{E}_{k+k'}(f \otimes f') = \mathcal{E}_k(f)\,\mathcal{E}_{k'}(f'), \quad \forall f \in \mathcal{F}_k, f' \in \mathcal{F}_{k'}.$$

We say that the evaluator is symmetric if $\mathcal{E}$ is symmetric in the variables z_i, i.e., $\mathcal{E}(f \circ \tau) = \mathcal{E}(f)$ for any permutation τ of variables.

EXAMPLE 2.8. For a Laurent series $f(z) = \sum_{i \geq -k} a_i z^i$ at $z = 0$, let $\mathrm{fp}|_{z=0}$ be the projection operator $\mathrm{fp}|_{z=0}(f) = a_0$. Define a map $\mathcal{E}^0$ on a type $\mathcal{I}$ filtered algebra $\mathcal{F}_\otimes$, in particular on $\mathcal{M}_\otimes$, by taking

$$(2) \qquad \mathcal{E}_k^0(f) := \mathrm{fp}_{z_k=0} \left(\cdots \left(\mathrm{fp}_{z_1=0}\, f(\underline{z}) \right) \cdots \right), \quad f \in \mathcal{F}_k.$$

Then $\mathcal{E}^0$ is a generalized evaluator. Indeed, it is clearly linear and coincides with the evaluation at zero on holomorphic functions. It is compatible with the filtration since for any f in $\mathcal{F}_k$ seen as a function in $k + 1$ variables which is constant in the $(k+1)$-th variable, we have

$$\begin{aligned} \mathcal{E}_{k+1}^0(f) &= \mathrm{fp}_{z_{k+1}=0} \left(\cdots \left(\mathrm{fp}_{z_1=0}\, f(z_1, \cdots, z_k) \right) \cdots \right) \\ &= \mathrm{fp}_{z_k=0} \left(\cdots \left(\mathrm{fp}_{z_1=0}\, f(z_1, \cdots, z_k) \right) \cdots \right) \\ &= \mathcal{E}_k^0(f). \end{aligned}$$

---

[6]Our definition differs from Speer's insofar as unlike Speer (see definition in Section 3 of [**31**]), we do not assume either symmetry or uniform analyticity. We introduce symmetry as an extra requirement on the evaluator and drop uniform analyticity altogether.

It is compatible with the tensor product. Indeed, adopting the notations of the above definition we have

$$
\begin{aligned}
\mathcal{E}^0_{k+k'}(f \otimes f') &= \mathrm{fp}_{z_{k+k'}=0} \left(\cdots \left(\mathrm{fp}_{z_1} (f \otimes f')(z_1, \cdots, z_k, \cdots z_{k+k'}) \right) \cdots \right) \\
&= \mathrm{fp}_{z_{k+k'}=0} \left(\cdots \left(\mathrm{fp}_{z_1=0} (f(z_1, \cdots, z_k) f'(z_{k+1}, \cdots z_{k+k'})) \right) \cdots \right) \\
&= \mathrm{fp}_{z_k=0} \left(\cdots \left(\mathrm{fp}_{z_1=0} f(z_1, \cdots, z_k) \right) \cdots \right) \\
&\quad \cdot \mathrm{fp}_{z_{k+k'}=0} \left(\cdots \left(\mathrm{fp}_{z_{k+1}=0} f'(z_{k+1}, \cdots, z_{k+k'}) \right) \cdots \right) \\
&= \mathcal{E}^0_k(f) \, \mathcal{E}^0_{k'}(f').
\end{aligned}
$$

EXAMPLE 2.9. Fix a $k \geq 1$ and $\tau \in \Sigma_k$. Use the same notations as in the last example and define

$$
\mathcal{E}^0_\tau(f) := \cdots \mathrm{fp}_{z_{k+1}=0} \left(\mathrm{fp}_{z_{\sigma(k)}=0} \left(\cdots \left(\mathrm{fp}_{z_{\sigma(1)}=0} f(\underline{z}) \right) \cdots \right) \right), \quad f \in \mathcal{F}.
$$

Then we obtain a generalized evaluator—the permutated version of $\mathcal{E}^0$.

Also

$$
(3) \qquad \mathcal{E}^{sym}_k(f) := \frac{1}{k!} \sum_{\sigma \in \Sigma_k} \mathrm{fp}_{z_{\sigma(k)}=0} \left(\cdots \left(\mathrm{fp}_{z_{\sigma(1)}=0} f(\underline{z}) \right) \cdots \right), \quad f \in \mathcal{F}_k,
$$

yields a symmetrized generalized evaluator $\mathcal{E}^{sym}$ on $\mathcal{F}$.

As discussed in [17,29,30,31], generalized evaluators provide a nice renormalization scheme if the regularized homomorphisms take values in $\mathcal{M}_\otimes$. In this case, renormalization by generalized evaluators is simpler than renormalization by Birkhoff factorizations. But in general lack of compatibility of the product with the tensor product introduces technical difficulties when renormalizing, which are common to both Birkhoff-Hopf renormalization and renormalization by generalized evaluators.

In this paper, in addition to the case discussed in [17], we discuss more applications of generalized evaluators in later sections.

2.2. Generalized Evaluators from Birkhoff Factorizations

Renormalization by generalized evaluators and renormalization by Birkhoff factorizations are closely related. The following construction via renormalization by Birkhoff factorization yields another type of generalized evaluator on subalgebras of the filtered algebra $\mathcal{B}_\otimes$.

Let $\mathcal{K}_\otimes$ be a filtered subalgebra of the filtered algebra $\mathcal{B}_\otimes$ equipped with a connected filtered Hopf algebra structure such that $\Delta(\mathcal{K} \cap \mathrm{Hol}(\mathbb{C})) \subset \mathrm{Hol}(\mathbb{C}) \otimes \mathrm{Hol}(\mathbb{C})$. The diagonal map

$$
\begin{aligned}
\delta : \mathbb{C} &\to \mathbb{C}^\infty \\
(4) \qquad\qquad z &\mapsto z(1, 1, \cdots, 1, \cdots) = (z, z, \cdots, z, \cdots)
\end{aligned}
$$

induces an algebra homomorphism

$$
\begin{aligned}
\delta^* : \mathcal{B}_\otimes &\to \mathrm{Mer}(\mathbb{C}), \\
(5) \qquad\qquad f &\mapsto (\delta^* f : z \mapsto f \circ \delta(z)).
\end{aligned}
$$

Therefore, we have an algebra homomorphism

$$\delta^* : \mathcal{K}_\otimes \hookrightarrow \mathcal{B}_\otimes \to \mathrm{Mer}(\mathbb{C}),$$

which in turn induces via renormalization by Birkhoff factorization (with the Rota-Baxter algebra $(\mathrm{Mer}(\mathbb{C}), \mathrm{id-fp})$, where $\mathrm{id-fp}$ is taking the pole part of a Laurent series) an algebra homomorphism

$$\delta^*_+ : \mathcal{K}_\otimes \to \mathrm{Hol}(\mathbb{C}).$$

This further gives rise to a map on $\mathcal{K}_\otimes$ defined by

$$\mathcal{E}_\delta : \mathcal{K}_\otimes \to \mathbb{C},$$
$$f \mapsto \delta^*_+(f)(0).$$

PROPOSITION 2.10. $\mathcal{E}_\delta$ is a generalized evaluator on $\mathcal{K}_\otimes$.

PROOF. By construction, $\mathcal{E}_\delta$ is linear. It coincides with the ordinary evaluation on holomorphic functions, since the map δ^*_+ coincides with δ^* on holomorphic functions because of the assumption $\Delta(\mathcal{K} \cap \mathrm{Hol}(\mathbb{C})) \subset \mathrm{Hol}(\mathbb{C}) \otimes \mathrm{Hol}(\mathbb{C})$. The map $\mathcal{E}_\delta$ defines a character, i.e., it is compatible with the tensor product, since by construction δ^*_+ is an algebra homomorphism on $\mathcal{K}_\otimes$.   □

This construction can be adapted to more general settings. First let us introduce some concepts.

Let (R, P) be a commutative Rota-Baxter algebra, i.e., a commutative algebra R with a linear map $P : R \to R$ such that

$$P(x)P(y) = P(xP(y)) + P(P(x)y) + \lambda P(xy)$$

for some weight λ. Let us assume that $P^2 = P$ which implies that $\lambda = -1$. We also set

$$R_- = \mathbf{k} + P(R), \quad R_+ = \mathbf{k} + (\mathrm{id} - P)(R),$$

which are both (unitary) subalgebras of R. Then the algebraic Birkhoff decomposition of $\phi : \mathcal{H} \to R$ gives

$$\phi = \phi_-^{\star(-1)} \star \phi_+$$

with $\phi_\pm : \mathcal{H} \to R_\pm$.

DEFINITION 2.11. Let $\mathcal{H}$ be a connected filtered Hopf algebra and (R, P) a Rota-Baxter algebra. Let $\phi : \mathcal{H} \to R$ be an algebra homomorphism. The renormalization ϕ_+ is called **consistent** if $\phi_+(x) = \phi_+(y)$ whenever $\phi(x) = \phi(y)$.

DEFINITION 2.12. Let $\mathcal{H}$ be a connected filtered Hopf algebra and $(\mathrm{Mer}(\mathbb{C}), P)$ a Rota-Baxter algebra. Let $\phi : \mathcal{H} \to \mathrm{Mer}(\mathbb{C})$ be an algebra homomorphism. The renormalization ϕ_+ is called **holomorphic** if $\phi_+(\mathcal{H}) \subset \mathrm{Hol}(\mathbb{C})$, and $\phi_+(x) = \phi(x)$ whenever $\phi(x) \in \mathrm{Hol}(\mathbb{C})$.

DEFINITION 2.13. An algebra homomorphism $\phi : \mathcal{H} \to Mer(\mathbb{C})$ has a **holomorphic lifting** if there exists a filtered subalgebra $\mathcal{K}_\otimes$ of $\mathcal{M}_\otimes$, a (filtered) algebra homomorphism $\Phi : \mathcal{H} \to \mathcal{K}_\otimes$ and an algebra homomorphism $\eta : \mathcal{K}_\otimes \to Mer(\mathbb{C})$ that give the following commutative diagram:

$$
\begin{array}{ccc}
\mathcal{H} & \xrightarrow{\ \Phi\ } & \mathcal{K}_\otimes \\
\| & & \downarrow{\scriptstyle \eta} \\
\mathcal{H} & \xrightarrow{\ \phi\ } & Mer(\mathbb{C})
\end{array}
$$

with $\eta(Hol(\mathbb{C}^\infty)) \subset Hol(\mathbb{C})$ and $\eta(f)(0) = f(0)$ for $f \in Hol(\mathbb{C}^\infty)$.

THEOREM 2.14. *Let $\mathcal{H}$ be a connected filtered Hopf algebra and let*

$$\phi : \mathcal{H} \to Mer(\mathbb{C})$$

be an algebra homomorphism with a holomorphic lifting given by a $\Phi : \mathcal{H} \to \mathcal{K}_\otimes$. If the Rota-Baxter algebra $(Mer(\mathbb{C}), P)$ makes the renormalization ϕ_+ holomorphic and consistent, then there is a generalized evaluator $\mathcal{E}$ on $Im\,\Phi$ such that

$$\mathcal{E} \circ \Phi(x) = \phi_+(x)(0).$$

In other words, renormalizing via Birkhoff factorization in this setting amounts to renormalizing via the generalized evaluator $\mathcal{E}$ on the subalgebra $Im\,\Phi$.

PROOF. By definition, the generalized evaluator $\mathcal{E}$ satisfies the following requirement:

$$\Phi(x) = f \implies \mathcal{E}(f) = \phi_+(x)(0).$$

Since the renormalization is holomorphic, taking the value at 0 makes sense, and if $\Phi(x) = \Phi(y) = f$, then $\phi(x) = \phi(y)$. By the consistency of ϕ_+, we have $\phi_+(x) = \phi_+(y)$. Therefore $\mathcal{E}$ is well defined on $Im\Phi$.

Let us now check that $\mathcal{E}$ is a character. If $f = \Phi(x)$ and $g = \Phi(y)$, then $f \otimes g = \Phi(x) \otimes \Phi(y) = \Phi(xy)$. So

$$\mathcal{E}(f \otimes g) = \mathcal{E} \circ \Phi(xy) = \phi_+(xy) = \phi_+(x)\phi_+(y) = \mathcal{E}(f)\mathcal{E}(g).$$

The character $\mathcal{E}$ coincides with the ordinary evaluation at 0 on holomorphic functions. If $f = \Phi(x)$ is holomorphic, then because of its holomorphic lifting property, $\phi(x)$ is holomorphic. Since ϕ_+ is a holomorphic renormalization, we have

$$\mathcal{E}(f) = \phi_+(x)(0) = \phi(x)(0) = \Phi(x)(0) = f(0)$$

Thus $\mathcal{E}$ is a generalized evaluator on $Im\Phi$. □

2.3. Comparison of Birkhoff-Hopf Renormalization Schemes

Having compared Birkhoff-Hopf renormalization with renormalization via generalized evaluators, we now compare different Birkhoff-Hopf renormalization schemes.

The following proposition describes the behavior of $\phi_\pm$ as $\mathcal{H}$ and R vary and could be proved by an induction on the filtration of $\mathcal{H}$.

PROPOSITION 2.15. *Let* $\mathcal{H} = \oplus_{n \geq 0} \mathcal{H}_n$ *be a connected filtered Hopf algebra and* (R, P) *be a commutative Rota-Baxter algebra with* $P^2 = P$. *Let* $\phi : \mathcal{H} \to R$ *be an algebra homomorphism.*

(1) *Let* (R', P') *be a commutative Rota-Baxter algebra with* $P'^2 = P'$ *and let* $\chi : (R, P) \to (R', P')$ *be a Rota-Baxter algebra homomorphism. Then we have*

$$\text{(6)} \qquad (\chi \circ \phi)_{\pm} = \chi \circ \phi_{\pm}.$$

(2) *Let* $\mathcal{H}' = \oplus_{n \geq 0} \mathcal{H}'_n$ *be a connected filtered (cograded) Hopf algebra and let* $f : \mathcal{H}' \to \mathcal{H}$ *be a homomorphism of filtered cograded Hopf algebras. Then we have*

$$\text{(7)} \qquad (\phi \circ f)_{\pm} = \phi_{\pm} \circ f.$$

We next apply these general relations to a special situation.

When an algebra R of complex-valued meromorphic functions around $z_0 \in \mathbb{C}$ is equipped with the projection $I - P$ to a subspace R_+ of the subalgebra of holomorphic functions at z_0, we obtain a regularized evaluation at the point z_0:

$$\text{ev}_{z_0}^P : R \to \mathbb{C},$$
$$f \mapsto (I - P)(f)(z_0).$$

When R_+ coincides with the algebra of holomorphic functions at z_0, implementing the map $\text{ev}_{z_0}^P$ amounts to a minimal substraction scheme, which restricts to an algebra morphism on the range of $I - P$. This applies in particular to the case when (R, P) is the Rota-Baxter algebra of meromorphic functions of one variable around zero with the projection to the pole part and with $z_0 = 0$. This serves as a motivation for the following setup. For a connected filtered Hopf algebra $\mathcal{H}$ and a commutative Rota-Baxter algebra (R, P) with $P^2 = P$, a **renormalization scheme** is defined to be a pair of algebra homomorphisms $\phi : \mathcal{H} \to R$ and $\text{ev}^P : R_+ \to \mathbf{k}$. The following direct consequence of Proposition 2.15 can be used to compare two renormalization schemes.

PROPOSITION 2.16. *Let* $\mathcal{H}$ *be a connected filtered Hopf algebra. For* $i = 1, 2$, *let* (R_i, P_i) *be two Rota-Baxter algebras with* $P_i^2 = P_i$. *Let* $\phi_i : \mathcal{H} \to R_i$ *and* $\text{ev}^{P_i} : R_{i,+} \to \mathbf{k}$ *be two renormalization schemes. If there is a Rota-Baxter algebra homomorphism* $\chi : R_1 \to R'_2$ *such that* $\chi \circ \phi_1 = \phi_2$ *and* $\text{ev}^{P_2} \circ \chi = \text{ev}^{P_1}$, *then*

$$\text{(8)} \qquad \text{ev}^{P_1} \circ \phi_{1,+} = \text{ev}^{P_2} \circ \phi_{2,+}.$$

That is, the renormalized values from the two renormalization schemes are the same.

See Corollary 3.3 for the application of this proposition to the renormalization of Riemann integrals.

3. Renormalization of Integrals Indexed by Rooted Trees

We formulate a framework to discuss integrals indexed by rooted trees in the algebraic contexts of Hopf algebra and operated algebras. The related algebraic concepts are reviewed in the first subsection and are applied to integrals in the section.

3.1. Hopf Algebra and Operated Algebra of Decorated Rooted Trees

We recall the rich structure of decorated trees first as a Hopf algebra and then as a free operated algebra.

3.1.1. Hopf Algebra of Decorated Trees

We recall the Hopf algebra of decorated planar rooted trees [13] whose commutative (non-planar) version was introduced in [4,22,28] as a toy model of the Hopf algebra of Feynman graphs in QFT to study renormalization of perturbative QFT. It is also related to the Hopf algebras of rooted trees of Grossman-Larson [14] and of Loday-Ronco [24].

A rooted tree is a connected and simply connected set of vertices and oriented edges such that there is precisely one distinguished vertex, called the root, with no incoming edge. A planar rooted tree is a rooted tree with a fixed embedding into the plane.

Let $\mathcal{T}$ be the set of (isomorphic classes of) planar rooted trees. Let $\mathcal{F}$ be the free monoid generated by $\mathcal{T}$ and let $\mathcal{H}_{\mathcal{T}}$ be the noncommutative polynomial algebra over $\mathbb{C}$ generated by $\mathcal{T}$: $\mathcal{H}_{\mathcal{T}} = \mathbb{C}\langle \mathcal{T} \rangle$. Elements in $\mathcal{F}$ are called (rooted planar) forests. The **depth** $\mathrm{d}(F)$ of a forest is the length of the longest path from one of its roots to the leaves.

Let Ω be a set. Let $\mathcal{T}(\Omega)$ be the set of planar rooted trees with vertices decorated by elements in Ω. Similarly define $\mathcal{F}(\Omega)$ to be the free monoid generated by $\mathcal{T}(\Omega)$ and $\mathcal{H}_{\mathcal{T}(\Omega)}$ to be the noncommutative polynomial algebra over $\mathbb{Q}$ generated by $\mathcal{T}(\Omega)$.

We next define a coalgebra structure on $\mathcal{H}_{\mathcal{T}(\Omega)}$. When Ω is a singleton, we recover the case of undecorated trees. A subforest of a tree T consists of a set of vertices of T together with their descendants and edges connecting all these vertices.

The **coproduct** is then defined as follows. Let $\mathcal{F}_T$ be the set of subforests of the decorated rooted tree $T \in \mathcal{T}(\Omega)$, including the empty subforest, identified with 1, and the full tree. Define

$$\Delta(T) = \sum_{F \in \mathcal{F}_T} F \otimes (T/F).$$

The quotient T/F is obtained by removing the subforest F and edges connecting the subforest to the rest of the tree. We use the convention that if F is the empty subforest, then $T/F = T$, and if $F = T$, then $T/F = 1$. We then extend Δ to $\mathcal{H}_{\mathcal{T}(\Omega)}$ by multiplicity with the convention $\Delta(1) = 1 \otimes 1$. Then Δ is an algebra homomorphism. Also define

$$\varepsilon : \mathcal{H}_{\mathcal{T}(\Omega)} \to \mathbf{k}$$

by $\varepsilon(T) = 0$ for $T \in \mathcal{T}$, $\varepsilon(1) = 1$ and extend by multiplicity. Further, define the degree of a tree and a forest (i.e., monomial in $\mathbb{C}[\mathcal{T}(\Omega)]$) to be its number of vertices. We then obtain a connected graded Hopf algebra as in [4,22].

For a given $\omega \in \Omega$, let $B_{\omega}^+ : \mathcal{H}_{\mathcal{T}(\Omega)} \to \mathcal{H}_{\mathcal{T}(\Omega)}$ be the linear map given by taking the product $T_1 \cdots T_k$ of k trees $T_1, \cdots, T_k$ to the tree T consisting of a new vertex decorated by ω, subtrees T_i and an edge from the new vertex to the root of each T_i. The map B_{ω}^+ is also called the **grafting operator** in combinatorics. Then [4]

$$\Delta B_{\omega}^+ = B_{\omega}^+ \otimes 1 + (\mathrm{id}_H \otimes B_{\omega}^+)\Delta.$$

3.1.2. Universal Property of Decorated Rooted Forests

We next characterize the set of decorated rooted forests by a universal property.

DEFINITION 3.1. An **operated semigroup** is a semigroup U together with an operator $\alpha : U \to U$. A morphism between operated semigroups (U, α) and (V, β) is a semigroup homomorphism $f : U \to V$ such that $f \circ \alpha = \beta \circ f$.

When a semigroup is replaced by a monoid we obtain the concept of a **operated monoid**. Let **k** be a commutative ring. We similarly define the concepts of a **operated k-algebra** or **operated nonunitary k-algebra**.

More generally, for a set Ω, an Ω-**operated semigroup** is a semigroup U together with a family of operators $\alpha_\omega : U \to U$ indexed by $\omega \in \Omega$. Other concepts related to structures with one operator similarly generalize to those related to structures with multiple operators.

THEOREM 3.2. ([15]) *With the grafting operators $B_\omega^+ : \mathcal{H}_{\mathcal{T}(\Omega)} \to \mathcal{H}_{\mathcal{T}(\Omega)}$ (denoted by $\lfloor \; \rfloor_\omega$ in [15]),*

(1) $\mathcal{F}(\Omega)$ *is the initial object in the category of Ω-operated monoids. More precisely, for any Ω-operated monoid $(M, P_\omega, \omega \in \Omega)$ where $P_\omega : M \to M$ is a set map, there is a unique homomorphism $f : \mathcal{F}(\Omega) \to M$ of Ω-mapped monoids.*

(2) $\mathcal{H}_{\mathcal{T}(\Omega)} = \oplus_{F \in \mathcal{F}(\Omega)} \mathbb{Q} \, F$ *is the initial object in the category of Ω-operated algebras. More precisely, for any Ω-operated algebra $(A, P_\omega, \omega \in \Omega)$, where $P_\omega : A \to A$ is a linear map, there is a unique homomorphism $f : \mathcal{H}_{\mathcal{T}(\Omega)} \to A$ of Ω-operated algebras.*

3.2. Operated Algebra of Integrals

The above framework of operated algebras can be applied to integrals in various generalities.

3.2.1. Nested Families of Integrals with One Primitive Integral

In the simplest case, for a given function $f_0(x, c)$, consider the family $\mathcal{G}$ of (possibly formal) integrals indexed by $\mathcal{F}$ recursively defined by

(1) $\phi(\bullet)(c) = \int_0^\infty f_0(x, c) dx$;

(2) $\phi(T_1 \cdots T_k)(c) = \phi(T_1) \cdots \phi(T_k)$ for $T_1 \cdots T_k \in \mathcal{F}$;

(3) $\phi(B^+(F))(c) = \int_0^\infty \phi_F(x) f_0(x, c) dx$ for $F \in \mathcal{F}$.

Thus

$$\phi : \mathcal{H}_{\mathcal{T}} \to \mathbb{Q}\mathcal{G}, F \mapsto \phi(F)(c)$$

is the unique morphism of operated algebras, where the operator on $\mathbb{Q}\mathcal{G}$ is given by $I(f)(c) := \int_0^\infty f(x) f_0(x, c) dx$. We can think of $\mathcal{G}$ as a family of nested integrals "spanned" by one primitive integral $\int_0^\infty f_0(x, c) dx$.

3.2.2. Nested Families of Integrals with Multiple Primitive Integrals

More generally, we can consider a nested family of integrals with multiple primitive integrals. Let Ω be a set and let $f_\omega(x, c)$, $\omega \in \Omega$, be a family of functions indexed by Ω. Let A be a suitable algebra of functions. For $\omega \in \Omega$, define an operator

$$I_\omega : A \to A, I_\omega(f)(c) = \int_0^\infty f_\omega(x, c) f(x) dx.$$

Then A is an Ω-operated algebra. Thus by the universal property of $\mathcal{H}_{\mathcal{T}(\Omega)}$ in Theorem 3.2, there is a unique homomorphism

$$\phi : \mathcal{H}_{\mathcal{T}(\Omega)} \to A$$

of Ω-operated algebras. In particular, this means that
  (1) $\phi(\bullet_\omega) = \int_0^\infty f_\omega(x,c)dx$;
  (2) $\phi(T_1 \cdots T_k) = \phi(T_1) \cdots \phi(T_k)$;
  (3) $\phi(B_\omega^+(F)) = I_\omega(\phi(F))$.

3.2.3. Regularized Families of Integrals

The integrals in the previous cases might be divergent and thus need regulators to make them convergent. Introducing multiple regulators can also be put in the context of operated algebras.

For $\omega \in \Omega$, define an operator

$$I_\omega(f)(c) : \varepsilon \mapsto \int_0^\infty f_\omega(x,c;\varepsilon)f(x)dx.$$

If $I_\omega(f)(c)$ lies in A, then A is an Ω-operated algebra. Thus by the universal property of $\mathcal{H}_{\mathcal{T}(\Omega)}$ in Theorem 3.2, there is a unique homomorphism

$$\phi_{bu,+}^{reg} : \mathcal{H}_{\mathcal{T}(\Omega)} \to A$$

of Ω-operated algebras. In particular, this means that
  (1) $\phi_{bu}^{reg}(\bullet_\omega)(c;\varepsilon) = \int_0^\infty f_\omega(x,c;\varepsilon)f(x)dx$;
  (2) $\phi_{bu}^{reg}(T_1 \cdots T_k)(c;\varepsilon_1,\cdots,\varepsilon_{|T_1\cdots T_k|})$
        $= \phi_{bu}^{reg}(T_1)(c;\varepsilon_1,\cdots,\varepsilon_{|T_1|}) \cdots \phi_{bu}^{reg}(T_k)(c;\varepsilon_{|T_1\cdots T_{k-1}|+1},\cdots,\varepsilon_{|T_1\cdots T_k|})$;
  (3) $\phi_{bu}^{reg}(B_\omega^+(F))(c;\varepsilon_1,\cdots,\varepsilon_{|F|+1}) = I_\omega(\phi_{bu}^{reg}(F)(c;\varepsilon_1,\cdots,\varepsilon_{|F|}))$.
This setup will be applied in Section 4.

Setting $\varepsilon_i = \varepsilon$ yields a homomorphism

$$\phi_{bu}^{reg} : \mathcal{H}_{\mathcal{T}(\Omega)} \to A$$

of Ω-operated algebras. In particular, this means that
  (1) $\phi_{bu}^{reg}(\bullet_\omega)(c;\varepsilon) = \int_0^\infty f_\omega(x,c;\varepsilon)f(x)dx$;
  (2) $\phi_{bu}^{reg}(T_1 \cdots T_k)(c;\varepsilon) = \phi_{bu}^{reg}(T_1)(c;\varepsilon) \cdots \phi_{bu}^{reg}(T_k)(c;\varepsilon)$;
  (3) $\phi_{bu}^{reg}(B_\sigma^+(F))(c;\varepsilon) = I_\omega(\phi_{bu}^{reg}(F)(c;\varepsilon))$.
The following corollary of Proposition 2.16 shows how some of the regulators are related.

COROLLARY 3.3. *Suppose the regulations (algebra homomorphisms)* $\phi : \mathcal{H}_{\mathcal{F}} \to \mathbb{C}\mathcal{G} \to \mathbb{C}[\varepsilon^{-1},\varepsilon]]$ *(resp.* $\phi' : \mathcal{H}_{\mathcal{F}} \to \mathbb{C}\mathcal{G} \to \mathbb{C}[\varepsilon^{-1},\varepsilon]]$*) are given by a regulator* $\sigma(x;\varepsilon)$ *(resp.* $\sigma(x;k\varepsilon)$ *for a constant* $k \in \mathbb{R}$*). Then the renormalized values from* ϕ *agree with the renormalized values from* ϕ'*.*

PROOF. We note that ϕ and ϕ' are related by $\phi' = \chi\phi$, where $\chi : \mathbb{C}[\varepsilon^{-1},\varepsilon]] \to \mathbb{C}[\varepsilon^{-1},\varepsilon]]$ is defined by $\chi(g(\varepsilon)) = g(k\varepsilon)$. Since χ is an automorphism on the Rota-Baxter algebra $\mathbb{C}[\varepsilon^{-1},\varepsilon]]$ with the usual projection operator and is compatible with the evaluator $ev_0^P : \mathbb{C}[[\varepsilon]] \to \mathbb{C}$, the corollary follows from Proposition 2.16.   □

4. Renormalizations on Trees

We revisit the toy model first considered by Connes and Kreimer [4] and further explored in [16]. As in Section 3.2.1, we define a family of nested integrals by one primitive integral $1/(x+c)$, $c > 0$, i.e., for each rooted forest F, we define a formal integral $f_F(c)$ by the recursive structure of rooted forests as

$$f_\bullet(c) = \int_0^\infty \frac{dx}{x+c},$$

and assume that $f_F(c)$ have been defined for all rooted forests F with $0 \leq d(F) \leq k$, and let T be a rooted tree with $d(T) = k+1$. Then $T = B_+(\overline{F})$ for a rooted forest $\overline{F}$ with $d(F) = k$ and we define

$$f_T(c) = \int_0^\infty f_{\overline{F}}(x)\frac{dx}{x+c}.$$

In general, let F be a rooted forest with $d(F) = k+1$. It decomposes in a unique way as

$$F = T_1 \cdots T_n,$$

where $T_i, 1 \leq i \leq n$, are rooted trees with $d(T_i) \leq k+1$ and we set

$$f_F(c) = \prod_{i=1}^n f_{T_i}(c).$$

Here are formal expressions corresponding to some rooted forests:

$$f_{\bullet,\bullet}(c) = \left(\int_0^\infty \frac{dx}{x+c}\right)^2, \quad f_{\centerdot}^{\centerdot}(c) = \int_0^\infty \left(\int_0^\infty \frac{dx_1}{x_1+x}\right)\frac{dx}{x+c},$$

$$f(c) = \int_0^\infty \left(\int_0^\infty \frac{dx_1}{x_1+x} \int_0^\infty \frac{dx_2}{x_2+x}\right)\frac{dx}{x+c}.$$

Thus we have the formal map

$$\phi : \mathcal{F} \to \{f_F \mid F \in \mathcal{F}\}$$

sending $F \in \mathcal{F}$ to the divergent integral f_F.

4.1. Riesz Regularization

To get renormalized values of these integrals, we first work in the framework of Section 3.2.3 and apply a Riesz type regularization (see below) which amounts to replacing the integrand $f(x)$ by $f(x)x^{-\varepsilon}$. We do this again recursively as we define $f_F(c)$, up to replacing du by $u^{-\varepsilon}du$ for each variable u. Thus we define

$$f_\bullet(c; \varepsilon) = \int_0^\infty \frac{x^{-\varepsilon}dx}{x+c}$$

and for a rooted tree $T = B_+(\overline{F})$, we define recursively

$$f_T(c; \varepsilon) = \int_0^\infty f_{\overline{F}}(x; \varepsilon)\frac{x^{-\varepsilon}dx}{x+c}.$$

In view of Corollary 3.3, such a regularization might not be as special as it appears, since a change of variable $\varepsilon \mapsto k\varepsilon$ for $0 < k \in \mathbb{R}$ does not change the renormalized values.

LEMMA 4.1. *For any forest* F, $f_F(c; \varepsilon)$ *is convergent for* $\varepsilon \in \mathbb{C}$ *with* $\mathrm{Re}(\varepsilon) \in (0, \frac{1}{|F|})$, *where* $|F|$ *is the degree (i.e., the number of vertices) of* F.

PROOF. For $0 < a < 1$ we have

$$\int_0^\infty \frac{x^{-a} dx}{x + 1} = B(1 - a, a),$$

where $B(x, y) = \int_0^\infty \frac{t^{x-1}}{(1+t)^{x+y}} \, dt$ is the beta function defined for positive x and y. Applying the identity

$$B(x, y) \, B(x + y, 1 - y) = \frac{\pi}{x \, sin(\pi y)}$$

to $x = 1 - a$ and $y = a$ yields

$$\int_0^\infty \frac{x^{-a} dx}{x + 1} = \frac{\pi}{sin(a\pi)}.$$

Hence,

$$f_\bullet(c; \varepsilon) = \frac{\pi}{c^\varepsilon \sin(\pi\varepsilon)}$$

for $\mathrm{Re}(\varepsilon) \in (0, 1)$. Since $\int_0^\infty \frac{x^{-a} dx}{x+c} = c^{-a} \int_0^\infty \frac{x^{-a} dx}{x+1}$, iterating this computation yields

$$(9) \qquad f_F(c; \varepsilon) = c^{-|F|\varepsilon} \prod_{i=1}^{|F|} \frac{\pi}{sin(k_i \pi \varepsilon)},$$

with positive integers $k_i \leq |F|$. Thus the equation holds for $\mathrm{Re}(\varepsilon) \in (0, \frac{1}{|F|})$.   □

Therefore $f_F(c; \varepsilon)$ can be analytically continued to a convergent Laurent series, still denoted by $f_F(c; \varepsilon)$, in $\mathbb{C}[\varepsilon^{-1}, \varepsilon]]$. For a fixed value of c, we now define an algebra homomorphism

$$(10) \qquad \phi^{reg} : \mathcal{H}_\mathcal{T} \to \mathbb{C}[\varepsilon^{-1}, \varepsilon]], \quad \phi^{reg}(F) = f_F(c; \varepsilon), \quad F \in \mathcal{F}.$$

This homomorphism is also compatible with the operated algebra structure on $\mathcal{H}_\mathcal{T}$ given by the grafting and the operated algebra structure on the corresponding integrals given by the integral operator

$$f(c; \varepsilon) \mapsto \int_0^\infty f(x; \varepsilon) \frac{x^{-\varepsilon} \, dx}{x + c}.$$

4.1.1. Birkhoff-Hopf Renormalized Integrals

We now make use of the Hopf algebra $\mathcal{H}_\mathcal{T}$ and apply the algebraic Birkhoff decomposition to the morphism ϕ^{reg} to build an algebra homomorphism

$$\phi_+^{reg} : \mathcal{H}_\mathcal{T} \to \mathbb{C}[[\varepsilon]].$$

We then define the **renormalized value** of the formal integral $f_F(c)$ to be

$$\bar{f}_F(c) := \lim_{\varepsilon \to 0} \phi_+^{reg}(F)(c; \varepsilon).$$

For example,

$$f_\bullet(c;\varepsilon) = \int_0^\infty \frac{x^{-\varepsilon}\,dx}{x+c} = \frac{1}{\varepsilon} - \ln c + \left(\frac{\pi^2}{6} + \frac{(\ln c)^2}{2}\right)\varepsilon + o(\varepsilon).$$

Hence the renormalized value of $f_\bullet(c)$ is $\bar{f}_\bullet(c) = -\ln c$.

REMARK 4.2. We saw that a linear transformation $\varepsilon \to k\varepsilon$ does not modify the renormalized value. In contrast, introducing second order powers of ε (higher order powers will not contribute to the renormalized value of $\bar{f}_\bullet(c;\varepsilon)(c)$) by the transformation

$$\tau_\mu : \varepsilon \mapsto \varepsilon + \mu\varepsilon^2 + o(\varepsilon^2)$$

introduces a constant term arising from the pole part

$$\frac{1}{\varepsilon + \mu\varepsilon^2} = \frac{1}{\varepsilon}(1 - \mu\varepsilon + o(\varepsilon)) = -\mu + \frac{1}{\varepsilon} + o(1),$$

so that the new renormalized value reads:

$$\mathrm{fp}_{\varepsilon=0}\bar{f}_\bullet(c;\tau_\mu(\varepsilon)) = -\ln(e^\mu c).$$

Hence, a transformation by a diffeomorphism τ_μ induces a rescaling of c by e^μ:

$$\mathrm{fp}_{\varepsilon=0}f_\bullet(c;\tau_\mu(k\varepsilon)) = \mathrm{fp}_{\varepsilon=0}f_\bullet(e^\mu c;\varepsilon) = \mathrm{fp}_{\varepsilon=0}\left(e^{-\varepsilon\mu}f_\bullet(c;\varepsilon)\right), \quad \forall k > 0, \quad \forall \mu > 0.$$

Algebraic Birkhoff decomposition provides an inductive procedure to compute higher iterated integrals:

$$\phi_+^{reg}(\mathbf{!})(c;\varepsilon) = (\mathrm{id} - P)\big(\phi^{reg}(\mathbf{!})(c;\varepsilon) - P(\phi^{reg}(\bullet)(c;\varepsilon))\phi^{reg}(\bullet)(c;\varepsilon)\big),$$

which is the power series part of $\phi^{reg}(\mathbf{!})(c;\varepsilon) - P(\phi^{reg}(\bullet)(c;\varepsilon))\phi^{reg}(\bullet)(c;\varepsilon)$. Since

$$\phi^{reg}(\mathbf{!})(c;\varepsilon) = \frac{1}{2\varepsilon^2} - \frac{\ln c}{\varepsilon} + ((\ln c)^2 + \frac{5}{12}\pi^2) - \frac{\ln c}{6}(4(\ln c)^2 + 5\pi^2)\varepsilon + o(\varepsilon),$$

we obtain

$$\bar{f}_{\mathbf{!}}(c) = \frac{1}{2}(\ln c)^2 + \frac{\pi^2}{4}.$$

By similar computations, we obtain

$$\bar{f}_{\mathbf{!}}(c) = -\frac{1}{6}(\ln c)^3 - \frac{5}{12}\pi^2(\ln c)$$

$$\bar{f}_\wedge(c) = -\frac{1}{3}(\ln c)^3 - \frac{1}{3}\pi^2 \ln c$$

$$\bar{f}_{\mathbf{!}}(c) = \frac{1}{24}(\ln c)^4 + \frac{7}{24}\pi^2(\ln c)^2 + \frac{5}{32}\pi^4.$$

The renormalized values in this case are quite simple, as we can see in the following proposition.

PROPOSITION 4.3. *Let* $F \in \mathcal{F}$ *be a forest whose number of vertices is denoted by* $|F|$. *Then* $\bar{f}_F(c)$ *is a homogeneous expression of degree* $|F|$ *in* $\mathbb{Q}[\ln c, \pi^2]$, *where* $\ln c$ *and* π *are both assigned the degree 1.*

PROOF. We know that (from (9))

$$f_F(c) = c^{-|F|\varepsilon} \prod_{i=1}^{|F|} \frac{\pi}{\sin(k_i \pi \varepsilon)}$$

and

$$\frac{1}{\sin(x)} = \frac{1}{x} + a_1 x + \sum_{k \geq 1} a_{2k+1} x^{2k+1},$$

where $a_i \in \mathbb{Q}$. So

$$\frac{\pi}{\sin(k_i \pi x)} = \sum_{k \geq 0} c_{2k-1} \pi^{2k} x^{2k-1}$$

with all coefficients in $\mathbb{Q}$. Therefore (by induction),

$$\prod_{i=1}^{|F|} \frac{\pi}{\sin(k_i \pi \varepsilon)} = \sum_{k \geq 0} c_{2k-|F|} \pi^{2k} \varepsilon^{2k-|F|},$$

where the c_i's are $\mathbb{Q}$-coefficients which depend on k_j's.

So if we let $\ln c$ and π be of degree 1, then

$$f_F(c) = c^{-|F|\varepsilon} \prod_{i=1}^{|F|} \frac{\pi}{\sin(k_i \pi \varepsilon)} = e^{-|F|(\ln c)\varepsilon} \prod_{i=1}^{|F|} \frac{\pi}{\sin(k_i \pi \varepsilon)} = \sum_{k \geq -|F|} f_k \varepsilon^k,$$

where f_k is a homogeneous polynomial in $\mathbb{Q}[\ln c, \pi^2]$ with degree $k + |F|$; in particular, the constant term is of degree $|F|$.

Since $|F| = |F'| + |F/F'|$ if F' is a subforest of F, the contribution from every counter term has degree $|F|$, so the proposition follows by implementing a renormalization process. $\square$

4.1.2. Renormalization by Generalized Evaluators

As an alternative to renormalization by Birkhoff-Hopf factorization, we now implement renormalization by generalized evaluators, which corresponds to a blow-up of the previous construction. Define $\phi_{bu}^{reg}(F)(c; \varepsilon_1, \cdots, \varepsilon_{|F|})$ recursively by

(1) $\phi_{bu}^{reg}(\bullet)(c; \varepsilon_1) = \int_0^\infty \frac{x^{-\varepsilon_1} dx}{x+c}$;

(2) $\phi_{bu}^{reg}(T_1 \cdots T_k)(c; \varepsilon_1, \cdots, \varepsilon_{|T_1 \cdots T_k|})$
$= \phi_{bu}^{reg}(T_1)(c; \varepsilon_1, \cdots, \varepsilon_{|T_1|}) \phi_{bu}^{reg}(T_2)(c; \varepsilon_{|T_1|+1}, \cdots, \varepsilon_{|T_1 T_2|})$
$\cdots \phi_{bu}^{reg}(T_k)(c; \varepsilon_{|T_1 \cdots T_{k-1}|+1}, \cdots, \varepsilon_{|T_1 \cdots T_k|})$;

(3) $\phi_{bu}^{reg}(B^+(F))(c; \varepsilon_1, \cdots, \varepsilon_{|F|+1}) = \int_0^\infty \phi_{bu}^{reg}(F)(x; \varepsilon_1, \cdots, \varepsilon_{|F|}) \frac{x^{-\varepsilon_{|F|+1}}}{x+c} dx$.

REMARK 4.4. Note that a rescaling $c \mapsto e^\mu c$

$$\phi_{bu}^{reg}(B^+(F))(e^\mu c; \varepsilon_1, \cdots, \varepsilon_{|F|+1})$$

$$= e^{(\varepsilon_1 + \cdots + \varepsilon_{|F|+1})\mu} \int_0^\infty \phi_{bu}^{reg}(F)(e^\mu x; \varepsilon_1, \cdots, \varepsilon_{|F|}) \frac{x^{-\varepsilon_{|F|+1}}}{x+c} dx$$

induces a rescaling of the integration variable x by e^μ. There is an analogy between the parameter $\lambda = e^\mu$ and the renormalization group parameter in physics; since the renormalization group parameter is the dimensionless parameter $\frac{p^2}{\lambda^2}$ expressed in terms of the external momentum p, a closer analogy would require choosing $\frac{c}{\lambda}$ as the actual parameter with c seen as external parameter.

If we decorate the forest F by variables $\varepsilon_1, \cdots, \varepsilon_{|F|}$ such that
(1) every tree has a set of variables, and if $i > j$, then the indices of variables for tree i are all bigger than the indices of variables for tree j, and
(2) the root of every tree always has the largest index among the indices of variables in that tree,

then by an easy induction, we have the following lemma.

LEMMA 4.5. *For any forest F,*

$$\phi_{bu}^{reg}(F)(c; \varepsilon_1, \cdots, \varepsilon_{|F|}) = c^{-\varepsilon_1 - \cdots - \varepsilon_{|F|}} \prod_v \frac{\pi}{\sin(\pi e_v)},$$

where the product is over all vertices v, and e_v is the sum of variables associated to v and leaves of subtrees with root v.

EXAMPLE 4.6.

$$\phi_{bu}^{reg}(\mathbf{!})(c; \varepsilon_1, \varepsilon_2) = \int_0^\infty \left(\int_0^\infty \frac{x_1^{-\varepsilon_1} \, dx_1}{x_1 + x} \right) \frac{x^{-\varepsilon_2} \, dx}{x + c}$$

$$= \int_0^\infty \frac{\pi}{\sin(\varepsilon_1 \pi)} \frac{x^{-\varepsilon_2 - \varepsilon_1} \, dx}{x + c}$$

$$= c^{-(\varepsilon_1 + \varepsilon_2)} \frac{\pi^2}{\sin(\varepsilon_1 \pi) \sin((\varepsilon_1 + \varepsilon_2) \pi)}$$

and similarly,

$$\phi_{bu}^{reg}(\mathbf{!})(c; \varepsilon_1, \varepsilon_2, \varepsilon_3) = c^{-(\varepsilon_1 + \varepsilon_2 + \varepsilon_3)} \frac{\pi^3}{\sin(\varepsilon_1 \pi) \sin((\varepsilon_1 + \varepsilon_2) \pi) \sin((\varepsilon_1 + \varepsilon_2 + \varepsilon_3) \pi)}.$$

Now we can take any generalized evaluator $\mathcal{E}$ and get the renormalized value

$$\phi_{\mathcal{E}}(F)(c) = \mathcal{E} \left(\varepsilon \mapsto \phi_{bu}^{reg}(F)(c; \varepsilon) \right).$$

In particular, we can take the evaluator $\mathcal{E}^0$ in Example 2.8 or its permutation versions, or even the symmetric one. Because these evaluators take 0 values for any rational functions in $\mathcal{B}$ with homogeneous degree not 0, although the renormalized values may be different, they nevertheless all lie in $\mathbb{Q}[\ln c, \pi^2]$.

EXAMPLE 4.7. Since

$$c^{-\varepsilon} \frac{\pi}{\sin(\pi \varepsilon)} = \frac{1}{\varepsilon} - \ln c + \left(\frac{\pi^2}{6} + \frac{(\ln c)^2}{2} \right) \varepsilon + o(\varepsilon),$$

we have

$$\phi_{bu}^{reg}(\mathbf{!})(c; \varepsilon_1, \varepsilon_2) = \frac{\pi}{\sin(\pi \varepsilon_1)} \left(\frac{1}{\varepsilon_1 + \varepsilon_2} - \ln c + \left(\frac{\pi^2}{6} + \frac{(\ln c)^2}{2} \right) (\varepsilon_1 + \varepsilon_2) + o(\varepsilon_1 + \varepsilon_2) \right),$$

so that applying the symmetrized generalized evaluator we get

$$\mathcal{E}_2^{sym} \circ \phi_{bu}^{reg}(\mathbf{!})(c) = \mathcal{E}_2^{sym} \left(c^{-(\varepsilon_1 + \varepsilon_2)} \frac{\pi^2}{\sin(\varepsilon_1 \pi) \sin((\varepsilon_1 + \varepsilon_2) \pi)} \right)$$

$$= \frac{1}{2} \mathrm{fp}_{\varepsilon_1 = 0} \left(c^{-\varepsilon_1} \frac{\pi^2}{\sin(\varepsilon_1 \pi) \sin(\varepsilon_1 \pi)} \right)$$

$$+ \frac{1}{2} \mathrm{fp}_{\varepsilon_1=0} \left(\frac{\pi}{\sin(\varepsilon_1 \pi)} \right) \mathrm{fp}_{\varepsilon_2=0} \left(c^{-\varepsilon_2} \frac{\pi}{\sin(\varepsilon_2 \pi)} \right)$$

$$= \frac{\pi^2}{4} + \frac{(\ln c)^2}{2}.$$

By an argument similar to that of Proposition 4.3, we obtain the following result.

PROPOSITION 4.8. *Let* $F \in \mathcal{F}$ *be a forest. Then* $\phi_\varepsilon(F)(c)$ *is a homogeneous element of degree* $|F|$ *in* $\mathbb{Q}[\ln c, \pi^2]$, *where* $\ln c$ *and* π *are both assigned the degree 1.*

REMARK 4.9. Consequently, a rescaling $c \mapsto e^\mu c$ gives rise to a polynomial expression in μ of degree $|F|$.

4.2. "Covariant" Regularization

Instead of the regulator applied in the last section, we can apply another regulator defined by the following initial condition and recursion, close to covariant methods used in [**31**] in the context of Feynman diagrams, where the Green function is taken to a power $\varepsilon + 1$:

$$(11) \qquad f_\bullet^*(c; \varepsilon) := \int_0^\infty \frac{dx}{(x + c)^{\varepsilon+1}}$$

$$(12) \qquad f_{B^+(\bar{F})}^*(c; \varepsilon) := \int_0^\infty f_{\bar{F}}^*(x; \varepsilon) \frac{dx}{(x + c)^{\varepsilon+1}}.$$

We have similar results as before. But in this case the renormalized values are more interesting. For $\mathrm{Re}(x) > 0$ and $\mathrm{Re}(y) > 0$, the beta function gives

$$B(x, y) = \int_0^\infty \frac{t^{x-1}}{(1 + t)^{x+y}} \, dt = \frac{\Gamma(x) \Gamma(y)}{\Gamma(x + y)}.$$

So we have

$$f_\bullet^*(c; \varepsilon) = \frac{c^{-\varepsilon}}{\varepsilon} = c^{-\varepsilon} B(1, \varepsilon)$$

and

$$f_{\mathbf{i}}^*(c; \varepsilon) = c^{-2\varepsilon} B(1, \varepsilon) B(1 - \varepsilon, 2\varepsilon).$$

By induction, we show that for the ladder tree L_k of length k, we have

$$f_{L_k}^*(c; \varepsilon) = c^{-k\varepsilon} \prod_{i=0}^{k-1} B(1 - i\varepsilon, (i + 1)\varepsilon)$$

and, for a general forest F,

$$f_F^*(c; \varepsilon) = c^{-|F|\varepsilon} \prod_{i=1}^{|F|} B(1 - k_i \varepsilon, (k_i + 1)\varepsilon)$$

with integers $k_i \leq |F|$, giving a series convergent for $\mathrm{Re}(\varepsilon) \in (0, 1/|F|)$.

As before, $f_F^*(c; \varepsilon)$ can be analytically continued to a convergent Laurent series, and we hence define an algebra homomorphism

$$(13) \qquad \phi^{*,reg} : \mathcal{H}_{\mathcal{T}} \to \mathbb{C}[\varepsilon^{-1}, \varepsilon]], \quad \phi^{*,reg}(F) = f_F^*(c; \varepsilon), \quad F \in \mathcal{F}.$$

4.2.1. Birkhoff-Hopf Renormalization

Applying the algebraic Birkhoff decomposition, we obtain an algebra homomorphism

$$\phi_+^{*,reg} : \mathcal{H}_{\mathcal{T}} \to \mathbb{C}[[\varepsilon]].$$

We then define the **renormalized value** of the formal integral $f_F^*(c)$ to be

$$\bar{f}_F^*(c) := \lim_{\varepsilon \to 0} \phi_+^{*,reg}(F)(c; \varepsilon).$$

LEMMA 4.10. *In* $\mathbb{Q}[\pi^2, \zeta(2n+1), n \geq 1]$, *if we let* π *have degree 1 and* $\zeta(2n+1)$, $n \geq 1$, *have degree* $2n + 1$, *then*

$$B(1 - k\varepsilon, (k+1)\varepsilon) = \frac{1}{k+1}\varepsilon^{-1} + \sum c_i \varepsilon^i,$$

where $c_i \in \mathbb{Q}[\pi^2, \zeta(2n+1), n \geq 1]$ *is homogeneous and has degree* $i + 1$.

By this lemma and a similar argument as for Proposition 4.3, we obtain

PROPOSITION 4.11. *For any* $F \in \mathcal{F}$, $\bar{f}_F^*(c)$ *is a homogeneous polynomial of degree* $|F|$ *in* $\mathbb{Q}[\ln c, \pi^2, \zeta(2n+1), n \geq 1]$, *where* $\ln c$ *and* π *have degree 1 and* $\zeta(2n+1)$, $n \geq 1$, *has degree* $2n + 1$.

Assuming the well-known conjecture [1] that the set $\{\pi, \zeta(2n+1), n \geq 1\}$ is algebraic independent, then the polynomial in this proposition is unique. It would be interesting to understand the rational coefficients of these polynomials. The first few polynomials are

$$\bar{f}_\bullet^*(c) = -\ln c$$

$$\bar{f}_{\mathfrak{l}}^*(c) = \frac{1}{2}(\ln c)^2 + \frac{1}{6}\pi^2$$

$$\bar{f}_{\mathfrak{l}}^*(c) = -\frac{1}{6}(\ln c)^3 - \frac{1}{3}\pi^2 \ln c - \frac{1}{3}\zeta(3)$$

$$\bar{f}_\wedge^*(c) = -\frac{1}{3}\pi^2 \ln c$$

$$\bar{f}_{\mathfrak{l}}^*(c) = \frac{1}{24}(\ln c)^4 + \frac{13}{120}\pi^4 + \frac{1}{3}(\ln c)\zeta(3) + \frac{1}{4}(\ln c)^2\pi^2$$

$$\bar{f}_{\mathfrak{l}}^*(c) = -(1/5)\zeta(5) - (2/9)\zeta(3)\pi^2 - (1/120)(\ln c)^5 - (1/9)(\ln c)^3\pi^2$$

$$- (19/90)(\ln c)\pi^4 - (1/6)(\ln c)^2\zeta(3).$$

4.2.2. Renormalization by Evaluators

Once again, we can implement the alternative renormalization method by generalized evaluators. Recursively define $\phi_{bu}^{*;reg}(F)(c; \varepsilon_1, \cdots, \varepsilon_{|F|})$ by

(1) $\phi_{bu}^{*;reg}(\bullet)(c; \varepsilon_1) = \int_0^\infty \frac{dx}{(x+c)^{\varepsilon_1+1}};$

(2) $\phi_{bu}^{*;reg}(T_1 \cdots T_k)(c; \varepsilon_1, \cdots, \varepsilon_{|T_1 \cdots T_k|})$

$$= \phi_{bu}^{*,reg}(T_1)(c; \varepsilon_1, \cdots, \varepsilon_{|T_1|}) \phi_{bu}^{*,reg}(T_2)(c; \varepsilon_{|T_1|+1}, \cdots, \varepsilon_{|T_1T_2|})$$
$$\cdots \phi_{bu}^{*,reg}(T_k)(c; \varepsilon_{|T_1 \cdots T_{k-1}|+1}, \cdots, \varepsilon_{|T_1 \cdots T_k|});$$

(3) $\phi_{bu}^{*,reg}(B^+(F))(c; \varepsilon_1, \cdots, \varepsilon_{|F|+1}) = \displaystyle\int_0^\infty \frac{\phi_{bu}^{*,reg}(F)(x; \varepsilon_1, \cdots, \varepsilon_{|F|}) dx}{(x+c)^{\varepsilon_{|F|+1}+1}}.$

Then the structure for $\phi_{bu}^{*,reg}(F)$ is very clear:

$$\phi_{bu}^{*,reg}(F)(c; \varepsilon_1, \cdots, \varepsilon_{|F|}) = c^{-\varepsilon_1 - \cdots - \varepsilon_{|F|}} \prod_v B(1 - d_v, e_v),$$

where the product is over all vertices v, d_v is the sum of variables associated to leaves of subtrees with root v and e_v is the sum of variables associated to v and leaves of subtrees with root v.

Now we can take any generalized evaluator $\mathcal{E}$ and get the renormalized value

$$\phi_{\mathcal{E}}^*(F)(c) = \mathcal{E}\left(\varepsilon \mapsto \phi_{bu}^{*,reg}(F)(c; \varepsilon)\right).$$

Similarly, the following result holds.

PROPOSITION 4.12. *For any $F \in \mathcal{F}$, $\phi_{\mathcal{E}}^*(F)(c)$ is a homogeneous polynomial of degree $|F|$ in $\mathbb{Q}[\ln c, \pi^2, \zeta(2n+1), n \geq 1]$, where $\ln c$ and π have degree 1 and $\zeta(2n+1)$, $n \geq 1$, has degree $2n+1$.*

REMARK 4.13. Consequently, here again, a rescaling $c \mapsto e^\mu c$ gives rise to a polynomial expression in μ of degree $|F|$.

5. Renormalizations on Trees Decorated by Symbols

In this section, we want to renormalize integrals associated with trees decorated by classical pseudodifferential symbols. We refer the reader to [17,27] for the notion of pseudodifferential symbols with constant coefficients supported in $\mathbb{R}_+$. Among these symbols we find **classical pseudodifferential** or polyhomogeneous symbols which are pseudodifferential symbols $\sigma \in C^\infty(\mathbb{R}_+)$ with the following asymptotic expansion:

$$\sigma(x) = \sum_{j=0}^\infty \sigma_{\alpha-j}(x) \chi(x), \quad \forall x \in \mathbb{R}_+, \quad \sigma_{\alpha-j}(tx) = t^{\alpha-j}\sigma(x), \quad \forall t > 0,$$

where χ is a smooth function which is one outside the interval $(0,1)$ and which vanishes in a neighborhood of 0. The complex number α is the order of the symbol and classical symbols of any order form an algebra under the ordinary product of functions, denoted by $CS(\mathbb{R}_+)$.

EXAMPLE 5.1. For any positive number c, the map $x \mapsto \frac{1}{x+c}$ lies in $CS(\mathbb{R}_+)$, hence so does the map

$$x \mapsto \frac{\sigma(x)}{x+c}$$

define a symbol in $CS(\mathbb{R}_+)$ for any $\sigma \in CS(\mathbb{R}_+)$.

For a given positive number c, the map

$$\mathcal{I} : \sigma \mapsto \int_0^\infty \sigma(x)(x+c)^{-1} dx,$$

which is well defined on symbols $\sigma \in CS(\mathbb{R}_+)$ whose order α have negative real part, is a priori ill defined on general symbols. We build a meromorphic extension of this map to all symbols via holomorphic families of classical symbols.

5.1. Integrals on Non-integer Order Symbols

Having discussed two special cases of regularization in the last section, let us now address more general cases. We first come back to the regularized function $f_\bullet(c; \varepsilon)$ studied previously, setting for convenience $\alpha := -\varepsilon$.

LEMMA 5.2. *Let $c > 0$ and set $\rho_\alpha(x) := x^\alpha$ for $x > 0$ and $c \in \mathbb{C}$. The map*

$$c \mapsto \mathcal{I}(\rho_\alpha)(c) := \int_0^\infty x^\alpha \, (x + c)^{-1} \, dx$$

can be extended to a homogeneous function of degree α

$$\overline{\mathcal{I}}(\rho_\alpha)(c) := \fint_0^\infty x^\alpha \, (x + c)^{-1} \, dx = c^\alpha \, \Gamma(\alpha + 1) \, \Gamma(-\alpha),$$

which when seen as a function of α and for a fixed parameter c, is a meromorphic map on the whole complex plane with simple poles at integers given by

$$Res_{\alpha=k}\overline{\mathcal{I}}(\rho_\alpha)(c) = (-1)^k \, c^k, \quad Res_{\alpha=-k-1}\overline{\mathcal{I}}(\rho_\alpha)(c) = (-1)^k \, c^{-k-1}, \quad \forall k \in \mathbb{Z}_{\geq 0}.$$

Note that $\fint_0^\infty x^\alpha \, (x + c)^{-1} \, dx$ is just a notation for the function extending the function defined by $\int_0^\infty x^\alpha \, (x + c)^{-1} \, dx$.

PROOF. For $\operatorname{Re}(\alpha) \in (-1, 0)$, $\mathcal{I}(\rho_\alpha)(c)$ is convergent, and by adequate changes of variables one shows that

$$\int_0^\infty x^\alpha \, (x + c)^{-1} = c^\alpha \int_0^1 (1 - y)^\alpha \, y^{-1-\alpha} \, dy$$

$$= c^\alpha \, B(\alpha + 1, -\alpha)$$

$$= c^\alpha \, \Gamma(\alpha + 1) \, \Gamma(-\alpha)$$

where, as before, $B(a, b) = \int_0^1 t^{a-1} \, (1 - t)^{b-1} \, dt = \frac{\Gamma(a) \, \Gamma(b)}{\Gamma(a+b)}$ stands for the beta function.

Since the gamma function Γ extends to a meromorphic function which we denote here by $\overline{\Gamma}$ with simple poles at non-positive integers $-k \in \mathbb{Z}_{\leq 0}$ and residues $Res_{z=-k}\overline{\Gamma}(z) = \frac{(-1)^k}{k!}$, it follows that $\alpha \mapsto \mathcal{I}(\rho_\alpha)(c)$ can be extended to a meromorphic function $\alpha \mapsto \overline{\mathcal{I}}(\rho_\alpha)(c)$ with simple poles in $\mathbb{Z}$ described as in the lemma as a result of the fact that $\Gamma(k) = (k - 1)!$. $\square$

Until this point, we addressed both singularities at 0 and ∞. From now on we focus on the singularity at ∞ by introducing a smooth cut-off function χ which rules out singularities at zero. Thus χ is a smooth cut-off function on $\mathbb{R}_+$ which vanishes in a small neighborhood of zero and is one outside the interval $(0, 1)$.

LEMMA 5.3. *For any positive constant c the map*

$$\alpha \mapsto \mathcal{I}(\chi \, \rho_\alpha)(c) := \int_0^\infty \chi(x) \, x^\alpha \, (x + c)^{-1} \, dx,$$

which is well defined on the half plane $Re(\alpha) < 0$ uniquely extends to a meromorphic map

$$\alpha \mapsto \overline{\mathcal{I}}(\chi \rho_\alpha)(c) := \int_0^\infty \chi(x) \, x^\alpha \, (x+c)^{-1} \, dx$$

on the whole complex plane with simple poles at non-negative integers given by

$$Res_{\alpha=k}\overline{\mathcal{I}}(\chi \rho_\alpha)(c) = (-1)^k \, c^k, \quad \forall k \in \mathbb{Z}_{\geq 0}.$$

The map $c \mapsto \overline{\mathcal{I}}(\chi \rho_\alpha)(c)$ is well defined at $\alpha \notin \mathbb{Z}_{\geq 0}$ and gives rise to a symbol in $CS(\mathbb{R}_+)$ of order α.

PROOF. Let us first assume that $Re(\alpha) < 0$. Introducing the cut-off function χ, we write

(14) $$\mathcal{I}(\chi \rho_\alpha)(c) = \mathcal{I}(\rho_\alpha)(c) - \mathcal{I}((\chi(x) - 1) \rho_\alpha)(c).$$

Let us first study the expression $\mathcal{I}((\chi(x) - 1) \rho_\alpha)(c)$. Since $\chi(x) - 1$ is identically one in a neighborhood $(0, \varepsilon)$ of zero and vanishes outside the interval $(0, 1)$, we have

$$\mathcal{I}((\chi(x) - 1) \rho_\alpha)(c) = -\int_0^\varepsilon x^\alpha \, (x+c)^{-1} \, dx + \int_\varepsilon^1 (\chi(x) - 1) \, x^\alpha \, (x+c)^{-1} \, dx.$$

The second integral on the right hand side extends to a holomorphic function in α as an integral of a holomorphic function $(\chi(x) - 1) \, x^\alpha \, (x+c)^{-1}$ over a compact interval, whereas the first integral extends to a holomorphic function on the half plane $Re(\alpha) > -1$ since $x^\alpha \, (x+c)^{-1} \sim_0 c^{-1} x^\alpha$. Iterated integrations by parts shows that the map $\int_0^\varepsilon x^\alpha \, (x+c)^{-1} \, dx$ actually extends to a meromorphic map $\overline{\mathcal{I}}(\chi\rho_\alpha)(c) = \int_0^\varepsilon x^\alpha \, (x+c)^{-1} \, dx$ with poles at the discrete set of points $-1 - k, k \in \mathbb{Z}_+$. Moreover we have

$$Res_{\alpha=-k-1}\overline{\mathcal{I}}(\rho_\alpha)(c) = Res_{\alpha=-k-1} \int_0^\varepsilon x^\alpha \, (x+c)^{-1} \, dx$$

$$= Res_{\alpha=-k-1} \int_0^\varepsilon (1 - \chi)(x) \, x^\alpha \, (x+c)^{-1} \, dx.$$

Thus for a fixed parameter c, replacing $\overline{\mathcal{I}}(\rho_\alpha)(c)$ by $\overline{\mathcal{I}}(\chi \rho_\alpha)(c) = \overline{\mathcal{I}}(\rho_\alpha)(c) + \overline{\mathcal{I}}((\chi(x) - 1) \rho_\alpha)(c)$ gets rid of divergences at zero as a function of α and rules out the poles $-k - 1, k \in \mathbb{Z}_{\geq 0}$. Only the poles $\alpha = k \in \mathbb{Z}_{\geq 0}$ at non-negative integers remain. Consequently, a complex number $\alpha \notin \mathbb{Z}_{\geq 0}$ is not a pole, so that the map $c \mapsto \overline{\mathcal{I}}(\chi(x) \rho_\alpha)(c)$ is well defined.

For $c > 1$, the map $y \mapsto (1 - \chi)(c\,y)$ has support in $[0, 1)$ and

$$\overline{\mathcal{I}}(\chi \rho_\alpha)(c) = c^\alpha \int_0^\infty \chi(y\,c) \, y^\alpha (y+1)^{-1} \, dy$$

$$= c^\alpha \int_0^\infty y^\alpha (y+1)^{-1} \, dy - c^\alpha \int_0^1 (1 - \chi(y\,c)) \, y^\alpha (y+1)^{-1} \, dy$$

$$= c^\alpha \int_0^\infty y^\alpha (y+1)^{-1} \, dy + c^{\alpha+1} \, \chi'(y_0\,c) \int_0^1 y^\alpha (y+1)^{-1} \, dy$$

for some $y_0 \in (0, 1)$. Since χ' has compact support in $[0, 1]$ so does the map $c \mapsto \chi'(c\,y_0)$ have compact support and it is therefore smoothing. Since $\alpha \notin \mathbb{Z}_{\geq 0}$, the extended integral $K = \int_0^\infty y^\alpha \, (y+1)^{-1} \, dy$ is well defined, so that we can write

$$\overline{\mathcal{I}}(\chi \, \rho_\alpha)(c) = K\chi(c) \, c^\alpha + \left(K(1-\chi(c)) \, c^\alpha + c^{\alpha+1} \, \chi'(y_0 \, c) \int_0^1 y^\alpha (y+1)^{-1} \, dy \right)$$
$$= K\chi(c) \, c^\alpha + \rho(c),$$

where $\rho(c) := \left(K(1-\chi(c)) \, c^\alpha + c^{\alpha+1} \, \chi'(y_0 \, c) \int_0^1 y^\alpha (y+1)^{-1} \, dy \right)$ is a smoothing symbol. Thus the map $c \mapsto \overline{\mathcal{I}}(\chi \, \rho_\alpha)(c)$ defines a classical symbol of order α.   □

A symbol $\sigma \in CS^{\notin \mathbb{Z}}(\mathbb{R}_+)$ of order α has the following asymptotic expansion:

$$\sigma(x) \sim_{x \to +\infty} \sum_{j=0}^\infty a_{\alpha-j} \, x^{\alpha-j} \, \chi(x),$$

for some real numbers $a_{\alpha-j}, j \in \mathbb{Z}_{\geq 0}$. Indeed, a general classical symbol of order α reads

$$\sigma(x) \sim_{x \to +\infty} \sum_{j=0}^\infty \sigma_{\alpha-j}(x) \, \chi(x),$$

where $\sigma_{\alpha-j}(x)$ is positively homogeneous of degree $\alpha - j$. The above description then follows from setting $a_{\alpha-j} := \sigma_{\alpha-j}(1)$.

Combining the $\mathcal{I}(\chi\sigma_{\alpha-j})$ for some symbol σ of non-integer order leads to the following result.

PROPOSITION 5.4. *Let* $CS^{\notin \mathbb{Z}}(\mathbb{R}_+)$ *denote the set of non-integer order classical symbols on* $\mathbb{R}_+$. *The map*

$$\mathcal{I} : \sigma \mapsto \left(c \mapsto \int_0^\infty \sigma(x) \, (x+c)^{-1} \, dx \right),$$

which is well defined for symbols of order α *with negative real part, extends to a map denoted by* $\overline{\mathcal{I}} : \sigma \mapsto \int_0^\infty \sigma(x) \, (x+c)^{-1} \, dx$, *which stabilizes* $CS^{\notin \mathbb{Z}}(\mathbb{R}_+)$ *and sends a symbol of order* α *to a symbol with the same order.*

PROOF. On the one hand, by Lemma 5.3, each of the maps $\mathcal{I}(\chi \, \rho_{\alpha-j}) : c \mapsto \int_0^\infty x^{\alpha-j} \, (x+c)^{-1} \, \chi(x) dx$ is well defined whenever $\alpha \notin \mathbb{Z}_{\geq 0}$. On the other hand, for N chosen large enough so that $\sigma^{(N)} := \sigma - \sum_{j=0}^{N-1} a_{\alpha-j} \, x^{\alpha-j} \, \chi(x)$ lies in $L^1(\mathbb{R}_+)$, we set by linearity

$$\mathcal{I}(\sigma)(c) = \sum_{j=0}^{N-1} a_{\alpha-j} \, \mathcal{I}(\chi \, \rho_{\alpha-j})(c) + \int_0^\infty \frac{\sigma^{(N)}(x)}{x+c} \, dx.$$

This definition is clearly independent of the choice of N.

Again by Lemma 5.3 we know that each extended integral $\overline{\mathcal{I}}(\chi \, \rho_{\alpha-j})$ defines a symbol of order $\alpha - j$, hence their sum defines a classical symbol of order α so that $\overline{\mathcal{I}}(\sigma)$ also defines a classical symbol of order α.   □

It is useful to view $\overline{\mathcal{I}}(\sigma)$ as the extended integral of a symbol depending on a parameter c, in which case it is natural to differentiate with respect to c. Whenever the symbol σ has negative order we have

$$\partial_c^j \mathcal{I}(\sigma)(c) = \int_0^\infty \partial_c^j \frac{\sigma(x)}{x+c} \, dx = (-1)^j j! \int_0^\infty \frac{\sigma(x)}{(x+c)^{(j+1)}} \, dx.$$

For a general symbol σ of non-integer order α, the symbols $\frac{\sigma(x)}{(x+c)^{(j+1)}}$ have non-integer order $\alpha - j$ as a result of Proposition 5.4. On the other hand, for $N > \text{Re}(\alpha)$, the derivative $\partial_c^j \sigma$ has negative order and a Taylor expansion in c at zero yields

$$\overline{\mathcal{I}}(\sigma)(c) = \sum_{j=0}^{N-1} (-1)^j c^j \int_0^\infty \frac{\sigma(x)}{x^{(j+1)}}\, dx$$

$$(15) \qquad\qquad + (-1)^N c^N \int_0^1 (1-t)^{N-1} \int_0^\infty \frac{\sigma(x)}{(x+tc)^{(j+1)}}\, dx\, dt,$$

where $\int_0^\infty \frac{\sigma(x)}{x^{(j+1)}}\, dx$ is defined as in Proposition 5.4.

5.2. Meromorphic Extensions of Integrals of Symbols

The methods we use in this section are close to the ones implemented in [26] in the context of nested integrals.

We need the notion of holomorphic family of symbols taken from [20,21] but previously introduced by Guillemin under the name of gauged symbols.

A family $\{f(z)\}_{z \in \Omega}$ in a topological vector space $\mathcal{A}$ which is parameterized by a complex domain Ω is holomorphic at $z_0 \in \Omega$ if the corresponding function $f : \Omega \to \mathcal{A}$ admits a Taylor expansion in a neighborhood N_{z_0} of z_0

$$f(z) = \sum_{k=0}^\infty f^{(k)}(z_0) \frac{(z - z_0)^k}{k!}$$

which is convergent, uniformly on compact subsets in a neighborhood of z_0 (i.e., locally uniformly), with respect to the topology on $\mathcal{A}$. The vector spaces of functions we consider here are $C^\infty(\mathbb{R}_+)$ (in view of the smoothing part) and the space of classical pseudodifferential symbols on $\mathbb{R}_+$.

DEFINITION 5.5. Let Ω be a domain of $\mathbb{C}$. A family $(\sigma(z))_{z \in \Omega}$ is called a **holomorphic family of classical symbols** on $\mathbb{R}_+$ with constant coefficients parameterized by Ω if

(1) the map $z \mapsto \alpha(z)$, with $\alpha(z)$ the order of $\sigma(z)$, is holomorphic in z,

(2) $z \mapsto \sigma(z)$ is holomorphic as an element of $C^\infty(\mathbb{R}_+)$ and for each $z \in \Omega$, $\sigma(z) \sim \sum_{j=0}^\infty \chi\, \sigma(z)_{\alpha(z)-j}$ (for some smooth function χ which is identically one outside the unit ball and vanishes in a neighborhood of 0) lies in $CS^{\alpha(z)}(\mathbb{R}_+)$, and

(3) for any integer $N \geq 0$, the remainder term $\sigma_{(N)}(z) = \sigma(z) - \sum_{j=0}^{N-1} \sigma(z)_{\alpha(z)-j}$ is holomorphic in $z \in \Omega$ as an element of $C^\infty(\mathbb{R}_+)$ and its k-th derivative

$$\xi \mapsto \partial_z^k \sigma_{(N)}(z)\xi := \partial_z^k \left(\sigma_{(N)}(z)(x,\xi)\right)$$

is a symbol on $\mathbb{R}_+$ of order $\alpha(z) - N + \varepsilon$ for all $\varepsilon > 0$ locally uniformly in z, i.e., the k-th derivative $\partial_z^k \sigma_{(N)}(z)$ satisfies a uniform estimate

$$(17) \qquad |\partial_x^\alpha \partial_\xi^\beta \sigma(x,\xi)| \leq C_{K\alpha\beta}(1 + |\xi|)^{\text{Re}(a)-|\beta|},$$

where $\text{Re}(a)$ is the real part of a and $|\beta| = \beta_1 + \cdots + \beta_n$ in z on compact subsets in Ω.

Well-known regularization procedures, such as Riesz regularization in number theory and dimensional regularization in physics, arise from embedding a symbol into an appropriate holomorphic family of symbols.

DEFINITION 5.6. A **holomorphic regularization** is a map $\mathcal{R} : \sigma \mapsto \sigma(z)$ which takes a symbol $\sigma \in CS(\mathbb{R}_+)$ of order α to a holomorphic family of symbols $\sigma(z) \in CS(\mathbb{R}_+)$ of order $\alpha(z) = \alpha - q\,z$ for some positive q.

EXAMPLE 5.7. $\mathcal{R}(\sigma)(x) = \sigma(x)\,|x|^{-z}\,\chi(x)$, where χ is a smooth cut-off function around zero as above, defines a holomorphic regularization with $q = 1$ called Riesz regularization.

REMARK 5.8. To simplify the presentation, in the following we set $q = 1$, an assumption which holds for both Riesz and dimensional regularization.

In the following theorem we consider meromorphic families

$$\sigma(z) = \frac{\tau(z)}{\prod_{j=1}^{k}(z - z_j)^{m_j}}$$

of symbols with poles at a fixed set $z_j, j = 1, \cdots, k$, of complex numbers of order $m_j, j = 1, \cdots, k$, where $\tau(z)$ is a holomorphic family of symbols and $m_j, j = 1, \cdots, k$, are non-negative integers.

THEOREM 5.9. *Given a holomorphic (resp. meromorphic) family of classical symbols $\sigma(z)$ in $CS(\mathbb{R}_+)$ of order $\alpha(z) = \alpha - z$ (resp. with a pole at zero of order m), the map $c \mapsto \overline{\mathcal{I}}(\sigma(z))(c)$ defines a meromorphic family of classical symbols of order $\alpha(z)$ with a simple pole (resp. a pole of order $m + 1$) at zero whenever $\alpha \in \mathbb{Z}_{\geq 0}$. The residue at zero is given by a polynomial symbol in c:*

$$Res_{z=0}\overline{\mathcal{I}}(\sigma(z))(c) = \sum_{j=0}^{\alpha}(-1)^{\alpha-j}\,a_{\alpha-j}\,c^{\alpha-j},$$

where as before we have written $\sigma \sim \sum_{j=0}^{\infty} a_{\alpha-j}\,x^{\alpha-j}$.

PROOF. We present two proofs, both rather elementary but which reflect two different renormalization approaches, the first one by direct expansion and the second one by derivation with respect to the parameter c.

Proof 1. The symbol $\sigma(z)$ has order $\alpha(z) = \alpha - z$ so that

$$\overline{\mathcal{I}}(\sigma(z))(c) = \sum_{j=0}^{N-1} a_{\alpha(z)-j}\overline{\mathcal{I}}(\chi\,\rho_{\alpha(z)-j})(c) + \int_0^{\infty} \frac{\sigma^{(N)}(z)(x)}{x+c}\,dx$$

defines a symbol of order $\alpha(z)$. By Lemma 5.3, this actually defines a meromorphic family of symbols which has a simple pole (resp. of order $m+1$) at $z = 0$ whenever $\alpha - j \in \mathbb{Z}_{\geq 0}$ for some non-negative integer j, i.e., whenever $\alpha(0) \in \mathbb{Z}_{\geq 0}$.

When starting from a holomorphic family and for $\alpha \in \mathbb{Z}_{\geq 0}$, we know by Lemma 5.3 that the pole at $z = 0$ reads

$$\operatorname{Res}_{z=0}\overline{\mathcal{I}}(\sigma(z))(c) = \sum_{j=0}^{\alpha} a_{\alpha-j}\operatorname{Res}_{z=0}\overline{\mathcal{I}}(\chi\,\rho_{\alpha(z)-j})(c)$$

$$= \sum_{j=0}^{\alpha} a_{\alpha-j}\operatorname{Res}_{z=0}\overline{\mathcal{I}}(\rho_{\alpha(z)-j})(c)$$

$$= \sum_{j=0}^{\alpha}(-1)^{\alpha-j}\,a_{\alpha-j}\,c^{\alpha-j},$$

which is polynomial in c. In the meromorphic case higher order poles arise, in which case the ordinary residue turns out not to be a relevant quantity.

Proof 2.   We now embed a general symbol $\sigma \in CS(\mathbb{R}_+)$ of order α in a holomorphic family $\sigma(z)$ of order $\alpha - z$. When applied to $\sigma(z)$ the Taylor expansion (15) yields a holomorphic part corresponding to the remainder term, whereas the residues arise from the first N terms:

$$\overline{\mathcal{I}}(\sigma(z))(c) = \sum_{j=0}^{N-1}(-1)^j c^j \int_0^\infty \frac{\sigma(z)(x)}{x^{j+1}}\,dx$$

$$+ (-1)^N c^N \int_0^1 (1-t)^{N-1} \int_0^\infty \frac{\sigma(z)(x)}{x+tc^{j+1}}\,dx\,dt.$$

When $\alpha \in \mathbb{Z}_{\geq 0}$, we can take $N = \alpha + 1$ so that the poles at $z = 0$ are given by the polynomial expression

$$\sum_{j=0}^{\alpha}(-1)^j c^j \operatorname{Res}_{z=0} \int_0^\infty \frac{\sigma(z)(x)}{(x+c)^{j+1}}\,dx$$

$$= \sum_{j=0}^{\alpha}(-1)^j c^j \operatorname{Res}_{z=0} \int_1^\infty \frac{\sigma(z)(x)}{x^{j+1}}\,dx$$

$$= \sum_{j=0}^{\alpha}(-1)^j c^j \sum_{i=0}^{\infty} a_{\alpha-i}\operatorname{Res}_{z=0} \int_1^\infty x^{\alpha-i-j-1-z}\,dx$$

$$= \sum_{j=0}^{\alpha}(-1)^j c^j \sum_{i=0}^{\alpha-j+1} a_{\alpha-i}\,\delta_{\alpha-j-i}dx$$

$$= \sum_{j=0}^{\alpha}(-1)^{\alpha-j}c^{\alpha-j}\,a_{\alpha-j},$$

where as before we have written $\sigma \sim \sum_{i=0}^{\infty} a_{\alpha-i}\,x^{\alpha-i}$. This coincides with the polynomial we previously derived. $\square$

We infer from Proposition 5.4 that for symbols $\sigma_1, \cdots \sigma_k$ with orders $\alpha_1, \cdots, \alpha_k$ such that $\alpha_1 + \cdots + \alpha_i \notin \mathbb{Z}_{\geq 0}$ for any $i \in \{1, \cdots, k\}$, the iterated extended integrations make sense as integrals extended to non-integer order symbols. Setting

$\vec{\sigma} := (\sigma_1, \cdots \sigma_k)$ we infer that

$$\overline{\mathcal{I}}_k(\vec{\sigma})(c) := \int_0^\infty \frac{\sigma_k(x_k)}{x_k + c} \, dx_k \int_0^\infty \frac{\sigma_{k-1}(x_{k-1})}{x_{k-1} + x_k} \, dx_{k-1} \cdots \int_0^\infty \frac{\sigma_1(x_1)}{x_1 + x_2} \, dx_1$$

defines a symbol of order $\alpha_1 + \cdots + \alpha_k$.

In the general case, we replace the symbols σ_i by holomorphic families $\sigma_i(z_i)$ of order $\alpha_i - z_i$, thereby introducing poles on hyperplanes which pass through zero whenever $\alpha_1 + \cdots + \alpha_i \in \mathbb{Z}_{\geq 0}$ for some $i \in \{1, \cdots, k\}$.

PROPOSITION 5.10. *Let $\vec{\sigma}(\vec{z}) := (\sigma_1(z_1), \cdots \sigma_k(z_k))$ be a k-tuple of* **holomorphic** *families of classical symbols of order $\alpha_1(z_1) = \alpha_1 - z_1, \cdots, \alpha_k(z_k) = \alpha_k - z_k$. Then integrals of the type*

$$\overline{\mathcal{I}}_k(\vec{\sigma}(\vec{z}))(c)$$

$$:= \int_0^\infty \frac{\sigma_k(z_k)(x_k)}{x_k + c} \, dx_k \int_0^\infty \frac{\sigma_{k-1}(z_{k-1})(x_{k-1})}{x_{k-1} + x_k} \, dx_{k-1} \cdots \int_0^\infty \frac{\sigma_1(z_1)(x_1)}{x_1 + x_2} \, dx_1$$

give rise to meromorphic symbols $c \mapsto \overline{\mathcal{I}}(\vec{\sigma}(\vec{z})(c))$ with simple poles at zero located on hyperplanes $z_1 + \cdots + z_i = 0$ with i running from 1 to k.

PROOF. We proceed by induction on k.

The first step $k = 1$ follows from Theorem 5.9 applied to the holomorphic family $\sigma(z)$. To prove the induction step, assuming the result holds for $k > 1$, we write

$$\overline{\mathcal{I}}_{k+1}(\sigma_1, \cdots, \sigma_{k+1})(c) := \int_0^\infty \frac{\overline{\mathcal{I}}_k(\sigma_1, \cdots, \sigma_k)(x_{k+1})\sigma_{k+1}(z_{k+1})(x_{k+1})}{x_{k+1} + c} \, dx_{k+1}.$$

We then apply Theorem 5.9 to the meromorphic family

$$x_{k+1} \mapsto \overline{\mathcal{I}}_k(\sigma_1, \cdots, \sigma_k)(x_{k+1})\sigma_{k+1}(z_{k+1})(x_{k+1}),$$

which by the induction assumption has order $\vec{\alpha}(\vec{z}) := \alpha_1(z_1) + \cdots + \alpha_k(z_k)$ and simple poles located on hyperplanes $z_1 + \cdots + z_i = 0$, $i \in \{1, \cdots, k\}$. We infer that the map $c \mapsto \overline{\mathcal{I}}_{k+1}(\sigma_1, \cdots, \sigma_{k+1})(c)$ defines a meromorphic family of symbols of order $\vec{\alpha}(\vec{z}) + \alpha_{k+1}(z_{k+1}) = \alpha_1(z_1) + \cdots + \alpha_{k+1}(z_{k+1}) = \sum_{i=1}^{k+1}(\alpha_i - z_i)$ whose poles at zero are simple and located on hyperplanes $z_1 + \cdots z_i = 0$ with $i \in \{1, \cdots, k+1\}$. □

5.3. Renormalizations

The above results can be applied to renormalization on rooted trees in two layers. First, given a symbol σ and holomorphic regularization $\mathcal{R} : \sigma \mapsto \sigma(z)$, this construction yields a homomorphism from the algebra $\mathcal{H}_{\mathcal{T}}$ of rooted trees to the algebra of meromorphic functions. Second, given a holomorphic map $\mathcal{R} : CS(\mathbb{R}_+) \times \Omega \to CS(\mathbb{R}_+)$, it gives a homomorphism from the Hopf algebra of rooted tress decorated by element in $CS(\mathbb{R}_+)$ to the algebra of meromorphic functions.

Given a symbol σ and holomorphic regularization $\mathcal{R} : \sigma \mapsto \sigma(z)$, it defines a homomorphism

$$\phi^{\mathcal{R}} : \mathcal{H}_{\mathcal{T}} \to \mathcal{B}_\otimes$$

recursively by

(1) $\phi^{\mathcal{R}}(\bullet)(c; \varepsilon_1) = \overline{\mathcal{I}}(\sigma(\varepsilon_1)(c);$

(2) $\phi^{\mathcal{R}}(T_1 \cdots T_k)(c; \varepsilon_1, \cdots, \varepsilon_{|T_1 \cdots T_k|})$

$$= \phi^{\mathcal{R}}(T_1)(c; \varepsilon_1, \cdots, \varepsilon_{|T_1|}) \phi^{\mathcal{R}}(T_2)(c; \varepsilon_{|T_1|+1}, \cdots, \varepsilon_{|T_1 T_2|})$$
$$\cdots \phi^{\mathcal{R}}(T_k)(c; \varepsilon_{|T_1 \cdots T_{k-1}|+1}, \cdots, \varepsilon_{|T_1 \cdots T_k|});$$

(3) $\phi^{\mathcal{R}}(B^+(F))(c; \varepsilon_1, \cdots, \varepsilon_{|F|+1}) = \int_0^\infty \phi^{\mathcal{R}}(F)(x; \varepsilon_1, \cdots, \varepsilon_{|F|}) \frac{\sigma(\varepsilon_{|F|+1})(x)dx}{x+c}$.

Therefore, we can apply either a generalized evaluator or Birkhoff decomposition (in this case, it is equivalent to a generalized evaluator) to get renormalized values of rooted trees. But in general the calculation of these values is hard.

Given a holomorphic map $\mathcal{R} : CS(\mathbb{R}_+) \times \Omega \to CS(\mathbb{R}_+)$, it defines a homomorphism

$$\phi^{\mathcal{R}} : \mathcal{H}_{\mathcal{T}(CS(\mathbb{R}_+))} \to \mathcal{B}_{\otimes}$$

recursively by

(1) $\phi_{bu}^{\mathcal{R}}(\bullet_\sigma)(c; \varepsilon_1) = \overline{\mathcal{I}}(\vec{\sigma}(\varepsilon_1)(c));$

(2) $\phi_{bu}^{\mathcal{R}}(T_1 \cdots T_k)(c; \varepsilon_1, \cdots, \varepsilon_{|T_1 \cdots T_k|})$
$$= \phi_{bu}^{\mathcal{R}}(T_1)(c; \varepsilon_1, \cdots, \varepsilon_{|T_1|}) \phi_{bu}^{\mathcal{R}}(T_2)(c; \varepsilon_{|T_1|+1}, \cdots, \varepsilon_{|T_1 T_2|})$$
$$\cdots \phi_{bu}^{\mathcal{R}}(T_k)(c; \varepsilon_{|T_1 \cdots T_{k-1}|+1}, \cdots, \varepsilon_{|T_1 \cdots T_k|});$$

(3) $\phi_{bu}^{\mathcal{R}}(B_\sigma^+(F))(c; \varepsilon_1, \cdots, \varepsilon_{|F|+1}) = \int_0^\infty \phi_{bu}^{\mathcal{R}}(F)(x; \varepsilon_1, \cdots, \varepsilon_{|F|}) \frac{\sigma(\varepsilon_{|F|+1})(x)dx}{x+c}$.

Then either a generalized evaluator or Birkhoff decomposition will give us a character:

$$\bar{\phi}_{bu}^{\mathcal{R}} : \mathcal{H}_{\mathcal{T}(CS(\mathbb{R}_+))} \to \mathbb{C}.$$

REMARK 5.11. We expect a re-parametrization $z_i \mapsto \tau_\mu(z_i)$ (common to all the z_i's) as in Remark 4.2 to give rise to a polynomial expression in μ of degree $|F|$ and hope to investigate this issue in terms of the renormalization group in a forthcoming paper.

References

[1] Y. André, *Une Introduction aux Motifs*, Société Mathématique de France, Paris, 2004.

[2] N. Berline and M. Vergne, *Local Euler-Maclaurin formula for polytopes*, arXiv:math/0507256.

[3] S. Bloch, H. Esnault and D. Kreimer, *Motives associated to graph polynomials*, Comm. Math. Phys. **267** (2006), 181-225.

[4] A. Connes, D. Kreimer, *Hopf algebras, renormalisation and noncommutative geometry*, Comm. Math. Phys. **199** (1988), 203-242.

[5] A. Connes and D. Kreimer, *Renormalization in quantum field theory and the Riemann-Hilbert problem. I. The Hopf algebra structure of graphs and the main theorem*, Comm. Math. Phys. **210** (2000), 249-273.

[6] A. Connes and D. Kreimer, *Renormalization in quantum field theory and the Riemann-Hilbert problem. II. The β-function, diffeomorphisms and the renormalization group*, Comm. Math. Phys. **216** (2001), 215-241.

[7] A. Connes and M. Marcolli, *From physics to number theory via noncommutative geometry, Part II: Renormalization, the Riemann-Hilbert correspondence, and motivic Galois theory*, Frontiers in Number Theory, Physics, and Geometry, II, Springer-Verlag, 2006, pp. 617-713.

[8] C. De Concini, C. Procesi, *Wonderful models of subspace arrangements*, Selecta Mathematica, New Series **1** (1995), 459-494.

[9] K. Ebrahimi-Fard, L. Guo and D. Kreimer, *Spitzer's identity and the algebraic Birkhoff decomposition in pQFT*, J. Phys. A: Math. Gen. **37** (2004), 11037-11052.

[10] K. Ebrahimi-Fard, L. Guo and D. Manchon, *Birkhoff type decompositions and the Baker-Campbell-Hausdorff recursion*, Comm. Math. Phys. **267** (2006), 821-845.

[11] A. Einstein, *Quantum mechanics and reality (Quanten-Mechanik und Wirklichkeit)*, Dialectica **2** (1948), 320-324.

[12] E. M. Feichtner, *De Concini-Procesi wonderful arrangement models, A discrete point of view*, MSRI Publications **52** (2005), 333-360.

[13] L. Foissy, *Les algèbres de Hopf des arbres enracinés décorés. I*, Bull. Sci. Math. **126** (2002), 193-239.

[14] R. Grossman and R. G. Larson, *Hopf-algebraic structure of families of trees*, J. Algebra **126** (1989), 184-210.

[15] L. Guo, *Operated semigroups, Motzkin paths and rooted tree*, J. Algebraic Combinatorics **29** (2009), 35-62.

[16] L. Guo, *Algebraic Birkhoff decomposition and its applications*, Automorphic Forms and the Langlands Program, International Press, 2008, pp. 283-323.

[17] L. Guo, S. Paycha, B.-Y. Xie and B. Zhang, *Double shuffle relations and renormalization of multiple zeta values*, Geometry of Algebraic Cycles, Clay Mathematics Institute and Amer. Math. Soc., 2010, pp. 145-188.

[18] L. Guo and B. Zhang, *Renormalization of multiple zeta values*, J. Algebra **319** (2008), 3770-3809.

[19] L. Guo and B. Zhang, *Differential algebraic Birkhoff decomposition and renormalization of multiple zeta values*, J. Number Theory **128** (2008), 2318-2339.

[20] M. Kontsevich and S. Vishik, *Geometry of determinants of elliptic operators*, Progress in Mathematics **131** (1994), 173–197.

[21] M. Kontsevich and S. Vishik, *Determinants of elliptic pseudodifferential operators*, Max Planck Preprint (1994).

[22] D. Kreimer, *On the Hopf algebra of perturbative quantum field theory*, Adv. Theo. Math. Phys. **2** (1998), 303-334.

[23] M. Lesch and M. Pflaum, *Traces on algebras of parameter dependent pseudodifferential operators and the eta-invariant*, Trans. Amer. Math. Soc. **352** (2000), 4911-4936.

[24] J.-L. Loday and M. O. Ronco, *Hopf algebra of the planar binary trees*, Adv. Math. **139** (1998), 293-309.

[25] D. Manchon, *Hopf algebras in renormalisation*, Handbook of Algebra (M. Hazewinkel, ed.) **5** (2008), 365-427, arXiv:math. QA/0408405.

[26] D. Manchon and S. Paycha, *Shuffle relations for regularised integrals of symbols*, Comm. Math. Phys. **270** (2007), 13-31.

[27] D. Manchon and S. Paycha, *Chen sums of symbols on $\mathbb{R}^n$ and renormalised multizeta values*, Int. Math. Reas. Not. (to appear), arXiv:math/0702135.

[28] R. Mills, *Tutorial on infinities in QED*, Renormalization: From Lorentz to Landau (and beyond), Springer-Verlag, 1993.

[29] S. Paycha, *Renormalised multiple integrals of symbols with linear constraints*, Comm. Math. Phys. **286** (2009), 495-540.

[30] S. Paycha *Renormalised multiple sums and integrals with constraints: A comparative study*, preprint (2008)..

[31] E. Speer, *Analytic renormalisation*, Jour. Math. Phys. **9** (1968), 1404-1410.

DEPARTMENT OF MATHEMATICS AND COMPUTER SCIENCE, RUTGERS UNIVERSITY, NEWARK, NJ 07102, USA.
E-mail address: `liguo@rutgers.edu`

LABORATOIRE DE MATHÉMATIQUES APPLIQUÉES, UNIVERSITÉ BLAISE PASCAL (CLERMONT II), COMPLEXE UNIVERSITAIRE DES CÉZEAUX, 63177 AUBIÈRE CEDEX, FRANCE.
E-mail address: `sylvie.paycha@math.univ-bpclermont.fr`

YANGTZE CENTER OF MATHEMATICS, SICHUAN UNIVERSITY, CHENGDU, 610064, P. R. CHINA.
E-mail address: `zhangbin@scu.edu.cn`

Absolute Modular Forms

Shin-ya Koyama and Nobushige Kurokawa

ABSTRACT. We introduce the notion of absolute modular forms and give examples expressed in terms of higher derivatives of the multiple sine functions. We also prove a certain identity between them, which suggests a graded structure of the space of absolute modular forms of weight 3.

1. Introduction

The absolute modular forms are a new kind of modular form. These forms are functions of the semi-lattice

$$\mathbf{Z}_{\geq 0}\omega_1 + \cdots + \mathbf{Z}_{\geq 0}\omega_r.$$

The absolute modular group is identified as

$$(1.1) \qquad GL_r(\mathbf{F}_1) = S_r = \mathrm{Aut}(\mathbf{Z}_{\geq 0}\omega_1 + \cdots + \mathbf{Z}_{\geq 0}\omega_r).$$

We refer to [**KOW,CCM**] for mathematics over $\mathbf{F}_1$.

Concretely speaking, we say that a (holomorphic) function f on

$$\mathcal{D}_r = \left\{ (u_1, \cdots, u_r) \in \mathbf{C}^r \left| \begin{array}{l} u_1, \cdots, u_r \text{ and } 1 \text{ belong to one side} \\ \text{with respect to a line crossing } 0 \end{array} \right. \right\}$$

is an absolute modular form of weight k, if it satisfies the following two conditions:

(1) $f(u_1, \cdots, u_r)$ is symmetric, and
(2) $f(\frac{1}{u_1}, \frac{u_2}{u_1}, \cdots, \frac{u_r}{u_1}) = u_1^k f(u_1, u_2, \cdots, u_r)$.

The Eisenstein series

$$\mathcal{E}_k(u_1, \cdots, u_r) = \sideset{}{'}\sum_{n_1, \cdots, n_{r+1} \geq 0} (n_1 u_1 + \cdots + n_r u_r + n_{r+1})^{-k} \qquad (k > r+1)$$

is a typical example. It is analogous to the classical modular forms, but a remarkable difference is that the sum is taken over the semi-lattice

$$\mathbf{Z}_{\geq 0}\omega_1 + \cdots + \mathbf{Z}_{\geq 0}\omega_r,$$

while the classical sum was taken over the whole lattice.

We find that the domains $\mathcal{D}_r$ for $r = 1, 2$ are given by

$$\mathcal{D}_1 = \mathbf{C} \setminus \mathbf{R}_{\leq 0} = \{u \in \mathbf{C} \mid -\pi < \arg(u) < \pi\}$$

and

$$\mathcal{D}_2 = \{(u, v) \in \mathcal{D}_1^2 \mid -\pi < \arg(u) - \arg(v) < \pi\}.$$

We recall that the ordinary modular forms have two formulations: a function on the upper half plane and a function on the lattice $\mathbf{Z}\omega_1 + \mathbf{Z}\omega_2$. We usually identify them as a function in $\tau = \omega_2/\omega_1$ on the upper (or lower) half plane.

The group (1.1) is analogous to the modular group $GL_2(\mathbf{Z}) = \mathrm{Aut}(\mathbf{Z}\omega_1 + \mathbf{Z}\omega_2)$. From this viewpoint, we introduce another domain

$$\mathbf{D}_r = \left\{(\omega_1, \cdots, \omega_r) \in \mathbf{C}^r \,\middle|\, \begin{array}{l} \omega_1, \cdots, \omega_r \text{ belong to one side} \\ \text{with respect to a line crossing } 0 \end{array}\right\}$$

so that we regard a function F on $\mathbf{D}_r$ as an absolute modular form, if it satisfies the following two conditions

(1) $F(\omega_1, \cdots, \omega_r)$ is symmetric, and
(2) $F(\omega_1, \cdots, \omega_r)$ is homogeneous of degree $-k$:

$$F(c\omega_1, \cdots, c\omega_r) = c^{-k} F(\omega_1, \cdots, \omega_r)$$

for $c \in \mathbf{C} \setminus \{0\}$.

A function $f : \mathcal{D}_r \to \mathbf{C}$ is associated to a function $F : \mathbf{D}_{r+1} \to \mathbf{C}$ under

$$f(u_1, \cdots, u_r) = F(u_1, \cdots, u_r, 1)$$

and

$$F(\omega_1, \cdots, \omega_{r+1}) = \omega_{r+1}^{-k} f\left(\frac{\omega_1}{\omega_{r+1}}, \cdots, \frac{\omega_r}{\omega_{r+1}}\right).$$

It is easy to see that the conditions (1) and (2) on f and F are equivalent under this correspondence.

In this paper we deal with functions f on $\mathcal{D}_r$ of weight $k \geq 0$, which equivalently are the functions F on $\mathbf{D}_{r+1}$. Our chief concern is the following two constructions:

(a) $\mathcal{S}_k(u_1, \cdots, u_r) = S_{r+1}^{(k)}(0, (u_1, \cdots, u_r, 1))$ for $k \geq 0$, and
(b) $\mathcal{E}_k(u_1, \cdots, u_r) = \zeta_{r+1}(k, (u_1, \cdots, u_r, 1))$ for $k \geq 0$.

The notation is as follows: $S_r(x, (\omega_1, \cdots, \omega_r))$ is the multiple sine function defined as

$$S_r(x, (\omega_1, \cdots, \omega_r))$$

$$= \coprod_{n_1, \cdots, n_r \geq 0} (n_1\omega_1 + \cdots + n_r\omega_r + x) \left(\coprod_{m_1, \cdots, m_r \geq 1} (m_1\omega_1 + \cdots + m_r\omega_r - x)\right)^{(-1)^{r-1}}$$

with the symbol $\coprod$ being the regularized product due to Deninger [D]:

$$\coprod_\lambda \lambda = \exp\left(-\frac{d}{ds}\sum_\lambda \lambda^{-s}\bigg|_{s=0}\right).$$

Using the multiple gamma function

$$\Gamma_r(x, (\omega_1, \cdots, \omega_r)) = \left(\coprod_{n_1, \cdots, n_r \geq 0} (n_1\omega_1 + \cdots + n_r\omega_r + x)\right)^{-1},$$

we can express the multiple sine function as

$$S_r(x, (\omega_1, \cdots, \omega_r)) = \Gamma_r(x, (\omega_1, \cdots, \omega_r))^{-1}\Gamma_r(\omega_1 + \cdots + \omega_r - x, (\omega_1, \cdots, \omega_r))^{(-1)^r}.$$

We refer to [**K1,K2,KK1**] for the theory of multiple sine functions; see the excellent survey by Manin [**M**].

The other notation, $\zeta_r(s, (\omega_1, \cdots, \omega_r))$, is a kind of multiple version of the Riemann (or Hurwitz) zeta function

$$\zeta_r(s, (\omega_1, \cdots, \omega_r)) = \sum_{n_1, \cdots, n_r \geq 0}{}' (n_1 \omega_1 + \cdots + n_r \omega_r)^{-s}.$$

We remark that $\mathcal{E}_k(u_1, \cdots, u_r)$, whose original definition was (1.1) for $k > r + 1$, is extended to $k \leq r + 1$ by the analytic continuation of $\zeta_r(k, (u_1, \cdots, u_r, 1))$. This enables us to consider those with smaller weights such as $\mathcal{E}_0(u_1, \cdots, u_r)$.

It would be suggestive to regard $\mathcal{E}_k$ and $\mathcal{S}_k$ as the (generalized) "Eisenstein series" and "cusp forms," respectively. The present state of our experience on absolute modular forms is primitive. We must postpone developing the general theory.

The absolute modular form is also called the Stirling modular form. It was originated in an old paper of Barnes [**B**] (p. 397), where a function $\rho_r(\omega_1, \cdots, \omega_r)$ was called the "Stirling modular form" associated to the multiple gamma function. It was named after the Stirling asymptotic formula for $n!$ as $n \to \infty$ (cf. [**K3,K4,K5,KK3,Ko**]). The Stirling modular form $\rho_r(\omega_1, \cdots, \omega_r)$ appears as the normalization factor of the multiple gamma functions. Indeed Barnes defined the multiple gamma function $\Gamma_r^B(x, (\omega_1, \cdots, \omega_r))$ so that it satisfied the normalization

$$\lim_{x \to 0} \frac{\Gamma_r^B(x, (\omega_1, \cdots, \omega_r))^{-1}}{x} = 1.$$

It is related to our $\Gamma_r(x, (\omega_1, \cdots, \omega_r))$ by

$$\Gamma_r^B(x, (\omega_1, \cdots, \omega_r)) = \Gamma_r(x, (\omega_1, \cdots, \omega_r)) \rho_r(\omega_1, \cdots, \omega_r).$$

We see that $\rho_r(\omega_1, \cdots, \omega_r)$ is a complicated function and it is rather difficult to make a general theory containing it. Indeed $\rho_r(\omega_1, \cdots, \omega_r)$ satisfies condition (1) above, but unfortunately we need a homogeneous function $k = k(\omega_1, \cdots, \omega_r)$ for condition (2). The function $\rho_r(\omega_1, \cdots, \omega_r)$ is important and we know the following result proved in [**K2**]:

$$S_r'(0, \boldsymbol{\omega}) = \begin{cases} \rho_r(\boldsymbol{\omega})^2 \prod_{j=1}^{r-1} P_j(\boldsymbol{\omega})^{(-1)^{j-1}} & \text{if } r \text{ is odd,} \\ \prod_{j=1}^{r-1} P_j(\boldsymbol{\omega})^{(-1)^{j-1}} & \text{if } r \text{ is even,} \end{cases}$$

where

$$P_j(\boldsymbol{\omega}) = \prod_{1 \leq i_1 < i_2 < \cdots < i_j \leq r} \rho_r(\omega_{i_1}, \cdots, \omega_{i_j})$$

for $\boldsymbol{\omega} = (\omega_1, \cdots, \omega_r)$.

We notice the following neat expression for $\rho_r(\omega_1, \cdots, \omega_r)$, which was not mentioned in [**B**]:

$$\rho_r(\omega_1, \cdots, \omega_r) = \prod_{n_1, \cdots, n_r \geq 0}{}' (n_1 \omega_1 + \cdots + n_r \omega_r)$$
$$= \exp\left(-\zeta_r'(0, (\omega_1, \cdots, \omega_r))\right).$$

We recall that the Kronecker's Jugendtraum gives a strong motivation for the study of absolute modular forms and functions. It is a famous problem to find such a function that generates abelian extensions (class fields) of a given algebraic number field by its division values. The work in the rational number field case by Kronecker and the real quadratic field case by Shintani [**S**] suggests our looking at the extension

$$\mathbf{Q}\left(S_r\left(\frac{\omega_1 + \cdots + \omega_r}{N}, (\omega_1, \cdots, \omega_r)\right), \omega_1, \cdots, \omega_r\right)$$

over $\mathbf{Q}(\omega_1, \cdots, \omega_r)$. This function

$$S_r\left(\frac{\omega_1 + \cdots + \omega_r}{N}, (\omega_1, \cdots, \omega_r)\right)$$

on $\mathbf{D}_r$ (or the function $S_{r+1}\left(\frac{u_1+\cdots+u_r+1}{N}, (u_1, \cdots, u_r, 1)\right)$ on $\mathcal{D}_r$) is a typical absolute modular function (or an absolute modular form of weight 0).

In this paper we calculate $\mathcal{S}_k$ and $\mathcal{E}_k$ with k small and $r = 1, 2$. The following four theorems concern the case of $r = 1$.

THEOREM 1.
$$S_1(u) = \frac{2\pi}{\sqrt{u}}.$$

THEOREM 2.
$$S_2(u) = \begin{cases} \dfrac{8\pi^2 i}{\sqrt{u}}\left(\dfrac{1}{u}\mathcal{E}_1\left(-\dfrac{1}{u}\right) - E_1(u)\right) & \text{if} \quad \mathrm{Im}(u) > 0 \\[3mm] -\dfrac{8\pi^2 i}{\sqrt{u}}\left(\dfrac{1}{u}E_1\left(\dfrac{1}{u}\right) - E_1(-u)\right) & \text{if} \quad \mathrm{Im}(u) < 0, \end{cases}$$

where
$$E_1(\tau) = -\frac{1}{4} + \sum_{n=1}^{\infty} d(n)e^{2\pi i n\tau}$$

for $\mathrm{Im}(\tau) > 0$ with $d(n) = \sum_{d|n} 1$.

THEOREM 3.
$$\mathcal{E}_0(u) = \frac{1}{12}\left(u + \frac{1}{u} - 9\right).$$

THEOREM 4.
$$S_3(u) = \frac{3}{4}S_2(u)^2 S_1(u)^{-1} - \frac{3}{8}(4\mathcal{E}_0(u) + 3)S_1(u)^3.$$

Among others, Theorem 2 shows an unexpected relation of one-variable absolute modular forms of weight 2 and an Eisenstein-like series of weight 1. Note that $1/\sqrt{u}$ is of weight 1 as in Theorem 1. Theorem 4 indicates a graded structure of the space of absolute modular forms of weight 3. Our proof shows that this identity comes from the (quasi-)modularity of

$$E_2(\tau) = -\frac{1}{24} + \sum_{n=1}^{\infty} \sigma(n)e^{2\pi i n\tau}$$

with
$$\sigma(n) = \sum_{d|n} d.$$

Here we are using the notation

$$E_k(\tau) = \frac{\zeta(1-k)}{2} + \sum_{n=1}^{\infty} \sigma_{k-1}(n)e^{2\pi in\tau}$$

with

$$\sigma_{k-1}(n) = \sum_{d \mid n} d^{k-1}.$$

In this paper we do not deal with the "cuspidality." At present we think that the cuspidality of $\mathcal{S}_k(u)$ means that $\mathcal{S}_k(\infty) = 0$. This vanishing is obvious for $\mathcal{S}_1(u)$ from Theorem 1. It is not so difficult to show $\mathcal{S}_2(\pm i\infty) = 0$ and $\mathcal{S}_3(\pm i\infty) = 0$ from Theorems 2 and 4, respectively. We hope to report on them in a forthcoming paper.

Our results have some applications. For example, we see from Theorem 2 and $\mathcal{S}_2(1) = -4\pi$ that the following corollary holds.

COROLLARY 1.

$$\lim_{\substack{\tau \to 1 \\ \mathrm{Im}(\tau)>0}} \left(E_1\left(-\frac{1}{\tau}\right) - \tau E_1(\tau) \right) = -\frac{1}{2\pi i}.$$

The reason it is possible to prove such results is that the domain $\mathcal{D}_1$ contains the positive real numbers $\mathbf{R}_{>0}$ (see Figure 1). This indicates a merit of absolute modular forms. We refer to [**K3,K5,KK3**] for related results.

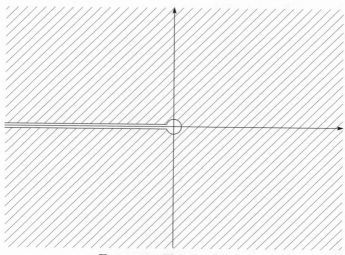

FIGURE 1 The domain $\mathcal{D}_1$.

The following theorems are the results of calculations in the case of $r = 2$.

THEOREM 5.

$$\mathcal{E}_0(u,v) = \frac{1}{24}\left(u + \frac{1}{u} + v + \frac{1}{v} + \frac{u}{v} + \frac{v}{u} - 21 \right).$$

THEOREM 6.

$$S_1(u,v) = \frac{\rho_3(u,v,1)^2 \rho_1(u)\rho_1(v)\rho_1(1)}{\rho_2(u,v)\rho_2(u,1)\rho_2(v,1)}$$

$$= \frac{(2\pi)^{\frac{3}{2}}}{\sqrt{uv}} \cdot \frac{\rho_3(u,v,1)^2}{\rho_2(u,v)\rho_2(u,1)\rho_2(v,1)}.$$

2. Proofs of Theorems 1 and 2

Since we have seen these results in other contexts essentially (see [K2], for example), we give concise proofs here.

2.1. Proof of Theorem 1

It is sufficient to show that

$$S_2'(0,(\omega_1,\omega_2)) = \frac{2\pi}{\sqrt{\omega_1\omega_2}}.$$

The periodicity proved in [KK1]

$$S_2(x,(\omega_1,\omega_2)) = S_2(x+\omega_2,(\omega_1,\omega_2))S_1(x,\omega_1)$$

with

$$S_1(x,\omega) = 2\sin\left(\frac{\pi x}{\omega}\right)$$

implies

$$S_2'(0,(\omega_1,\omega_2)) = S_2(\omega_2,(\omega_1,\omega_2))\frac{2\pi}{\omega_1}.$$

Here,

$$S_2(\omega_2,(\omega_1,\omega_2)) = \frac{\Gamma_2(\omega_1,(\omega_1,\omega_2))}{\Gamma_2(\omega_2,(\omega_1,\omega_2))}$$

$$= \lim_{x\to 0}\frac{\Gamma_2(\omega_1+x,(\omega_1,\omega_2))}{\Gamma_2(\omega_2+x,(\omega_1,\omega_2))}$$

$$= \lim_{x\to 0}\frac{\Gamma_2(x,(\omega_1,\omega_2))\Gamma_1(x,\omega_2)^{-1}}{\Gamma_2(x,(\omega_1,\omega_2))\Gamma_1(x,\omega_1)^{-1}}$$

$$= \lim_{x\to 0}\frac{\Gamma_1(x,\omega_1)}{\Gamma_1(x,\omega_2)}$$

$$= \lim_{x\to 0}\frac{\frac{\Gamma(\frac{x}{\omega_1})}{\sqrt{2\pi}}\omega_1^{\frac{x}{\omega_1}-\frac{1}{2}}}{\frac{\Gamma(\frac{x}{\omega_2})}{\sqrt{2\pi}}\omega_2^{\frac{x}{\omega_2}-\frac{1}{2}}}$$

$$= \sqrt{\frac{\omega_1}{\omega_2}}.$$

Thus

$$S_2'(0,(\omega_1,\omega_2)) = \frac{2\pi}{\sqrt{\omega_1\omega_2}}.$$

□

2.2. Proof of Theorem 2

A formula of Shintani [**S**] (Proposition 5) shows that

$$
\mathrm{Cot}_2(x,(1,\tau)) = \pi i \left(\frac{x}{\tau} - \frac{1}{2}\left(1 + \frac{1}{\tau}\right)\right)
$$

$$
- 2\pi i \sum_{n=0}^{\infty} \sum_{m=1}^{\infty} e^{2\pi i n m \tau} e^{2\pi i m x} + \frac{2\pi i}{\tau} \sum_{n=1}^{\infty} \sum_{m=1}^{\infty} e^{-\frac{2\pi i n m}{\tau}} e^{\frac{2\pi i m x}{\tau}},
$$

where the double cotangent function Cot_2 is defined as

$$
\mathrm{Cot}_2(x,(1,u)) = \frac{S_2'(x,(1,u))}{S_2(x,(1,u))}.
$$

Hence

$$
\mathrm{Cot}_2(\tau,(1,\tau)) = \frac{\pi i}{2}\left(1 - \frac{1}{\tau}\right) - 2\pi i \sum_{n=1}^{\infty}\sum_{m=1}^{\infty} e^{2\pi i n m \tau} + \frac{2\pi i}{\tau}\sum_{n=1}^{\infty}\sum_{m=1}^{\infty} e^{-\frac{2\pi i n m}{\tau}}
$$

$$
= \frac{\pi i}{2}\left(1 - \frac{1}{\tau}\right) - 2\pi i \left(E_1(\tau) + \frac{1}{4}\right) + \frac{2\pi i}{\tau}\left(E_1\left(-\frac{1}{\tau}\right) + \frac{1}{4}\right)
$$

$$
= \frac{2\pi i}{\tau}\left(E_1\left(-\frac{1}{\tau}\right) - \tau E_1(\tau)\right).
$$

Now we will prove that

$$
S_2''(0,(1,\tau)) = \frac{4\pi}{\sqrt{\tau}}\mathrm{Cot}_2(\tau,(1,\tau)).
$$

This is obtained from the periodicity

$$
S_2(x,(1,\tau)) = S_2(x+\tau,(1,\tau))S_1(x)
$$

with $S_1(x) = 2\sin(\pi x)$. In fact, differentiating this periodicity relation twice leads to

$$
S_2''(x,(1,\tau))
$$
$$
= S_2''(x+\tau,(1,\tau))S_1(x) + 2S_2'(x+\tau,(1,\tau))S_1'(x) + S_2(x+\tau,(1,\tau))S_1''(x).
$$

Thus we obtain the desired relation

$$
S_2''(0,(1,\tau)) = 2S_2'(\tau,(1,\tau))S_1'(0)
$$
$$
= 4\pi S_2'(\tau,(1,\tau))
$$
$$
= \frac{4\pi}{\sqrt{\tau}} \cdot \frac{S_2'(\tau,(1,\tau))}{S_2(\tau,(1,\tau))}
$$
$$
= \frac{4\pi}{\sqrt{\tau}}\mathrm{Cot}_2(\tau,(1,\tau)),
$$

where we used the fact

$$
S_2(\tau,(1,\tau)) = \frac{1}{\sqrt{\tau}}
$$

proved in [**KK2**]. Thus we have

$$
\mathcal{S}_2(\tau) = S_2''(0,(1,\tau)) = \frac{8\pi^2 i}{\tau\sqrt{\tau}}\left(E_1\left(-\frac{1}{\tau}\right) - \tau E_1(\tau)\right).
$$

This proves Theorem 2 for $\text{Im}(u) > 0$. The case $\text{Im}(u) < 0$ is given by the reflection. ☐

3. Proof of Theorem 3

We prove that

$$\zeta_2(0, (\omega_1, \omega_2)) = \frac{1}{12}\left(\frac{\omega_2}{\omega_1} + \frac{\omega_1}{\omega_2} - 9\right).$$

For this we recall the Riemann-Mellin type integral expression for $\zeta_2(s, (\omega_1, \omega_2))$ (cf. Barnes [**B**]). It says that

$$\zeta_2(s, (\omega_1, \omega_2)) = \frac{1}{\Gamma(s)}\int_0^\infty \Theta(t)t^{s-1}dt$$

in $\text{Re}(s) > 2$ with

$$\Theta(t) = {\sum_{n_1,n_2\geq 0}}' e^{-(n_1\omega_1 + n_2\omega_2)t}$$

$$= \frac{1}{(1 - e^{-t\omega_1})(1 - e^{-t\omega_2})} - 1$$

$$= \frac{e^{t\omega_1} + e^{t\omega_2} - 1}{(e^{t\omega_1} - 1)(e^{t\omega_2} - 1)}.$$

Let

$$\Theta(t) = \frac{a_{-2}}{t^2} + \frac{a_{-1}}{t} + a_0 + a_1 t + \cdots$$

be the Laurent expansion around $t = 0$. An easy calculation shows that

$$a_{-2} = \frac{1}{\omega_1\omega_2}$$

$$a_{-1} = \frac{1}{2}\left(\frac{1}{\omega_1} + \frac{1}{\omega_2}\right)$$

$$a_0 = \frac{1}{12}\left(\frac{\omega_2}{\omega_1} + \frac{\omega_1}{\omega_2} - 9\right).$$

To make an analytic continuation of $\zeta_2(s, (\omega_1, \omega_2))$ in $\text{Re}(s) > -1$, we split the integral into three parts:

$$\zeta_2(s, (\omega_1, \omega_2)) = \frac{1}{\Gamma(s)}\int_1^\infty \Theta(t)t^{s-1}dt + \frac{1}{\Gamma(s)}\int_0^1 \left(\Theta(t) - \frac{a_{-2}}{t^2} - \frac{a_{-1}}{t} - a_0\right)t^{s-1}dt$$

$$+ \frac{1}{\Gamma(s)}\int_0^1 \left(\frac{a_{-2}}{t^2} + \frac{a_{-1}}{t} + a_0\right)t^{s-1}dt.$$

The first term is holomorphic for all $s \in \mathbf{C}$, since the integral is absolutely convergent. The second term is holomorphic in $\text{Re}(s) > -1$ since

$$\Theta(t) - \frac{a_{-2}}{t^2} - \frac{a_{-1}}{t} - a_0 = O(t)$$

as $t \to 0$. The third term is

$$\frac{1}{\Gamma(s)}\left(\frac{a_{-2}}{s-2} + \frac{a_{-1}}{s-1} + \frac{a_0}{s}\right),$$

which is meromorphic in $s \in \mathbf{C}$. Hence the above expression gives an analytic continuation of $\zeta_2(s, (\omega_1, \omega_2))$ in $\mathrm{Re}(s) > -1$. Moreover this calculation shows that $\zeta_2(s, (\omega_1, \omega_2))$ is holomorphic at $s = 0$ and its value is given by $\zeta_2(0, (\omega_1, \omega_2)) = a_0$. Thus

$$\zeta_2(0, (\omega_1, \omega_2)) = \frac{1}{12}\left(\frac{\omega_2}{\omega_1} + \frac{\omega_1}{\omega_2} - 9\right).$$

□

4. Proof of Theorem 4

We prove the following facts (a):

$$S_2'''(0, (1, \tau)) = \frac{6\pi}{\sqrt{\tau}}\left(\mathrm{Cot}_2'(\tau, (1, \tau)) + \mathrm{Cot}_2(\tau, (1, \tau))^2 - \frac{\pi^2}{3}\right),$$

and (b):

$$\mathrm{Cot}_2'(\tau, (1, \tau)) = \frac{\pi^2}{6}\left(1 - \frac{1}{\tau^2}\right).$$

Before proving (a) and (b) we show that Theorem 4 follows from them. In fact (a) asserts that

$$\begin{aligned}
S_3(\tau) &= S_2'''(0, (1, \tau)) \\
&= \frac{6\pi}{\sqrt{\tau}}\mathrm{Cot}_2'(\tau, (1, \tau)) + \frac{6\pi}{\sqrt{\tau}}\mathrm{Cot}_2(\tau, (1, \tau))^2 - \frac{2\pi^3}{\sqrt{\tau}}.
\end{aligned}$$

Recall the formula

$$\begin{aligned}
\mathrm{Cot}_2(\tau, (1, \tau)) &= \frac{\sqrt{\tau}}{4\pi}S_2''(0, (1, \tau)) \\
&= \frac{\sqrt{\tau}}{4\pi}S_2(\tau),
\end{aligned}$$

which was proved in Section 2.2. Then by (b) we have

$$\begin{aligned}
S_3(\tau) &= \frac{\pi^3}{\sqrt{\tau}}\left(1 - \frac{1}{\tau^2}\right) + \frac{3\sqrt{\tau}}{8\pi}S_2(\tau)^2 - \frac{2\pi^3}{\sqrt{\tau}} \\
&= -\frac{1}{8}\left(\frac{2\pi}{\sqrt{\tau}}\right)^3\left(\tau + \frac{1}{\tau}\right) + \frac{3}{4}\left(\frac{2\pi}{\sqrt{\tau}}\right)^{-1}S_2(\tau)^2 \\
&= -\frac{1}{8}S_1(\tau)^3(12\mathcal{E}_0(\tau) + 9) + \frac{3}{4}S_1(\tau)^{-1}S_2(\tau)^2
\end{aligned}$$

as desired.

4.1. Proof of (a)

The definition

$$\mathrm{Cot}_2(x, (1, \tau)) = \frac{S_2'(x, (1, \tau))}{S_2(x, (1, \tau))}$$

implies

$$\mathrm{Cot}_2'(x, (1, \tau)) = \frac{S_2''(x, (1, \tau))}{S_2(x, (1, \tau))} - \frac{S_2'(x, (1, \tau))^2}{S_2(x, (1, \tau))^2}.$$

Hence

$$\text{Cot}'_2(\tau, (1, \tau)) = \frac{S''_2(\tau, (1, \tau))}{S_2(\tau, (1, \tau))} - \left(\frac{S'_2(\tau, (1, \tau))}{S_2(\tau, (1, \tau))}\right)^2$$

$$= \frac{S''_2(\tau, (1, \tau))}{S_2(\tau, (1, \tau))} - \text{Cot}_2(\tau, (1, \tau))^2.$$

Now we show that

$$S''_2(\tau, (1, \tau)) = \frac{1}{6\pi}\left(S'''_2(0, (1, \tau)) + \frac{2\pi^3}{\sqrt{\tau}}\right).$$

We obtain this by differentiating the periodicity relation

$$S_2(x, (1, \tau)) = S_2(x + \tau, (1, \tau))S_1(x)$$

three times at $x = 0$. In fact the identity

$$S'''_2(x, (1, \tau)) = S'''_2(x + \tau, (1, \tau))S_1(x) + 3S''_2(x + \tau, (1, \tau))S'_1(x)$$
$$+ 3S'_2(x + \tau, (1, \tau))S''_1(x) + S_2(x + \tau, (1, \tau))S'''_1(x)$$

gives

$$S'''_2(0, (1, \tau)) = 6\pi S''_2(\tau, (1, \tau)) - 2\pi^3 S_2(\tau, (1, \tau))$$

$$= 6\pi S''_2(\tau, (1, \tau)) - \frac{2\pi^3}{\sqrt{\tau}},$$

since $S_1(x) = 2\sin(\pi x)$ and $S_2(\tau, (1, \tau)) = \frac{1}{\sqrt{\tau}}$. Thus

$$\text{Cot}'_2(\tau, (1, \tau)) = \frac{\sqrt{\tau}}{6\pi}S'''_2(0, (1, \tau)) + \frac{\pi^2}{3} - \text{Cot}_2(\tau, (1, \tau))^2.$$

This gives (a).   □

4.2. Proof of (b)

We use the method of the proof of Theorem 2. In this case we start from

$$\text{Cot}'_2(x, (1, \tau)) = \frac{\pi i}{\tau} + 4\pi^2 \sum_{n=0}^{\infty}\sum_{m=1}^{\infty} me^{2\pi inm\tau}e^{2\pi imx} - \frac{4\pi^2}{\tau^2}\sum_{n=1}^{\infty}\sum_{m=1}^{\infty} me^{-\frac{2\pi inm}{\tau}}e^{\frac{2\pi imx}{\tau}}.$$

Then we have

$$\text{Cot}'_2(\tau, (1, \tau)) = \frac{\pi i}{\tau} + 4\pi^2 \sum_{n=1}^{\infty}\sum_{m=1}^{\infty} me^{2\pi inm\tau} - \frac{4\pi^2}{\tau^2}\sum_{n=1}^{\infty}\sum_{m=1}^{\infty} me^{-\frac{2\pi inm}{\tau}}$$

$$= \frac{\pi i}{\tau} + 4\pi^2\left(E_2(\tau) + \frac{1}{24}\right) - \frac{4\pi^2}{\tau^2}\left(E_2\left(-\frac{1}{\tau}\right) + \frac{1}{24}\right)$$

$$= 4\pi^2\left(E_2(\tau) - \frac{1}{\tau^2}E_2\left(-\frac{1}{\tau}\right)\right) + \frac{\pi i}{\tau} + \frac{\pi^2}{6}\left(1 - \frac{1}{\tau^2}\right)$$

$$= \frac{\pi^2}{6}\left(1 - \frac{1}{\tau^2}\right),$$

where we used the modularity of $E_2(\tau)$:

$$E_2\left(-\frac{1}{\tau}\right) = \tau^2 E_2(\tau) - \frac{\tau}{4\pi i}.$$

                                                                               □

REMARK. Similar calculations show that

$$\mathcal{S}_5 = \frac{5}{2}\mathcal{S}_4\mathcal{S}_2\mathcal{S}_1^{-1} - \frac{15}{16}\mathcal{S}_2^4\mathcal{S}_1^{-3} + \frac{15}{16}(4\mathcal{E}_0 + 3)\mathcal{S}_2^2\mathcal{S}_1 + \frac{1}{192}(144\mathcal{E}_0^2 + 216\mathcal{E}_0 + 89)\mathcal{S}_1^5.$$

5. Proof of Corollary 1

Theorem 2 implies

$$\lim_{\substack{\tau \to 1 \\ \mathrm{Im}(\tau) > 0}} \left(E_1\left(-\frac{1}{\tau}\right) - \tau E_1(\tau) \right) = \frac{1}{8\pi^2 i} \mathcal{S}_2(1)$$

$$= \frac{1}{8\pi^2 i} S_2''(0, (1, 1)),$$

and Theorem 4(1) of [**K2**] asserts that

$$S_2''(0, (1, 1)) = -4\pi.$$

Hence we get Corollary 1. □

6. Proof of Theorem 5

We calculate $\zeta_3(0, (\omega_1, \omega_2, \omega_3))$. Exactly similarly to the case of $\zeta_2(0, (\omega_1, \omega_2))$ treated in Section 3 we have

$$\zeta_3(0, (\omega_1, \omega_2, \omega_3)) = a_0 = a_0(\omega_1, \omega_2, \omega_3),$$

where a_0 is the constant term of the Laurent expansion of

$$\Theta(t) = \frac{1}{(1 - e^{-t\omega_1})(1 - e^{-t\omega_2})(1 - e^{-t\omega_3})} - 1$$

around $t = 0$:

$$\Theta(t) = \frac{a_{-3}}{t^3} + \frac{a_{-2}}{t^2} + \frac{a_{-1}}{t} + a_0 + a_1 t + \cdots.$$

The direct calculation shows that

$$a_0 = \frac{\omega_1\omega_2^2 + \omega_2\omega_3^2 + \omega_3\omega_1^2 + \omega_1^2\omega_2 + \omega_2^2\omega_3 + \omega_3^2\omega_1 - 21\omega_1\omega_2\omega_3}{24\omega_1\omega_2\omega_3}.$$

Hence

$$\mathcal{E}_0(u, v) = \zeta_3(0, (u, v, 1))$$

$$= \frac{1}{24}\left(u + \frac{1}{u} + v + \frac{1}{v} + \frac{u}{v} + \frac{v}{u} - 21\right).$$

□

7. Proof of Theorem 6

We calculated $S_3'(0, (\omega_1, \omega_2, \omega_3))$ in [**K2**] as

$$S_3'(0, (\omega_1, \omega_2, \omega_3)) = \frac{\rho_3(\omega_1, \omega_2, \omega_3)^2 \rho_1(\omega_1)\rho_2(\omega_2)\rho_3(\omega_3)}{\rho_2(\omega_1, \omega_2)\rho_2(\omega_2, \omega_3)\rho_2(\omega_3, \omega_1)}$$

with

$$\rho_1(\omega) = \sqrt{\frac{2\pi}{\omega}}.$$

Hence

$$\mathcal{S}_1(u,v) = S_3'(0,(u,v,1))$$
$$= \frac{(2\pi)^{\frac{3}{2}}}{\sqrt{uv}} \cdot \frac{\rho_3(u,v,1)^2}{\rho_2(u,v)\rho_2(u,1)\rho_2(v,1)}.$$

$\square$

References

[B]   Barnes, B.W., *On the theory of the multiple gamma function*, Trans. Cambridge Philos. Soc. **19** (1904), 374–425.

[CCM]   Connes, A., Consani, C., and Marcolli, M., *Fun with* $\mathbf{F}_1$, J. Number Theory **129** (2009), 1532–1561.

[D]   Deninger, C., *Local L-factors of motives and regularized determinant*, Invent. Math. **107** (1992), 135–150.

[K1]   Kurokawa, N., *Gamma factors and Plancherel measures*, Proc. Japan Acad. **68A** (1992), 256–260.

[K2]   Kurokawa, N., *Derivatives of multiple sine functions*, Proc. Japan Acad. **80A** (2004), 65–69.

[K3]   Kurokawa, N., *Limit values of Eisenstein series and multiple cotangent functions*, J. Number Theory **128** (2008), 1775–1783.

[K4]   Kurokawa, N., *The central value of the triple sine function*, Kodai Math. J. **32** (2009), 159–171.

[K5]   Kurokawa, N., *Automorphy of the principal Eisenstein series of weight 1: an application of the double sine function*, Kodai Math J. **32** (2009), 391–403.

[KK1]   Kurokawa, N., and Koyama, S., *Multiple sine functions*, Forum Math. **15** (2003), 839–876.

[KK2]   Koyama, S., and Kurokawa, N., *Zeta functions and normalized multiple sine functions*, Kodai Mathematical Journal **28** (2005), 534–550.

[KK3]   Koyama, S., and Kurokawa, N., *Multiple Eisenstein series and multiple cotangent functions*, J. Number Theory **128** (2008), 1769–1774.

[Ko]   Koyama, S., *Division values of multiple sine functions*, Kodai Mathematical Journal **32** (2009), 1–51.

[KOW]   Kurokawa, N., Ochiai, H. and Wakayama, M., *Absolute derivations and zeta function*, Doc. Math. (extra volume) (2003), 565–584.

[M]   Manin, Yu. I., *Lectures on zeta functions and motives (according to Deninger and Kurokawa)*, Asterisque **228** (1995), 121–163.

[S]   Shintani, T., *On a Kronecker limit formula for real quadratic fields*, J. Fac. Sci. Univ. Tokyo IA **24** (1977), 167–199.

INSTITUTE OF MATHEMATICAL SCIENCES, EWHA WOMAN'S UNIVERSITY, DAEHYUN-DONG 11-1, SEDAEMOON-KU, 120-750, SEOUL, SOUTH KOREA.
Current address: Department of Biomedical Engineering, Toyo University, 2100 Kujirai, Kawagoe-shi 350-8585, Japan.
E-mail address: koyama@toyonet.toyo.ac.jp

DEPARTMENT OF MATHEMATICS, TOKYO INSTITUTE OF TECHNOLOGY, 2-12-1, OH-OKAYAMA, MEGURO-KU, TOKYO 152-8551, JAPAN.
E-mail address: kurokawa@math.titech.ac.jp

Absolute Zeta Functions and Absolute Tensor Products

Shin-ya Koyama and Nobushige Kurokawa

ABSTRACT. We first survey the absolute tensor product. In particular we find that zeta functions which are meromorphic of order r are expected to have Euler factors expressed in terms of the elliptic gamma function of order $r - 1$. We make some observations and give evidence for this phenomenon. We then generalize the absolute Weil type zeta function of Deitmar-Koyama-Kurokawa to noncommutative versions. We obtain the determinant expressions and give many examples.

1. Zeta Functions of Tensor Products over $\mathbf{F}_1$

We recall the construction of the absolute tensor product, which is called the Kurokawa tensor product by Manin [M]. We refer to [KK3] and [KW] for details. Let

$$Z_j(s) = \prod_{\rho \in \mathbf{C}} (s - \rho)^{m_j(\rho)}$$

$$= \exp\left(-\frac{\partial}{\partial w}\bigg|_{w=0} \sum_{\rho \in \mathbf{C}} \frac{m_j(\rho)}{(s-\rho)^w}\right)$$

be "zeta functions" expressed as regularized products, where

$$m_j \;:\; \mathbf{C} \to \mathbf{Z}$$

denotes the multiplicity function for $j = 1, \dots, r$. The absolute tensor product $(Z_1 \otimes \cdots \otimes Z_r)(s)$ is defined as

$$(Z_1 \otimes \cdots \otimes Z_r)(s) = \prod_{\rho_1, \cdots, \rho_r \in \mathbf{C}} (s - (\rho_1 + \cdots + \rho_r))^{m(\rho_1, \cdots, \rho_r)}$$

with

$$m(\rho_1, \cdots, \rho_r) = m_1(\rho_1) \cdots m_r(\rho_r) \times \begin{cases} 1 & \operatorname{Im}(\rho_j) \geq 0, \quad (j = 1, \dots, r) \\ (-1)^{r-1} & \operatorname{Im}(\rho_j) < 0, \quad (j = 1, \dots, r) \\ 0 & \text{otherwise.} \end{cases}$$

This definition originates from [K2]. We refer to the excellent survey by Manin [M]. The notation of the regularized product is due to Deninger [Den]. The notion of such an infinite determinant has its origins in Ray-Singer [RS]. It is also a standard tool in mathematical physics (Hawking [**H**]). See [**HKW**] concerning the needed regularized products. The absolute tensor product was studied by Schröter [**S**] as the "Kurokawa tensor product."

We are especially interested in the case of Hasse zeta functions $Z_j(s) = \zeta(s, A_j)$ for commutative rings $A_1, \dots, A_r$ of finite type over **Z**. We recall that the Hasse zeta function $\zeta(s, A)$ of a commutative ring A is defined to be

$$\zeta(s, A) = \prod_{\mathbf{m}}(1 - N(\mathbf{m})^{-s})^{-1}$$

$$= \exp\left(\sum_{\mathbf{m}} \sum_{k=1}^{\infty} \frac{1}{k} N(\mathbf{m})^{-ks}\right),$$

where $\mathbf{m}$ runs over maximal ideals of A and $N(\mathbf{m}) = \#(A/\mathbf{m})$. This is also written as

$$\zeta(s, A) = \exp\left(\sum_{p:\text{primes}} \sum_{m=1}^{\infty} \frac{|\text{Hom}_{\text{ring}}(A, \mathbf{F}_{p^m})|}{m} p^{-ms}\right).$$

For simplicity we write

$$\zeta(s, A_1 \otimes_{\mathbf{F}_1} \cdots \otimes_{\mathbf{F}_1} A_r) := \zeta(s, A_1) \otimes \cdots \otimes \zeta(s, A_r).$$

We find that $\zeta(s, A_1 \otimes_{\mathbf{F}_1} \cdots \otimes_{\mathbf{F}_1} A_r)$ has the following additive structure on zeros and poles: if the condition $\zeta(s_j, A_j) = 0$ or ∞ holds for $j = 1, \dots, r$, and $\text{Im}(s_j)$ $(j = 1, \dots, r)$ have the same signature, then

$$\zeta(s_1 + \cdots + s_r, A_1 \otimes_{\mathbf{F}_1} \cdots \otimes_{\mathbf{F}_1} A_r) = 0 \text{ or } \infty,$$

where $s_1, \dots, s_r$ may be any mixed combination of zeros and poles.

Such an additive structure was crucial in the study of Hasse zeta functions of positive characteristic (congruence zeta functions) pursued by Grothendieck [**G**] and Deligne [**D**], where Euler products were important to restrict the region of zeros and poles for our reaching the analog of the Riemann Hypothesis (the Weil Conjecture). Indeed, Deligne considered the tensor product $A_1 \otimes_{\mathbf{F}_p} \cdots \otimes_{\mathbf{F}_p} A_r$ of $\mathbf{F}_p$-algebras A_j. The singularities of its Hasse-Weil zeta function $\zeta^{HW}(s, A_1 \otimes_{\mathbf{F}_p} \cdots \otimes_{\mathbf{F}_p} A_r)$ has zeros and poles at a sum of zeros or poles of $\zeta^{HW}(s, A_j)$ over $j = 1, 2, \dots, r$. In particular, when $A_1 = \cdots = A_j =: A$, this additive structure implies that if $\zeta^{HW}(\rho, A) = 0, \infty$, then $\zeta^{HW}(r\rho, A^{\otimes r}) = 0, \infty$. For proving the RH-type property stated as $|\text{Re}(\rho) - \frac{k}{2}| = 0$ for some $k = 0, 1, 2, 3, \dots, 2\dim A$, he considered the "trivial" zero-free region given by the Euler product for $\zeta^{HW}(s, A^{\otimes r})$. It follows that $|\text{Re}(r\rho) - \frac{rk}{2}| \le \frac{1}{2}$, which equivalently is $|\text{Re}(\rho) - \frac{k}{2}| \le \frac{1}{2r}$. Since this holds for any positive integer r, he obtained the conclusion by letting $r \to \infty$. Grothendieck's determinant expression for the Hasse-Weil zeta function was crucial to Deligne's method.

We expect that our multiple zeta functions also have Euler products of the following form:

$$\zeta(s, A_1) \otimes \cdots \otimes \zeta(s, A_r) = \prod_{(\mathbf{m}_1, \dots, \mathbf{m}_r)} H_{(\mathbf{m}_1, \dots, \mathbf{m}_r)}(N(\mathbf{m}_1)^{-s}, \dots, N(\mathbf{m}_1)^{-s}),$$

where $\mathbf{m}_i$ runs over the maximal ideals of A_i and $H_{(\mathbf{m}_1, \dots, \mathbf{m}_r)}(T_1, \dots, T_r)$ is a power series in $(T_1, \dots, T_r)$ of the constant term 1 with a possible degeneration at

$(\mathbf{m}_1, \ldots, \mathbf{m}_r)$, where $N(\mathbf{m}_i) = N(\mathbf{m}_j)$ for some $i \neq j$. Here we mean by degeneration logarithmic degeneration (power series with log terms). More generally we expect that the multiple zeta function $Z_1(s) \otimes \cdots \otimes Z_r(s)$ has an Euler product

$$Z_1(s) \otimes \cdots \otimes Z_r(s) = \prod_{(p_1, \ldots, p_r) \in P_1 \times \cdots \times P_r} H_{(p_1, \ldots, p_r)}(N(p_1)^{-s}, \ldots, N(p_r)^{-s})$$

when each zeta function $Z_j(s)$ has an Euler product

$$Z_j(s) = \prod_{p \in P_j} H_p^j(N(p)^{-s})$$

and a functional equation; here $H_p^j(T)$ is a power series in T and $H_{(p_1, \ldots, p_r)}(T_1, \ldots, T_r)$ is a power series in $(T_1, \ldots, T_r)$ with a possible degeneration at $(p_1, \ldots, p_r)$, where $N(p_i) = N(p_j)$ for some $i \neq j$.

In [**KK3**] we investigated the absolute tensor product $\zeta(s, \mathbf{F}_p) \otimes \zeta(s, \mathbf{F}_q)$ for primes p and q by using a signed double Poisson summation formula, where $\zeta(s, \mathbf{F}_p) = (1 - p^{-s})^{-1}$. In other words we constructed a zeta function having zeros (or poles) at sums of poles of $\zeta(s, \mathbf{F}_p)$ and those of $\zeta(s, \mathbf{F}_q)$. We state the results as follows. The proofs and some refinements are in [**KK3**] and [**KK4**]. For simplicity we use the notation $F(s) \cong G(s)$ for functions $F(s)$ and $G(s)$ to indicate that $F(s) = e^{Q(s)} G(s)$ for some polynomial $Q(s)$.

THEOREM 1.1. *Let p and q be distinct prime numbers. Define the function $\zeta_{p,q}(s)$ in $\mathrm{Re}(s) > 0$ as follows:*

$$\zeta_{p,q}(s) := \exp\left(-\frac{i}{2} \sum_{n=1}^{\infty} \frac{\cot\left(\pi n \frac{\log p}{\log q}\right)}{n} p^{-ns} - \frac{i}{2} \sum_{n=1}^{\infty} \frac{\cot\left(\pi n \frac{\log q}{\log p}\right)}{n} q^{-ns}\right.$$

$$\left. -\frac{1}{2} \sum_{n=1}^{\infty} \frac{1}{n} p^{-ns} - \frac{1}{2} \sum_{n=1}^{\infty} \frac{1}{n} q^{-ns}\right).$$

Then the function $\zeta_{p,q}(s)$ has the following properties:

(0) *It converges absolutely in $\mathrm{Re}(s) > 0$.*

(1) *The function $\zeta_{p,q}(s)$ has an analytic continuation to all $s \in \mathbf{C}$ as a meromorphic function of order two.*

(2) *All zeros and poles of $\zeta_{p,q}(s)$ are simple and located at*

$$s = 2\pi i \left(\frac{m}{\log p} + \frac{n}{\log q}\right),$$

where (m, n) is either a pair of nonnegative integers or a pair of negative integers. Indeed it gives a zero or pole depending on whether they are nonnegative or negative.

(3) *We have the identification*

$$\zeta_{p,q}(s) \cong \zeta(s, \mathbf{F}_p) \otimes \zeta(s, \mathbf{F}_q).$$

(4) *The function $\zeta_{p,q}(s)$ satisfies a functional equation:*

$$\zeta_{p,q}(-s) = \zeta_{p,q}(s)^{-1} (pq)^{\frac{s}{2}} (1 - p^{-s})(1 - q^{-s})$$

$$\times \exp\left(\frac{i \log p \log q}{4\pi} s^2 - \frac{\pi i}{6}\left(\frac{\log q}{\log p} + \frac{\log p}{\log q} + 3\right)\right).$$

When $p = q$ the result is as follows.

THEOREM 1.2. *Let*

$$\zeta_{p,p}(s) := \exp\left(\frac{i}{2\pi}\sum_{n=1}^{\infty}\frac{1}{n^2}p^{-ns} - \left(1 - \frac{i\log p}{2\pi}s\right)\sum_{n=1}^{\infty}\frac{1}{n}p^{-ns}\right)$$

in $\mathrm{Re}(s) > 0$. *Then the function* $\zeta_{p,p}(s)$ *has the following properties:*

(0) *It converges absolutely in* $\mathrm{Re}(s) > 0$.

(1) *The function* $\zeta_{p,p}(s)$ *has an analytic continuation to all* $s \in \mathbf{C}$ *as a meromorphic function of order two.*

(2) *All zeros and poles of* $\zeta_{p,p}(s)$ *are located at*

$$s = \frac{2\pi in}{\log p},$$

which gives a zero or pole of order $|n+1|$, *depending on whether* n *is a nonnegative or negative integer.*

(3) *We have the identification*

$$\zeta_{p,p}(s) \cong \zeta(s, \mathbf{F}_p) \otimes \zeta(s, \mathbf{F}_p).$$

(4) *The function* $\zeta_{p,p}(s)$ *satisfies a functional equation:*

$$\zeta_{p,p}(-s) = \zeta_{p,p}(s)^{-1}p^s(1 - p^{-s})^2\exp\left(\frac{i(\log p)^2}{4\pi}s^2 - \frac{5\pi i}{6}\right).$$

The proofs of Theorems 1.1 and 1.2 were done by constructing the signed double Poisson summation formula, and the theory of multiple sine functions developed in [**KK2**] is essential.

From our viewpoint, it is very interesting to see the nature of $\prod_{p,q}\zeta_{p,q}(s)$. Unfortunately, however, it does not converge even for sufficiently large $\mathrm{Re}(s)$. Our "α-version" $\zeta_{p,q}^{\alpha}(s)$ treated below remedies the situation. In passing we find the analyticity of the diagonal Euler product, as stated in the following theorem.

THEOREM 1.3. *Let*
$$Z(s) = \prod_p \zeta_{p,p}(s).$$

Then $Z(s)$ *is absolutely convergent in* $\mathrm{Re}(s) > 1$, *and it has an analytic continuation with singularities to* $\mathrm{Re}(s) > 0$ *with the natural boundary* $\mathrm{Re}(s) = 0$.

Later in [**KK4**] we constructed "the double Riemann zeta function" $\zeta(s, \mathbf{Z}) \otimes \zeta(s, \mathbf{Z})$ by establishing the signed double explicit formula which is stated in the following theorem. It generalizes the signed double Poisson summation formula used in the proofs of Theorems 1.1 and 1.2.

For simplicity put $\xi(s) = \hat{\zeta}(s + \frac{1}{2}) = \hat{\zeta}(s + \frac{1}{2}, \mathbf{Z})$ for $\hat{\zeta}(s) = \Gamma_{\mathbf{R}}(s)\zeta(s)$ with $\Gamma_{\mathbf{R}}(s) = \pi^{-s/2}\Gamma(s/2)$. The functional equation of $\zeta(s)$ is written as $\xi(s) = \xi(-s)$. We recall that nontrivial zeros of $\zeta(s)$ are zeros in the strip $|\mathrm{Re}(s - \frac{1}{2})| < 1/2$. We denote by $\frac{1}{2} + i\gamma$ such a zero, where γ is a complex number in $-1/2 < \mathrm{Im}(\gamma) < 1/2$.

Hereafter let $h(t)$ be an odd regular function in $|\mathrm{Re}(t)| < 1$ satisfying $h(t) = O(|t|^{-3})$ as $|t| \to \infty$. We put $H_{\alpha}(t) := h(2\alpha + it)$ and

$$\widetilde{H}(u) := \int_{-\infty}^{\infty} H(t)e^{itu}\,dt.$$

THEOREM 1.4. *Let $1/2 < \alpha < 1$. We have*

$$\sum_{\mathrm{Re}(\gamma_1),\mathrm{Re}(\gamma_2)>0} H_0(\gamma_1 + \gamma_2) = \sum_{\substack{p,q \\ p \neq q}} \mathcal{H}^\alpha_{p,q} + \sum_p \mathcal{H}^\alpha_{p,p} + \sum_p \mathcal{H}^\alpha_{p,\infty} + \mathcal{H}^\alpha_{\infty,\infty} + \mathcal{H}^\alpha_0,$$

where the sum in the left hand side is taken over pairs $(\frac{1}{2}+i\gamma_1, \frac{1}{2}+i\gamma_2)$ of nontrivial zeros of the Riemann zeta function, the sum in the right hand side is taken over pairs of distinct primes p, q or primes p, and we define for pairs of distinct primes p, q

$$\mathcal{H}^\alpha_{p,q} = \frac{i}{4\pi^2} \sum_{m,n} \frac{\log p \log q}{\log(p^m q^n)} \frac{1}{p^{m(\alpha+\frac{1}{2})} q^{n(\alpha+\frac{1}{2})}} \left(\widetilde{H}_0(-m\log p) + \widetilde{H}_0(-n\log q) \right)$$

$$+ \frac{i}{4\pi^2} \sum_{m,n} \frac{\log p \log q}{\log(\frac{p^m}{q^n})} \frac{1}{p^{m(\alpha+\frac{1}{2})} q^{n(\alpha+\frac{1}{2})}} \left(\widetilde{H}_\alpha(-m\log p) - \widetilde{H}_\alpha(-n\log q) \right),$$

and for a prime p,

$$\mathcal{H}^\alpha_{p,p} = \frac{i}{4\pi^2} \sum_{m,n} \frac{\log p}{(m+n)p^{(m+n)(\alpha+\frac{1}{2})}} \left(\widetilde{H}_0(-m\log p) + \widetilde{H}_0(-n\log p) \right)$$

$$+ \frac{i}{4\pi^2} \sum_{m \neq n} \frac{\log p}{(m-n)p^{(m+n)(\alpha+\frac{1}{2})}} \left(\widetilde{H}_\alpha(-m\log p) - \widetilde{H}_\alpha(-n\log p) \right)$$

$$+ \frac{1}{4\pi^2} (\log p)^2 \sum_{m=1}^\infty p^{-2m(\alpha+\frac{1}{2})} \widetilde{tH_\alpha(t)}(-m\log p),$$

$$\mathcal{H}^\alpha_{p,\infty} = -\frac{1}{2\pi^2} \sum_{m=1}^\infty \frac{\log p}{p^{m(\alpha+\frac{1}{2})}} \int_{-\infty}^\infty p^{-imt} H_\alpha(t) \int_0^t p^{imt'} \frac{\Gamma'_{\mathbf{R}}}{\Gamma_{\mathbf{R}}}\left(\alpha + \frac{1}{2} + it'\right) dt' dt$$

$$- \frac{1}{2\pi^2} \sum_{m=1}^\infty \frac{\log p}{p^{m(\alpha+\frac{1}{2})}} \int_{-\infty}^\infty H_0(t) \int_0^t \mathrm{Re}\left(p^{im(t-t')} \frac{\Gamma'_{\mathbf{R}}}{\Gamma_{\mathbf{R}}}\left(\alpha + \frac{1}{2} + it'\right)\right) dt' dt,$$

$$\mathcal{H}^\alpha_{\infty,\infty} = \frac{1}{4\pi^2} \int_{-\infty}^\infty H_\alpha(t) \int_0^t \frac{\Gamma'_{\mathbf{R}}}{\Gamma_{\mathbf{R}}}\left(\alpha + \frac{1}{2} + it_1\right) \frac{\Gamma'_{\mathbf{R}}}{\Gamma_{\mathbf{R}}}\left(\alpha + \frac{1}{2} + i(t-t_1)\right) dt_1 dt$$

$$+ \frac{1}{4\pi^2} \int_{-\infty}^\infty H_0(t) \int_0^t \frac{\Gamma'_{\mathbf{R}}}{\Gamma_{\mathbf{R}}}\left(\alpha + \frac{1}{2} + it_1\right) \frac{\Gamma'_{\mathbf{R}}}{\Gamma_{\mathbf{R}}}\left(\alpha + \frac{1}{2} - i(t-t_1)\right) dt_1 dt,$$

and

$$\mathcal{H}^\alpha_0 = -\frac{\alpha}{\pi} \int_0^\pi \sum_{\mathrm{Re}(\gamma_1)>0} h(i\gamma_1 + \alpha e^{i\theta}) \frac{\xi'}{\xi}(\alpha e^{i\theta}) e^{i\theta} d\theta$$

$$- \frac{\alpha^2}{4\pi^2} \int_0^\pi \int_0^\pi h(\alpha e^{i\theta_1} + \alpha e^{i\theta_2}) \frac{\xi'}{\xi}(\alpha e^{i\theta_1}) \frac{\xi'}{\xi}(\alpha e^{i\theta_2}) e^{i(\theta_1+\theta_2)} d\theta_1 d\theta_2,$$

where $m, n \in \mathbf{Z}$, $m, n \geq 1$.

Notice that only pairs of zeros in the upper (or lower) half plane are counted in the left hand side of Theorem 1.4. The method of Cramér [C] is important in the proof.

For defining the (p, q)-Euler factors, we put

(1.1) $$h(t) = \frac{1}{(t+s)^2} - \frac{1}{(t-s)^2}$$

in Theorem 1.4. We denote by p, q any (finite or infinite) places. We call $\zeta_{p,q}^\alpha(s)$ a (p, q)-Euler factor of the double Riemann zeta function $\zeta(s, \mathbf{Z}) \otimes \zeta(s, \mathbf{Z})$ if and only if it holds that

$$\zeta_{p,q}^\alpha(s+1) = \exp\left(\iint \mathcal{H}_{p,q}^\alpha(s) ds ds\right).$$

We also denote the remainder factor $\zeta_0^\alpha(s)$ by

$$\zeta_0^\alpha(s+1) = \exp\left(\iint \mathcal{H}_0^\alpha(s) ds ds\right).$$

Note that these definitions imply some ambiguity emerging from the integral constants, that is, the factor $\exp(Q(s))$ with $Q(s)$ a polynomial with $\deg Q \leq 2$. The double Riemann zeta function $\zeta(s, \mathbf{Z}) \otimes \zeta(s, \mathbf{Z})$ is expressed by an Euler product over the pairs of places (p, q). In the following theorem we denote the dilogarithm of order r by $\mathrm{Li}_r(u) = \sum_{n=1}^\infty \frac{u^n}{n^r}$ $(r = 1, 2, 3, \dots)$.

THEOREM 1.5. *The (p, q)-Euler factors of the double Riemann zeta function $\zeta(s, \mathbf{Z}) \otimes \zeta(s, \mathbf{Z})$ are described as follows.*

(1) *For distinct prime numbers p and q, we put*

$$\zeta_{p,q}^\alpha(s) = \exp\left(\frac{1}{\pi i} \sum_{m,n} \frac{(\log p)(\log q)}{(m \log p)^2 - (n \log q)^2}\right.$$
$$\left(\frac{\cosh(m\alpha \log p)}{q^{n(\alpha+\frac{1}{2})}} p^{-m(s-\frac{1}{2})} + \frac{n \log q}{m \log p} \frac{\sinh(m\alpha \log p)}{q^{n(\alpha+\frac{1}{2})}} p^{-m(s-\frac{1}{2})}\right.$$
$$\left.\left.- \frac{m \log p}{n \log q} \frac{\sinh(n\alpha \log q)}{p^{m(\alpha+\frac{1}{2})}} q^{-n(s-\frac{1}{2})} - \frac{\cosh(n\alpha \log q)}{p^{m(\alpha+\frac{1}{2})}} q^{-n(s-\frac{1}{2})}\right)\right)$$

in $\mathrm{Re}(s) > \alpha + \frac{1}{2}$, where the sum is taken over all pairs of all positive integers m and n. Then it is a (p, q)-Euler factor of the double Riemann zeta function, and it has an analytic continuation to the entire plane.

(2) *For a prime number p, a (p, p)-Euler factor is given as follows in $\mathrm{Re}(s) > \alpha + \frac{1}{2}$:*

$$\zeta_{p,p}^\alpha(s)$$
$$= \exp\left(\frac{2}{\pi i} \sum_{m \neq n} \frac{p^{-m(s-\frac{1}{2})-n(\alpha+\frac{1}{2})}}{m^2 - n^2} \left(\cosh(m\alpha \log p) + \frac{n}{m} \sinh(m\alpha \log p)\right)\right.$$
$$\left. + \frac{1}{2\pi i}\left((\log p)(s-1-2\alpha)\log(1-p^{-s}) - \mathrm{Li}_2\left(p^{-s}\right) + \mathrm{Li}_2\left(p^{-s-2\alpha}\right)\right)\right).$$

It has an analytic continuation to the entire plane.

(3) *The (p, ∞)-factor $\zeta_{p,\infty}^\alpha(s)$ of the double Riemann zeta function has an analytic continuation to the entire plane, and moreover $\prod_p \zeta_{p,\infty}^\alpha(s)$ has an analytic continuation to the entire plane with possible singularities at*

$s = \frac{1}{2} - 2k \pm \alpha$ $(k \geq 0)$, $1 - 2k$ $(k \geq 0)$, $-2k$ $(k \geq 1)$, $\rho - 2k$ $(k \geq 0)$, $\frac{1}{2} + \rho \pm \alpha$, $\frac{3}{2} \pm \alpha$ with ρ any nontrivial zero of $\zeta(s)$.

(4) The (∞, ∞)-factor $\zeta_{\infty,\infty}^{\alpha}(s)$ of the double Riemann zeta function is analytic with possible singularities at $s = -2n$, $-2n + \alpha + \frac{1}{2}$ with $n = 0, 1, 2, \ldots$.

THEOREM 1.6. *The remaining factor $\zeta_0^{\alpha}(s)$ of the double Riemann zeta function is an analytic function on $\mathbf{C}$ with possible singularities at*

$$s = \rho + \frac{1}{2} + \operatorname{sgn}(\operatorname{Im}(\rho))\alpha e^{i\theta}$$

with $0 \leq \theta \leq \pi$ for any nontrivial zero ρ of $\zeta(s)$ and at s belonging to $|s - 1| \leq 2\alpha$.

In the next theorem we use the half Riemann zeta function $\zeta_+(s)$ studied in [**HKW**] and the multiple gamma function $\Gamma_r(s)$ of Barnes [**Bar**].

THEOREM 1.7. *The Euler product for the double Riemann zeta function*

$$\zeta(s, \mathbf{Z}) \otimes \zeta(s, \mathbf{Z}) \cong \left(\prod_{p,q} \zeta_{p,q}^{\alpha}(s) \right) \zeta_0^{\alpha}(s) \left(\frac{\prod_{m=1}^{\infty} \zeta_+(s + 2m)}{\zeta_+(s - 1)} \right)^2 \Gamma_2\left(\frac{s}{2}\right)^{-1} \Gamma_1(s)^2 s(s-2)$$

is absolutely convergent in $\operatorname{Re}(s) > \alpha + \frac{3}{2}$, where (p, q) runs through pairs of all (finite or infinite) places. It has an analytic continuation (with singularities) to the entire plane and satisfies a functional equation between s and $2 - s$.

Later Akatsuka [**Ak**] successfully eliminated the parameter α from the double Riemann zeta function. He obtained the (p, q)-Euler factor

$$\zeta_{p,q}(s) = \exp\left(\frac{1}{\pi i} \sum_{p} \sum_{m=1}^{\infty} \sum_{q} \sum_{\substack{n=1 \\ q^n \neq p^m}}^{\infty} \frac{p^{-m(s-1)} q^{-n} \log p}{n(m \log p - n \log q)} \right.$$

$$\left. - \frac{1}{\pi i} \sum_{p} \sum_{m=1}^{\infty} \sum_{q} \sum_{n=1}^{\infty} \frac{p^{-ms} q^{-n} \log p}{n(m \log p + n \log q)} \right),$$

and he proved that the double Euler product

$$\zeta^{\otimes 2}(s) = \prod_{p,q} \zeta_{p,q}(s)$$

satisfies a similar property as in Theorem 1.5.

For $r \geq 0$ and $x, q_j \in \mathbf{C}$ $(j = 1, \ldots, r)$ we define the *elliptic gamma function* of order r by

$$(1.2) \qquad G_r(x; q_1, \ldots, q_r) := \prod_{n_1, \ldots, n_r \geq 0} (1 - x q_1^{n_1} \cdots q_r^{n_r}).$$

We conventionally put $G_0(x) = 1 - x$. The product in (1.2) is absolutely convergent when $|q_j| < 1$ for $j = 1, \ldots, r$.

The function (1.2) was originally dealt with by Appell [**A**]. He actually considered

$$O_q(x; \omega_1, \dots, \omega_r) = \prod_{n_1, \dots, n_r \geq 0} (1 - q^{n_1 \omega_1 + \dots + n_r \omega_r + x})$$

$$= G_r(q^x; q^{\omega_1}, \dots, q^{\omega_r})$$

with $\mathrm{Re}(\omega_j) > 0$ $(j = 1, \dots, r)$ and $0 < q < 1$.

The elliptic gamma function (1.2) has another expression (Kurokawa-Wakayama [**KW**], Proposition 2.1):

$$(1.3) \qquad G_r(x; q_1, \dots, q_r) = \exp\left(-\sum_{m=1}^{\infty} \frac{x^m}{m(1 - q_1^m) \cdots (1 - q_r^m)} \right).$$

The sum in (1.3) is absolutely convergent for suitable $q_j \in \mathbf{C}$ outside $|q_j| < 1$. For example, it is valid for $q_j = e^{2\pi i \alpha_j}$ with $\alpha_j \in \mathbf{R} \setminus \mathbf{Z}$ such that either $\alpha_j \in (\overline{\mathbf{Q}} \setminus \mathbf{Q}) \cap \mathbf{R}$ (Roth) or $\alpha_j = \log p / \log q$ with p and q distinct primes (Baker).

In [**KW**], Theorem 1.1, Kurokawa and Wakayama proved the following theorem.

THEOREM 1.8. *Let $p_1, \dots, p_r$ be distinct prime numbers. Then for $\mathrm{Re}(s) > 0$,*

$$\zeta(s, \mathbf{F}_{p_1}) \otimes \cdots \otimes \zeta(s, \mathbf{F}_{p_r})$$

$$= \left(G_{r-1}\left(p_1^{-s}; \exp\left(\frac{2\pi i \log p_1}{\log p_2} \right), \dots, \exp\left(\frac{2\pi i \log p_1}{\log p_r} \right) \right) \times \cdots \right.$$

$$\left. \times G_{r-1}\left(p_r^{-s}; \exp\left(\frac{2\pi i \log p_r}{\log p_1} \right), \dots, \exp\left(\frac{2\pi i \log p_r}{\log p_{r-1}} \right) \right) \right)^{(-1)^r}.$$

Here, G_{r-1} is considered using (1.3) and the convergence comes from a delicate transcendency result of Baker [**B**] Theorem 3.1.

In particular, when $r = 2$, we have for distinct prime numbers p and q

$$\zeta(s, \mathbf{F}_p) \otimes \zeta(s, \mathbf{F}_q) = G_1\left(p^{-s}; \exp\left(\frac{2\pi i \log p}{\log q} \right) \right) G_1\left(q^{-s}; \exp\left(\frac{2\pi i \log q}{\log p} \right) \right).$$

As $\zeta(s, \mathbf{F}_{p_1}) \otimes \cdots \otimes \zeta(s, \mathbf{F}_{p_r})$ is a meromorphic function of order r, we reach the following expectation.

Expectation. Zeta functions which are meromorphic of order r have Euler products with Euler factors being expressed in terms of G_{r-1}.

EXAMPLE 1.9. [Dedekind zeta functions $(r = 1)$] Let K be a number field. The Dedekind zeta function of K is

$$\zeta_K(s) = \prod_p (1 - N(p)^{-s})^{-1} = \prod_p G_0(N(p))^{-1},$$

where p runs through all maximal ideals of O_K and $N(p) = |O_K/(p)|$.

EXAMPLE 1.10. [Selberg zeta functions for $SL(2, \mathbf{R})$ $(r = 2)$] Let $\Gamma \subset SL(2, \mathbf{R})$ be a discrete subgroup. The Selberg zeta function of Γ is

$$Z_\Gamma(s) = \prod_{n=0}^{\infty} \prod_p (1 - N(p)^{-s-n}) = \prod_p G_1(N(p)^{-s}; N(p)^{-1}),$$

where p runs through all primitive hyperbolic conjugacy classes of Γ and $N(p)$ is the larger square of the eigenvalues of p.

EXAMPLE 1.11. [Selberg zeta functions for $SL(2, \mathbf{C})$ $(r = 3)$] Let $\Gamma \subset SL(2, \mathbf{C})$ be a discrete subgroup. The Selberg zeta function of Γ is

$$Z_\Gamma(s) = \prod_p \prod_{\substack{k \geq 0, \ l \geq 0 \\ k \equiv l \ (\mathrm{mod} \ \nu_p)}} (1 - a(p)^{-2k} \overline{a(p)}^{-2l} N(p)^{-s}),$$

where p runs through certain hyperbolic conjugacy classes of Γ, the eigenvalues of p are denoted by $a(p)$ and $\overline{a(p)}$ with $|a(p)| > 1$, and $N(p) = |a(p)|^2$. The symbol ν_p is defined as $\nu_p = |(\Gamma_p)_{\mathrm{tor}}|$, the order of the torsion of the centralizer of p in Γ. For hyperbolic classes p such that $\nu_p = 1$, the Euler factor has the form

$$\prod_{l=0}^{\infty} \prod_{k=0}^{\infty} (1 - a(p)^{-2k} \overline{a(p)}^{-2l} N(p)^{-s}) = G_2(N(p)^{-s}; a(p)^{-2}, \overline{a(p)}^{-2}).$$

2. Absolute Weil Zeta Functions

For a group G we denote $\mathbf{F}_1[G] = G \cup \{0\}$, where 0 is an element satisfying $0 \cdot g = g \cdot 0 = 0$ for any $x \in G$. For a positive integer $m \geq 1$, we denote

$$\begin{aligned} \mathbf{F}_{1^m} &= \mathbf{F}_1[\boldsymbol{\mu}_m] \\ &= \{0\} \cup \boldsymbol{\mu}_m \\ &= \{0\} \cup \{z \in \mathbf{C} \mid z^m = 1\}. \end{aligned}$$

We define an $\mathbf{F}_1$-algebra as a multiplicative monoid with 0.

Let $1 \leq r \in \mathbf{Z}$. For a $\mathbf{Z}$-algebra A, we recall that the noncommutative Hasse zeta function of A is

$$(2.1) \qquad \zeta_{\mathbf{Z}}^r(s, A) = \exp\left(\sum_{p: \text{ prime}} \sum_{m=1}^{\infty} \frac{|\mathrm{Hom}(A, M_r(\mathbf{F}_{p^m}))|}{m} p^{-ms} \right),$$

where $M_r(\mathbf{F}_{p^m})$ is the ring of matrices over $\mathbf{F}_{p^m}$ of size r, and Hom is the set of ring homomorphisms.

When $r = 1$ and A is a commutative finitely generated $\mathbf{Z}$-algebra, the definition (2.1) coincides with the usual Hasse zeta function. Namely, we have an expression

$$\zeta_{\mathbf{Z}}^1(s, A) = \prod_{\substack{M \subset A \\ \text{maximal ideal}}} (1 - N(M)^{-s})^{-1}$$

with $N(M) = |A/M|$. When $r \geq 2$ and A is a commutative finitely generated $\mathbf{Z}$-algebra, Kurokawa [K] and Fukaya [F] study some analogous versions of (2.1).

When A is an $\mathbf{F}_1$-algebra, we analogously define

(2.2) $$\zeta_{\mathbf{F}_1}^r(s, A) = \exp\left(\sum_{m=1}^{\infty} \frac{|\mathrm{Hom}(A, M_r(\mathbf{F}_{1^m}))|}{m} e^{-ms}\right),$$

where Hom is the set of monoid homomorphisms.

When $r = 1$ and A is a finitely generated abelian monoid, which is an $\mathbf{F}_1$-algebra, (2.2) was studied in [**DKK**] and [**KKK**]. The explicit form and the functional equation of $\zeta_{\mathbf{F}_1}(s, A)$ are given by the following proposition and theorem.

PROPOSITION 2.1. ([**DKK**], Proposition 2.1) *Assume* $A = \mathbf{F}_1[G] = G \cup \{0\}$ *with G a finitely generated multiplicative abelian group of rank r. Put*

$$G \cong \mathbf{Z}^r \times \boldsymbol{\mu}_{n_1} \times \boldsymbol{\mu}_{n_2} \times \cdots \times \boldsymbol{\mu}_{n_k}$$

with $n_1 | n_2 | \cdots | n_k$. Then

$$\zeta_{\mathbf{F}_1}(s, A) = \begin{cases} \displaystyle\prod_{\underline{d}|\underline{n}} \left(1 - e^{-|\underline{d}|s}\right)^{-\varphi(\underline{d})d_1^{k-2}\cdots d_k^{-1}} & \text{for} \quad r = 0 \\ \displaystyle\prod_{\underline{d}|\underline{n}} \exp\left(\frac{g_r(e^{-|\underline{d}|s})\varphi(\underline{d})d_1^{r+k-1}\cdots d_k^{r-1}}{(1-e^{-|\underline{d}|s})^r}\right) & \text{for} \quad r \geq 1, \end{cases}$$

where the notation $\underline{d}|\underline{n}$ means that the product is over all tuples

$$\underline{d} = (d_1, \ldots, d_k) \in \mathbf{N}^k$$

such that $d_1 | n_1, d_2 | \frac{n_2}{d_1}, \ldots, d_k | \frac{n_k}{d_1 \cdots d_{k-1}}$. Further, we put

$$|\underline{d}| = d_1 \cdots d_k$$

and

$$\varphi(\underline{d}) = \varphi(d_1) \cdots \varphi(d_k).$$

THEOREM 2.2. ([**KKK**], Theorem 2)

(1) *Assume* $A = \mathbf{F}_1[G] = G \cup \{0\}$ *with G a finite abelian group of order n. Put*

$$G \cong \boldsymbol{\mu}_{n_1} \times \boldsymbol{\mu}_{n_2} \times \cdots \times \boldsymbol{\mu}_{n_k}$$

with $n_1 | n_2 | \cdots | n_k$ and $n = n_1 \cdots n_k$. Then

(2.3) $$\zeta_{\mathbf{F}_1}(s, A) = \det\left(1 - \Phi_A e^{-s}\right)^{-1/n}$$

and the following functional equation holds:

$$\zeta_{\mathbf{F}_1}(-s, A) = w(A)e^{-ns}\zeta_{\mathbf{F}_1}(s, A),$$

where $w(A)$ is a complex number of modulus 1 satisfying

$$w(A) = (-1)^n \det(\Phi_A)^{\frac{1}{n}},$$

and where we define the absolute Frobenius operator Φ_A on the group

$$G^{(2)} = \boldsymbol{\mu}_{n_1^2} \times \boldsymbol{\mu}_{n_2^2} \times \cdots \times \boldsymbol{\mu}_{n_k^2}$$

as

(2.4) $$\Phi_A(\alpha_1, \ldots, \alpha_k) = (\alpha_1^{n_1+1}, \ldots, \alpha_k^{n_k+1}).$$

(2) *Assume $A = \mathbf{F}_1[G] = G \cup \{0\}$ with G a finitely generated free abelian group of rank 1. Put*

$$G \cong \mathbf{Z} \times \boldsymbol{\mu}_{n_1} \times \boldsymbol{\mu}_{n_2} \times \cdots \times \boldsymbol{\mu}_{n_k}$$

with $n_1|n_2|\cdots|n_k$. Then the following functional equation holds:

$$\zeta_{\mathbf{F}_1}(-s, A) = \zeta_{\mathbf{F}_1}(s, A)^{-1} \prod_{d|n} e^{-\varphi(\underline{d})d_1^{k-1}\cdots d_k}.$$

(3) *If $A = \mathbf{F}_1[G] = G \cup \{0\}$ with G is a finitely generated abelian group of rank $r \geq 2$, the following functional equation holds:*

$$\zeta_{\mathbf{F}_1}(-s, A) = \zeta_{\mathbf{F}_1}(s, A)^{(-1)^r}.$$

Above all, the determinant expression (2.3) is crucial. By this we proved an analog of the Riemann hypothesis as well as a tensor structure of the zeta functions ([**KKK**]). We generalize (2.3) to $r \geq 2$ in the following theorem.

THEOREM 2.3. *Assume all prime divisors of n are bigger than r. In other words, assume that*

$$p \,|\, n \quad \Longrightarrow \quad p > r.$$

Let $A = \mathbf{F}_1[G] = G \cup \{0\}$ with G a finite abelian group of order n. Then $\zeta_{\mathbf{F}_1}^r(s, A)$ has an Euler product

$$(2.5) \qquad \zeta_{\mathbf{F}_1}^r(s, A) = \prod_{d|n}(1 - e^{-ds})^{-\frac{1}{d}\left(d^r - \sum_{d'|d}\varphi(d')^r\right)}$$

and a determinant expression

$$(2.6) \qquad \zeta_{\mathbf{F}_1}^r(s, A) = \det\left(1 - \Phi_A^{\otimes r}e^{-s}\right)^{-1/n^r},$$

where Φ_A is defined by (2.4) which is a square matrix of size n^2, and $\Phi_A^{\otimes r}$ denotes the Kronecker tensor power which is a square matrix of size n^{2r}.

PROOF. Put

$$G \cong \boldsymbol{\mu}_{n_1} \times \boldsymbol{\mu}_{n_2} \times \cdots \times \boldsymbol{\mu}_{n_k}$$

with $n_1|n_2|\cdots|n_k$ and $n = n_1 \cdots n_k$. We first claim

$$(2.7) \qquad \mathrm{Hom}(A, M_r(\mathbf{F}_{1^m})) = (n_1, m)^r \cdots (n_k, m)^r.$$

Then (2.5) follows from a direct calculation. For proving (2.7), it suffices to prove that

$$\mathrm{Hom}(\boldsymbol{\mu}_l, M_r(\mathbf{F}_{1^m})) = (l, m)^r,$$

by considering the image of a generator ζ_l of each $\boldsymbol{\mu}_l$ $(l = n_1, \ldots, n_k)$.

Since the image of ζ_l has to be invertible, it holds that

$$|\mathrm{Hom}(\boldsymbol{\mu}_l, M_r(\mathbf{F}_{1^m}))| = |\mathrm{Hom}(\boldsymbol{\mu}_l, GL_r(\mathbf{F}_{1^m}))|.$$

Here $GL_r(\mathbf{F}_{1^m})$ is the group of invertible matrices in the form of a permutation matrix with the entry 1 replaced by any element in $\boldsymbol{\mu}_m$.

Denote by α the image of ζ_l by a homomorphism. By the assumption, α is a diagonal matrix with diagonal entries in $\boldsymbol{\mu}_m$, because the order of α must be equal to l, not divisible by any positive integer less than l. Any element

$$\alpha \in \begin{pmatrix} \boldsymbol{\mu}_m & \cdots & 0 \\ \vdots & \ddots & \vdots \\ 0 & \cdots & \boldsymbol{\mu}_m \end{pmatrix}$$

with $\alpha^l = 1$ can be an image. Thus

$$|\mathrm{Hom}(\boldsymbol{\mu}_l, M_2(\mathbf{F}_{1^m}))| = \left| \left\{ \alpha \in \begin{pmatrix} \boldsymbol{\mu}_m & \cdots & 0 \\ \vdots & \ddots & \vdots \\ 0 & \cdots & \boldsymbol{\mu}_m \end{pmatrix} : \alpha^l = 1 \right\} \right| = (l, m)^r.$$

This proves (2.7). Then we have

$$\zeta_{\mathbf{F}_1}^r(s, A) = \exp\left(\sum_{m=1}^{\infty} \frac{|\mathrm{Hom}(A, M_r(\mathbf{F}_{1^m}))|}{m} e^{-ms} \right)$$

$$= \exp\left(\sum_{m=1}^{\infty} \frac{(n_1, m)^r \cdots (n_k, m)^r}{m} e^{-ms} \right).$$

For proving (2.6), it suffices to show that

$$(2.8) \qquad \sum_{m=1}^{\infty} \frac{n^r (n_1, m)^r \cdots (n_k, m)^r}{m} e^{-ms} = \sum_{m=1}^{\infty} \frac{\mathrm{tr}((\Phi_A^{\otimes r})^m)}{m} e^{-ms}$$

by taking the logarithm of (2.6). From a general identity $\mathrm{tr}((A \otimes B)^m) = \mathrm{tr}(A^m \otimes B^m) = \mathrm{tr}(A^m)\mathrm{tr}(B^m)$, we have

$$\mathrm{tr}((\Phi_A^{\otimes r})^m) = \mathrm{tr}((\Phi_A^m)^r).$$

On the other hand, it holds by [**KKK**], Lemma 1, that

$$\mathrm{tr}(\Phi_A^m) = n \, |\mathrm{Hom}(A, \boldsymbol{\mu}_m)| = n(n_1, m) \cdots (n_k, m).$$

These identities lead to (2.8). $\qquad \square$

EXAMPLE 2.4. For any prime $p > r$, it holds that

$$\zeta_{\mathbf{F}_1}^r(s, \mathbf{F}_{1^p}) = (1 - e^{-s})^{-1} (1 - e^{-ps})^{-\frac{p^r - 1}{p}}.$$

By this theorem, it is easily shown that $\zeta_{\mathbf{F}_1}^r(s, A)$ satisfies an analog of the Riemann hypothesis and the tensor structure concerning zeros and poles. The proofs are the same as in [**KKK**], Theorem 3. The functional equation is also obtained as follows.

COROLLARY 2.5. *The zeta function in the preceding theorem satisfies the functional equation*

$$\zeta_{\mathbf{F}_1}^r(-s, A) = \zeta_{\mathbf{F}_1}^r(s, A)(-1)^{n^r} \det(\Phi_A)^{rn^{r-2}} e^{-n^r s}.$$

PROOF. In [**KKK**], Proposition 1, we obtained the functional equation of the zeta function $\zeta_\sigma(s) = \det(1 - \sigma e^{-s})^{-1}$ of a finite permutation matrix σ over N elements as

$$\zeta_\sigma(-s) = (-1)^N \det(\sigma) e^{-Ns} \zeta_\sigma(s).$$

Identifying $\Phi_A^{\otimes r}$ with a permutation matrix over n^{2r} elements, we compute by the theorem that

$$\zeta_{\mathbf{F}_1}^r(-s, A) = \zeta_{\Phi_A^{\otimes r}}(-s)^{\frac{1}{n^r}}$$

$$= \left(\zeta_{\Phi_A^{\otimes r}}(s)(-1)^{n^{2r}} \det(\Phi_A^{\otimes r}) e^{-n^{2r}s} \right)^{\frac{1}{n^r}}$$

$$= \left(\zeta_{\Phi_A^{\otimes r}}(s)(-1)^{n^{2r}} \det(\Phi_A)^{rn^{2(r-1)}} e^{-n^{2r}s} \right)^{\frac{1}{n^r}}$$

$$= \zeta_{\mathbf{F}_1}^r(s, A)^{\frac{1}{n^r}} (-1)^{n^r} \det(\Phi_A)^{rn^{r-2}} e^{-n^r s}.$$

$\square$

Investigating the cases when n has small prime divisors is an interesting problem. It seems that this situation is much more complicated. We give some examples below.

EXAMPLE 2.6. When $A = \mathbf{F}_1[\boldsymbol{\mu}_2 \times \boldsymbol{\mu}_2]$, we have

$$\zeta_{\mathbf{F}_1}^2(s, A) = (1 - e^{-s})^{-25} \exp\left(\frac{10 - 36e^{-s} + 9e^{-2s} + 18e^{-3s}}{(1 - e^{-s})^2} \right).$$

PROOF. The result follows from

$$|\mathrm{Hom}(A, M_2(\mathbf{F}_{1^m}))| = \begin{cases} (m+5)^2, & m \geq 3 \\ (m+2)^2, & m = 1, 2. \end{cases}$$

$\square$

EXAMPLE 2.7. When $A = \mathbf{F}_{1^3}$, we have

$$\zeta_{\mathbf{F}_1}^3(s, A) = (1 - e^{-s})^{-1} (1 - e^{-3s})^{-\frac{20}{3}} \exp\left(3 \frac{e^{-3s}(1 + 2e^{-3s})}{(1 - e^{-3s})^2} \right).$$

PROOF. The result follows from

$$|\mathrm{Hom}(A, M_3(\mathbf{F}_{1^m}))| = \begin{cases} 1, & 3 \nmid m \\ 27 + 2^2, & 3 \mid m. \end{cases}$$

$\square$

EXAMPLE 2.8. When $A = \mathbf{F}_{1^2}$, we have

$$\zeta_{\mathbf{F}_1}^2(s, A) = (1 - e^{-s})^{-5} \exp\left(\frac{-4 + 3e^{-s} + 3e^{-2s}}{2(e^{-s} - 1)} \right).$$

PROOF. The result follows from

$$|\text{Hom}(A, M_2(\mathbf{F}_{1^m}))| = \begin{cases} m+5, & m \geq 3 \\ m+2, & m = 1, 2. \end{cases}$$

$\square$

EXAMPLE 2.9. For $A = \mathbf{F}_{1^2}$, we have

$$\zeta_{\mathbf{F}_1}^r(s, A) = \exp\left(\sum_{l=1}^{\infty} \frac{r!}{(2l-1)!} \sum_{k=0}^{[r/2]} \frac{e^{-2ls}}{2^k(r-2k)!} \left(\frac{(2l)^k}{2l-1}e^s + \frac{(2l+1)^k}{(2l)^2}2^{2(l-k)}\right)\right).$$

PROOF. The result follows from

$$|\text{Hom}(A, M_r(\mathbf{F}_{1^m}))| = \begin{cases} \displaystyle\sum_{k=0}^{[r/2]} \frac{r!(m+1)^k}{2^k(r-2k)!}, & 2 \nmid m \\ \displaystyle\sum_{k=0}^{[r/2]} \frac{r!(m+1)^k}{2^k(r-2k)!}2^{m-2k}, & 2 \mid m. \end{cases}$$

$\square$

In what follows we give some examples for the case when A is infinite. We define the Euler polynomials $g_r(T) \in \mathbb{Z}[T]$ $(r = 1, 2, 3, \dots)$ by

$$(2.9) \qquad g_1(T) = T \quad \text{and} \quad g_{r+1}(T) = \sum_{k=1}^{r} \binom{r}{k-1}(T-1)^{r-k}g_k(T).$$

For example,

$$g_1(T) = T$$
$$g_2(T) = T$$
$$g_3(T) = T^2 + T$$
$$g_4(T) = T^3 + 4T^2 + T.$$

In [**DKK**], Lemma 2.2, we proved the following lemma.

LEMMA 2.10. For $r = 1, 2, 3, \dots$, we have

$$\sum_{\nu=1}^{\infty} \nu^{r-1}T^{\nu} = \frac{g_r(T)}{(1-T)^r},$$

where $g_r(T) \in \mathbb{Z}[T]$ is defined by (2.9).

By using this lemma, we obtain the following.

EXAMPLE 2.11. When $A = \mathbf{F}_1[T^{\pm}]$, the explicit form of $\zeta_{\mathbf{F}_1}^r(s, A)$ is given by

$$\zeta_{\mathbf{F}_1}^r(s, A) = \exp\left(\frac{g_r(e^{-s})r!}{(1-e^{-s})^r}\right).$$

It satisfies the functional equation

$$\zeta_{\mathbf{F}_1}^r(-s, A) = \zeta_{\mathbf{F}_1}^r(s, A)^{(-1)^r}.$$

PROOF. An element in $\text{Hom}(A, M_r(\mathbf{F}_{1^m}))$ is determined by the image of T, which must be invertible. Thus

$$|\text{Hom}(A, M_r(\mathbf{F}_{1^m}))| = |GL_r(\mathbf{F}_{1^m})| = m^r r!.$$

Then we compute

$$\zeta^r_{\mathbf{F}_1}(s, A) = \exp\left(\sum_{m=1}^{\infty} \frac{m^r r!}{m} e^{-ms}\right)$$

$$= \exp\left(\sum_{m=1}^{\infty} m^{r-1} e^{-ms}\right)^{r!}$$

$$= \exp\left(\frac{g_r(e^{-s})r!}{(1 - e^{-s})^r}\right).$$

□

EXAMPLE 2.12. When $A = \mathbf{F}_1[T]$, the explicit form of $\zeta^r_{\mathbf{F}_1}(s, A)$ is given by

$$\zeta^r_{\mathbf{F}_1}(s, A) = (1 - e^{-s})^{-1} \exp\left(\sum_{k=1}^{r} \binom{r}{k} r^k \frac{g_k(e^{-s})}{(1 - e^{-s})^k}\right).$$

PROOF. An element in $\text{Hom}(A, M_r(\mathbf{F}_{1^m}))$ is determined by the image of T, which must belong to $M_r(\mathbf{F}_{1^m}) = \text{End}(\mathbf{F}^r_{1^m})$. Thus

$$|\text{Hom}(A, M_r(\mathbf{F}_{1^m}))| = |\text{End}(\mathbf{F}^r_{1^m})| = (rm + 1)^r.$$

Then

$$\zeta^r_{\mathbf{F}_1}(s, A) = \exp\left(\sum_{m=1}^{\infty} \frac{(rm + 1)^r}{m} e^{-ms}\right)$$

$$= \exp\left(\sum_{m=1}^{\infty} \frac{1}{m} \sum_{k=0}^{r} \binom{r}{k} (rm)^k e^{-ms}\right)$$

$$= \exp\left(\sum_{m=1}^{\infty} \left(\frac{1}{m} + \sum_{k=1}^{r} \binom{r}{k} r^k m^{k-1}\right) e^{-ms}\right)$$

$$= (1 - e^{-s})^{-1} \exp\left(\sum_{k=1}^{r} \binom{r}{k} r^k \sum_{m=1}^{\infty} m^{k-1} e^{-ms}\right)$$

$$= (1 - e^{-s})^{-1} \exp\left(\sum_{k=1}^{r} \binom{r}{k} r^k \frac{g_k(e^{-s})}{(1 - e^{-s})^k}\right).$$

□

References

[Ak]   H. Akatsuka, *The double explicit formula and the double Riemann zeta function*, Comm. Number Theory and Phys. **3** (2009), 1-35.

[A]   P. Appell, *Sur une classe de fonctions analogues aux fonctions Eulériennes*, Math. Ann. **19** (1882), 84-102.

[B]   A. Baker *Transcendental number theory*, Cambridge University Press, 1975.

[Bar] E.W. Barnes, *On the theory of the multiple gamma function*, Trans. Cambridge Philos. Soc. **19** (1904), 374-425.

[C]   H. Cramér, *Studien über die Nullstellen der Riemannschen Zetafunktion*, Math. Z. **4** (1919), 104-130.

[D]   P. Deligne, *La conjecture de Weil (I)*, Inst. Hautes Études Sci. Publ. Math. **43** (1974), 273-307.

[DKK] A. Deitmar, S. Koyama and N. Kurokawa, *Absolute zeta functions*, Proc. Japan Acad. **84A** (2008), 138-142.

[Den] C. Deninger, *Local L-factors of motives and regularized determinants*, Invent. Math. **107** (1992), 135-150.

[F]   T. Fukaya, *Hasse zeta functions of non-commutative rings*, J. Algebra **208** (1998), 304-342.

[H]   S.W. Hawking, *Zeta function regularization of path integrals in curved spacetime*, Comm. Math. Phys. **55** (1977), 133-148.

[HKW] M. Hirano, N. Kurokawa and M. Wakayama, *Half zeta functions*, J. Ramanujan Math. Soc. **18** (2003), 195-209.

[G]   A. Grothendieck *Séminaire de Géométrie Algébrique du Bois-Marie 1965-66 (SGA 5)*, Lecture Notes in Math. **589**, Springer-Verlag, 1977.

[KKK] S. Kim, S. Koyama and N. Kurokawa, *The Riemann hypothesis and functional equations for zeta functions over* $\mathbf{F}_1$, Proc. Japan Acad. **85A** (2009), 75-80.

[K]   N. Kurokawa, *Zeta functions of categories*, Proc. Japan Acad. **72** (1996), 221-222.

[K2]   N. Kurokawa, *Multiple zeta functions: an example*, Adv. Stud. Pure Math. **21** (1992), 219-226.

[KK1] N. Kurokawa and S. Koyama, *Normalized double sine functions*, Proc. Japan Acad. **79** (2003), 14-18.

[KK2] N. Kurokawa and S. Koyama, *Multiple sine functions*, Forum Math. **15** (2003), 839-876.

[KK3] S. Koyama and N. Kurokawa, *Multiple zeta functions: the double sine function and the signed double Poisson summation formula*, Compositio Math. **140** (2004), 1176-1190.

[KK4] S. Koyama and N. Kurokawa, *Multiple Euler products*, Proc. St. Petersburg Math. Soc. **11** (2005), 123-166 (in Russian). English translation is published in AMS Translations, Series 2 (2006), 101-140.

[KW]   N. Kurokawa and M. Wakayama, *Absolute tensor products*, International Math. Res. Notices, **5** (2004), 249-260.

[M]   Yu. I. Manin, *Lectures on zeta functions and motives (according to Deninger and Kurokawa)*, Asterisque **228** (1995), 121-163.

[RS]   D.B. Ray and I.M. Singer, *Analytic torsion for complex manifolds*, Ann. Math. **98** (1973), 154-177.

[S]   M. Schröter, *Über Kurokawa-Tensorprodukte von L-Reihen*, Schriftenreihe des Mathematischen Instituts der Universität Münster, 3. Serie **18** (1996), 240 pp.

INSTITUTE OF MATHEMATICAL SCIENCES, EWHA WOMAN'S UNIVERSITY, DAEHYUN-DONG 11-1, SEDAEMOON-KU, 120-750, SEOUL, SOUTH KOREA.
Current address: Department of Biomedical Engineering, Toyo University, 2100 Kujirai, Kawagoe-shi 350-8585, Japan.
E-mail address: koyama@toyonet.toyo.ac.jp

DEPARTMENT OF MATHEMATICS, TOKYO INSTITUTE OF TECHNOLOGY, 2-12-1, OH-OKAYAMA, MEGURO-KU, TOKYO 152-8551, JAPAN.
E-mail address: kurokawa@math.titech.ac.jp

Mapping $\mathbb{F}_1$-land:
An Overview of Geometries over
the Field with One Element

Javier López Peña and Oliver Lorscheid

ABSTRACT. This paper gives an overview of the various approaches towards $\mathbb{F}_1$-geometry. In the first part, we review all known theories in literature so far, which are: Deitmar's $\mathbb{F}_1$-schemes, Toën and Vaquié's $\mathbb{F}_1$-schemes, Haran's F-schemes, Durov's generalized schemes, Soulé's varieties over $\mathbb{F}_1$ as well as his and Connes-Consani's variations of this theory, Connes and Consani's $\mathbb{F}_1$-schemes, the author's torified varieties and Borger's Λ-schemes. In the second part, we tie up these different theories by describing functors between the different $\mathbb{F}_1$-geometries, which partly rely on the work of others, partly describe work in progress and partly add new insights in the field. This leads to a commutative diagram of $\mathbb{F}_1$-geometries and functors between them that connects all the reviewed theories. We conclude the paper by reviewing the second author's constructions that lead to a realization of Tits' idea about Chevalley groups over $\mathbb{F}_1$.

Introduction

The birth of the field with one element took place with Jacques Tits' paper [**Ti57**] in 1956, in which he indicated that the analogy between the symmetric group S_n and the Chevalley group $\mathrm{GL}_n(\mathbb{F}_q)$ (as observed by Robert Steinberg in [**St51**] in 1951) should find an explanation by interpreting S_n as a Chevalley group over the "field of characteristic one." Though Tits' idea was not further developed by him or by his contemporary mathematicians, it is one of the guiding thoughts in the development of $\mathbb{F}_1$-geometry today. For a recent treatment of Steinberg's paper, see [**So09**], and for an overview of what geometric objects over $\mathbb{F}_q$ should become if $q = 1$, see [**C04**].

It took more than 35 years before the field of one element reoccurred in the mathematical literature. In an unpublished note ([**KS**]), Mikhail Kapranov and Alexander Smirnov developed the philosophy that the set of the n-th roots of unity should be interpreted as $\mathbb{F}_{1^n}$, a field extension of $\mathbb{F}_1$ in analogy to the field extension $\mathbb{F}_{p^n}$ of $\mathbb{F}_p$. A scheme that contains the n-th roots of unity should be thought of as a scheme over $\mathbb{F}_{1^n}$. The tensor product $\mathbb{F}_{1^n} \otimes_{\mathbb{F}_1} \mathbb{Z}$ should be the group ring $\mathbb{Z}[\mathbb{Z}/n\mathbb{Z}]$.

In the early nineties, Christoph Deninger published his studies ([**D91**], [**D92**], [**D94**], etc.) on motives and regularized determinants. In his paper [**D92**], Deninger gave a description of conditions on a category of motives that would admit a translation of Weil's proof of the Riemann hypothesis for function fields of projective curves over finite fields $\mathbb{F}_q$ to the hypothetical curve $\overline{\operatorname{Spec}\mathbb{Z}}$. In particular, he showed that the following formula would hold:

$$2^{-1/2}\pi^{-s/2}\Gamma(\frac{s}{2})\zeta(s) =$$

$$\frac{\det_\infty\left(\frac{1}{2\pi}(s-\Theta)\Big|H^1(\overline{\operatorname{Spec}\mathbb{Z}},\mathcal{O}_T)\right)}{\det_\infty\left(\frac{1}{2\pi}(s-\Theta)\Big|H^0(\overline{\operatorname{Spec}\mathbb{Z}},\mathcal{O}_T)\right)\det_\infty\left(\frac{1}{2\pi}(s-\Theta)\Big|H^2(\overline{\operatorname{Spec}\mathbb{Z}},\mathcal{O}_T)\right)}$$

where $\det_\infty$ denotes the regularized determinant, Θ is an endofunctor that comes with the category of motives and $H^i(\overline{\operatorname{Spec}\mathbb{Z}},\mathcal{O}_T)$ are certain proposed cohomology groups. This description combines with Nobushige Kurokawa's work on multiple zeta functions ([**Ku92**]) giving hope that there exist some motives h^0 ("*the absolute point*"), h^1 and h^2 ("*the absolute Tate motive*") with zeta functions

$$\zeta_{h^w}(s) = \det_\infty\left(\frac{1}{2\pi}(s-\Theta)\Big|H^w(\overline{\operatorname{Spec}\mathbb{Z}},\mathcal{O}_T)\right)$$

for $w = 0, 1, 2$. Deninger computed that $\zeta_{h^0}(s) = s/2\pi$ and that $\zeta_{h^2}(s) = (s-1)/2\pi$. It was Yuri Manin who proposed in [**Ma95**] the interpretation of h^0 as $\operatorname{Spec}\mathbb{F}_1$ and the interpretation of h^2 as the affine line over $\mathbb{F}_1$. The quest for a proof of the Riemann hypothesis was from now on a main motivation to look for a geometry over $\mathbb{F}_1$. Kurokawa continued his work on zeta functions in the spirit of $\mathbb{F}_1$-geometry in the collaboration [**KOW02**] with Hiroyuki Ochiai and Masato Wakayama, as well as in [**Ku05**].

In 2004, Christophe Soulé proposed a first definition of an algebraic variety over $\mathbb{F}_1$ in [**So04**], based on the observation that the extension of the base field of a scheme can be characterized by a universal property. His suggestion for a variety over $\mathbb{F}_1$ is an object involving a functor, a complex algebra, a $\mathbb{Z}$–scheme and certain morphisms and natural transformations such that a corresponding universal property is satisfied. Shortly after that, many different approaches to $\mathbb{F}_1$-geometry arose.

Anton Deitmar reinterpreted in [**D05**] the notion of a fan as given by Kazuya Kato in [**Ka94**] as a scheme over $\mathbb{F}_1$. He calculated zeta functions for his $\mathbb{F}_1$-schemes ([**D06**]) and showed that the $\mathbb{F}_1$-schemes whose base extension to $\mathbb{C}$ are complex varieties correspond to toric varieties ([**D07**]). Bertrand Töen and Michel Vaquié associated to any symmetric monoidal category with certain additional properties a category of geometric objects. In the case of the category of sets with the monoidal structure given by the cartesian product, the geometric objects are locally representable functors on monoids, which are $\mathbb{F}_1$-schemes in the sense of Töen and Vaquié. Florian Marty further developed the theory providing the notions of Zariski open objects ([**Ma07**]) and smooth morphisms ([**Ma09b**]) in this context.

Haran (see [**Ha07**]) proposed using certain categories modeled over finite sets as a replacement for rings, and actually produced a candidate for the compactification $\overline{\operatorname{Spec}\mathbb{Z}}$ of $\operatorname{Spec}\mathbb{Z}$ in his framework. Nikolai Durov developed in [**Du07**] an extension of classical algebraic geometry within a categorical framework that essentially implied replacing rings by a certain type of monad. As a byproduct of his theory he obtained a definition of $\mathbb{F}_1$ and an algebraic geometry attached to it. See [**Fr09**] for a summary.

In 2008, Alain Connes, Katia Consani and Matilde Marcolli showed ([**CCM09**]) that the Bost-Connes system defined in [**BC95**] admits a realization as a geometric object in the sense of Soulé. The suggestion of Soulé to consider a variation of the functor in his approach and other ideas led Connes and Consani to the variation of Soulé's approach as presented in [**CC08**]. That paper contains a first contribution to Tits' idea of Chevalley groups over $\mathbb{F}_1$, namely, Connes and Consani define them as varieties over $\mathbb{F}_{1^2}$. Soulé himself wrote only later a text with his originally suggested modification. It can be found in this volume ([**So09**]).

Yuri Manin proposed in [**Ma08**] a notion of analytic geometry over $\mathbb{F}_1$. The key idea is that one should look for varieties having "enough cyclotomic points." This idea relates to Kazuo Habiro's notion ([**Ha04**]) of the cyclotomic completion of a polynomial ring, which finds the interpretation as the ring of analytic functions on the set of all roots of unity. Matilde Marcolli presents in [**Ma09**] an alternative model to the BC-system for the noncommutative geometry of the cyclotomic tower. For that purpose, she uses the multidimensional analogs of the Habiro ring defined by Manin and constructs a class of multidimensional BC-endomotives satisfying a particular universality property. Marcolli's endomotives turn out to be closely related to Λ-rings, in the sense of [**BS08**].

Aiming at unifying the different notions of varieties over $\mathbb{F}_1$ (after Soulé and Connes-Consani) as well as establishing new examples of $\mathbb{F}_1$-varieties, the authors of this text introduced in [**LL**] the notion of torified varieties. Of particular interest are Grassmann varieties, which are shown in [**LL**] to be torified varieties and to provide candidates for $\mathbb{F}_1$-varieties. However, these candidates fail to satisfy the constraints of Soulé's and Connes-Consani $\mathbb{F}_1$-geometries. Independently of this work, Connes and Consani introduced in [**CC09**] a new notion of scheme over $\mathbb{F}_1$, which simplified the previous approaches by Soulé and themselves by merging Deitmar's and Toën-Vaquies viewpoints into it. We will show in this paper that this notion is closely related to torified varieties. The second author showed in [**Lo09**] that Connes-Consani's new notion of $\mathbb{F}_1$-geometry is indeed suitable to realize Tits' original ideas on Chevalley groups over $\mathbb{F}_1$.

Another promising notion of $\mathbb{F}_1$-geometry was given recently by James Borger in [**Bo09**], who advocates the use of Λ-ring structures as the natural descent data from $\mathbb{Z}$ to $\mathbb{F}_1$.

The aim of this paper is to give an overview of the new land of geometries over the field with one element. First, we review briefly the different developments and building bricks of $\mathbb{F}_1$-geometries. Second, we tie up and lay paths and bridges between the different $\mathbb{F}_1$-geometries by describing functors between them. This will finally lead to a commutative diagram of $\mathbb{F}_1$-geometries and functors between them that can be considered as a first map of $\mathbb{F}_1$-land.

The paper is organized as follows. In the first part, we review the different notions of $\mathbb{F}_1$-geometries. We describe a theory in detail where it seems important to gain an insight into its technical nature, but we will refrain from a detailed treatment in favor of a rough impression where technicalities lead too far for a brief account. In these cases, we provide the reader with references to the literature where the missing details can be found.

The following approaches towards $\mathbb{F}_1$-geometries are reviewed: Deitmar's $\mathbb{F}_1$-schemes (which we call $\mathcal{M}$-schemes in the following) are described in Section 1.1, Toën and Vaquié's $\mathbb{F}_1$-schemes in Section 1.2, Haran's $\mathbb{F}$-schemes in Section 1.3 and Durov's generalized Schemes in Section 1.4. In Section 1.5, we give a taste of Soulé's definition of a variety over $\mathbb{F}_1$ and describe his and Connes-Consani's variations on this notion. In Section 1.6, we present Connes and Consani's new proposal of $\mathbb{F}_1$-geometry including the notion of an $\mathcal{M}_0$-scheme. In Section 1.7, we review our definition of a torified variety. Finally, we give in Section 1.8 an insight into Borger's Λ-schemes.

In the second part, we review and construct functors between the categories introduced in the first part of the paper. The very central objects of $\mathbb{F}_1$-geometries are toric varieties. As we will see, these can be realized in all considered $\mathbb{F}_1$-geometries. We begin in Section 2.1 by recalling the definition of a toric variety and following Kato ([**Ka94**]) to show that toric varieties are $\mathbb{F}_1$-schemes after Deitmar, i.e., $\mathcal{M}$-schemes. In Section 2.2, we describe Florian Marty's work on comparing Deitmar's and Toën-Vaquié's notions of $\mathbb{F}_1$-schemes.

In Section 2.3, we lay the path from $\mathcal{M}$-schemes to $\mathcal{M}_0$-schemes and Connes-Consani's $\mathbb{F}_1$-schemes. In Section 2.4, we recall from [**Ha07**] Haran's notion of $\mathbb{F}$–schemes and that $\mathcal{M}$-schemes and $\mathcal{M}_0$-schemes are particular examples. In Section 2.5, we see that the same result holds true for Durov's generalized schemes (with 0) in place of $\mathbb{F}$-schemes. In Section 2.6, we review Peter Arndt's work in progress [**A**] comparing Haran's and Durov's approaches towards $\mathbb{F}_1$-geometry. In Section 2.7, we refer to Borger's paper [**Bo09**] to establish $\mathcal{M}$-schemes as Λ-schemes.

In Section 2.8, we recall from our previous paper [**LL**] that toric varieties are affinely torified. In Section 2.9, we give an idea of why affinely torified varieties define varieties over $\mathbb{F}_1$ (after Soulé resp. Connes and Consani). In Section 2.10, we extend the definition of a torified variety to a generalized torified scheme in order to show that the idea behind the notion of a torified variety and an $\mathbb{F}_1$-scheme after Connes and Consani are the same. All these categories and functors between them will be summarized in the diagram in Figure 1 of Section 2.11.

We conclude the paper with a review of the realization of Tits' ideas on Chevalley groups over $\mathbb{F}_1$ by the second author in Section 2.12.

Note: The present overview was written during the summer of 2009. Since then, several new relevant works in the area have appeared. Although the content of these papers is not used in the present work, the interested reader might also profit from looking at [**CC09b**], [**Ha09**], [**LB09**], [**Le09**], [**CC10**].

Acknowledgments. The authors thank the organizers and participants of the Nashville conference on $\mathbb{F}_1$-geometry for a very interesting event and for numerous inspiring discussions that finally lead to an overview as presented in the present paper. In particular, the authors thank Peter Arndt, Pierre Cartier, Javier Fresán, Florian Marty, Andrew Salch and Christophe Soulé for useful discussions and openly sharing pieces of their unpublished works with us. The authors thank Susama Agarwala, Snigdhayan Mahanta, Jorge Plazas and Bora Yakinoglu for complementing the authors' research on $\mathbb{F}_1$ with many other valuable aspects. Javier López Peña was supported by the EU Marie-Curie fellowship PIEF-GA-2008-221519 at Queen Mary University of London. Oliver Lorscheid was supported by the Max-Planck-Institut für Mathematik in Bonn.

1. The Building Bricks of $\mathbb{F}_1$-geometries

In the first part of the paper, we present an introduction to several different approaches to $\mathbb{F}_1$-geometry. Some of the original definitions have been reformulated in order to unify notation, simplify the exposition or make similarities with other notions apparent. In what follows—unless explicitly mentioned otherwise—all rings will be assumed to be commutative rings with 1. Monoids are commutative semi-groups with 1 and will be written multiplicatively. Schemes are understood to be schemes over $\mathbb{Z}$, and by variety we will mean a reduced scheme of finite type.

1.1. $\mathcal{M}$-schemes after Kato and Deitmar

In [**D05**], Deitmar proposes the definition of a geometry over the field with one element by following the ideas of Kato (see [**Ka94**]) of mimicking classical scheme theory but using the category $\mathcal{M}$ of commutative monoids (abelian multiplicative semigroups with 1) in place of the usual category of (commutative and unital) rings. To a large extent, this idea leads to a theory that is formally analogous to algebraic geometry.

Given a monoid A, an **ideal** of A is a subset $\mathfrak{a}$ such that $\mathfrak{a}A \subseteq \mathfrak{a}$. An ideal $\mathfrak{a}$ is **prime** if whenever $xy \in \mathfrak{a}$, then either $x \in \mathfrak{a}$ or $y \in \mathfrak{a}$ (or equivalently, if the set $A \setminus \mathfrak{a}$ is a submonoid of A). The set $\operatorname{spec} A$ of all the prime ideals of the monoid A can be endowed with the **Zariski topology**, for which a set $V \subseteq \operatorname{spec} A$ is closed if there is an ideal $\mathfrak{a} \subseteq A$ such that $V = V(\mathfrak{a}) = \{\mathfrak{q} \in \operatorname{spec} A \mid \mathfrak{a} \subseteq \mathfrak{q}\}$. A commutative monoid always contains a **minimal prime ideal**, the empty set, and a *unique* **maximal prime ideal**, $\mathfrak{m}_A := A \setminus A^\times$, where $A^\times$ is the group of units of A. For any submonoid $S \subseteq A$ the **localization of A at S** is defined by $S^{-1}A = \{\frac{a}{s} \mid a \in A, s \in S\}/\sim$, where $a/s \sim b/t$ if there exists some $u \in S$ such that $uta = usb$. The localization $\operatorname{Quot} A := A^{-1}A$ is called the **total fraction monoid** of A. The monoid A is **integral** (or **cancellative**) if the natural map $A \to \operatorname{Quot} A$ is injective. For each prime ideal $\mathfrak{p} \subseteq A$ we construct the **localization at $\mathfrak{p}$** as $A_\mathfrak{p} := (A \setminus \mathfrak{p})^{-1}A$.

Let $X = \operatorname{spec} A$ be endowed with the Zariski topology. Given an open set $U \subseteq X$, define

$$\mathcal{O}_X(U) := \left\{ s : U \to \coprod_{\mathfrak{p} \in U} A_\mathfrak{p} \,\middle|\, s(\mathfrak{p}) \in A_\mathfrak{p} \text{ and } s \text{ is loc. a quotient of elements in } A \right\},$$

where s is locally a quotient of elements in A if $s(\mathfrak{q}) = a/f$ for some $a, f \in A$ with $f \notin \mathfrak{q}$ for all $\mathfrak{q}$ in some neighborhood of $\mathfrak{p}$. We call $\mathcal{O}_X$ the **structure sheaf** of X. The **stalks** are $\mathcal{O}_{X,\mathfrak{p}} := \varinjlim_{\mathfrak{p} \in U} \mathcal{O}_X(U) \cong A_\mathfrak{p}$. Taking global sections yields $\mathcal{O}_X(X) = \Gamma(\operatorname{spec} A, \mathcal{O}_X) \cong A$.

A **monoidal space** is a pair $(X, \mathcal{O}_X)$, where X is a topological space and $\mathcal{O}_X$ is a sheaf of monoids. A **morphism of monoidal spaces** is a pair $(f, f^\#)$, where $f : X \to Y$ is a continuous map and $f^\# : \mathcal{O}_Y \to f_*\mathcal{O}_X$ is a morphism of sheaves. The morphism $(f, f^\#)$ is **local** if for each $x \in X$, we have $(f_x^\#)^{-1}(\mathcal{O}_{X,x}^\times) = \mathcal{O}_{Y,f(x)}^\times$. For every monoid A and $X = \operatorname{spec} A$, the pair $(X, \mathcal{O}_X)$ is a monoidal space, called an **affine $\mathcal{M}$-scheme**, and every morphism of monoids $\varphi : A \to B$ induces a local morphism between the corresponding monoidal spaces. A monoidal space $(X, \mathcal{O}_X)$ is called a $\mathcal{M}$**-scheme** (over $\mathbb{F}_1$) if for all $x \in X$ there is an open set U containing x such that $(U, \mathcal{O}_{X|U})$ is an affine $\mathcal{M}$-scheme.

Let $\mathbb{Z}[A]$ denote the semigroup ring of A. The base extension functor $- \otimes_{\mathbb{F}_1} \mathbb{Z}$ that sends $\operatorname{spec} A$ to $\operatorname{Spec} \mathbb{Z}[A]$ has a right-adjoint given by the forgetful functor from rings to monoids ([**D05**], Theorem 1.1). Both functors extend to functors between $\mathcal{M}$-schemes and schemes over $\mathbb{Z}$ ([**D05**], Section 2.3). We will often write $X_{\mathbb{Z}}$ for $X \otimes_{\mathbb{F}_1} \mathbb{Z}$.

1.2. Schemes over $\mathbb{F}_1$ in the Sense of Toën and Vaquié

Toën and Vaquié introduce in their paper [**TV08**] $\mathbb{F}_1$-schemes as functors on monoids that can be covered by representable functors. To avoid confusion with the other notions of $\mathbb{F}_1$-schemes in the present text, we will call the $\mathbb{F}_1$-schemes after Toën and Vaqiué's TV-schemes. We will explain the notion of a TV-scheme in this section.

As in the previous section, we let $\mathcal{M}$ be the category of monoids. Then the category of affine TV-schemes is by definition the opposite category $\mathcal{M}^{op}$. If A is a monoid in $\mathcal{M}$, then we write $\operatorname{Spec}_{TV} A$ for the same object in the opposite category. The category of presheaves $\mathcal{P}\mathrm{sh}(\mathcal{M}^{op})$ on $\mathcal{M}^{op}$ consists of functors $\mathcal{M}^{op} \to \mathcal{S}\mathrm{ets}$ as objects and natural transformations between them as morphisms. The affine TV-scheme $X = \operatorname{Spec}_{TV} A$ can be regarded as a presheaf $X : \mathcal{M}^{op} \to \mathcal{S}\mathrm{ets}$ by sending $Y = \operatorname{Spec}_{TV} B$ to $X(Y) = \operatorname{Hom}(Y, X) = \operatorname{Hom}(A, B)$. This defines an embedding of categories $\mathcal{M}^{op} \hookrightarrow \mathcal{P}\mathrm{sh}(\mathcal{M}^{op})$.

For a monoid A, let A-Mod be the category of sets with A-action together with equivariant maps. A homomorphism $f : A \to B$ of monoids is *flat* if the induced functor

$$- \otimes_A B : \ A\text{-Mod} \ \longrightarrow \ B\text{-Mod}$$

commutes with finite limits.

Let $X = \operatorname{Spec}_{TV} A$ and $Y = \operatorname{Spec}_{TV} B$. A morphism $\varphi : Y \to X$ in $\mathcal{M}^{op}$ is called *Zariski open* if the dual morphism $f : A \to B$ is an flat epimorphism of finite presentation. A *Zariski cover* of $X = \operatorname{Spec}_{TV} A$ is a collection $\{X_i \to X\}_{i \in I}$ of Zariski open morphisms in $\mathcal{M}^{op}$ such that there is a finite subset $J \subset I$ with the property that the functor

$$\prod_{j \in J} - \otimes_A A_j : \ A\text{-Mod} \ \longrightarrow \ \prod_{j \in J} A_j\text{-Mod}$$

is conservative. This defines the *Zariski topology* on $\operatorname{Spec} A$. The category $\mathcal{S}\mathrm{h}(\mathcal{M}^{op})$ of sheaves is the full subcategory of $\mathcal{P}\mathrm{sh}(\mathcal{M}^{op})$ whose objects satisfy the sheaf axiom for the Zariski topology. Toën and Vaquié show that $\operatorname{Spec} A$ is indeed a sheaf for every monoid A.

Let $\mathcal{F} : \mathcal{M}^{op} \to \mathcal{S}\mathrm{ets}$ be a subsheaf of an affine TV-scheme $X = \operatorname{Spec} A$. Then $F \subset X$ is *Zariski open* if there exists a family $\{X_i \to X\}_{i \in I}$ of Zariski open morphisms in $\mathcal{M}^{op}$ such that $\mathcal{F}$ is the direct image of

$$\coprod_{i \in I} X_i \to X$$

as a sheaf. Let $\mathcal{F}$ be a subsheaf of a sheaf $\mathcal{G} : \mathcal{M}^{op} \to \mathcal{S}\mathrm{ets}$. Then $\mathcal{F} \subset \mathcal{G}$ is *Zariski open* if for all affine TV-schemes $X = \operatorname{Spec} A$, the natural transformation $\mathcal{F} \times_{\mathcal{G}} X \to X$ is a monomorphism in $\mathcal{S}\mathrm{h}(\mathcal{M}^{op})$ with Zariski open image.

Let $\mathcal{F} : \mathcal{M} \to \mathcal{S}\mathrm{ets}$ be a functor. An *open cover of $\mathcal{F}$* is a collection $\{X_i \hookrightarrow \mathcal{F}\}_{i \in I}$ of Zariski open subfunctors such that

$$\coprod_{i \in I} X_i \to \mathcal{F}$$

is an epimorphism in $\mathcal{S}\mathrm{h}(\mathcal{M}^{op})$.

A *TV-scheme* is a sheaf $\mathcal{F} : \mathcal{M}^{\mathrm{op}} \to \mathcal{S}\mathrm{ets}$ that has an open cover by affine TV-schemes.

Given an affine TV-scheme $X = \mathrm{Spec}_{TV} A$, we define its base extension to $\mathbb{Z}$ as the scheme $X_{\mathbb{Z}} = \mathrm{Spec}\,\mathbb{Z}[A]$. This extends to a base extension functor

$$- \otimes_{\mathbb{F}_1} \mathbb{Z} : \{\text{TV-schemes}\} \longrightarrow \{\text{schemes}\}.$$

REMARK. In their paper [**TV08**], Toën and Vaquié define schemes w.r.t. any *cosmos*, that is, a symmetric closed-monoidal category that admits small limits and small colimits. It is possible to speak of monoids in such a category. In the case of the category $\mathcal{S}\mathrm{ets}$, we obtain $\mathcal{M}$ as monoids and the schemes w.r.t. $\mathcal{S}\mathrm{ets}$ are TV-schemes as described above. The category of $\mathbb{Z}$-modules yields commutative rings with 1 as monoids, and the schemes w.r.t. the category of $\mathbb{Z}$-modules are schemes in the usual sense. There are several other interesting categories obtained by this construction that can be found in [**TV08**]; also see Remark 1.6.1.

1.3. Haran's Non-additive Geometry

In [**Ha07**], Haran proposes an "absolute geometry" in which the field with one element and rings over $\mathbb{F}_1$ are realized as certain categories.

Haran's definition of the field with one element is the category whose objects are pointed finite sets and morphisms are maps $\varphi : X \to Y$ such that $f_{|X \setminus \varphi^{-1}(0_Y)}$ is injective (where 0_Y is the base point of Y).

Neglecting the base point yields an equivalence of this category with the category $\mathbb{F}$ of finite sets together with partial bijections, i.e., the morphism set from X to Y is

$$\mathbb{F}_{Y,X} = \mathrm{Hom}_{\mathbb{F}}(X, Y) := \{\varphi : U \xrightarrow{\sim} V \mid U \subseteq X, V \subseteq Y\}.$$

The category $\mathbb{F}$ is endowed with two functors $\oplus$ (disjoint union) and $\otimes$ (cartesian product). These functors induce two structures $(\mathbb{F}, \oplus, [0] = \varnothing)$ and $(\mathbb{F}, \otimes, [1] = \{*\})$ of symmetric monoidal categories on $\mathbb{F}$. An $\mathbb{F}$-**ring** is a category A with objects $\mathrm{Ob}(A) = \mathrm{Ob}(\mathbb{F})$ being finite sets, endowed with a faithful functor $\mathbb{F} \to A$ that is the identity on objects, and two functors $\oplus, \otimes : A \times A \to A$ extending the ones in $\mathbb{F}$ and each making A into a symmetric monoidal category. For example, every ring together with the matrix algebras with coefficients in the ring is an $\mathbb{F}$-ring. This embeds the category of rings into the category of $\mathbb{F}$-rings.

An **ideal** of an $\mathbb{F}$-ring A is a collection of subsets $\{I_{Y,X} \subseteq A_{Y,X}\}_{X,Y \in |\mathbb{F}|}$ that is closed under the operations $\circ$, $\oplus$, $\otimes$, i.e.,

$$A \circ I \circ A \subseteq I, \qquad A \otimes I \subseteq I \quad \text{and} \quad I \oplus I \subseteq I.$$

An ideal $\mathfrak{a} \subseteq A$ is **homogeneous** if it is generated by $\mathfrak{a}_{[1],[1]}$. A subset $\mathfrak{A} \subseteq A_{[1],[1]}$ is an **H-ideal** if for all $a_1, \ldots, a_n \in \mathfrak{A}$, $b \in A_{[1],[n]}$, $b' \in A_{[n],[1]}$ we have $b \circ (a_1 \oplus \cdots \oplus a_n) \circ b' \in \mathfrak{A}$. If $\mathfrak{a}$ is a homogeneous ideal, then $\mathfrak{a}_{[1],[1]}$ is an H-ideal, and conversely every H-ideal $\mathfrak{A}$ generates a homogeneous ideal $\mathfrak{a}$ such that $\mathfrak{a}_{[1],[1]} = \mathfrak{A}$. An H-ideal $\mathfrak{p} \subseteq A_{[1],[1]}$ is **prime** if its complement $A_{[1],[1]} \setminus \mathfrak{p}$ is multiplicatively closed. The set $\mathrm{Spec}\,A$ of all prime H-ideals can be endowed with the Zariski topology in the usual way. Localization of an $\mathbb{F}$-ring A with respect to a multiplicative subset $S \subseteq A_{[1],[1]}$ also works exactly as classical localization theory for commutative rings, with the localization functor having all the nice properties we might expect. We will denote by $A_{\mathfrak{p}} = \left(A_{[1],[1]} \setminus \mathfrak{p}\right)^{-1} A$ the localization of A with respect to the complement of a prime H-ideal $\mathfrak{p}$.

As in Deitmar's geometry, we can use this localization theory to build structure sheaves. Let A be an $\mathbb{F}$-ring and $U \subseteq \operatorname{Spec} A$ an open set. For $X, Y \in |\mathbb{F}|$ let $\mathcal{O}_A(U)_{Y,X}$ denote the set of functions

$$s : U \longrightarrow \coprod_{\mathfrak{p} \in U} (A_{\mathfrak{p}})_{Y,X}$$

such that $s(\mathfrak{p}) \in (A_{\mathfrak{p}})_{Y,X}$ and s is locally a fraction, i.e. for all $\mathfrak{p} \in U$, there is a neighborhood V of $\mathfrak{p}$ in U, an $a \in A_{Y,X}$ and an $f \in A_{[1],[1]} \setminus \coprod_{\mathfrak{q} \in V} \mathfrak{q}$ such that $s(\mathfrak{q}) = a/f$ for all $\mathfrak{q} \in V$. This construction yields a sheaf $\mathcal{O}_A$ of $\mathbb{F}$-rings, which is the **structure sheaf** of A. For each H-prime $\mathfrak{p}$ the stalk $\mathcal{O}_{A,\mathfrak{p}} := \varinjlim_{\mathfrak{p} \in U} \mathcal{O}_A(U)$ is isomorphic to $A_{\mathfrak{p}}$. Moreover, the $\mathbb{F}$-ring $\Gamma(\operatorname{Spec} A, \mathcal{O}_A)$ of global sections is isomorphic to A.

An $\mathbb{F}$-**ringed space** is a pair $(X, \mathcal{O}_X)$, where X is a topological space and $\mathcal{O}_X$ is a sheaf of $\mathbb{F}$-rings. An $\mathbb{F}$-ringed space $(X, \mathcal{O}_X)$ is $\mathbb{F}$-*locally ringed* if for each $p \in X$ the stalk $\mathcal{O}_{X,p}$ is local, i.e., contains a unique maximal H-ideal $\mathfrak{m}_{X,p}$. A **(Zariski)** $\mathbb{F}$-**scheme** is an $\mathbb{F}$-locally ringed space $(X, \mathcal{O}_X)$ such that there is an open covering $X = \bigcup_{i \in I} U_i$ for which the canonical maps $P_i : (U_i, \mathcal{O}_{X|U_i}) \to \operatorname{Spec} \mathcal{O}_X(U_i)$ are isomorphisms of $\mathbb{F}$-locally ringed spaces. The category of $\mathbb{F}$-**schemes** is the category of inverse systems (or pro-objects) in the category of Zariski $\mathbb{F}$-schemes.

For $\mathbb{F}$-rings A, B, C and morphisms $A \to B$ and $A \to C$, we can construct the *(relative) tensor product* $B \otimes_A C$. Using this construction, it follows that the category of (Zariski) $\mathbb{F}$-schemes contains fibered sums. In particular, we can take $A = \mathbb{F}$ and $C = \mathbb{F}(\mathbb{Z})$ in order to obtain an extension of scalars from $\mathbb{F}$ to $\mathbb{F}(\mathbb{Z})$; however, the category of $\mathbb{F}(\mathbb{Z})$-algebras is not equivalent to the category of rings (this is pretty much due to the existence of $\mathbb{F}$-rings which are not matrix rings; see [**A**] for details), so this functor does not provide an extension of scalars from $\mathbb{F}$-schemes to the usual schemes.

The embedding of the category of rings inside the category of $\mathbb{F}$-rings shows that every scheme can be regarded as an $\mathbb{F}$-scheme. A similar construction allows production of an $\mathbb{F}$-ring out of a monoid, providing a relation with $\mathcal{M}$-schemes (see Section 2.4 for further details). It is worth noting that all the examples mentioned here are what Haran calls *rings of matrices*; there are examples (see [**Ha07**] §2.3, Examples 3 and 5) of more exotic $\mathbb{F}$-rings that are not rings of matrices. Haran succeeds (see [**Ha07**], §6.3) in defining the completion $\overline{\operatorname{Spec} \mathbb{Z}}$ of the spectrum of $\mathbb{Z}$, which is one step in Deninger's program to prove the Riemann hypothesis.

1.4. Durov's Generalized Schemes

In [**Du07**], Durov introduces a generalization of classical algebraic geometry, into which Arakelov geometry fits naturally. As a byproduct of his theory of generalized rings, he obtains a model for the field with one element and a notion of a geometry over $\mathbb{F}_1$. Here, we outline Durov's construction briefly. For full details, see [**Du07**] or consider [**Fr09**] for a summary.

Any ring R can be realized as the endofunctor in the category of sets that maps a set X to R^X, the free R-module generated by X. The ring multiplication and unit translate into properties of this functor, making it into a **monad** that commutes with filtered direct limits (see for instance [**St72**] or [**MC98**], Chapter VI). Motivated by this fact, Durov defines a **generalized (commutative) ring** as a monad in the category of sets that is commutative (see [**Du07**], §5.1, for the

precise notion of commutativity) and commutes with filtered direct limits. If the
set $A([0]) = A(\varnothing)$ is not empty, A is said to *admit a constant*, or we say that A
is a *generalized ring with zero* (see [**Du07**], §5.1). There is a natural notion of
a **module** over a monad, which allows construction of the category of modules
over any generalized ring A. Every generalized ring A has an underlying monoid
$|A| := A([1])$, so we can define a **prime ideal** of A as any A-submodule $\mathfrak{p}$ of $|A|$
such that the complement $|A| \setminus \mathfrak{p}$ is a multiplicative system. The set Spec A of all
the prime ideals in A can be endowed with the Zariski topology in the usual way.

The notions of localization, presheaves and sheaves of generalized rings are
defined analogously to the usual theory of schemes resp. to the theory of the former
sections. We can talk about **generalized ringed spaces** $(X, \mathcal{O}_X)$ consisting of
a topological space X with a sheaf of generalized rings $\mathcal{O}_X$. A generalized ringed
space is **local** if for every $p \in X$ the generalized ring $\mathcal{O}_{X,p}$ has a unique maximal
ideal. Every generalized ring defines a locally generalized ringed space (Spec $A, \mathcal{O}_A$),
which is called the **generalized affine scheme** associated to A. A **generalized
scheme** is then a locally generalized ringed space which is locally isomorphic to
a generalized affine scheme. In this setting, the category of $\mathbb{F}_1$-schemes consists
precisely of those generalized schemes $(X, \mathcal{O}_X)$ such that the set $\Gamma(X, \mathcal{O}_X(\mathbf{0}))$ is
not empty, also called **generalized schemes with zero** (see [**Du07**], §6.5.6).

Examples of generalized schemes in the sense of Durov include the usual
schemes via the aforementioned realization of a ring as a generalized ring, as well as
schemes over the spectrum of monoids or semi-rings, in particular in Durov's theory it is possible to speak about schemes over the natural numbers, the completion
$\mathbb{Z}_\infty$ or the tropical semi-ring $\mathbb{T}$. As Haran does in the context of his non-additive
geometry, Durov defines the completion $\overline{\mathrm{Spec}\,\mathbb{Z}}$ of the spectrum of $\mathbb{Z}$ as a generalized scheme with zero. However, Durov's construction forces the fibered product
$\overline{\mathrm{Spec}\,\mathbb{Z}} \times_{\mathbb{F}_1} \overline{\mathrm{Spec}\,\mathbb{Z}}$ to be isomorphic again to $\overline{\mathrm{Spec}\,\mathbb{Z}}$, making it unsuitable for pur-
suing Deninger's program.

1.5. Varieties over $\mathbb{F}_1$ in the Sense of Soulé and Their Variations

The first suggestion of a category that realizes a geometry over $\mathbb{F}_1$ was given
by Soulé in [**So04**]. This notion underwent several variations ([**So09**], [**CC08**])
that finally lead to the notion of $\mathbb{F}_1$-schemes as given by Connes and Consani in
[**CC09**] (see the following section). We do not give the formal definitions from
[**So04**], [**So09**] and [**CC08**] in all completeness, but try to give an overview of how
ideas developed.

1.5.1. S-varieties

We begin with the notion of varieties over $\mathbb{F}_1$ as given in [**So04**]. To avoid
confusion, we will call these varieties over $\mathbb{F}_1$ *S-varieties*. For a precise definition,
see [**So04**] or [**LL**].

An *affine S-variety* X consists of
(1) a functor $\underline{X} : \{\text{finite flat rings}\} \to \{\text{finite sets}\}$,
(2) a complex (not necessarily commutative) algebra $\mathcal{A}_X$,
(3) an affine scheme $X_{\mathbb{Z}} = \mathrm{Spec}\, A$ of finite type,
(4) a natural transformation $\mathrm{ev}_X : \underline{X} \Rightarrow \mathrm{Hom}(\mathcal{A}_X, - \otimes_{\mathbb{Z}} \mathbb{C})$,
(5) an inclusion of functors $\underline{\iota} : \underline{X} \Rightarrow \mathrm{Hom}(A, -)$ and
(6) an injection $\iota_{\mathbb{C}} : A \otimes_{\mathbb{Z}} \mathbb{C} \hookrightarrow \mathcal{A}_X$

such that the diagram

(1.5.1.1)
$$\begin{CD}
\underline{X}(R) @>{\iota(R)}>> \operatorname{Hom}(A, R) \\
@V{\operatorname{ev}_X(R)}VV @VV{-\otimes_{\mathbb{Z}}\mathbb{C}}V \\
\operatorname{Hom}(\mathcal{A}_X, R \otimes_{\mathbb{Z}} \mathbb{C}) @>{\iota_{\mathbb{C}}^*}>> \operatorname{Hom}(A \otimes_{\mathbb{Z}} \mathbb{C}, R \otimes_{\mathbb{Z}} \mathbb{C})
\end{CD}$$

commutes for all finite flat rings R and such that a certain universal property is satisfied. This universal property characterizes $X_{\mathbb{Z}}$ together with $\underline{\iota}$ and $\iota_{\mathbb{C}}$ as the unique extension of the triple $(\underline{X}, \mathcal{A}_X, \operatorname{ev}_X)$ to an affine S-variety. We define $X_{\mathbb{Z}}$ as the *base extension of X to $\mathbb{Z}$*.

A *morphism $X \to Y$ between affine S-varieties* consists of
(1) a natural transformation $\underline{X} \to \underline{Y}$,
(2) a $\mathbb{C}$-linear map $\mathcal{A}_Y \to \mathcal{A}_X$ and
(3) a morphism $X_{\mathbb{Z}} \to Y_{\mathbb{Z}}$ of schemes
such that they induce a morphism between the diagrams (1.5.1.1) corresponding to X and Y.

The idea behind the definition of global S-varieties is to consider functors on affine S-varieties. More precisely, an *S-variety X* consists of
(1) a functor $\underline{X} : \{\text{affine S-varieties}\} \to \{\text{finite sets}\}$,
(2) a complex algebra $\mathcal{A}_X$,
(3) a scheme $X_{\mathbb{Z}}$ of finite type with global sections A,
(4) a natural transformation $\operatorname{ev}_X : \underline{X} \Rightarrow \operatorname{Hom}(\mathcal{A}_X, \mathcal{A}_{(-)})$,
(5) an inclusion of functors $\underline{\iota} : \underline{X} \Rightarrow \operatorname{Hom}((-)_{\mathbb{Z}}, X_{\mathbb{Z}})$ and
(6) an injection $\iota_{\mathbb{C}} : A \otimes_{\mathbb{Z}} \mathbb{C} \hookrightarrow \mathcal{A}_X$
such that the diagram

$$\begin{CD}
\underline{X}(V) @>{\iota(V)}>> \operatorname{Hom}(V_{\mathbb{Z}}, X_{\mathbb{Z}}) \\
@V{\operatorname{ev}_X(V)}VV @VV{\text{global sections}}V \\
\operatorname{Hom}(\mathcal{A}_X, \mathcal{A}_V) @>{\iota_{\mathbb{C}}^*}>> \operatorname{Hom}(A \otimes_{\mathbb{Z}} \mathbb{C}, \mathcal{A}_V)
\end{CD}$$

commutes for all affine S-varieties V and such that a certain universal property is satisfied. This universal property plays the same role as the universal property in the definition of an affine S-variety. We define $X_{\mathbb{Z}}$ as the base extension of X to $\mathbb{Z}$. Morphisms of S-varieties are defined analogously to the affine case.

A remarkable property of these categories is that the dual of the category of finite flat rings embeds into the category of affine S-varieties and that the category of affine S-varieties embeds into the category of S-varieties. The essential image of the latter embedding are those S-varieties whose base extension to $\mathbb{Z}$ is an affine scheme. Furthermore, a system of affine S-varieties ordered by inclusions that is closed under intersections can be glued to an S-variety.

In his paper, Soulé constructs for every smooth toric variety an S-variety whose base extension to $\mathbb{Z}$ is isomorphic to the toric variety. This result was extended in [**LL**] (Theorem 3.11) (also see Section 2.9).

The paper contains a definition of a zeta function for those S-varieties X that admit a polynomial counting function, i.e., that the function $q \mapsto \#X_{\mathbb{Z}}(\mathbb{F}_q)$ on prime powers q is given by a polynomial in q with integer coefficients. This provides, up to a factor 2π, a first realization of Deninger's motivic zeta functions of h^0 and h^2 (as in the Introduction) as zeta functions of the *absolute point* $\operatorname{Spec}\mathbb{F}_1$ and the *affine line over* $\mathbb{F}_1$.

1.5.2. S^*-varieties

In [**So09**], Soulé describes the first modification of this category. The idea is to exchange the functor on finite flat rings by a functor on finite abelian groups. Correspondingly, the functors $\operatorname{Hom}(\mathcal{A}_X, -\otimes_\mathbb{Z}\mathbb{C})$ and $\operatorname{Hom}(A,-)$ in the definition of an affine S-variety are replaced by $\operatorname{Hom}(\mathcal{A}_X,\mathbb{C}[-])$ and $\operatorname{Hom}(A,\mathbb{Z}[-])$, respectively, where $\mathbb{Z}[D]$ and $\mathbb{C}[D]$ are the group algebras of a finite abelian group D. We call the objects that satisfy the definition of an affine S-variety with these changes *affine S^*-varieties*. Morphisms between affine S^*-varieties and the base extension to $\mathbb{Z}$ is defined analogously to the case of affine S-varieties. Also the definition of S^*-*varieties* and morphisms between S^*-varieties are defined in complete analogy to the case of S-varieties.

Soulé proves similar results for S^*-varieties as for S-varieties: affine S^*-varieties are S^*-varieties. A system of affine S^*-varieties ordered by inclusions that is closed under intersections can be glued to an S^*-variety. Every smooth toric variety has a model as an S^*-variety (see also Section 2.9). The definition of zeta functions transfers to this context.

1.5.3. CC-varieties

In [**CC08**], Connes and Consani modify the notion of an S^*-variety in the following way. They endow the functor $\underline{X}$ from finite abelian groups to finite sets with a grading, i.e., for every finite abelian group D, the set $\underline{X}(D)$ has a decomposition $\underline{X}(D) = \coprod_{n\in\mathbb{N}}\underline{X}^n(D)$ into a disjoint union of sets. They then exchange the complex algebra by a complex variety. We call the objects that satisfy the definition of an affine S^*-variety with these changes *affine CC-varieties*. Morphisms between affine CC-varieties and the base extension to $\mathbb{Z}$ are defined analogously to the case of affine S-varieties and S^*-varieties.

The category of affine CC-varieties is embedded into a larger category that plays the role of the category of locally ringed spaces in the theory of schemes. *CC-varieties* are defined as those objects in the larger category that admit a cover by affine CC-varieties.

The definition of zeta functions transfers to this context. Connes and Consani show certain examples of CC-schemes, where the zeta function can be read off from the graded functor of the CC-scheme.

The main application of this paper is the construction of models of split reductive groups as CC-varieties over $\mathbb{F}_{1^2}$. This is a first construction of objects in $\mathbb{F}_1$-geometry that contributes to Tits' ideas from [**Ti57**]. These results were extended in [**LL**]; in particular, they also hold in the context of S-varieties and S^*-varieties (see Section 2.9). However, these categories are not suitable for defining a group law for any split reductive group over $\mathbb{F}_1$ (see [**CC08**], p. 25, and [**LL**], Section 6.1). For how this can be done in the context of a CC-scheme, see the next section.

1.6. Schemes over $\mathbb{F}_1$ in the Sense of Connes and Consani

In their paper [**CC09**], Connes and Consani merge Soulé's idea and its variations with Kato's resp. Deitmar's monoidal spaces and Toën and Vaquié's sheaves on monoids. Roughly speaking, a scheme over $\mathbb{F}_1$ in the sense of Connes and Consani, which we call a *CC-scheme* to avoid confusion, is a triple consisting of a locally representable functor on monoids, a scheme and an evaluation map. Unlike Kato/Deitmar and Toën-Vaquié, all monoids are considered together with a base point as in Haran's theory ([**Ha07**])—the locally representable functor is a functor from $\mathcal{M}_0$ (see Section 1.3) to $\mathcal{S}$ets.

We make these notions precise. First of all, we can reproduce all steps in the construction of $\mathcal{M}$-schemes as in Section 1.1 to define $\mathcal{M}_0$-schemes. Namely, an *ideal* of a monoid A with 0 is a subset $I \subset A$ containing 0 such that $IA \subset I$. A *prime ideal* is an ideal $\mathfrak{p} \subset A$ such that $A-\mathfrak{p}$ is a submonoid of A. As in Section 1.1, we can define *localizations* and the *Zariski topology* on $\mathrm{Spec}_{\mathcal{M}_0} A = \{$prime ideals of $A\}$, the *spectrum of A*.

A *monoidal space with* 0 is a topological space together with a sheaf in $\mathcal{M}_0$. Together with its Zariski topology and its localizations, $\mathrm{Spec}\,A$ is a monoidal space with 0. A *geometric $\mathcal{M}_0$-scheme* is a monoidal space with 0 that has an open covering by spectra of monoids with 0.

Similar to the definition in Section 1.2, $\mathcal{M}_0$-schemes are defined as locally representable functors on $\mathcal{M}_0$. In detail, an *$\mathcal{M}_0$-functor* is a functor from $\mathcal{M}_0$ to $\mathcal{S}$ets. Every monoid A with 0 defines an $\mathcal{M}_0$-functor $X = \mathrm{Spec}_{\mathcal{M}_0} A$ by sending a monoid B with 0 to the set $X(B) = \mathrm{Hom}(\mathrm{Spec}_{\mathcal{M}_0} B, \mathrm{Spec}_{\mathcal{M}_0} A) = \mathrm{Hom}(A, B)$. An *affine $\mathcal{M}_0$-scheme* is a $\mathcal{M}_0$-functor that is isomorphic to $\mathrm{Spec}_{\mathcal{M}_0} A$ for some monoid A with 0.

A subfunctor $Y \subset X$ of an $\mathcal{M}_0$-functor X is called an *open subfunctor* if for all monoids A with 0 and all morphisms $\varphi : Z = \mathrm{Spec}_{\mathcal{M}_0} A \to X$, there is an ideal I of A such that for all monoids B with 0 and for all $\rho \in Z(B) = \mathrm{Hom}(A, B)$,

$$\varphi(\rho) \in Y(B) \subset X(B) \iff \rho(I)B = B.$$

An *open cover of X* is a collection $\{X_i \to X\}_{i \in I}$ of open subfunctors such that the map $\coprod_{i \in I} X_i(H) \to X(H)$ is surjective for every monoid H that is a union of a group with 0. An *$\mathcal{M}_0$-scheme* is an $\mathcal{M}_0$-functor that has an open cover by affine $\mathcal{M}_0$-schemes.

If X is an $\mathcal{M}_0$-scheme, then the *stalk at x* is $\mathcal{O}_{X,x} = \lim_{\to} X(U)$, where U runs through all open neighborhoods of x. A *morphism of $\mathcal{M}_0$-schemes* $\varphi : X \to Y$ is a natural transformation of functors that is *local*, i.e., for every $x \in X$ and $y = \varphi(x)$, the induced morphism $\varphi_x^\sharp : \mathcal{O}_{Y,y} \to \mathcal{O}_{X,x}$ between the stalks satisfies $(\varphi_x^\sharp)^{-1}(\mathcal{O}_{X,x}^\times) = \mathcal{O}_{Y,y}^\times$.

Connes and Consani claim that the $\mathcal{M}_0$-functor $\mathrm{Hom}(\mathrm{Spec}_{\mathcal{M}_0} -, X)$ is a $\mathcal{M}_0$-scheme for every geometric $\mathcal{M}_0$-scheme X and that, conversely, every $\mathcal{M}_0$-scheme is of this form ([**CC09**], Proposition 3.16). For this reason, we shall make no distinction between geometric $\mathcal{M}_0$-schemes and $\mathcal{M}_0$-schemes from now on.

The association

$$A \to \left(\mathbb{Z}[A]/1 \cdot 0_A - 0_{\mathbb{Z}[A]} \right),$$

where 0_A is the zero of A and $0_{\mathbb{Z}[A]}$ is the zero of $\mathbb{Z}[A]$ extends to the base extension functor

$$- \otimes_{\mathcal{M}_0} \mathbb{Z} : \{\mathcal{M}_0 - \text{schemes}\} \longrightarrow \{\text{schemes}\}.$$

We denote the base extension of an $\mathcal{M}_0$-scheme X to $\mathbb{Z}$ by $X_{\mathbb{Z}} = X \otimes_{\mathcal{M}_0} \mathbb{Z}$.

The ideas from Sections 1.5.1–1.5.3 now find a simplified form. A *CC-scheme* is a triple $(\widetilde{X}, X, \mathrm{ev}_X)$, where $\widetilde{X}$ is an $\mathcal{M}_0$-scheme, X is a scheme, viewed as a functor on the category of rings, and $\mathrm{ev}_X : \widetilde{X}_{\mathbb{Z}} \Rightarrow X$ is a natural transformation that induces a bijection $\mathrm{ev}_X(k) : \widetilde{X}_{\mathbb{Z}}(k) \to X(k)$ for every field k. The natural notion of a *morphism between CC-schemes* $(\widetilde{Y}, Y, \mathrm{ev}_Y)$ *and* $(\widetilde{X}, X, \mathrm{ev}_X)$ is a pair $(\widetilde{\varphi}, \varphi)$, where $\widetilde{\varphi} : \widetilde{Y} \to \widetilde{X}$ is a morphism of $\mathcal{M}_0$-schemes and $\varphi : Y \to X$ is a morphism of schemes such that the diagram

$$
\begin{array}{ccc}
\widetilde{X}_{\mathbb{Z}} & \xrightarrow{\widetilde{\varphi}_{\mathbb{Z}}} & \widetilde{Y}_{\mathbb{Z}} \\
{\scriptstyle \mathrm{ev}_X} \downarrow & & \downarrow {\scriptstyle \mathrm{ev}_Y} \\
X & \xrightarrow{\varphi} & Y
\end{array}
$$

commutes. The base extension of $\mathcal{X} = (\widetilde{X}, X, \mathrm{ev}_X)$ to $\mathbb{Z}$ is $\mathcal{X}_{\mathbb{Z}} = X$.

REMARK 1.6.1. The definition of an $\mathcal{M}_0$-scheme is in the line of thought of Toën and Vaquié's theory as described in Section 1.2. Indeed, monoid objects in the category of pointed sets are monoids with 0. It seems to be likely, but it is not obvious, that the notion a scheme w.r.t. the category of pointed sets are $\mathcal{M}_0$-schemes. Andrew Salch is working on making this precise. Furthermore, Salch constructs a cosmos such that monoid objects on this category correspond to triples $(\widetilde{X}, X, e_X)$, where $\widetilde{X}$ is an $\mathcal{M}_0$-functor, X is a functor on rings and $e_X : \widetilde{X}_{\mathbb{Z}} \Rightarrow X$ is a natural equivalence.

It is, however, questionable whether CC-schemes have an interpretation as schemes w.r.t. some category in the sense of Toën and Vaquié since it is not clear what an affine CC-scheme should be. See Remark 2.10.4 for a more detailed explanation.

1.7. Torified Varieties

Torified varieties and schemes were introduced by the authors of this text in [**LL**] in order to establish examples of varieties over $\mathbb{F}_1$ in the sense of Soulé (see Section 1.5.1) and Connes-Consani (see Section 1.5.3).

A **torified scheme** is a scheme X endowed with a **torification** $e_X : T \to X$, i.e., $T = \coprod_{i \in I} \mathbb{G}_m^{d_i}$ is a disjoint union of split tori and e_X is a morphism of schemes such that the restrictions $e_X{}_{|\mathbb{G}_m^{d_i}}$ are immersions and $e_X(k) : T(k) \to X(k)$ is a bijection for every field k. Examples of torified schemes include (see in [**LL**], §1.3, or [**Lo09**]):

(1) Toric varieties, with torification given by the orbit decomposition.
(2) Grassmann and Schubert varieties, where the torification is induced by the Schubert cell decomposition.
(3) Split reductive groups, with torification coming from the Bruhat decomposition.

Let X and Y be torified schemes with torifications $e_X : T = \coprod_{i \in I} \mathbb{G}_m^{d_i} \to X$ and $e_Y : S = \coprod_{j \in J} \mathbb{G}_m^{f_j} \to Y$, respectively. A **torified morphism** $(X, T) \to (Y, S)$ consists of a pair of morphisms $\varphi : X \to Y$ and $e_\varphi : T \to S$ such that the diagram

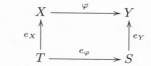

commutes and such that for every $i \in I$ there is a $j \in J$ such that the restriction $e_{\varphi | \mathbb{G}_m^{d_i}} : \mathbb{G}_m^{d_i} \to \mathbb{G}_m^{f_j}$ is a morphism of algebraic groups.

A torified scheme (X, T) is **affinely torified** if there is an affine open cover $\{U_j\}$ of X that respects the torification, i.e., such that for each j there is a subset $T_j = \coprod_{i \in I_j} \mathbb{G}_m^{d_j}$ such that the restriction $e_{X | T_j}$ is a torification of U_j. A torified morphism is **affinely torified** if there is an affine open cover $\{U_j\}$ of X respecting the torification and such that the image of each U_j is an affine subscheme of Y. An *(affinely) torified variety* is an (affinely) torified scheme that is reduced and of finite type over $\mathbb{Z}$.

Toric varieties and split reductive groups are examples of affinely torified varieties (see [**LL**], §1.3), while the torifications associated to Grassmann and Schubert varieties are in general not affine.

1.8. Λ-schemes after Borger

An approach in a different vein to all the other $\mathbb{F}_1$-geometries is Borger's notion of a Λ-scheme (see [**Bo09**]). A Λ-scheme is a scheme with an additional decoration, which is interpreted as descent datum to $\mathbb{F}_1$. In order to give a quick impression, we restrict ourselves to introducing only flat Λ-schemes in this paper.

A *flat Λ-scheme* is a flat scheme X together with a family $\Phi = \{\varphi_p : X \to X\}_{p \text{ prime}}$ of pairwise commuting endomorphisms of X such that the diagram

$$\begin{array}{ccc} X \otimes_{\mathbb{Z}} \mathbb{F}_p & \xrightarrow{\text{Frob}_p} & X \otimes_{\mathbb{Z}} \mathbb{F}_p \\ \downarrow & & \downarrow \\ X & \xrightarrow{\varphi_p} & X \end{array}$$

commutes for every prime number p. Here, $\text{Frob}_p : X \otimes_{\mathbb{Z}} \mathbb{F}_p \to X \otimes_{\mathbb{Z}} \mathbb{F}_p$ is the Frobenius morphism of the reduction of X modulo p.

The base extension of this flat Λ-scheme is the scheme X. In particular every reduced scheme X that has a family Φ as described above is flat.

Examples of Λ-schemes are toric varieties. For more examples, consider [**Bo09**] (Section 2) or Section 2.7 of this paper.

2. Paths and Bridges

In this part of the paper, we review and construct various functors between the different $\mathbb{F}_1$-geometries. In some central cases, we will describe functors in detail, in other cases, we will cite a reference. We will also describe some work in progress by Arndt and Marty on such functors. All these functors are displayed in Figure 1 in Section 2.11.

2.1. Toric Varieties as $\mathcal{M}$-schemes

In this section, we recall the definition of a toric variety, describe the reformulation of toric varieties in terms of a fan of monoids as done by Oda ([**O88**]) and relate them to $\mathcal{M}$-schemes as done by Kato ([**Ka94**]). For more details on toric varieties, consider [**Fu93**] and [**Ma00**].

A *(strictly convex and rational) cone (in $\mathbb{R}^n$)* is an additive semigroup $\tau \subset \mathbb{R}^n$ of the form $\tau = \sum_{i \in I} t_i \mathbb{R}_{\geq 0}$, where $\{t_i\}_{i \in I} \subset \mathbb{Q}^n \subset \mathbb{R}^n$ is a linearly independent

set. A *face* σ *of* τ is a cone of the form $\sigma = \sum_{i \in J} t_i \mathbb{R}_{\geq 0}$ for some subset $J \subset I$. A *fan* is a non-empty collection Δ of cones in $\mathbb{R}^n$ such that
  (1) each face of a cone $\tau \in \Delta$ is in Δ and
  (2) for all cones $\tau, \sigma \in \Delta$, the intersection $\tau \cap \sigma$ is a face of both τ and σ.
In particular, the face relation makes Δ a partially ordered set.

If τ is a cone in $\mathbb{R}^n$, we define A_τ as the intersection $\tau^\vee \cap \mathbb{Z}^n$ of the dual cone $\tau^\vee$ with the lattice $\mathbb{Z}^n \subset \mathbb{R}^n$. Since the generators t_i of τ have rational coordinates, A_τ is a finitely generated (additively written) monoid. An inclusion $\sigma \subset \tau$ of cones induces an open embedding of schemes $\operatorname{Spec} \mathbb{Z}[A_\sigma] \hookrightarrow \operatorname{Spec} \mathbb{Z}[A_\tau]$. A *toric variety (of dimension n)* is a scheme X together with a fan Δ (in $\mathbb{R}^n$) such that

$$X \simeq \varinjlim_{\tau \in \Delta} \operatorname{Spec} \mathbb{Z}[A_\tau].$$

A *morphism* $\psi : \Delta \to \Delta'$ *of fans* is map $\widetilde{\psi}$ between partially ordered sets together with a direct system of semigroup morphisms $\psi_\tau : \tau \to \widetilde{\psi}(\tau)$ (with respect to the inclusion of cones). The dual morphisms induce monoid homomorphisms $\psi_\tau^\vee : A_{\widetilde{\psi}(\tau)} \to A_\tau$. Taking the direct limit over the system of scheme morphisms $\operatorname{Spec} \mathbb{Z}[\psi_\tau^\vee] : \operatorname{Spec} \mathbb{Z}[A_\tau] \to \operatorname{Spec} \mathbb{Z}[A_{\widetilde{\psi}(\tau)}]$ yields a morphism $\varphi : X \to X'$ between toric varieties. Such a morphism is called a *toric morphism*.

Note that a toric variety is determined by its fan and a toric morphism between toric varieties is determined by the morphism between the fans. This leads to a completely combinatorial description of the category of toric varieties in terms of monoids as follows (see [**Ka94**], Section 9).

Recall the definition of the quotient group $\operatorname{Quot} A$ and of an integral monoid from Section 1.1. The monoid A is *saturated* if it is integral and if for all $a \in A$, $b \in \operatorname{Quot} A$ and $n > 0$ such that $b^n = a$, we have that $b \in A$. A *fan in* $\mathbb{Z}^n$ is a collection Δ of (additive) submonoids $A \subset \mathbb{Z}^n$ such that
  (1) all $A \in \Delta$ are finitely generated and saturated, $A^\times = 1$ and $\mathbb{Z}^n / \operatorname{Quot} A$ is torsion-free,
  (2) for all $A \in \Delta$ and $\mathfrak{p} \in \operatorname{Spec}_{\mathcal{M}} A$, we have $A \setminus \mathfrak{p} \in \Delta$, and
  (3) for all $A, B \in \Delta$, we have $A \cap B = A \setminus \mathfrak{p} = B \setminus \mathfrak{q}$ for some $\mathfrak{p} \in \operatorname{Spec}_{\mathcal{M}} A$ and $\mathfrak{q} \in \operatorname{Spec}_{\mathcal{M}} B$.
In particular, Δ is a diagram of monoids via the inclusions $A \setminus \mathfrak{p} \hookrightarrow A$.

This yields the following alternative description: a toric variety is a scheme X together with a fan Δ in $\mathbb{Z}^n$ such that

$$X \simeq \varinjlim_{A \in \Delta} \operatorname{Spec} \mathbb{Z}[A].$$

Note that $\operatorname{Spec} \mathbb{Z}[A]$ is the base extension of the $\mathcal{M}$-scheme $\operatorname{Spec}_{\mathcal{M}} A$ to $\mathbb{Z}$, and that for

$$\widetilde{X} = \varinjlim_{A \in \Delta} \operatorname{Spec}_{\mathcal{M}} A,$$

$\widetilde{X}_{\mathbb{Z}} \simeq X$. We introduce some further definitions to state the following theorem, where we follow ideas from [**D07**] (Section 4).

An $\mathcal{M}$-scheme is *connected* if its topological space is connected. An $\mathcal{M}$-scheme is *integral/of finite type/of exponent* 1 if all its stalks are integral/of finite type/ (multiplicatively) torsion-free.

THEOREM 2.1.1. ([**LL**], Theorem 4.1) *The association as described above extends to a functor*

$$\mathcal{K} : \{toric\ varieties\} \longrightarrow \{\mathcal{M}\text{-schemes}\}$$

that induces an equivalence of categories

$$\mathcal{K} : \{toric\ varieties\} \xrightarrow{\sim} \left\{ \begin{array}{l} connected\ integral\ \mathcal{M}\text{-schemes} \\ of\ finite\ type\ and\ of\ exponent\ 1 \end{array} \right\}$$

with $- \otimes_{\mathbb{F}_1} \mathbb{Z}$ *being its inverse.*

2.2. Comparison of $\mathcal{M}$-schemes with TV-schemes

The category of D-schemes is equivalent to the category of TV-schemes. In the second arXiv version of Florian Marty's paper [**Ma07**], one finds partial results, in particular for the affine case.

Namely, it is possible to associate to each affine TV-scheme $X = \operatorname{Spec}_{TV} A$ a topological space Y whose locale is the locale of X (defined by the Zariski open subsheaves as introduced in Section 1.2) and such that Y is homeomorphic to the topological space $\operatorname{Spec}_{\mathcal{M}} A$ as considered in Section 1.1. More precisely, this construction yields a functor

$$\mathcal{M}_1 : \{\text{affine TV-schemes}\} \longrightarrow \{\text{affine }\mathcal{M}\text{-schemes}\}$$

that is an equivalence of categories with inverse $\mathcal{M}_2 : \operatorname{Spec}_{\mathcal{M}} A \mapsto \operatorname{Hom}(A, -)$. This equivalence of categories was extended to an equivalence

$$\{\text{TV-schemes}\} \underset{\mathcal{M}_2}{\overset{\mathcal{M}_1}{\rightleftarrows}} \{\mathcal{M}\text{-schemes}\}$$

by Alberto Vezzani in his master thesis "The geometry over the field with one element" at Universit degli Studi di Milano in 2010. A publication is in preparation ([**Ve10**]).

2.3. $\mathcal{M}$-schemes as $\mathcal{M}_0$- and CC-schemes

The category of $\mathcal{M}$-schemes embeds into the category of CC-schemes. We proceed in two steps.

First, we construct a functor from $\mathcal{M}$-schemes to $\mathcal{M}_0$-schemes. Consider the fully faithful functor $\mathcal{M} \xrightarrow{+0} \mathcal{M}_0$ that associates to a monoid A the monoid $A_{+0} = A \cup \{0\}$ with 0 whose multiplication is extended by $0 \cdot a = 0$ for every $a \in A_{+0}$. This induces a faithful functor

$$\{\text{monoidal spaces}\} \xrightarrow{+0} \{\text{monoidal spaces with 0}\}$$

that we denote by the same symbol, by composing the structure sheaf $\mathcal{O}_X$ of a monoidal space X with $\mathcal{M} \xrightarrow{+0} \mathcal{M}_0$. We reason in the following that this functor restricts to a functor from the category of $\mathcal{M}$-schemes to the category of $\mathcal{M}_0$-schemes.

The prime ideals of a monoid A (see Section 1.1) coincide with the prime ideals of A_{+0} (see Section 1.6) since we required that a prime ideal of a monoid with 0 contain 0. Recall that localizations, the Zariski topology, monoidal spaces (without and with 0) for $\mathcal{M}$-schemes and $\mathcal{M}_0$-schemes are defined in complete analogy. Thus the above functor restricts to a faithful functor

$$\{\mathcal{M}\text{-schemes}\} \xrightarrow{+0} \{\mathcal{M}_0\text{-schemes}\}.$$

Second, we define the functor

$$\mathcal{F} : \{\mathcal{M}_0\text{-schemes}\} \longrightarrow \{\text{CC-schemes}\}$$

as sending an $\mathcal{M}_0$-scheme X to the CC-scheme $(X, X_{\mathbb{Z}}, \mathrm{id}_{X_{\mathbb{Z}}})$. A morphism $\varphi : Y \to X$ of $\mathcal{M}_0$-schemes defines the morphism $(\varphi, \varphi_{\mathbb{Z}})$ of CC-schemes. The condition that

$$
\begin{array}{ccc}
Y_{\mathbb{Z}} & \xrightarrow{\varphi_{\mathbb{Z}}} & X_{\mathbb{Z}} \\
\mathrm{id}_{Y_{\mathbb{Z}}} \downarrow & & \downarrow \mathrm{id}_{X_{\mathbb{Z}}} \\
Y_{\mathbb{Z}} & \xrightarrow{\varphi_{\mathbb{Z}}} & X_{\mathbb{Z}}
\end{array}
$$

commutes is trivially satisfied and shows that $\mathcal{F}$ is faithful. The composition of the functors $+0$ and $\mathcal{F}$ yields a functor

$$\{\mathcal{M}\text{-schemes}\} \longrightarrow \{\text{CC-schemes}\}.$$

2.4. $\mathcal{M}$- and $\mathcal{M}_0$-schemes as $\mathbb{F}$-schemes

Let M be a monoid with 0. We can define an $\mathbb{F}$-ring $\mathbb{F}\langle M \rangle$ by setting $\mathbb{F}\langle M \rangle_{Y,X} = \mathrm{Hom}_{\mathbb{F}\langle M \rangle}(X, Y)$, that is, the set of $Y \times X$ matrices with values in M with at most one non-zero entry in every row and column. This is an $\mathbb{F}$-ring with composition defined by matrix multiplication. This is possible because all sums that occur in the product of two matrices with at most one non-zero entry in every row and column range over at most one term. This construction yields a faithful functor

$$\mathbb{F}\langle - \rangle : \mathcal{M}_0 \longrightarrow \{\mathbb{F}\text{-rings}\}$$

(see [**Ha07**], §2.3, Example 2), which has a right adjoint functor, namely, the functor $-_{[1],[1]}$ that takes any $\mathbb{F}$-ring A to the monoid $A_{[1],[1]} = \mathrm{Hom}_A([1], [1])$. Composition with the functor $+0$ from Section 2.3 yields a faithful functor from $\mathcal{M}$ to the category of $\mathbb{F}$-rings, which admits a right adjoint given by composing $-_{[1],[1]}$ with the forgetful functor.

Since ideals, localization and gluing in the category of $\mathbb{F}$-rings are defined in terms of the underlying monoid $A_{[1],[1]}$, which is in complete analogy to the construction of $\mathcal{M}_0$-schemes, we obtain a faithful functor

$$\mathbb{F}\langle - \rangle : \{\mathcal{M}_0\text{-schemes}\} \longrightarrow \{\mathbb{F}\text{-schemes}\}.$$

Composing with $+0$ yields a faithful functor from the category of $\mathcal{M}$-schemes to the category of $\mathbb{F}$-schemes.

2.5. $\mathcal{M}_0$-schemes as Generalized Schemes with Zero

In this section we mention some relations between the category of $\mathcal{M}$-schemes and that of Durov's generalized schemes with zero. What follows is based on a work in progress by Peter Arndt (see [**A**]).

Associated to any monoid with zero M, we can construct the monad $T_M : \mathcal{S}ets \to \mathcal{S}ets$ that takes any set X into $T_M(X) := (M \times X)/\sim$, where we identify all the elements of the form $(0, x)$ and assume that if $X = \varnothing$ the quotient consists of one element (the empty class). This monad is algebraic since cartesian products commute with filtered colimits, so we have a functor T_- from $\mathcal{M}_0$ to the category of generalized rings with 0. The functor T_- has as a right adjoint, namely, the functor $|-|$ that associates to a generalized ring A its underlying monoid $|A| := A([1])$.

The functor T_- commutes with localizations because localizations in generalized rings are defined in terms of localizations of the underlying monoids. Thus it naturally extends to an embedding of categories

$$T_- : \{\mathcal{M}_0\text{-schemes}\} \longrightarrow \{\text{Generalized schemes with } 0\}.$$

2.6. Relation between Durov's Generalized Rings with Zero and Haran's $\mathbb{F}$-rings

In this section, we explain the relation between the categories of generalized rings with zero defined by Durov and that of $\mathbb{F}$-rings defined by Haran, following some remarks by Durov (see [**Du07**], §5.3.25) and a work in progress by Peter Arndt (see [**A**]).

Given a generalized ring with zero T, we can construct the $\mathbb{F}$-ring $T^{\mathcal{D}}$ defined by the sets of morphisms

$$T^{\mathcal{D}}_{Y,X} = \mathrm{Hom}_{\mathcal{S}\mathrm{ets}}(TX, TY)$$

for all the set morphisms between TX and TY. This construction yields a functor

$$\mathcal{D} : \{\text{Generalized rings with } 0\} \longrightarrow \{\mathbb{F}\text{-rings}\}$$

admitting a left adjoint $\mathcal{A}$, where for every $\mathbb{F}$-ring R the monad $\mathcal{A}_R$ is defined by $\mathcal{A}_R([n]) := R_{[1],[n]}$.

It is worth noting that the functor $\mathcal{D}$ that sends generalized rings to $\mathbb{F}$-rings is not monoidal. This is due to the fact that the product $\odot$ in the category of $\mathbb{F}$-rings is not compatible with the tensor product of generalized rings (see [**Ha07**], Remark 7.19, and [**A**] for further details).

It seems to be likely that the above pair of functors lift to the corresponding categories of schemes, providing an adjunction

$$\{\text{Generalized schemes with } 0\} \underset{\mathcal{A}}{\overset{\mathcal{D}}{\rightleftarrows}} \{\mathbb{F}\text{-schemes}\}.$$

One of the interesting applications of the functor $\mathcal{A}$ is that we can define a base change functor from $\mathbb{F}$-rings to usual rings by composing $\mathcal{A}$ with Durov's base change functor. This extension of scalars is left adjoint to the inclusion of rings into $\mathbb{F}$-rings. Complete details on this extension of scalars functor and its properties will be given in [**A**].

2.7. $\mathcal{M}$-schemes as Λ-schemes

In [**Bo09**], Borger describes different examples of (flat) Λ-schemes. We will recall his constructions briefly to explain why the category of $\mathcal{M}$-schemes embeds into the category of Λ-schemes.

Given a monoid A, we can endow the (flat) scheme $X = \mathrm{Spec}\,\mathbb{Z}[A]$ with the following Λ-scheme structure $\Phi = \{\varphi_p : X \to X\}_{p \text{ prime}}$: for each prime p, the endomorphism $\varphi_p : X \to X$ is induced by the algebra map $\mathbb{Z}[A] \to \mathbb{Z}[A]$ given by $a \to a^p$ for every $a \in A$. Every morphism of monoids $A \to B$ therefore induces a morphism of the associated Λ-schemes $\mathrm{Spec}\,\mathbb{Z}[B] \to \mathrm{Spec}\,\mathbb{Z}[A]$.

Borger remarks in [**Bo09**] (Section 2.3) that all small colimits and limits of Λ-schemes exist and that they commute with the base extension to $\mathbb{Z}$. Every $\mathcal{M}$-scheme X has an open cover by affine $\mathcal{M}$-schemes X_i. By definition of the base extension to $\mathbb{Z}$ this yields an open cover of $X_{\mathbb{Z}}$ by $X_{i,\mathbb{Z}}$. Since the $X_{i,\mathbb{Z}}$ have the structure of a Λ-scheme and $X_{\mathbb{Z}}$ is the limit over the $X_{i,\mathbb{Z}}$ and their intersections, X inherits the structure of a Λ-scheme. This yields the functor

$$\mathcal{B} : \{\mathcal{M}\text{-schemes}\} \longrightarrow \{\Lambda\text{-schemes}\}.$$

Note, however, that $\mathcal{B}$ is not essentially surjective, as certain quotients of Λ-schemes are Λ-schemes that are not induced by an $\mathcal{M}$-scheme. See [**Bo09**] (Sections 2.5 and 2.6) for details.

REMARK. As mentioned to us by James Borger, there is also a functor

$$\mathcal{B}_0 : \{\mathcal{M}_0\text{-schemes}\} \longrightarrow \{\Lambda\text{-schemes}\}$$

constructed essentially in the same way as $\mathcal{B}$, except that in the construction of the monoid algebra $\mathbb{Z}[A]$ one needs to identify the zero element of A with the zero of $\mathbb{Z}$. The triangle formed by the functors $+0$, $\mathcal{B}$ and $\mathcal{B}_0$ is obviously commutative.

2.8. Toric Varieties and Affinely Torified Varieties

As was mentioned in Section 1.7, the orbit decomposition $\coprod_{\tau \in \Delta} T_\tau = \operatorname{Spec} \mathbb{Z}[A_\tau^\times] \longrightarrow X$ provides an affine torification of a toric variety X with fan Δ (for details, see [**LL**], §1.3.3). Moreover, every toric morphism between toric varieties induces a morphism between the corresponding affinely torified varieties.

In other words, we obtain an embedding of categories

$$\iota : \{\text{Toric varieties}\} \longrightarrow \{\text{Affinely torified varieties}\}.$$

2.9. Relation between Affinely Torified Varieties, S-varieties and Their Variations

The relation between affinely torified varieties and S-varieties or CC-varieties is established in [**LL**] (§2.2, §3.3). Theorems 3.11 and 2.10 in [**LL**] provide embeddings $\mathcal{S}$ and $\mathcal{L}$ of the category of affinely torified varieties into the category of S-varieties resp. CC-varieties.

In order to show that the above functors define varieties over $\mathbb{F}_1$ in the sense of Soulé or Connes-Consani, it is necessary to show that the universal property holds, which boils down to proving that a certain morphism $\varphi_{\mathbb{C}} : X \to V$ of affine complex varieties is actually defined over the integers, where X is an affinely torified variety and V is an arbitrary variety. The main idea of the proof is to consider for each irreducible component of X the unique open torus T_i in the torification of X that is contained in the irreducible component (see [**LL**], Corollary 1.4). Since a split torus satisfies the universal property of an S-variety resp. a CC-variety (see [**So04**] and [**CC08**]), the morphism $\varphi_{\mathbb{C}|T_i}$ is defined over $\mathbb{Z}$ and extends to a rational function ψ on X defined over the integers with the locus of poles contained in the complement of T_i. But the extension to $\mathbb{C}$ of the rational function is $\varphi_{\mathbb{C}}$, which has no poles. Thus ψ cannot have a pole and is a morphism of schemes.

Note that this proof, which works for both S-varieties and CC-varieties, can also be adopted to S*-varieties. This yields an embedding $\mathcal{S}^*$ of the category of affinely torified varieties into the category of S*-varieties.

In [**LL**], we find the construction of a partial functor $\mathcal{F}_{CC \to S}$ mapping CC-varieties to "S-objects," which, however, do not have to satisfy the universal property of an S-variety. In a similar fashion, replacing the complex algebra $\mathcal{A}_X$ by the complexified scheme $X_{\mathbb{Z}} \otimes \mathbb{C}$, or conversely replacing the complex scheme $X_{\mathbb{C}}$ by the algebra of global sections, provides ways to compare the categories of CC-varieties and S*-varieties. But there are technical differences between the two categories that prevent this correspondence from defining a functor. Namely, the notions of gluing affine pieces are different and it is not clear how to define the grading for an CC-variety. Regarding the other direction, going from the complex scheme $X_{\mathbb{C}}$ to the algebra of global functions is a loss of information if the CC-variety is not affine. All these issues suggest that the reader should consider these categories as similar in spirit, but technically different.

2.10. Generalized Torified Schemes and CC-schemes

Torified schemes are closely connected to Connes and Consani's notion of $\mathbb{F}_1$-schemes. In this section, we introduce a generalization of the notion of a torified scheme which allows the sheaves to have multiplicative torsion and which provides an easier setting to compare the category of torified schemes with the category of CC-schemes.

A **generalized torified scheme** is a triple $\mathcal{X} = (\widetilde{X}, X, e_X)$, where $\widetilde{X}$ is a geometric $\mathcal{M}_0$-scheme, X is a scheme, and $e_X : \widetilde{X}_{\mathbb{Z}} \to X$ is a morphism of schemes such that for every field k the map $e_X(k) : \widetilde{X}_{\mathbb{Z}}(k) \to X(k)$ is a bijection. A **morphism of torified schemes** is a pair of morphisms $\widetilde{\varphi} : \widetilde{X} \to \widetilde{Y}$ (morphism of $\mathcal{M}_0$-schemes) and $\varphi : X \to Y$ (morphism of schemes) such that the diagram

$$\begin{array}{ccc} \widetilde{X}_{\mathbb{Z}} & \xrightarrow{\widetilde{\varphi}} & \widetilde{Y}_{\mathbb{Z}} \\ {\scriptstyle e_X}\downarrow & & \downarrow{\scriptstyle e_Y} \\ X & \xrightarrow{\varphi} & Y. \end{array}$$

commutes. We denote the natural inclusion of the category of affinely torified varieties into the category of generalized torified schemes by ι. We also have an obvious inclusion

$$\mathcal{F}' : \{\mathcal{M}_0\text{-schemes}\} \longrightarrow \{\text{generalized torified schemes}\}$$

constructed in the same fashion as the functor $\mathcal{F}$ in Section 2.3.

Let X be a scheme together with a torification $e_X : T = \coprod_i \mathbb{G}_m^{d_i} \longrightarrow X$. Then $T = \widetilde{X}_{\mathbb{Z}}$ for the $\mathcal{M}_0$-scheme $\widetilde{X} = \coprod_i \mathbb{G}_{m,\mathcal{M}_0}^{d_i}$. Using this description, the relation between (generalized) torified schemes and CC-schemes becomes apparent.

THEOREM 2.10.1. *The functor*

$$\mathcal{I} : \{\text{generalized torified schemes}\} \longrightarrow \{\text{CC-schemes}\}$$
$$(\widetilde{X}, X, e_X) \longmapsto (\underline{\widetilde{X}}(-), \underline{X}(-), \underline{e}_X)$$

is an equivalence of categories.

PROOF. First of all note that the objects of both categories satisfy the condition $\widetilde{X}_{\mathbb{Z}}(k) = X(k)$ for every field k. The theorem follows from the facts that every $\mathcal{M}_0$-scheme $\widetilde{X}(-)$ is represented by a geometric $\mathcal{M}_0$-scheme ([**CC09**], Proposition

3.16) and that a natural transformation of representable functors is induced by a morphism between the representing objects (Yoneda's lemma).

This theorem establishes toric varieties, Grassmann and Schubert varieties and split reductive groups as CC-schemes. We have an immediate equivalence $\mathcal{I} \circ \mathcal{F}' \simeq \mathcal{F}$.

REMARK 2.10.2. If $\mathcal{X} = (\widetilde{X}, X, e_X)$ is a generalized torified scheme, then $\mathcal{X}$ is torified if and only if there is an isomorphism $\widetilde{X} \cong \coprod_{x \in \widetilde{X}} \mathbb{G}_{m,\mathcal{M}_0}^{\mathrm{rk}x}$, where $\mathrm{rk}x$ is the rank of the group $\mathcal{O}_{\widetilde{X},x}^{\times}$, and if the restriction $e_{X \,|\, \mathbb{G}_m^{\mathrm{rk}x}}$ is an immersion for every $x \in \widetilde{X}$. As a consequence we obtain the following result.

COROLLARY 2.10.3. *Let $\mathcal{X} = (\widetilde{X}, X, e_X)$ be a CC-scheme. If $\widetilde{X}$ is integral and torsion-free and if $e_{X \,|\, \widetilde{Y}_{\mathbb{Z}}}$ is an immersion for every connected component $\widetilde{Y}$ of $\widetilde{X}$, then $X = \mathcal{X}_{\mathbb{Z}}$ is torified.*

REMARK 2.10.4. The torification of $\mathrm{Gr}(2,4)$ given by a Schubert decomposition is not affine (see [**LL**], Section 1.3.4), making it unlikely that the corresponding CC-scheme has an open cover by open sub-CC-schemes of the form $(\widetilde{X}, X, e_X)$ such that both $\widetilde{X}$ and X are affine. This is one reason for the flexibility of CC-schemes, and it hints that the category of CC-schemes is different from categories that are obtained by gluing affine pieces. In particular, see Remark 1.6.1.

2.11. The Map of $\mathbb{F}_1$-land

The functors that we described in the previous sections are illustrated in the commutative diagram (Figure 1) of categories and functors. Note that we place the category of schemes on the outside in order to have a better overview of the different categories. In Figure 1, all the extensions of scalars from $\mathbb{F}_1$ to $\mathbb{Z}$ are denoted by $- \otimes_{\mathbb{F}_1} \mathbb{Z}$, and the functor ι is the canonical inclusion of (affinely) torified schemes into generalized torified schemes. All the other functors have been defined in the previous sections. As we are dealing with categories and functors, the diagram is commutative only up to isomorphism. Some of the functors admit adjoints also described in the paper, that in many cases have been left out of the diagram for the sake of clarity. The dashed arrows corresponding to the functors $\mathcal{D}$ and $\mathcal{A}'$ represent a work in progress.

A few remarks about the commutativity (up to isomorphism) of the diagram are in order. Any path that starts at toric varieties and ends at schemes will always produce the toric variety itself, so all these possible paths are equivalent. Commutativity of subdiagrams involving S-varieties and CC-varieties was proved in [**LL**] and extends verbatim to S*-varieties. The commutativity of the triangle involving CC-schemes, generalized schemes with 0 and H-schemes (also involving the functor $\mathcal{A}$ adjoint of $\mathcal{D}$) is explained in Arndt's work [**A**]. The equivalence $(- \otimes_{\mathbb{F}_1} \mathbb{Z}) \circ \mathcal{B} \simeq (- \otimes_{\mathbb{F}_1} \mathbb{Z}) \circ \mathcal{M}_1$ follows immediately from the definitions.

2.12. Algebraic Groups over $\mathbb{F}_1$

The aim of Connes and Consani's paper [**CC08**] as described in Section 1.5.3 was to realize Tits' idea from [**Ti57**] of giving the Weyl group of a split reductive group scheme (see [**SGA3**], Exposé XIX, Def. 2.7) an interpretation as a "Chevalley group over $\mathbb{F}_1$." Connes and Consani obtained partial results, namely, that every

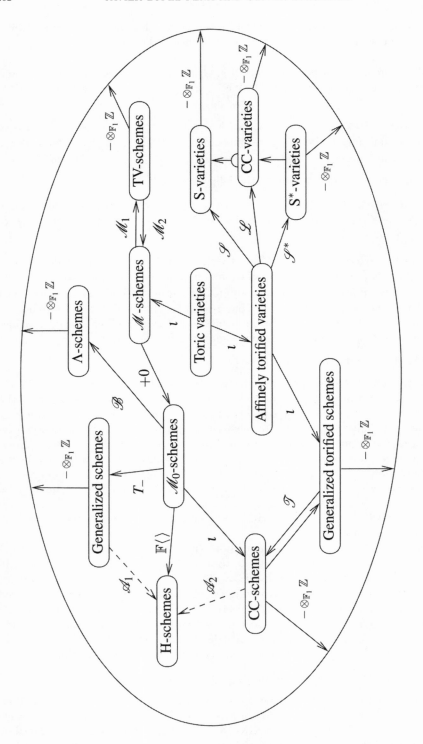

FIGURE 1   The map of $\mathbb{F}_1$-land.

split reductive group scheme G has a model (as a CC–variety) "over $\mathbb{F}_{1^2}$" and that the normalizer N of a maximal split torus T of G can be defined as a group object over $\mathbb{F}_{1^2}$. However, there is no model of G as an algebraic group over $\mathbb{F}_1$ or $\mathbb{F}_{1^2}$. For a further discussion of possibilities and limitation in this context, see [**LL**] (Section 6.1).

In the category of CC-schemes, the same reasons prevent split reductive group schemes (except for tori) from having as a model an algebraic group over $\mathbb{F}_1$. However, the second author of this paper showed that this category is flexible enough to invent new notions of morphisms that yield the desired results. We review the definitions and the main result from [**Lo09**].

Let $\widetilde{X}$ be an $\mathcal{M}_0$-scheme. Recall the definition of the rank $\mathrm{rk}x$ of $x \in \widetilde{X}$ from Remark 2.10.2. We define the sub-$\mathcal{M}_0$-scheme $\widetilde{X}^{\mathrm{rk}} \hookrightarrow \widetilde{X}$ as the disjoint union $\coprod \mathrm{Spec}_{\mathcal{M}_0} \mathcal{O}^{\times}_{\widetilde{X},x}$ over all points $x \in \widetilde{X}$ of minimal rank. A *strong morphism* between $\mathbb{F}_1$-schemes $(\widetilde{Y}, Y, e_Y)$ and $(\widetilde{X}, X, e_X)$ is a pair $(\widetilde{\varphi}, \varphi)$, where $\widetilde{\varphi} : \widetilde{Y}^{\mathrm{rk}} \to \widetilde{X}^{\mathrm{rk}}$ is a morphism of $\mathcal{M}_0$-schemes and $\varphi : Y \to X$ is a morphism of schemes such that the diagram

$$
\begin{array}{ccc}
\widetilde{Y}^{\mathrm{rk}}_{\mathbb{Z}} & \xrightarrow{\tilde{f}_{\mathbb{Z}}} & \widetilde{X}^{\mathrm{rk}}_{\mathbb{Z}} \\
{\scriptstyle e_Y}\downarrow & & \downarrow{\scriptstyle e_X} \\
Y & \xrightarrow{f} & X
\end{array}
$$

commutes.

The morphism $\mathrm{spec}\,\mathcal{O}^{\times}_{X,x} \to *_{\mathcal{M}_0}$ to the terminal object $*_{\mathcal{M}_0} = \mathrm{Spec}_{\mathcal{M}_0}\{0,1\}$ in the category of $\mathcal{M}_0$-schemes induces a morphism

$$
t_{\widetilde{X}} : \quad \widetilde{X}^{\mathrm{rk}} = \coprod_{x \in \widetilde{X}^{\mathrm{rk}}} \mathrm{spec}\,\mathcal{O}^{\times}_{X,x} \quad \longrightarrow \quad *_{\widetilde{X}} := \coprod_{x \in \widetilde{X}^{\mathrm{rk}}} *_{\mathcal{M}_0}.
$$

Given a morphism $\widetilde{\varphi} : \widetilde{Y}^{\mathrm{rk}} \to \widetilde{X}^{\mathrm{rk}}$ of $\mathcal{M}_0$-schemes, there is thus a unique morphism $t_{\widetilde{\varphi}} : *_{\widetilde{Y}} \to *_{\widetilde{X}}$ such that $t_{\widetilde{\varphi}} \circ t_{\widetilde{Y}} = t_{\widetilde{X}} \circ \widetilde{\varphi}$. Let X^{rk} denote the image of $e_X : \widetilde{X}^{\mathrm{rk}}_{\mathbb{Z}} \to X$. A *weak morphism* between $\mathbb{F}_1$-schemes $(\widetilde{Y}, Y, e_Y)$ and $(\widetilde{X}, X, e_X)$ is a pair $(\widetilde{\varphi}, \varphi)$, where $\widetilde{\varphi} : \widetilde{Y}^{\mathrm{rk}} \to \widetilde{X}^{\mathrm{rk}}$ is a morphism of $\mathcal{M}_0$-schemes and $\varphi : Y \to X$ is a morphism of schemes such that the diagram

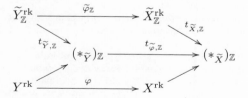

commutes.

If $\mathcal{X} = (\widetilde{X}, X, e_X)$ is an CC-scheme, then we define *the set of* $\mathbb{F}_1$*-points* $\mathcal{X}(\mathbb{F}_1)$ as the set of strong morphism from $(*_{\mathcal{M}_0}, \mathrm{Spec}\,\mathbb{Z}, \mathrm{id}_{\mathrm{Spec}\,\mathbb{Z}})$ to $(\widetilde{X}, X, e_X)$. An *algebraic group over* $\mathbb{F}_1$ is a group object in the category whose objects are $\mathbb{F}_1$-schemes and whose morphisms are weak morphisms. The base extension functor $- \otimes_{\mathbb{F}_1} \mathbb{Z}$ from this category to the category of schemes is given by sending $\mathcal{X} = (\widetilde{X}, X, e_X)$ to $\mathcal{X}_{\mathbb{Z}} = X$. On the other hand, the group law of an algebraic group $\mathcal{G}$ over $\mathbb{F}_1$ induces a group structure on the set $\mathcal{G}(\mathbb{F}_1)$. We realize Tits' idea in the following form.

THEOREM 2.12.1. ([**Lo09**], Theorem 7.9) *For every split reductive group scheme G with Weyl group W, there exists an algebraic group $\mathcal{G}$ over $\mathbb{F}_1$ such that $\mathcal{G}_{\mathbb{Z}} \simeq G$ as group schemes and $\mathcal{G}(\mathbb{F}_1) \simeq W$ as groups.*

References

[A]   P. Arndt, *Comparison of $\mathbb{F}_1$-geometries*, work in progress.

[Bo09]   J. Borger, *Λ-rings and the field with one element*, arXiv: 0906.3146v1 [math.NT], 2009.

[BS08]   J. Borger, B. de Smit, *Galois theory and integral models of Λ-rings*, Bull. Lond. Math. Soc. **40** (2008), 439–446.

[BC95]   J. Bost, A. Connes, *Hecke algebras, type III factors and phase transitions with spontaneous symmetry breaking in number theory*, Selecta Math. (N.S.) **1** (1995), 411–457.

[C04]   H. Cohn, *Projective geometry over $\mathbb{F}_1$ and the Gaussian binomial coefficients*, Am. Math. Monthly **111** (2004), 487–495.

[CC08]   A. Connes, C. Consani, *On the notion of geometry over $\mathbb{F}_1$*, arXiv: to appear in J. Algebraic Geometry.

[CC09]   A. Connes, C. Consani, *Schemes over $\mathbb{F}_1$ and zeta functions*, to appear in Compositio Math.

[CC09b] A. Connes, C. Consani, *Characteristic 1, entropy and the absolute point*, arXiv:0911. 3537v1 [math.AG], 2009.

[CC10]   A. Connes, C. Consani, *The hyperring of adèle classes*, arXiv:1001.4260v2 [math.AG], 2010.

[CCM09] A. Connes, C. Consani, M. Marcolli, *Fun with $\mathbb{F}_1$*, J. Number Theory **129** (2009), 1532–1561.

[D05]   A. Deitmar *Schemes over $\mathbb{F}_1$, number fields and function fields—two parallel worlds*, Progr. Math. **239** (2005).

[D06]   A. Deitmar, *Remarks on zeta functions and K-theory over $\mathbb{F}_1$*, Proc. Japan Acad. Ser. A Math. Sci. **82** (2006).

[D07]   A. Deitmar, *$\mathbb{F}_1$-schemes and toric varieties*, Contribs. Algebra Geometry **49** (2008), 517–525.

[SGA3]   M. Demazure, A. Grothendieck, *Séminaire de Géométrie Algébrique, Schémas en Groupes*, Lecture Notes in Mathematics 153. Springer-Verlag, 1962/64.

[D91]   C. Deninger, *On the Γ-factors attached to motives*, Invent. Math. **104** (1991), 245–261.

[D92]   C. Deninger, *Local L-factors of motives and regularized determinants*, Invent. Math. **107** (1992), 135–150.

[D94]   C. Deninger, *Motivic L-functions and regularized determinants*, in Motives (Seattle, WA, 1991), Proc. Sympos. Pure Math. 55, pp. 707–743, 1994.

[Du07]   N. Durov, *A new approach to Arakelov geometry*, PhD thesis, arXiv: 0704.2030v1 [math.AG], 2007.

[Fr09]   J. Fresán, *Compactification projective de Spec $\mathbb{Z}$ (d'après Durov)*, Master's thesis, arXiv: 0908.4059v1 [math.NT], 2009.

[Fu93]   W. Fulton, *Introduction to toric varieties*, Princeton University Press, 1993.

[Ha04]   K. Habiro, *Cyclotomic completions of polynomial rings*, Publ. RIMS Kyoto Univ. **40** (2004), 1127–1146.

[Ha07]   M. J. Shai Haran, *Non-additive geometry*, Compositio Math. **143** (2007), 618–688.

[Ha09]   M. J. Shai Haran, *Non-additive prolegomena (to any future arithmetic that will be able to present itself as a geometry)*, arXiv:0911.3522v1 [math.NT] 2009.

[KS]   M. Kapranov, A. Smirnov, *Cohomology determinants and reciprocity laws: number field case*, unpublished preprint.

[Ka94]   K. Kato, *Toric singularities*, Amer. J. Math. **116** (1994), 1073–1099.

[Ku92]   N. Kurokawa, *Multiple zeta functions: an example*, In Zeta functions in geometry (Tokyo, 1990), Adv. Stud. Pure Math. 21, pp. 219–226, 1992.

[Ku05]   N. Kurokawa, *Zeta functions over $\mathbb{F}_1$*, Proc. Japan Acad. Ser. A Math. Sci. **81** (2005), 180–184.

[KOW02] N. Kurokawa, H. Ochiai, M. Wakayama, *Absolute derivations and zeta functions*, Doc. Math. (Extra Volume: Kazuya Kato's Fiftieth Birthday) (2003), 565–584.

[LB09]   L. Le Bruyn, *(non)commutative f-un geometry*, arXiv:0909.2522v1 [math.RA], 2009.

[Le09]   P. Lescot, *Algèbre absolue*, arXiv:0911.1989v1 [math.RA], 2009.

[LL]   J. López Peña, O. Lorscheid, *Torified varieties and their geometries over* $\mathbb{F}_1$, to appear in Mathematische Zeitschrift.

[Lo09]   O. Lorscheid, *Algebraic groups over the field with one element*, arXiv:0907.3824v1 [math.AG], 2009.

[MC98]   S. Mac Lane *Categories for the working mathematician*, second edition, Springer-Verlag, 1998.

[Ma00]   V. Maillot, *Géométrie d'Arakelov des variétés toriques et fibrés en droites intégrables*, Mém. Soc. Math. Fr. (N.S.) **80** (2000).

[Ma95]   Y. Manin, *Lectures on zeta functions and motives (according to Deninger and Kurokawa)*, Astrisque **228** (1995), 121–163.

[Ma08]   Y. Manin, *Cyclotomy and analytic geometry over* $\mathbb{F}_1$, arXiv:0809.1564v1 [math.AG], 2008.

[Ma09]   M. Marcolli, *Cyclotomy and endomotives*, p–Adic Numbers, Ultrametric Analysis and Applications **1** (2009), 217–263.

[Ma07]   F. Marty, *Relative Zariski open objects*, arXiv:0712.3676v3 [math.AG], 2007.

[Ma09b]   F. Marty, *Smoothness in relative geometry*, arXiv:0812.1152v1 [math.AG], 2009.

[O88]   T. Oda, *Convex bodies and algebraic geometry*, Springer-Verlag, New York, 1988.

[So04]   C. Soulé, *Les variétés sur le corps à un élément*, Mosc. Math. J. **4** (2004), 217–244.

[So09]   C. Soulé, *Lectures on algebraic varieties over* $\mathbb{F}_1$, this volume.

[St51]   R. Steinberg, *A geometric approach to the representations of the full linear group over a Galois field*, Trans. Am. Math. Soc. **71** (1951), 274–282.

[St72]   R. Street, *The formal theory of monads*, J. Pure Appl. Alg. **2** (1972), 149–168.

[Ti57]   J. Tits, *Sur les analogues algébriques des groupes semi-simples complexes*, Colloque d'algèbre supérieure, tenu à Bruxelles du 19 au 22 décembre 1956 (1957), pp. 261–289.

[TV08]   B. Toën, M. Vaquié, *Au-dessous de Spec* $\mathbb{Z}$, J. of K-Theory (2008), 1–64.

[Ve10]   A. Vezzani, *Deitmar's versus Toën-Vaquié's schemes over* $\mathbb{F}_1$, preprint, 2010.

QUEEN MARY UNIVERSITY OF LONDON, SCHOOL OF MATHEMATICAL SCIENCES, MILE END ROAD, LONDON E14NS, UK.
E-mail address: jlopez@maths.qmul.ac.uk

MAX-PLANCK INSTITUT FÜR MATHEMATIK, VIVATSGASSE, 7. D-53111, BONN, GERMANY.
E-mail address: oliver@mpim-bonn.mpg.de

Lectures on Algebraic Varieties over $\mathbb{F}_1$

Christophe Soulé

Notes by David Penneys

ABSTRACT. In this paper we present the approach to varieties over the field with one element and their zeta functions as developed in [**So**], with minor changes.

1. Introduction

This paper is issued from a series of lectures given at the Seventh Annual Spring Institute on Noncommutative Geometry and Operator Algebras, 2009, at Vanderbildt University. They present a summary of the author's article [**So**], with a few modifications. These are made in order to take into account corrections and improvements due to A. Connes and C. Consani [**CC1,CC2**]. The notion of *affine gadget over* $\mathbb{F}_1$ introduced below (Definition 3.2) lies somewhere in between the notion of "truc" ([**So**], 3.1., Definition 1; see, however, (i) in [**So**]) and A. Connes and C. Consani's notion of "gadget over $\mathbb{F}_1$" [**CC1**]. We also added a discussion of the article of R. Steinberg [**St**] on the analogy between symmetric groups and general linear groups over finite fields.

Acknowledgments. I thank P. Cartier, A. Connes, C. Consani, J. López Peña and O. Lorscheid for helpful discussions.

2. Preliminaries

2.1. An Analogy

There is an analogy between the symmetric group Σ_n on n letters and the general linear group $GL(n, \mathbb{F}_q)$, where $q = p^k$ for a prime p. One of the first to write about this analogy was R. Steinberg in 1951 [**St**]. He used it to get a result in representation theory. This goes as follows.

For all $r \in \mathbb{N}$, define

$$[r] = q^{r-1} + q^{r-2} + \cdots + q + 1 = \frac{q^r - 1}{q - 1} \text{ and}$$

$$\{r\} = \prod_{i=1}^{r} [i].$$

Let $n \geq 1$ and $G = GL(n, \mathbb{F}_q)$. Let $\nu = (\nu_1, \ldots, \nu_n)$ be a partition of n, i.e.,

$$n = \sum_{i=1}^{n} \nu_i, \text{ where } 0 \leq \nu_1 \leq \nu_2 \leq \cdots \leq \nu_n.$$

Write every element of G as an n by n matrix of blocks of size $\nu_i \times \nu_j, 1 \leq i, j \leq n$. Consider the (parabolic) subgroup of such upper triangular matrices

$$G(\nu) = \left\{ g = \begin{pmatrix} * & * & * \\ 0 & * & * \\ 0 & 0 & * \end{pmatrix} \right\} \subset G.$$

One checks that

$$\#G/G(\nu) = \frac{\{n\}}{\prod_{i=1}^{n} \{\nu_i\}}.$$

Let

$$C(\nu) = \text{Ind}_{G(\nu)}^{G} \mathbf{1} = \mathbb{C}[G/G(\nu)]$$

be the induced representation of the trivial representation of $G(\nu)$.

THEOREM 2.1. *Let ν be a partition of n and $\lambda_i = \nu_i + i - 1$ for all $i \geq 1$. The virtual representation*

$$\Gamma(\nu) = \sum_{\kappa} \text{sgn}(\kappa_1, \ldots, \kappa_n) C(\lambda_1 - \kappa_1, \ldots, \lambda_n - \kappa_n)$$

is an irreducible representation of G (i.e., its character is the character of an irreducible representation), when $\kappa = (\kappa_1, \ldots, \kappa_n)$ runs over all $n!$ permutations of $0, 1, \ldots, n - 1$, with the convention that if $\lambda_i - \kappa_i < 0$ for some i, then $C(\lambda_1 - \kappa_1, \ldots, \lambda_n - \kappa_n) = 0$. Moreover, if $\Gamma(\mu) = \Gamma(\nu)$, then $\mu = \nu$.

To prove this result we consider the symmetric group $H = \Sigma_n$ and its subgroup

$$H(\nu) = \Sigma_{\nu_1} \times \cdots \times \Sigma_{\nu_n}.$$

Then

$$\#H/H(\nu) = \frac{n!}{\prod_{i=1}^{n} \nu_i!}.$$

Set

$$D(\nu) = \text{Ind}_{H(\nu)}^{H} \mathbf{1} = \mathbb{C}[H/H(\nu)]$$

and consider the virtual representation

$$\Delta(\nu) = \sum_{\kappa} \text{sgn}(\kappa_1, \ldots, \kappa_n) D(\lambda_1 - \kappa_1, \ldots, \lambda_n - \kappa_n).$$

THEOREM 2.2. (Frobenius, 1898) *$\Delta(\nu)$ is an irreducible representation. Moreover, $\Delta(\mu) = \Delta(\nu)$ implies $\mu = \nu$.*

The proof of Theorem 2.1 follows from this theorem and the following lemma.

LEMMA 2.3. *Let $x \mapsto \psi(\nu, x)$ be the character of $C(\nu)$ and $x \mapsto \varphi(\nu, x)$ the character of $D(\nu)$. Then for all μ, ν, we have*

$$\frac{1}{\#G} \sum_{x \in G} \psi(\nu, x)\psi(\mu, x) = \frac{1}{\#H} \sum_{x \in H} \varphi(\nu, x)\varphi(\mu, x).$$

PROOF. The left hand side (resp. the right hand side) of this equality is the number of double cosets of G (resp. H) modulo $G(\mu)$ and $G(\nu)$ (resp. $H(\mu)$ and $H(\nu)$). We have an inclusion $H \subset G$ such that $H(\mu) = G(\mu) \cap H$, and the Bruhat decomposition implies that the map

$$H(\mu)\backslash H/H(\nu) \to G(\mu)\backslash G/G(\nu)$$

is a bijection.

Let $\chi(\nu, x)$ be the character of $C(\nu)$. From the lemma and Frobenius' theorem we deduce that

$$\frac{1}{\#G} \sum_{x \in G} \chi(\nu, x)\overline{\chi(\mu, x)} = \delta_{\mu, \nu},$$

and, to get Theorem 2.1, it remains to check that $\chi(\nu, 1) > 0$.

2.2. The Field $\mathbb{F}_1$

In [**T**] Tits noticed that the analogy above extends to an analogy between the group $G(\mathbb{F}_q)$ of points in $\mathbb{F}_q$ of a Chevalley group scheme G and its Weyl group W. He had the idea that there should exist a "field of characteristic one" $\mathbb{F}_1$ such that

$$W = G(\mathbb{F}_1).$$

He showed, furthermore, that when q goes to 1, the finite geometry attached to $G(\mathbb{F}_q)$ becomes the finite geometry of the Coxeter group W.

Thirty five years later, Smirnov [**S**], and then Kapranov and Manin wrote about $\mathbb{F}_1$, viewed as the missing ground field over which number rings are defined. Since then several people have studied $\mathbb{F}_1$ and tried to define an algebraic geometry over it. Today, there are at least seven different definitions of such a geometry and a few studies comparing them.

3. Affine Varieties over $\mathbb{F}_1$

3.1. Schemes as Functors

We shall propose a definition for varieties over $\mathbb{F}_1$ based on three remarks. The first one is that schemes can be defined as covariant functors from rings to sets (satisfying some extra properties; see [**DG**]).

The second remark is that extensions of scalars can be defined in terms of functors. Namely, let k be a field, and let Ω be a k-algebra. If X is a variety over k, we denote by $X_\Omega = X \otimes_k \Omega$ ($= X \times_{\mathrm{Spec}(k)} \mathrm{Spec}(\Omega)$) its extension of scalars from k to Ω. Let $\underline{X}$ be the functor from k-algebras to sets defined by X and $\underline{X}_\Omega$ the functor from Ω-algebras to sets defined by X_Ω. Let β be the functor $_k\mathsf{Alg} \to {}_\Omega\mathsf{Alg}$ given by $R \mapsto R \otimes_k \Omega$.

PROPOSITION 3.1.
(1) *There is a natural transformation* $i\colon \underline{X} \to \underline{X}_\Omega \circ \beta$ *of functors* $_k\mathsf{Alg} \to \mathsf{Set}$. *For any k-algebra R the map* $\underline{X}(R) \to \underline{X}_\Omega(R_\Omega)$ *is injective.*
(2) *For any scheme S over Ω and any natural transformation* $\varphi\colon \underline{X} \to \underline{S} \circ \beta$, *there exists a unique algebraic morphism* $\varphi_\Omega\colon X_\Omega \to S$ *such that* $\varphi = \varphi_\Omega \circ i$. *In other words the following diagram is commutative:*

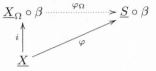

We deduce from this proposition that, if X is a variety over $\mathbb{F}_1$,
(1) X should determine a covariant functor $\underline{X}$ from $\mathbb{F}_1$-algebras to Set, and
(2) X should define a variety $X \otimes_{\mathbb{F}_1} \mathbb{Z}$ over $\mathbb{Z}$ by some universal property similar to the one in Proposition 3.1 (with $k = \mathbb{F}_1$ and $\Omega = \mathbb{Z}$).

3.2. A Definition

A third remark is that we know what should play the role of finite extensions of $\mathbb{F}_1$. According to both Kapranov-Smirnov [**KS**] and Kurokawa-Ochiai-Wakayama [**KOW**], the category of finite extensions of $\mathbb{F}_1$ is Ab_f, the category of finite abelian groups. If $D \in \mathsf{Ab}_f$, we define the extension of scalars of D from $\mathbb{F}_1$ to $\mathbb{Z}$ as the group-algebra $D \otimes_{\mathbb{F}_1} \mathbb{Z} = \mathbb{Z}[D]$. For example, $\mathbb{F}_{1^n} = \mathbb{Z}/n$, and

$$\mathbb{F}_{1^n} \otimes_{\mathbb{F}_1} \mathbb{Z} = \mathbb{Z}[T]/(T^n - 1)\,.$$

We now make the following definition.

DEFINITION 3.2. An *affine gadget over* $\mathbb{F}_1$ is a triple $X = (\underline{X}, \boldsymbol{A}_X, e_X)$ consisting of
(1) a covariant functor $\underline{X}\colon \mathsf{Ab}_f \to \mathsf{Set}$,
(2) a $\mathbb{C}$-algebra $\boldsymbol{A}_X$, and
(3) a natural transformation $e_X\colon \underline{X} \Rightarrow \mathrm{Hom}(\boldsymbol{A}_X, \mathbb{C}[-])$.

In other words, if $D \in \mathsf{Ab}_f$ and $P \in \underline{X}(D)$, we get a morphism of complex algebras $\boldsymbol{A}_X \to \mathbb{C}[D]$ that we write as $e_X(P)(f) = f(P) \in \mathbb{C}[D]$, the *evaluation* of $f \in \boldsymbol{A}_X$ at the point P.

EXAMPLE 3.3. Assume V is an affine algebraic variety over $\mathbb{Z}$. Then we can define an affine gadget $X = \mathcal{G}(V)$ as follows:
(1) $\underline{X}(D) = V(\mathbb{Z}[D])$,
(2) $\boldsymbol{A}_X = \Gamma(V_\mathbb{C}, \mathcal{O})$, and
(3) given $P \in V(\mathbb{Z}[D]) \subset V(\mathbb{C}[D])$ and $f \in \boldsymbol{A}_X$, then $f(P) \in \mathbb{C}[D]$ is the usual evaluation of the function f at P.

DEFINITION 3.4. A *morphism* of affine gadgets $\phi\colon X \to Y$ consists of
(1) a natural transformation $\underline{\phi}\colon \underline{X} \to \underline{Y}$, and
(2) a morphism of algebras $\phi^*\colon \boldsymbol{A}_Y \to \boldsymbol{A}_X$,
(3) which are compatible with evaluations, i.e., if $P \in \underline{X}(D)$ and $f \in \boldsymbol{A}_Y$, then $f(\underline{\phi}(P)) = (\phi^*(f))(P)$.

DEFINITION 3.5. An *immersion* is a morphism (ϕ, ϕ^*) such that both ϕ and ϕ^* are injective.

We can now define affine varieties over $\mathbb{F}_1$ as a special type of affine gadget.

DEFINITION 3.6. An *affine variety over* $\mathbb{F}_1$ is an affine gadget $X = (\underline{X}, \boldsymbol{A}_X, e_X)$ over $\mathbb{F}_1$ such that
(1) for any $D \in \mathsf{Ab}_f$, the set $\underline{X}(D)$ is finite;
(2) there exists an affine variety $X_{\mathbb{Z}} = X \otimes_{\mathbb{F}_1} \mathbb{Z}$ over $\mathbb{Z}$ and an immersion of affine gadgets $i \colon X \to \mathcal{G}(X_{\mathbb{Z}})$ [in particular, the points in the variety over $\mathbb{F}_1$ are points in $X_{\mathbb{Z}}$] satisfying the following universal property: for every affine variety V over $\mathbb{Z}$ and every morphism of affine gadgets $\varphi \colon X \to \mathcal{G}(V)$, there exists a unique algebraic morphism $\varphi_{\mathbb{Z}} \colon X_{\mathbb{Z}} \to V$ such that $\varphi = \mathcal{G}(\varphi_{\mathbb{Z}}) \circ i$, i.e., the diagram

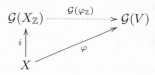

commutes.

3.3. Examples

EXAMPLE 3.7. Any finite abelian group D defines an affine variety over $\mathbb{F}_1$, denoted $\mathrm{Spec}(D)$: the functor $\underline{\mathrm{Spec}(D)}$ is the functor represented by D, the algebra is $\mathbb{C}[D]$, and the evaluation is the obvious one.

EXAMPLE 3.8. We define the multiplicative group $X = \mathbb{G}_m/\mathbb{F}_1$ as the triple $(\underline{X}, \boldsymbol{A}_X, e_X)$, where
(1) $\underline{X}(D) = D$,
(2) $\boldsymbol{A}_X$ is the algebra of continuous complex functions on the circle S^1, and
(3) if $P \in \underline{X}(D)$ and $f \in \boldsymbol{A}_X$, for every character $\chi \colon D \to \mathbb{C}^\times$, $f(P) \in \mathbb{C}[D]$ is such that $\chi(f(P)) = f(\chi(P))$.

PROPOSITION 3.9. $\mathbb{G}_m/\mathbb{F}_1$ *is an affine variety over* $\mathbb{F}_1$ *such that* $\mathbb{G}_m \otimes_{\mathbb{F}_1} \mathbb{Z} = \mathrm{Spec}(\mathbb{Z}[T, T^{-1}])$.

EXAMPLE 3.10. The affine line $\mathbb{A}^1/\mathbb{F}_1$ is defined as the triple $(\underline{X}, \boldsymbol{A}_X, e_X)$ by
(1) $\underline{X}(D) = D \amalg \{0\}$,
(2) $\boldsymbol{A}_X$ is the algebra of continuous functions on the closed unit disk which are holomorphic in the open unit disk, and
(3) if $P \in \underline{X}(D)$ and $f \in \boldsymbol{A}_X$, for any character $\chi \colon D \to \mathbb{C}^\times$, we have $\chi(f(P)) = f(\chi(P))$.

PROPOSITION 3.11. $\mathbb{A}^1/\mathbb{F}_1$ *is an affine variety over* $\mathbb{F}_1$ *with extension of scalars* $\mathbb{A}^1 \otimes_{\mathbb{F}_1} \mathbb{Z} = \mathrm{Spec}(\mathbb{Z}[T])$.

4. Varieties over $\mathbb{F}_1$

4.1. Definition

To get varieties over $\mathbb{F}_1$ (and not only affine ones), we proceed again by analogy with Proposition 3.1. Let $\mathsf{Aff}_{\mathbb{F}_1}$ be the category of affine varieties over $\mathbb{F}_1$ (a full subcategory of the category of affine gadgets).

DEFINITION 4.1. An *object over* $\mathbb{F}_1$ is a triple $X = (\underline{X}, \boldsymbol{\mathcal{A}}_X, e_X)$ consisting of
(1) a contravariant functor $\underline{X} \colon \mathsf{Aff}_{\mathbb{F}_1} \to \mathsf{Set}$,
(2) a $\mathbb{C}$-algebra $\boldsymbol{\mathcal{A}}_X$, and
(3) a natural transformation $e_X \colon \underline{X} \Rightarrow \mathrm{Hom}(\boldsymbol{\mathcal{A}}_X, \boldsymbol{\mathcal{A}}_-)$.

EXAMPLE 4.2. Assume V is an algebraic variety over $\mathbb{Z}$. Then we can define
an object $X = \mathcal{O}b(V)$ as follows:
(1) $\underline{X}(Y) = \mathrm{Hom}_{\mathbb{Z}}(Y_{\mathbb{Z}}, V)$,
(2) $\boldsymbol{\mathcal{A}}_X = \Gamma(V_{\mathbb{C}}, \mathcal{O})$, and
(3) given $u \in \mathrm{Hom}_{\mathbb{Z}}(Y_{\mathbb{Z}}, V)$ and $f \in \boldsymbol{\mathcal{A}}_X$, then $e_X(u)(f) = i^* u^*(f)$.

Morphisms and immersions of objects are defined as the corresponding notions
for affine gadgets. Finally, we have the following definition.

DEFINITION 4.3. A *variety over* $\mathbb{F}_1$ is an object $X = (\underline{X}, \boldsymbol{\mathcal{A}}_X, e_X)$ over $\mathbb{F}_1$ such
that
(1) for any $D \in \mathsf{Ab}_f$, the set $\underline{X}(\mathrm{Spec}(D))$ is finite;
(2) there exists a variety $X_{\mathbb{Z}} = X \otimes_{\mathbb{F}_1} \mathbb{Z}$ over $\mathbb{Z}$ and an immersion of objects
     $i \colon X \to \mathcal{O}b(X_{\mathbb{Z}})$ satisfying the following universal property: for every vari-
     ety V over $\mathbb{Z}$ and every morphism of objects $\varphi \colon X \to \mathcal{O}b(V)$, there exists a
     unique algebraic morphism $\varphi_{\mathbb{Z}} \colon X_{\mathbb{Z}} \to V$ such that $\varphi = \mathcal{O}b(\varphi_{\mathbb{Z}}) \circ i$.

4.2. Examples

Any affine variety X over $\mathbb{F}_1$ is also a variety over $\mathbb{F}_1$: $\underline{X}$ is the functor represented by X, $\boldsymbol{\mathcal{A}}_X$ and e_X are the obvious ones.

The following proposition (see [**So**], Proposition 5) allows one to define a variety
over $\mathbb{F}_1$ by glueing subvarieties.

PROPOSITION 4.4. *Let V be a variety over $\mathbb{Z}$ and $V = \bigcup_{i \in I} U_i$ a finite open
cover of V. Assume there is a finite family of varieties $X_i = (\underline{X}_i, \boldsymbol{\mathcal{A}}_i, e_i), i \in I$,
and $X_{ij} = (\underline{X}_{ij}, \boldsymbol{\mathcal{A}}_{ij}, e_{ij}), i \neq j$, and immersions $X_{ij} \to X_i$ and $X_i \to \mathcal{O}b(V)$ of
varieties over $\mathbb{F}_1$ such that*
(1) $X_{ij} = X_{ji}$ *and the composites $X_{ij} \to X_i \to \mathcal{O}b(V)$ and $X_{ij} \to X_j \to
     \mathcal{O}b(V)$ coincide;*
(2) *the maps $(X_{ij})_{\mathbb{Z}} \to (X_i)_{\mathbb{Z}}$ coincide with the inclusions $U_i \cap U_j \to U_i$, and
     the maps $X_i \to \mathcal{O}b(V)$ induce the inclusions $U_i \to V$.*
For any affine variety Y over $\mathbb{F}_1$ define

$$\underline{X}(Y) = \bigcup_i \underline{X}_i(Y)$$

(union in $\mathrm{Hom}_{\mathbb{Z}}(Y_{\mathbb{Z}}, V)$) and let

$$\boldsymbol{\mathcal{A}}_X = \left\{ (f_i) \in \prod_i \boldsymbol{\mathcal{A}}_i \;\middle|\; f_i|_{X_{ij}} = f_j|_{X_{ij}} \right\}.$$

*Then the object $X = (\underline{X}, \boldsymbol{\mathcal{A}}_X, e_X)$ (where e_X is the obvious evaluation) is a
variety over $\mathbb{F}_1$ and $X \otimes_{\mathbb{F}_1} \mathbb{Z}$ is canonically isomorphic to V.*

5. Zeta Functions

Let $X = (\underline{X}, \boldsymbol{A}_X, e_X)$ be a variety over $\mathbb{F}_1$. We make the following assumption.

ASSUMPTION. *There exists a polynomial* $N(x) \in \mathbb{Z}[x]$ *such that, for all* $n \geq 1$, $\#\underline{X}(\mathbb{F}_{1^n}) = N(n+1)$.

Consider the following series:

$$Z(q, T) = \exp\left(\sum_{r \geq 1} N(q^r)\frac{T^r}{r}\right).$$

Now take $T = q^{-s}$ to get a function of s and q. For every $s \in \mathbb{R}$, the function $Z(q, q^{-s})$ is meromorphic and has a pole at $q = 1$ of order $\chi = N(1)$. We let q go to 1 to get a zeta function over $\mathbb{F}_1$. We define

$$\zeta_X(s) = \lim_{q \to 1} Z(q, q^{-s})(q - 1)^\chi.$$

LEMMA 5.1. *If* $N(x) = \sum_{k=0}^{d} a_k x^k$, *then*

$$\zeta_X(s) = \prod_{k=1}^{d}(s - k)^{-a_k}.$$

PROOF. We may assume that $N(x) = x^k$. Then we have

$$Z(q, q^{-s}) = \exp\left(\sum_{r \geq 1} q^{kr}\frac{q^{-rs}}{r}\right) = \exp(-\log(1 - q^{k-s})) = \frac{1}{1 - q^{k-s}}.$$

Now we have that

$$\lim_{q \to 1} \frac{q - 1}{1 - q^{k-s}} = \frac{1}{s - k}.$$

For instance, if $X = \mathbb{G}_m/\mathbb{F}_1$, we get $\#\underline{X}(\mathbb{F}_{1^n}) = n = N(n+1)$ with $N(x) = x - 1$. Therefore

$$\zeta_X(s) = \frac{s}{s - 1}.$$

6. Toric Varieties over $\mathbb{F}_1$

6.1. Toric Varieties

Let $d \geq 1$, $N = \mathbb{Z}^d$, and $M = \mathrm{Hom}(N, \mathbb{Z})$. Let $N_\mathbb{R} = N \otimes_\mathbb{Z} \mathbb{R}$ and $M_\mathbb{R} = M \otimes_\mathbb{Z} \mathbb{R}$. We then have the duality pairing

$$\langle \cdot, \cdot \rangle \colon M_\mathbb{R} \times N_\mathbb{R} \to \mathbb{R}.$$

DEFINITION 6.1. A *cone* is a subset $\sigma \subset N_{\mathbb{R}}$ of the form

$$\sigma = \sum_{i \in I} \mathbb{R}_+ n_i,$$

where $(n_i)_{i \in I}$ is a finite family in N.

We define the *dual* and the *orthogonal* of σ by

$$\sigma^* = \{v \in M_{\mathbb{R}} \mid \langle v, x \rangle \geq 0 \text{ for all } x \in \sigma\} \text{ and}$$
$$\sigma^\perp = \{v \in M_{\mathbb{R}} \mid \langle v, x \rangle = 0 \text{ for all } x \in \sigma\},$$

respectively.

A cone is *strict* if it does not contain any line.

A *face* is a subset $\tau \subset \sigma$ such that there is a $v \in \sigma^*$ with $\tau = \sigma \cap v^\perp$.

DEFINITION 6.2. A *fan* is a finite collection $\Delta = \{\sigma\}$ of strict cones such that
(1) if $\sigma \in \Delta$, any face of σ is in Δ, and
(2) if $\sigma, \sigma' \in \Delta$, then $\sigma \cap \sigma'$ is a face of σ and σ'.

DEFINITION 6.3. Given Δ, we define a variety $\mathbb{P}(\Delta)$ over $\mathbb{Z}$ as follows. For all $\sigma \in \Delta$, consider the monoid $S_\sigma = M \cap \sigma^*$. Set

$$U_\sigma = \mathrm{Spec}(\mathbb{Z}[S_\sigma]).$$

If $\sigma \subset \tau$, we have $U_\sigma \subset U_\tau$. The variety $\mathbb{P}(\Delta)$ is obtained by glueing the affine varieties U_σ, $\sigma \in \Delta$, along the subvarieties $U_{\sigma \cap \tau}$.

We assume that Δ is *regular*, i.e., any $\sigma \in \Delta$ is spanned by a subset of a basis of N. We shall define a variety $X(\Delta)$ over $\mathbb{F}_1$ such that $X(\Delta) \otimes_{\mathbb{F}_1} \mathbb{Z} = \mathbb{P}(\Delta)$.

6.2. The Affine Case

First, let us fix $\sigma \in \Delta$. For any $m \in S_\sigma$, let $\chi^m : U_\sigma \to \mathbb{A}^1$ be the function defined by m. When D is a finite abelian group, we define $\underline{X}_\sigma(D) \subset U_\sigma(\mathbb{Z}[D])$ to be the set of points P such that for any $m \in S_\sigma$, $\chi^m(P) \in D \amalg \{0\}$.

Let

$$C_\sigma = \left\{ x \in U_\sigma(\mathbb{C}) \,\middle|\, |\chi^m(x)| \leq 1 \text{ for all } m \in S_\sigma \right\} \text{ and}$$

$$\overset{\circ}{C}_\sigma = \left\{ x \in C_\sigma \,\middle|\, |\chi^m(x)| < 1 \text{ for all } m \in S_\sigma \text{ with } \langle m, \sigma \rangle \neq 0 \right\}.$$

We define $\boldsymbol{A}_\sigma$ to be the ring of continuous functions $f : C_\sigma \to \mathbb{C}$ such that $f|_{\overset{\circ}{C}_\sigma}$ is holomorphic. Finally, if $P \in \underline{X}_\sigma(D)$, $f \in \boldsymbol{A}_\sigma$ and $\chi : D \to \mathbb{C}^\times$, we define $e_\sigma(P)$ by the formula $\chi(e_\sigma(P)(f)) = f(\chi(P))$.

The following is a generalization of Proposition 3.9. and Proposition 3.11.

PROPOSITION 6.4. *If σ is regular, then $X_\sigma = (\underline{X}_\sigma, \boldsymbol{A}_\sigma, e_\sigma)$ is an affine variety over $\mathbb{F}_1$ such that $X_\sigma \otimes_{\mathbb{F}_1} \mathbb{Z} = U_\sigma$.*

PROOF. Suppose $\{n_1, \ldots, n_d\}$ is a basis for N and that $\sigma = \mathbb{R}_+ n_1 + \cdots + \mathbb{R}_+ n_{d-r}$. Let $\{m_1, \ldots, m_d\}$ be the dual basis of M. Then

$$S_\sigma = \mathbb{N} m_1 + \cdots + \mathbb{N} m_{d-r} + \mathbb{Z} m_{d-r+1} + \cdots + \mathbb{Z} m_d = M \cap \sigma^*$$

and as $U_\sigma(\mathbb{C}) = \mathbb{C}^{d-r} \times (\mathbb{C}^\times)^r$, we have

$$C_\sigma = \{x \in U_\sigma(\mathbb{C}) \mid |x_1|, \ldots, |x_{d-r}| \le 1 \text{ and } |x_{d-r+1}| = \ldots = |x_r| = 1\} \text{ and }$$

$$\overset{\circ}{C}_\sigma = \{x \in U_\sigma(\mathbb{C}) \mid |x_1|, \ldots, |x_{d-r}| < 1 \text{ and } |x_{d-r+1}| = \ldots = |x_r| = 1\}.$$

Furthermore,

$$\underline{X}_\sigma(D) = (D \amalg \{0\})^{d-r} \times D^r.$$

Let V be an affine variety over $\mathbb{Z}$, and let $\varphi: X_\sigma \to \mathcal{G}(V)$ be a morphism of affine gadgets. We must find a $\varphi_\mathbb{Z}: U_\sigma \to V$ such that $\varphi = \mathcal{G}(\varphi_\mathbb{Z}) \circ i$. This is the same as a morphism from the algebra of functions on V to the algebra of functions on U_σ. Let $f \in \Gamma(V, \mathcal{O}_V)$. Then f induces a function $f_\mathbb{C}$ on the complex variety $V_\mathbb{C}$, and we may pull back this function to get a function on X_σ: $g_\mathbb{C} = \varphi^*(f_\mathbb{C}) \in \boldsymbol{A}_\sigma$. We must show that $g_\mathbb{C}$ is algebraic over $\mathbb{Z}$, i.e., that it comes from a $g \in \mathcal{O}(U_\sigma)$. Restrict $g_\mathbb{C}$ to $(S^1)^d$, and look at the Fourier expansion

$$g_\mathbb{C}(\exp(2\pi i\theta_1), \ldots, \exp(2\pi i\theta_d)) = \sum_{J \in \mathbb{Z}^d} c_J \exp(2\pi i(J \cdot \theta)), \text{ where } J \cdot \theta = \sum_{k=1}^d j_k \theta_k.$$

Since $g_\mathbb{C}$ is holomorphic on C_σ, we must have $c_J = 0$ when $j_k < 0$ for $1 \le k \le d-r$. We want to show that $g_\mathbb{C}$ is an integral polynomial in the first $d - r$ coordinates and an integral Laurent polynomial in the r remaining coordinates, i.e.,

$$g_\mathbb{C} \in \mathbb{Z}[T_1, \ldots, T_{d-r}, T_{d-r+1}^{\pm 1}, \ldots, T_d^{\pm 1}].$$

Let $n > 1$, and consider $D = (\mathbb{Z}/n)^d$. Then if

$$P_k = \underbrace{(0, \ldots, 0, 1, 0, \ldots, 0)}_{k\text{th slot is 1}},$$

we get a point $P = (P_1, \ldots, P_d) \in D^d \subset \underline{X}_\sigma(D)$. For $a = (a_k) \in D$, define $\chi_a: D \to \mathbb{C}^\times$ by

$$\chi_a(b) = \prod_{k=1}^d \exp\left(2\pi i \frac{a_k b_k}{n}\right).$$

Then, as φ commutes with evaluations, we get

$$\chi_a(e_\sigma(P)(g_\mathbb{C})) = g_\mathbb{C}(\chi_a(P)) = g_\mathbb{C}(\exp(2\pi i a_1/n), \ldots, (\exp(2\pi i a_d/n))$$
$$= \chi_a(f(\varphi(P))) = \chi_a(Q),$$

where $Q = f(\varphi(P)) \in f(V(\mathbb{Z}[D])) \subset \mathbb{Z}[D]$. The Fourier coefficients of $g_\mathbb{C}$ are given by the formula

$$c_J = \int_{(S^1)^d} g_\mathbb{C}(\exp(2\pi i\theta_1), \ldots, \exp(2\pi i\theta_k)) \exp(-2\pi i(J \cdot \theta)) d\theta_1 \cdots d\theta_d$$

$$= \lim_{n \to \infty} n^{-d} \sum_a g_\mathbb{C}(\exp(2\pi i a_1), \ldots, \exp(2\pi i a_k)) \exp(-2\pi i(J \cdot a)/n)$$

$$= \lim_{n \to \infty} n^{-d} \sum_a \chi_a(Q) \exp(-2\pi i(J \cdot a)/n).$$

But as $Q \in \mathbb{Z}[D]$ we must have, for every n,

$$n^{-d} \sum_a \chi_a(Q) \exp(-2\pi i(J \cdot a)/n) \in \mathbb{Z}.$$

Therefore $c_J \in \mathbb{Z}$, and $c_J = 0$ for almost all J, as desired.

6.3. The General Case

Let Δ be a regular fan. For every affine variety Y over $\mathbb{F}_1$ let

$$\underline{X}_\Delta(Y) = \bigcup_{\sigma \in \Delta} \mathrm{Hom}(Y, X_\sigma).$$

Define

$$C_\Delta = \bigcup_{\sigma \in \Delta} C_\sigma \subset \mathbb{P}(\Delta)(\mathbb{C}),$$

and let $\boldsymbol{\mathcal{A}}_\Delta$ be the algebra of continuous functions $f \colon C_\Delta \to \mathbb{C}$ such that, for all $\sigma \in \Delta$, the restriction of f to $\overset{\circ}{C}_\sigma$ is holomorphic. Finally, if $P \in \mathrm{Hom}(Y, X_\sigma) \subset \underline{X}_\Delta(Y)$ and $f \in \boldsymbol{\mathcal{A}}_\Delta$, define $e_\Delta(P)(f) = P^*(f) \in \boldsymbol{\mathcal{A}}_Y$.

The following is a consequence of Proposition 4.4 and Proposition 6.4.

THEOREM 6.5. *The object $X(\Delta) = (\underline{X}_\Delta, \boldsymbol{\mathcal{A}}_\Delta, e_\Delta)$ over $\mathbb{F}_1$ is a variety over $\mathbb{F}_1$ such that*

$$X(\Delta) \otimes_{\mathbb{F}_1} \mathbb{Z} = \mathbb{P}(\Delta).$$

REMARK 6.6. There exists $N(x) \in \mathbb{Z}[x]$ such that, for all $n \geq 1$, $\#X_\Delta(\mathbb{F}_{1^n}) = N(n+1)$.

7. Euclidean Lattices

Let Λ be a free $\mathbb{Z}$-module of finite rank, and $\| \cdot \|$ an Hermitian norm on $\Lambda \otimes_{\mathbb{Z}} \mathbb{C}$. We view $\overline{\Lambda} = (\Lambda, \infty \cdot \|)$ as a vector bundle on the complete curve $\mathrm{Spec}(\mathbb{Z}) \amalg \{\infty\}$. The finite pointed set

$$H^0(\mathrm{Spec}(\mathbb{Z}) \amalg \{\infty\}, \overline{\Lambda}) = \{s \in \Lambda \mid v_\infty(s) = -\log\|s\| \geq 0\} = \Lambda \cap B,$$

where $B = \{v \in \Lambda \otimes_{\mathbb{Z}} \mathbb{C} \mid \|v\| \leq 1\}$, is viewed as a finite-dimensional vector space over $\mathbb{F}_1$.

We can define an affine variety over $\mathbb{F}_1$ as follows. We let

$$\underline{X}(D) = \left\{ P = \sum_{v \in \Lambda \cap B} v \otimes \alpha_v \,\middle|\, \alpha_v \in D \right\} \subset \Lambda \otimes_{\mathbb{Z}} \mathbb{Z}[D].$$

If $\Lambda_0 \subset \Lambda$ is the lattice spanned by $\Lambda \cap B$ we consider

$$C = \left\{ v \in \Lambda_0 \otimes_{\mathbb{Z}} \mathbb{C} \,\middle|\, \|v\| \leq \mathrm{card}(V \cap B) \right\}$$

and we define $\boldsymbol{\mathcal{A}}_X$ as the algebra of continuous functions $f \colon C \to \mathbb{C}$ such that $f|_{\overset{\circ}{C}}$ is holomorphic. Finally, for each $D \in \mathsf{Ab}_f$, $P \in \underline{X}(D)$, $f \in \boldsymbol{\mathcal{A}}_X$, and $\chi \colon D \to \mathbb{C}^\times$, we define

$$\chi(f(P)) = f\left(\sum_{v \in \Lambda \cap B} \chi(a_v)v \right).$$

PROPOSITION 7.1.

(1) *The affine gadget* $X = (\underline{X}, \mathcal{A}_X, e_X)$ *is an affine variety over* $\mathbb{F}_1$ *such that* $X \otimes_{\mathbb{F}_1} \mathbb{Z} = \mathrm{Spec}(\mathrm{Symm}_{\mathbb{Z}}(\Lambda_0^*))$.

(2) *There is a polynomial* $N \in \mathbb{Z}[x]$ *such that, for all* $n \geq 1$, $\#X(\mathbb{F}_{1^n}) = N(2n + 1)$.

This proposition raises the question of whether there is a way to attach to $\overline{\Lambda}$ a torified variety in the sense of [**LL**].

References

[CC1]   A. Connes and C. Consani, *On the notion of Geometry over* $\mathbb{F}_1$, arXiv:0809.2926v2[math. AG] (2008).

[CC2]   A. Connes and C. Consani, *Schemes over* $\mathbb{F}_1$ *and zeta functions*, arXiv:0903.2024v2[math. AG] (2009).

[DG]   M. Demazure and P. Gabriel, *Introduction to algebraic geometry and algebraic groups*, North-Holland Mathematics Studies, Amsterdam, New York, Oxford: North-Holland. XIV **39** (1980), 358.

[KS]   M. Kapranov and A. Smirnov, *Cohomology determinants and reciprocity laws: number field case*, preprint.

[KOW]   N. Kurokawa, H. Ochiai and M. Wakayama, *Absolute derivations and zeta functions*, Documenta Math. (Extra Volume: Kazuya Kato's Fiftieth Birthday) (2003), 565–584.

[LL]   J. López Peña and O. Lorscheid, *Torified varieties and their geometries over* $\mathbb{F}_1$, arXiv: 0903.2173v2 [math.AG] (2009).

[S]   A. Smirnov, *Hurwitz inequalities for number fields*, St. Petersbg. Math. J. **4** (1993), 357–375. Translation from *Algebra Anal.* **4** (1992), 186–209.

[So]   C. Soulé, *Les variétés sur le corps à un élément*, Mosc. Math. J. **4** (2004), 217–244.

[St]   R. Steinberg, *A geometric approach to the representations of the full linear group over a Galois field*, Trans. Am. Math. Soc. **71** (1951), 274–282.

[T]   J. Tits, *Sur les analogues algébriques des groupes semi-simples complexes*, Colloque d'algèbre supérieure, Bruxelles (1956), 261–289. Centre Belge de Recherches Mathématiques, Gauthier-Villars (1957).

INSTITUT DES HAUTES ÉTUDES SCIENTIFIQUES, 35, ROUTE DE CHARTRES, 91440 BURES-SUR-YVETTE, FRANCE.

E-mail address: soule@ihes.fr

Transcendence of Values of Transcendental Functions at Algebraic Points

(Inaugural Monroe H. Martin Lecture and Seminar)

Paula Tretkoff

ABSTRACT. This paper is a transcript of the Inaugural Monroe H. Martin Lecture and Seminar given at Johns Hopkins University on February 23 and 24, 2009. In Part I (the lecture), we present classical and recent results on the transcendence of values of certain special functions of one variable at algebraic points. In Part II (the seminar), we describe some new results, jointly with Marvin D. Tretkoff, on the transcendence of values at algebraic points of hypergeometric functions of several variables. Note that the author's maiden name is Paula B. Cohen.

Part I: Transcendence of Values of Some Special Functions

In this part of our paper, we present some results on the transcendence of values of certain special functions of one variable at algebraic points. Recall that the algebraic numbers are the complex numbers satisfying a non-trivial polynomial relation with rational coefficients. We denote the field of rational numbers by $\mathbb{Q}$, and the field of algebraic numbers by $\overline{\mathbb{Q}}$. A transcendental number, α, is a complex number which is not algebraic. Therefore $P(\alpha) \neq 0$ for every polynomial $P \in \mathbb{Q}[x]$ with at least one non-zero coefficient.

The first examples of explicit transcendental numbers are due to Liouville in 1844 [**Lio1**], but their construction is rather artificial. Liouville showed that an irrational algebraic number cannot be too well approximated by rational numbers with denominators of relatively small size. He then constructed numbers that are so well approximated by rational numbers that they must be transcendental. An example is the number $\xi = \sum_{n=1}^{\infty} 10^{-n!}$. In 1874, Cantor gave another proof of the existence of transcendental numbers. He showed that the set of all algebraic numbers is countable, while the set of real numbers is uncountable. It follows that the set of transcendental numbers is uncountable. Somewhat paradoxically, it is usually very difficult to show that any given number is transcendental.

The ancient Greeks had asked whether it is possible to square the circle, that is, to construct, with compass and straight edge, a square with area equal to that of a given circle. If we set the radius of the circle equal to 1, this problem reduces to the construction of a segment of length $\sqrt{\pi}$. The impossibility of this construction is implied by the transcendence of π, which was proved by Lindemann in 1882 [**Lin**]. He used a method of Hermite, who proved the transcendence of e in 1873 [**Her**]. These were the first "naturally occurring" numbers proved to be transcendental. Weierstrass called Lindemann's work "one of the most beautiful theorems of arithmetic."

To mark the turn of the 20th century, Hilbert proposed a celebrated list of problems whose solution, he felt, would provide important goals for mathematicians during the following hundred years. These problems have inspired, and continue to inspire, important mathematical ideas. Hilbert said in 1900 that "Hermite's theorem on the arithmetic of the exponential function and their further development by Lindemann will undoubtedly remain an inspiration for mathematicians of future generations." The seventh of Hilbert's problems asked for a proof that α^β is transcendental for α, β algebraic with $\alpha \neq 0, 1$ and β irrational. This implies, for example, that $e^\pi = (e^{i\pi})^{-i} = (-1)^{-i}$ and $2^{\sqrt{2}}$ are transcendental. This generalizes a conjecture made by Euler in 1744 on the logarithms of rational numbers. Euler stated without proof that a number of the form $\log_a b$, where a and b are rational numbers, with b not equal to a rational power of a, must be a transcendental number. Hilbert's seventh problem was solved independently by Gelfond and Schneider in 1934. The Gelfond–Schneider theorem states that for any non-zero algebraic numbers α_1, α_2, β_1, β_2, with $\log \alpha_1$, $\log \alpha_2$ linearly independent over $\mathbb{Q}$, we have $\beta_1 \log \alpha_1 + \beta_2 \log \alpha_2 \neq 0$. In 1966, Baker obtained the analogous result for linear forms in arbitrarily many logarithms. This was a major breakthrough for which he was awarded the Fields Medal in 1970. The quantitative versions of Baker's results had a major impact on the study of diophantine equations, implying, in many examples, effectively computable bounds on the number and size of the solutions. For example, in 1909, Thue proved that the equation $F = F(x, y) = m$ has only a finite number of solutions in integers x, y. Here, F is an irreducible binary form of degree at least 3 with integer coefficients, and m is an integer. Thue used an ineffective strengthening of the 1844 theorem of Liouville mentioned earlier. Thue's method yielded an estimate for the number of solutions, but it did not furnish an estimate for the size of the solutions. Therefore, it did not enable one to actually solve the equations explicitly. In 1968, using lower bounds for linear forms in logarithms, Baker gave explicit bounds on the absolute values of the solutions of the Thue equation in terms of m and F. For details, see [**Bak**].

The proofs of these results share a philosophy of method originating in Hermite's work and still evident in many transcendence proofs to the present day. The evolution of this method, in increasingly sophisticated situations, has required breakthroughs and much additional mathematics. Despite this common thread, the description of any given transcendence proof is quite technical. We reproduce here a vague outline of the structure of a typical classical transcendence proof, transcribed from [**BuTu**], p. 10. To show a complex number α is transcendental, assume that it is algebraic and show that this leads to a contradiction. If α is algebraic, we have $P(\alpha) = 0$ for some non-zero polynomial $P = P(x)$ with integer coefficients. Using the integer coefficients of P, and some particular properties of α, construct an integer N. This step usually involves some techniques from algebra. Next, show that N is non-zero. This is usually the most difficult part of the proof. Then, show that $|N| < 1$. This step usually involves some analytic methods. To finish, we derive a contradiction, since we have constructed an integer N with $0 < |N| < 1$, and there is no such integer. Therefore our original premise that α is algebraic must be false, so that α must be transcendental.

There are many unsolved problems about the transcendence of everyday numbers that would seem to require new methods. For example, it is suspected, but unproved, that the following numbers are transcendental:

$$e\pi, \quad e + \pi, \quad \gamma, \quad \Gamma(1/5), \quad \zeta(2n+1),\, n \geq 1,\, n \in \mathbb{Z}.$$

Here, γ is Euler's constant, which is given by

$$\gamma = \lim_{n\to\infty} \left(1 + \frac{1}{2} + \frac{1}{3} + \ldots + \frac{1}{n} - \log n \right).$$

In 1755, Euler showed that for integers $n \geq 1$ the number $\zeta(2n)$ is a rational multiple of π^{2n}. Therefore, by Lindemann's result on the transcendence of π, the number $\zeta(2n)$ is transcendental. From the classical formula $\Gamma(s)\Gamma(1-s) = \pi/\sin(\pi s)$, it follows that $\Gamma(1/2)$ is transcendental. In the 1970's, using the theory of elliptic curves with complex multiplication, the Chudnovskys showed that the numbers $\Gamma(1/3)$ and $\Gamma(1/4)$ are transcendental. More recently, using modular forms, Nesterenko has shown the algebraic independence of π, e^π and $\Gamma(1/4)$.

A related challenge is to establish the irrationality of such numbers, a weaker property than transcendence. Again, there are many open problems and ample room for new methods. Euler showed in 1744 that e is irrational and Lambert showed in 1761 that π is irrational. It is not known whether $e\pi$, $e + \pi$, or γ is irrational. The irrationality of $\zeta(3)$ was established in 1978 by Apéry [**Ap**]. The irrationality of $\zeta(5)$ is unproved, although results of Rivoal and others show that infinitely many $\zeta(2n+1)$, $n \geq 1$, are irrational. There are now finer results, in particular Zudilin [**Zu**] has proved that one of $\zeta(5)$, $\zeta(7)$, $\zeta(9)$, or $\zeta(11)$ is irrational.

Many results in transcendental number theory can be formulated in terms of the *exceptional set* of a transcendental function, that is, *the set of algebraic arguments at which the function takes algebraic values*. For example, Lindemann showed in 1882 that the exceptional set of $\exp(z)$ is trivial:

$$\mathcal{E} = \{\alpha \in \overline{\mathbb{Q}} : \exp(\alpha) \in \overline{\mathbb{Q}}\} = \{0\}.$$

As $\exp(1) = e$ and $\exp(i\pi) = -1$, it follows that e and π are transcendental. Hilbert's seventh problem can be restated as the assertion that the exceptional set of z^β consist only of $\alpha = 0, 1$ when β is an algebraic irrational.

C.-L. Siegel made fundamental contributions to the study of the transcendence properties of various classes of functions generalizing the exponential function. In 1929, he introduced a general method for establishing the transcendence and algebraic independence of the values at algebraic points of a class of entire functions satisfying linear differential equations [**Si1**]. Siegel called these functions E-functions since the exponential function is in this class. Another example of an E-function is given by the normalized Bessel function:

$$K_\lambda(x) = \sum_{n=0}^{\infty} \frac{(-1)^n (\frac{1}{2}x)^{2n}}{n!(\lambda+1)\ldots(\lambda+n)},$$

where $\lambda \in \mathbb{Q}$, $\lambda \neq -1, -2, \ldots$. Siegel showed that if $\lambda \neq n + \frac{1}{2}$, then $K_\lambda(\alpha)$ and $K_\lambda'(\alpha)$ are algebraically independent for all $\alpha \in \overline{\mathbb{Q}}$, $\alpha \neq 0$. Siegel's method was later developed by Shidlovsky. At the same time, Siegel proposed studying power series, satisfying linear differential equations, that are reminiscent of a geometric series and have a finite radius of convergence. His E-function methods do not apply to them. He called these new functions G-functions and proposed studying their exceptional set. In particular, he asked whether this set is finite or infinite. Examples of G-functions are algebraic functions over $\mathbb{Q}$, for which the exceptional set is trivially all of $\overline{\mathbb{Q}}$, the k-th polylogarithm $L_k(x)$, and the classical Gauss hypergeometric function $F(a, b, c; x)$ with rational parameters a, b, c. Not much is known about the transcendence properties of the polylogarithm. For example, $L_k(1) = \zeta(k)$, and the transcendence of this number for odd $k \geq 3$ is unproved as mentioned earlier. We return to the example of $F(a, b, c; x)$ later. In 1932, Siegel [**Si2**] established the transcendence of at least one of any fundamental pair ω_1, ω_2 of periods of a Weierstrass elliptic function with algebraic invariants. Weierstrass elliptic functions can be viewed as generalizations of the exponential function. This raised the question as to whether every non-zero period is transcendental. Moreover, Siegel asked if the ratio $z = \omega_1/\omega_2$ is transcendental. This can be reinterpreted as the problem of determining the exceptional set of the classical elliptic modular function. Schneider solved these problems in 1937 [**Sch**].

We now describe Schneider's solution of Siegel's problem to determine the exceptional set of the classical elliptic modular function. This function $j = j(z)$ is defined on the upper half plane $\mathcal{H}$ consisting of complex numbers z with positive imaginary part. It is invariant with respect to the action of $\mathrm{SL}(2, \mathbb{Z})$ given by

$$z \mapsto \frac{az+b}{cz+d}, \qquad \begin{pmatrix} a & b \\ c & d \end{pmatrix} \in \mathrm{SL}(2, \mathbb{Z}).$$

Here, $\mathrm{SL}(2, \mathbb{Z})$ is the group of 2×2 matrices with integer entries and determinant equal to 1. In particular, $j(z) = j(z+1)$ for all $z \in \mathcal{H}$, and $j(z)$ is uniquely determined by the first two terms in its Fourier expansion:

$$j(z) = \exp(-2\pi i z) + 744 + \sum_{n=1}^{\infty} a_n \exp(2\pi i n z).$$

For all $n \geq 1$, the coefficient a_n turns out to be a positive integer. For example, $a_1 = 196884$, $a_2 = 21493760$. Schneider's celebrated result, proved in 1937, is that:

$$\mathcal{E} = \{z \in \mathcal{H} : z \in \overline{\mathbb{Q}} \text{ and } j(z) \in \overline{\mathbb{Q}}\} = \{z \in \mathcal{H} : [\mathbb{Q}(z) : \mathbb{Q}] = 2\}.$$

We can restate this result in terms of elliptic curves with complex multiplication. To every $z \in \mathcal{H}$ we associate the complex torus of dimension 1:

$$A_z = \mathbb{C}/(\mathbb{Z} + z\mathbb{Z}).$$

This torus has an underlying structure of a projective curve of genus $g = 1$ defined over the field $\mathbb{Q}(j(z))$, called an elliptic curve. Its endomorphism algebra $\mathrm{End}_0(A_z) = \mathrm{End}(A_z) \otimes_{\mathbb{Z}} \mathbb{Q}$ consists of multiplications by the numbers α preserving $\mathbb{Q} + z\mathbb{Q}$. It is not difficult to see that this algebra is either $\mathbb{Q}$ or $\mathbb{Q}(z)$. In the latter case, the number z must be imaginary quadratic, and we say that A_z has *complex multiplication* (CM) and that z is a *CM point*. Therefore, we may reformulate Schneider's result as saying:

$$\mathcal{E} = \{z \in \mathcal{H} : z \in \overline{\mathbb{Q}} \text{ and } j(z) \in \overline{\mathbb{Q}}\} = \{z \in \mathcal{H} : A_z \text{ has CM}\}.$$

That $j(z)$ is algebraic when A_z has CM was already known from the theory of elliptic curves. The hard part was to show that $j(z)$ is transcendental when z is algebraic and A_z does not have CM. The CM points are very important in number theory. The Kronecker-Weber theorem says that the finite abelian extensions of $\mathbb{Q}$ are obtained by adjoining the roots of unity $\exp(2\pi i\mathbb{Q})$. Hilbert's twelfth problem asks for a description of the abelian extensions of any number field in terms of special values of explicit functions. The theory of CM realizes this description for any imaginary quadratic field. For example, when z is imaginary quadratic, the field $\mathbb{Q}(z, j(z))$ is an extension of $\mathbb{Q}(z)$ with abelian Galois group. Moreover, we have very explicit information on the action of this Galois group on the number $j(z)$, which is an algebraic integer of degree the class number of $\mathbb{Q}(z)$. The study of abelian extensions of number fields is known as class field theory.

The proof of Schneider's theorem uses properties of doubly periodic functions and relies heavily on the group structure of elliptic curves. Schneider asked whether it is possible to prove his result using only the intrinsic properties of the elliptic modular function. This problem remains unsolved.

The generalization of Schneider's theorem to Siegel modular varieties follows from work of myself, Shiga and Wolfart in 1995. We briefly describe this result. Consider the Siegel upper half space of "genus" $g \geq 1$,

$$\mathcal{H}_g = \left\{ z \in M_g(\mathbb{C}) : z = z^t, \ \Im(z) \text{ positive definite} \right\}.$$

Notice that $\mathcal{H}_1 = \mathcal{H}$. The symplectic group with integer entries is given by

$$\mathrm{Sp}(2g, \mathbb{Z}) = \left\{ \gamma \in M_{2g}(\mathbb{Z}) : \gamma \begin{pmatrix} O_g & -I_g \\ I_g & O_g \end{pmatrix} \gamma^t = \begin{pmatrix} O_g & -I_g \\ I_g & O_g \end{pmatrix} \right\},$$

where I_g and O_g are the $g \times g$ identity and zero matrix, respectively. We have $\mathrm{Sp}(2, \mathbb{Z}) = \mathrm{SL}(2, \mathbb{Z})$. The group $\mathrm{Sp}(2g, \mathbb{Z})$ acts on the space $\mathcal{H}_g$ by

$$z \mapsto (Az + B)(Cz + D)^{-1}, \qquad z \in \mathcal{H}_g, \quad \gamma = \begin{pmatrix} A & B \\ C & D \end{pmatrix} \in \mathrm{Sp}(2g, \mathbb{Z}),$$

where A, B, C, D are in $M_g(\mathbb{Z})$. The quotient space for this action,

$$\mathcal{A}_g = \mathrm{Sp}(2g, \mathbb{Z}) \backslash \mathcal{H}_g,$$

is an analytic space parameterizing the complex isomorphism classes of (principally polarized) abelian varieties of dimension g. Indeed, to $z \in \mathcal{H}_g$, we associate the g-dimensional complex torus

$$A_z = \mathbb{C}^g / (\mathbb{Z}^g + z\mathbb{Z}^g),$$

which has the underlying structure of a projective commutative group variety, and is called an abelian variety. The space $\mathcal{A}_g$ has the underlying structure of a quasi-projective variety defined over $\overline{\mathbb{Q}}$, the Siegel modular variety. Two abelian varieties A, B related by a surjection $A \mapsto B$ with finite kernel are said to be isogenous, written $A \hat{=} B$. By the Poincaré irreducibility theorem, the abelian variety A_z is isogenous to a product of powers of simple mutually non-isogenous abelian varieties:

$$A_z \hat{=} A_1^{n_1} \times \ldots \times A_k^{n_k}, \qquad A_i \text{ simple}, \ A_i \hat{\neq} A_j, \ i \neq j.$$

The endomorphism algebra $\mathrm{End}_0(A) = \mathrm{End}(A) \otimes_{\mathbb{Z}} \mathbb{Q}$ of a simple abelian variety A is a division algebra over $\mathbb{Q}$ with positive involution. For $g > 1$, there are more possibilities for such division algebras than in the case $g = 1$. When A is simple and

$\mathrm{End}_0(A)$ is a number field $\mathbb{L}$ with $[\mathbb{L} : \mathbb{Q}] = 2 \dim(A)$, we say that A has complex multiplication (CM). In this case, the field $\mathbb{L}$ will be a totally imaginary quadratic extension of a totally real field. We say that A_z has CM when all its simple factors (in the above decomposition up to isogeny) have CM. The corresponding point $z \in \mathcal{H}_g$, or its $\mathrm{Sp}(2g, \mathbb{Z})$-orbit in $\mathcal{A}_g$, is said to be a CM or special point. The abelian varieties defined over $\overline{\mathbb{Q}}$ correspond to points of $\mathcal{A}_g(\overline{\mathbb{Q}})$, and include all CM abelian varieties. Moreover, the isomorphism classes of CM abelian varieties correspond to $\mathrm{Sp}(2g, \mathbb{Z})$-orbits of points in $\mathcal{H}_g \cap M_g(\overline{\mathbb{Q}})$. Conversely, we have the following generalization of Schneider's theorem, proved jointly by myself (using my maiden name, Paula B. Cohen) [Co1], Shiga and Wolfart [ShWo], using modern transcendence techniques, especially the results of Wüstholz [Wu2], [Wu3].

THEOREM 1. *Let* $J : \mathcal{H}_g \to \mathcal{A}_g(\mathbb{C})$ *be a holomorphic* $\mathrm{Sp}(2g, \mathbb{Z})$-*invariant map such that, for all CM points* z, *we have* $J(z) \in \mathcal{A}_g(\overline{\mathbb{Q}})$. *Then the exceptional set of* J, *given by*

$$\mathcal{E} = \{ z \in \mathcal{H}_g \cap M_g(\overline{\mathbb{Q}}) : J(z) \in \mathcal{A}_g(\overline{\mathbb{Q}}) \},$$

consists exactly of the CM points.

Therefore, the special values $J(z)$, $z \in \mathcal{H}_g \cap M_g(\overline{\mathbb{Q}})$, are "transcendental," that is, not in $\mathcal{A}_g(\overline{\mathbb{Q}})$, whenever z is not a CM point. Some refinements of this result were obtained by Derome [De]. As in the case $g = 1$, the CM points are important in number theory. One reason is that the action of the absolute Galois group on the torsion points of CM abelian varieties, and on the values of $J(z)$ at CM points, is well understood. This is a basic tool for studying the arithmetic of abelian varieties and modular forms.

A natural question to ask is how the image $\overline{\mathcal{E}}$ in $\mathcal{A}_g$ of the exceptional set $\mathcal{E}$ of J is distributed. We make the following conjecture, which by Theorem 1 is just a restatement of the André–Oort conjecture [An], [Oo]. A proof of this conjecture, assuming the Riemann hypothesis, has been announced by Klingler, Ullmo and Yafaev [KY], [UY].

CONJECTURE. *Let* Z *be an irreducible closed algebraic subvariety of* $\mathcal{A}_g$. *Then* $Z \cap \overline{\mathcal{E}}$ *is a Zariski dense subset of* Z *if and only if* Z *is a special subvariety of* $\mathcal{A}_g$.

The Siegel modular variety $\mathcal{A}_g$ is a moduli space for principally polarized abelian varieties of dimension g. If we impose some extra structure on these polarized abelian varieties, for example on their endomorphism ring, the corresponding moduli space is called a Shimura subvariety of $\mathcal{A}_g$. A special subvariety is an irreducible component of a Shimura subvariety or of its image under the correspondences coming from the action of $\mathrm{Sp}(2g, \mathbb{Q})$ on $\mathcal{H}_g$ (the Hecke correspondences).

We now turn to a special case of the problem on G-functions proposed by Siegel in 1929. Namely, we study the exceptional set of the classical Gauss hypergeometric function with rational parameters, focusing on criteria for this set to be finite or infinite. The methods we use are only relevant to this function, there being as yet no general method that applies to all G-functions. The Gauss hypergeometric differential equation with parameters a, b, c, $c \neq 0, -1, -2, \ldots$ is given by

$$x(1 - x)\frac{d^2 F}{dx^2} + (c - (a + b + 1)x)\frac{dF}{dx} - abF = 0$$

and has three regular singular points at $x = 0, 1, \infty$. The Gauss hypergeometric function $F = F(x) = F(a, b, c; x)$ is the solution of this differential equation given by the multi-valued function on $\mathbb{P}_1 \setminus \{0, 1, \infty\}$ with a branch defined for $|x| < 1$ by the series

$$F(x) = \sum_{n=0}^{\infty} \frac{(a, n)(b, n)}{(c, n)(1, n)} x^n, \qquad |x| < 1,$$

where $(w, 0) = 1$ and $(w, n) = w(w + 1) \ldots (w + n - 1)$, for $w \in \mathbb{C}$ and $n \geq 1$.

The solution space of the differential equation is two dimensional. A basis of the solution space at any point x_0 in $\mathbb{P}_1 \setminus \{0, 1, \infty\}$ changes into another such basis upon analytic continuation along closed loops in $\mathbb{P}_1 \setminus \{0, 1, \infty\}$ starting and ending at x_0. This gives rise to a representation of the fundamental group of $\mathbb{P}_1 \setminus \{0, 1, \infty\}$, with base point x_0, in the 2×2 matrices, whose image $\Gamma(a, b, c)$ in $\mathrm{PGL}(2, \mathbb{C})$ is called the monodromy group. If we change the choice of base point and solution space basis, we replace $\Gamma(a, b, c)$ by a conjugate group. If $\Gamma(a, b, c)$ is finite, then F is an algebraic function. The list of finite monodromy groups was found by Schwarz in 1873 [**Schw**]. In what follows, we assume that a, b, c, $c \neq 0, -1, -2, \ldots$, are rational and that $\Gamma(a, b, c)$ is infinite. To simplify the present discussion, from now on we make the additional implicit assumption that $0 < a < c$, $0 < b < c$, and $c < 1$. This ensures that the differential forms we introduce shortly are all holomorphic and that the monodromy group acts (not necessarily discontinuously) on $\mathcal{H}$. These assumptions are not very restrictive (see [**Wo**], [**CoWu**]). Under our assumptions, the series $F(x)$, $|x| < 1$, above has an analytic continuation to $\mathbb{C} \setminus [1, \infty)$ given by the Euler integral formula:

$$F = \frac{\int_1^{\infty} u^{b-c}(u - 1)^{c-a-1}(u - x)^{-b} du}{\int_1^{\infty} u^{-c}(u - 1)^{c-a-1} du}.$$

We assume that F is given by this analytic continuation in the discussion of the proof of Theorem 2 below. Our arguments are valid for any branch of F, since this only affects the choice of path of integration in the numerator of the above expression. In Theorem 2 below, by $F(a, b, c; x) \in \overline{\mathbb{Q}}$ we mean that some branch of the multi-valued function F takes an algebraic value at x. We have the following result.

THEOREM 2. *The exceptional set*

$$\mathcal{E} = \{x \in \overline{\mathbb{Q}} : F(a, b, c; x) \in \overline{\mathbb{Q}}\}$$

is infinite if and only if $\Gamma(a, b, c)$ is an arithmetic lattice in $PSL(2, \mathbb{R})$. Moreover, every element of the set $\mathcal{E}$ corresponds to a CM point in a certain moduli space $V(a, b, c)$ for abelian varieties.

An arithmetic group is a group commensurable with the integer points $G(\mathcal{O})$ of a linear algebraic group G defined over a number field F, where $\mathcal{O}$ is the ring of integers of F. A linear algebraic group is a group of matrices defined by algebraic conditions. Takeuchi [**Ta**] computed the complete list of arithmetic $\Gamma(a, b, c)$ in 1977. There are 85 such groups, up to conjugacy, and therefore infinitely many non-arithmetic $\Gamma(a, b, c)$. For example, the group $\mathrm{PSL}(2, \mathbb{Z})$ is arithmetic. It is the monodromy group of $F(\frac{1}{12}, \frac{5}{12}, \frac{1}{2}; x)$, which by Theorem 2 has an infinite exceptional set. The subgroup of $\mathrm{PSL}(2, \mathbb{R})$ generated by the fractional linear transformations given by the matrices

$$\begin{pmatrix} 0 & -1 \\ 1 & 0 \end{pmatrix}, \qquad \begin{pmatrix} 1 & \frac{1}{2}(1 + \sqrt{5}) \\ 0 & 1 \end{pmatrix}$$

is non-arithmetic. It is the monodromy group of $F(\frac{3}{20}, \frac{7}{20}, \frac{1}{2}; x)$, which has a finite exceptional set by Theorem 2. Wolfart [**Wo**] proved in 1988 that $\mathcal{E}$ is infinite when $\Gamma(a, b, c)$ is arithmetic. He was the first person to realize that there is a relation between the size of the exceptional set and the arithmetic properties of the monodromy group. Siegel seems to have been unaware of this relation. In the same paper, Wolfart claimed that $\mathcal{E}$ is finite when $\Gamma(a, b, c)$ is not arithmetic, but his proof contained a serious error discovered by Gubler. In 2002, Wüstholz and I (using my maiden name Cohen) [**CoWu**] showed that Wolfart's claim follows from a special case of the André–Oort conjecture in the theory of moduli varieties for abelian varieties (Shimura varieties). This special case was subsequently proved by Edixhoven and Yafaev [**EdYa**] in 2003. Together, these results give Theorem 2.

We briefly indicate some of the main ingredients of the proof of Theorem 2. Some of this will be revisited in Part 2 of the present paper. We mentioned above the Euler integral representation of $F(x)$. We can rewrite this as

$$F(x) = C \, \frac{\int_\gamma u^{b-c}(u-1)^{c-a-1}(u-x)^{-b}du}{\int_\gamma u^{-c}(u-1)^{c-a-1}du},$$

where $C \in \overline{\mathbb{Q}}^*$ and γ is a Pochhammer cycle (closed double loop) around $1, \infty$. The numerator is a period of a holomorphic differential form on the smooth projective curve $X(x)$ with affine model

$$w^N = u^{N(c-b)}(u-1)^{N(a+1-c)}(u-x)^{Nb},$$

where N is the least common denominator of $c - b$, $a + 1 - c$, b. This differential form and the curve $X(x)$ are defined over $\mathbb{Q}(x)$. To simplify our discussion, we assume N is a prime number (in Part II, we do not assume this). The denominator is a period of a holomorphic differential form on $X(0)$, both being defined over $\overline{\mathbb{Q}}$. When $x \in \mathcal{E}$, and $F(x) \neq 0$, the ratio of these non-zero periods is in $\overline{\mathbb{Q}}^*$ and the relevant differential forms and curves are defined over $\overline{\mathbb{Q}}$. (If $F(x) = 0$ then $x \notin \overline{\mathbb{Q}}$ [**Wo**].) We can associate to a smooth projective curve X an abelian variety, called its Jacobian, whose complex points are given by

$$\mathrm{Jac}(X) = H^{1,0}(X, \mathbb{C})^*/H_1(X, \mathbb{Z}).$$

Transcendence techniques imply that the only linear dependence relations between periods on abelian varieties are those induced by isogenies (assuming the base field $\overline{\mathbb{Q}}$ throughout). From this we deduce that, if $x \in \mathcal{E}$, then we have an isogeny of the form

$$\mathrm{Jac}(X(x)) \,\widehat{=}\, \mathrm{Jac}(X(0)) \times A.$$

The abelian variety A turns out to be independent of x. Moreover, both $\mathrm{Jac}(X(0))$ and A have CM, coming from the action of $\zeta = \exp(2\pi i/N)$ by the automorphism $(u, w) \mapsto (u, \zeta^{-1}w)$ of $X(x)$. The Shimura variety $V(a, b, c)$ of Theorem 2 is the smallest moduli space (Shimura variety) for the $\mathrm{Jac}(X(x))$ and this 1-parameter family determines a curve Z in $V(a, b, c)$. The exceptional set $\mathcal{E}$ is given by the intersection of Z with the moduli in $V(a, b, c)$ of abelian varieties isogenous to the fixed abelian variety $\mathrm{Jac}(X(0)) \times A$. By the special case of the André–Oort conjecture proved in [**EdYa**], this intersection is Zariski dense in Z if and only if Z is a special subvariety of $V(a, b, c)$. Finally, we observe that Z is a special subvariety of $V(a, b, c)$ if and only if $\Gamma(a, b, c)$ is arithmetic.

The monodromy groups of the higher dimensional hypergeometric functions, given by the Appell–Lauricella functions, were studied by Picard [**Pi**], Terada [**Te**], Deligne and Mostow [**DeMo**], [**Mo1**], [**Mo2**]. In particular, they determined the monodromy groups that act discontinuously on the complex n-ball for all $n \geq 1$, and, of those, which ones are arithmetic. The generalization of Theorem 2 to the exceptional set of these functions is due to Desrousseaux, M.D. Tretkoff, and myself [**DTT2**]. There, infinitude of the exceptional set is instead its Zariski density in the space of regular points of the system of partial differential equations of the Appell–Lauricella function. In the arithmetic case, the exceptional set again corresponds to certain CM points and we have Zariski density. In the non-arithmetic (which includes the non-discontinuous) case, when $n \geq 2$, the elements of the exceptional set no longer necessarily correspond to CM points, but rather may correspond to Shimura varieties of positive dimension. We can show that, in this situation, the exceptional set is not Zariski dense modulo a condition that arises naturally from the transcendence techniques and is related to unsolved problems on subvarieties of Shimura varieties. Recently, with M.D. Tretkoff, we improved the results of [**DTT2**] in the discontinuous case. Namely, the results are now unconditional, as we show in Part II of the present paper.

Part II: Transcendence Properties of Appell–Lauricella Functions

In this part we prove a new result, jointly with Marvin D. Tretkoff, and stated in Theorem 3 below, which is a partial improvement of our results with Desrousseaux in [**DTT2**]. Namely, we generalize Theorem 2 in Part I of the present paper to Appell–Lauricella hypergeometric functions of $n \geq 2$ variables, whose monodromy groups act discontinuously on the complex n-ball, denoted $\mathbb{B}_n$. Recall that $\mathbb{B}_n$ is the set of points $(x_0 : x_1 : \ldots : x_n) \in \mathbb{P}_n(\mathbb{C})$ satisfying $|x_1|^2 + |x_2|^2 + \ldots + |x_n|^2 < |x_0|^2$ with automorphism group $\mathrm{Aut}(\mathbb{B}_n) = \mathrm{PU}(1, n)$. When $n = 1$, we have a bijection $z \mapsto (z - i)/(z + i)$ from $\mathcal{H}$ to $\mathbb{B}_1$. This enables us to associate a subgroup of $\mathrm{PU}(1, 1)$ to each monodromy group $\Gamma(a, b, c)$ of $F(a, b, c; x)$ acting on $\mathcal{H}$. At the end of this paper, we show that we do indeed obtain new examples not appearing in [**DTT2**] for which Theorem 3 is valid. One of these examples corrects an error in the list of discontinuous monodromy groups for $n \geq 2$ given in [**Mo2**].

The Appell–Lauricella hypergeometric functions are multi-valued functions of $n \geq 1$ complex variables defined on the weighted configuration space of $n + 3$ distinct points in $\mathbb{P}_1$, given by

$$Q_n = \{(x_0, x_1, \ldots, x_{n+2}) \in \mathbb{P}_1^{n+3} : x_k \neq x_\ell, k \neq \ell\}/\mathrm{Aut}(\mathbb{P}_1),$$

where $\mathrm{Aut}(\mathbb{P}_1)$ acts diagonally. Using $\mathrm{Aut}(\mathbb{P}_1)$ to normalize the coordinates x_0, x_1, x_{n+2} to 0, 1, ∞, we can replace Q_n by

$$\mathcal{Q}_n = \{x = (x_2, \ldots, x_{n+1}) \in \mathbb{C}^n : x_i \neq 0, 1, x_i \neq x_j, i \neq j\}.$$

The space $\mathcal{Q}_n$ has a natural underlying quasi-projective variety structure. The weights are given by $n + 3$ numbers $\mu = \{\mu_i\}_{i=0}^{n+2}$. We assume throughout that the μ_i are all rational numbers satisfying the so-called ball $(n + 3)$-tuple condition (in [**Mo2**], it is called the disc $(n + 3)$-tuple condition) given by

$$\sum_{i=0}^{n+2} \mu_i = 2, \qquad 0 < \mu_i < 1, \ i = 0, \ldots, n + 2.$$

For each choice of μ, there is a system $\mathcal{H}_\mu$ of linear partial differential equations in the n complex variables $x_2, \ldots, x_{n+1}$, whose $(n+1)$-dimensional solution space gives rise to functions generalizing the classical hypergeometric functions of Part I. These functions are named after Gauss (when $n = 1$), after Appell (when $n = 2$) and after Lauricella (when $n \geq 3$). The space of regular points for $\mathcal{H}_\mu$ is $\mathcal{Q}_n$. When $n = 1$, we recover the discussion of Part 1 once we set $\mu_0 = c - b$, $\mu_1 = a + 1 - c$, $\mu_2 = b$, $\mu_3 = 1 - a$. The ball 4-tuple condition is implied by the inequalities $0 < a < c$, $0 < b < c$, $c < 1$ assumed in Part I. When $n = 2$, let $x = x_2$, $y = x_3$ and let a, b, b', c be such that $\mu_0 = c - b - b'$, $\mu_1 = a + 1 - c$, $\mu_2 = b$, $\mu_3 = b'$, $\mu_4 = 1 - a$. Let $F_\mu = F_\mu(x,y) = F(a,b,b',c;x,y)$ be the multi-valued function on $\mathcal{Q}_2$ given by the solution of $\mathcal{H}_\mu$ with a branch defined for $|x|, |y| < 1$ by the series

$$F(a,b,b',c;x,y) = \sum_{m,n} \frac{(a, m+n)(b, m)(b', n)}{(c, m+n)(1, m)(1, n)} x^m y^n, \qquad |x|, |y| < 1.$$

Notice that $F(a,b,b',c;x,0) = F(a,b,c;x)$, $|x| < 1$. The system $\mathcal{H}_\mu$ is given, in this case, by the three equations

$$x(1-x)\frac{\partial^2 F}{\partial x^2} + y(1-x)\frac{\partial^2 F}{\partial x \partial y} + (c - (a+b+1)x)\frac{\partial F}{\partial x} - by\frac{\partial F}{\partial y} - abF = 0,$$

$$y(1-y)\frac{\partial^2 F}{\partial y^2} + x(1-y)\frac{\partial^2 F}{\partial y \partial x} + (c - (a+b'+1)y)\frac{\partial F}{\partial y} - b'x\frac{\partial F}{\partial x} - ab'F = 0,$$

$$(x-y)\frac{\partial^2 F}{\partial x \partial y} - b'\frac{\partial F}{\partial x} + b\frac{\partial F}{\partial y} = 0.$$

Returning to the general case $n \geq 1$, for $x \in \mathbb{C}^n$ consider the differential form

$$\omega(\mu; x) = u^{-\mu_0}(u-1)^{-\mu_1} \prod_{i=2}^{n+1} (u - x_i)^{-\mu_i} du.$$

Let $F_\mu = F_\mu(x)$ be the multi-valued function on $\mathcal{Q}_n$ given by the solution of $\mathcal{H}_\mu$ with a branch defined for $|x_i| < 1$, $i = 2, \ldots, n+1$, by a holomorphic series with constant term equal 1, and analytic continuation given by the Euler integral representation

$$F_\mu(x) = \int_1^\infty \omega(\mu; x) \Big/ \int_1^\infty \omega(\mu; 0), \qquad x \in \mathcal{Q}_n.$$

We will mainly work with this branch of F_μ in what follows. As in the case $n = 1$ of Part I, our arguments are valid for any branch of the multi-valued function F_μ since this only affects the choice of path of integration in the numerator of the above expression. Notice that we may write the denominator of this last expression in terms of the classical beta function, namely,

$$\int_1^\infty \omega(\mu; 0) = B(1 - \mu_{n+2}, 1 - \mu_1) = \int_0^1 u^{-\mu_{n+2}}(1-u)^{-\mu_1} du.$$

If $\mu_1 + \mu_{n+2} = 1$, then $B(1 - \mu_{n+2}, 1 - \mu_1)$ is the product of a non-zero algebraic number and π. If $\mu_1 + \mu_{n+2} < 1$, then, up to multiplication by a non-zero algebraic number, $B(1 - \mu_{n+2}, 1 - \mu_1)$ is a non-zero period of a differential form of the second kind defined over $\overline{\mathbb{Q}}$. As $\omega(\mu; x)$ is holomorphic, by [**Wu2**], Theorem 5, for $\mu_1 + \mu_{n+2} \leq 1$, the number $F_\mu(x)$ is either zero or transcendental for all $x \in \mathcal{Q}_n \cap \overline{\mathbb{Q}}^n$.

We suppose from now on that $\mu_1 + \mu_{n+2} > 1$. Notice that, when $n = 1$, this corresponds to the condition $c < 1$, which was one of our assumptions in Part I.

We define the exceptional set $\mathcal{E}_\mu$ of F_μ to be

$$\mathcal{E}_\mu = \{x \in \mathcal{Q}_n \cap \overline{\mathbb{Q}}^n : F_\mu(x) \in \overline{\mathbb{Q}}^*\},$$

where, by $F_\mu(x) \in \overline{\mathbb{Q}}^*$, we mean that the value of some branch of F_μ at x is a non-zero algebraic number. For every $x \in \mathcal{Q}_n$, there will always be a branch of F_μ which does not vanish at x. However, if some branch vanishes at $x \in \mathcal{Q}_n$, we cannot assume some other branch takes a non-zero algebraic value at x. Wolfart's arguments in [**Wo**] showing that, when $n = 1$, the zeros of any branch of F_μ are transcendental, do not seem to readily extend to the case $n > 1$.

Let Γ_μ be the monodromy group of the system $\mathcal{H}_\mu$. The ball $(n+3)$-tuple condition ensures that we may assume Γ_μ acts on $\mathbb{B}_n$. Our new result is the following generalization of Theorem 2 in the case where Γ_μ acts discontinuously.

THEOREM 3. *Suppose that $\mu_1 + \mu_{n+2} > 1$ and that Γ_μ acts discontinuously on $\mathbb{B}_n$. Then the exceptional set $\mathcal{E}_\mu$ of F_μ is Zariski dense in $\mathcal{Q}_n$ if and only if Γ_μ is an arithmetic lattice in $PU(1, n)$.*

A lattice Γ in $PU(1, n)$ is a discrete subgroup of $PU(1, n)$ such that the quotient group $\Gamma \backslash PU(1, n)$ has finite Haar measure. A subgroup of $PU(1, n)$ acting discontinuously on $\mathbb{B}_n$ is a discrete subgroup of $PU(1, n)$. Conversely, as $PU(1, n)$ acts transitively on $\mathbb{B}_n$ with compact isotropy group, a discrete subgroup of $PU(1, n)$ acts discontinuously on $\mathbb{B}_n$. By [**Mo2**], Proposition 5.3, if the monodromy group Γ_μ associated to a ball $(n+3)$-tuple μ is a discrete subgroup of $PU(1, n)$, then it is a lattice in $PU(1, n)$. Therefore the Γ_μ acting discontinuously on $\mathbb{B}_n$ are precisely the Γ_μ that are lattices in $PU(1, n)$, and we shall use both descriptions in what follows.

When $n = 1$, there are infinitely many Γ_μ acting discontinuously on the unit disk. In fact, the Schwarz triangle groups $\Delta(p, q, r)$ with p, q, r either integers at least 2, or infinity, and $p^{-1} + q^{-1} + r^{-1} < 1$ provide examples (the numbers $|1 - \mu_i - \mu_j|^{-1}$, $i \neq j$, of the ball 4-tuple μ equal p, q or r). Up to conjugation, the subgroup $\Delta(p, q, r)$ of $PU(1, 1)$ is determined by the following presentation in terms of generators and relations:

$$\langle M_1, M_2, M_3 : M_1^p = M_2^q = M_3^r = M_1 M_2 M_3 = 1 \rangle,$$

with the corresponding relation being omitted if p, q or r is infinite. When $n = 1$, the property of being Zariski dense in $\mathcal{Q}_1 = \mathbb{P}_1 \setminus \{0, 1, \infty\}$ means being of infinite cardinality, and Theorem 2 implies Theorem 3 in that case. Picard studied the monodromy groups in the case $n = 2$. His work was made rigorous and extended to the case $n > 2$ by Terada [**Te**] and Deligne–Mostow [**DeMo**], who were mainly interested in finding examples of non-arithmetic Γ_μ acting discontinuously on higher dimensional spaces. When $n = 2$, there are 63 groups Γ_μ acting discontinuously, of which 19 are non-arithmetic. For $3 \leq n \leq 9$, there are 41 groups Γ_μ acting discontinuously, of which 1 is not arithmetic. For $n \geq 10$, no Γ_μ acts discontinuously.

The explicit list of all μ corresponding to these groups is given in [**Mo2**], p. 579 and pp. 584–586. (The entry 82 of the table on p. 586 of that reference is incorrect. It should be $(3/18, 3/18, 3/18, 12/18, 15/18)$. Moreover, the corresponding monodromy group is non-arithmetic, not arithmetic as claimed there. This gives an additional non-trivial non-arithmetic example covered by Theorem 3, as remarked before the case by case list at the end of this paper.)

We now prove Theorem 3. Those parts of the proof already found in [**DTT2**] are only summarized in what follows. We refer the reader to that reference for more details. For two non-zero complex numbers a and b, we write $a \sim b$ if $a/b \in \overline{\mathbb{Q}}^*$. We then say that a and b are proportional over $\overline{\mathbb{Q}}^*$.

Let $\mu = \{\mu_i\}_{i=0}^{n+2}$ be a ball $(n+3)$-tuple of rational numbers with $\mu_1 + \mu_{n+2} > 1$, and let N be the least common denominator of the μ_i. Let $K = \mathbb{Q}(\zeta)$, where $\zeta = \exp(2\pi i/N)$. For $s \in (\mathbb{Z}/N\mathbb{Z})^*$, let σ_s be the Galois embedding of K which maps ζ to ζ^s.

For $x \in \mathcal{Q}_n \cup \{0\}$, and f a divisor of N, let $X(f, \mu, x)$ be the projective non-singular curve with affine model

$$w^f = u^{N\mu_0}(u-1)^{N\mu_1} \prod_{i=2}^{n+1}(u - x_i)^{N\mu_i}.$$

The ball $(n+3)$-tuple condition ensures that $\omega(\mu; x)$, $x \in \mathcal{Q}_n$, is a holomorphic differential form on $X(N, \mu, x)$, and the condition $\mu_1 + \mu_{n+2} > 1$ ensures that $\omega(\mu; 0)$ is a holomorphic differential on $X(N, \mu, 0)$. For $x \in \mathcal{Q}_n \cup \{0\}$, let $T(N, \mu, x)$ be the connected component of the origin in the intersection over the proper divisors f of N of the kernel of the natural map between Jacobians: $\mathrm{Jac}(X(N, \mu, x)) \to \mathrm{Jac}(X(f, \mu, x))$.

The automorphism $(u, w) \mapsto (u, \zeta^{-1}w)$ of $X(N, \mu, x)$ induces an action of K on the holomorphic differential forms $H^0(T(N, \mu, x), \Omega)$ of $T(N, \mu, x)$. The principally polarized abelian variety $T_0 = T(N, \mu, 0)$ has dimension $\varphi(N)/2$. It is a power of a simple abelian variety with CM by a subfield of K. For $s \in (\mathbb{Z}/N\mathbb{Z})^*$, the dimension of the eigenspace of $H^0(T_0, \Omega)$ on which ζ acts by ζ^s is given by

$$r_s^{(0)} = 1 - \langle s(1 - \mu_1) \rangle - \langle s(1 - \mu_{n+2}) \rangle + \langle s(2 - \mu_1 - \mu_{n+2}) \rangle,$$

where $0 \le \langle x \rangle < 1$ denotes the fractional part of a real number x and $r_s^{(0)} + r_{N-s}^{(0)} = 1$. The number $B(1 - \mu_{n+2}, 1 - \mu_1)$ is proportional over $\overline{\mathbb{Q}}^*$ to a non-zero period of $\omega(\mu; 0)$. For $s \in (\mathbb{Z}/N\mathbb{Z})^*$, and $x \in \mathcal{Q}_n$, the dimension of the eigenspace of $H^0(T(N, \mu, x), \Omega)$ on which ζ acts by ζ^s is given by

$$r_s = -1 + \sum_{i=0}^{n+2} \langle s\mu_i \rangle,$$

and we have $r_s + r_{N-s} = n + 1$. Notice that $r_1 = r_1^{(0)} = 1$. The dimension of $T(N, \mu, x)$ is $(n+1)\varphi(N)/2$. When the number $\int_1^\infty \omega(\mu; x)$ is non-zero, it is proportional over $\overline{\mathbb{Q}}^*$ to a non-zero period of $\omega(\mu; x)$. For details, see [**DTT2**], §3.

For $x \in \mathcal{Q}_n \cap \overline{\mathbb{Q}}^n \cup \{0\}$, the principally polarized abelian variety $T(N, \mu, x)$ is defined over $\overline{\mathbb{Q}}$ as is the differential form $\omega(\mu; x)$. Moreover, when $x \in \mathcal{E}_\mu$, we have the relation between non-zero complex numbers

$$B(1 - \mu_1, 1 - \mu_{n+2}) \sim \int_1^\infty \omega(\mu; x),$$

which, as already remarked, implies the relation between non-zero periods

$$\int_\gamma \omega(\mu;0) \sim \int_\gamma \omega(\mu;x),$$

where γ is the cycle on each curve induced by the Pochhammer cycle between 1 and ∞.

As explained in [**ShWo**], Proposition 1, p. 6, and [**Co1**], it follows from [**Wu3**] that as these non-zero periods are proportional over $\overline{\mathbb{Q}}^*$, the abelian varieties $T(N, \mu, x)$ and T_0 must share a common simple factor B up to isogeny. By the arguments of [**Be**], §1, Example 3, the abelian variety T_0 is isogenous to B^s, the smallest power of B whose endomorphism algebra contains K. Moreover, $T(N, \mu, x)$ is isogenous to $B^s \times C \widehat{=} T_0 \times C$ for some abelian variety C whose endomorphism algebra contains K. The dimension of the eigenspace of $H^0(C, \Omega)$ on which ζ acts by ζ^s is given by $r_s^{(1)} = r_s - r_s^{(0)}$, where $r_s^{(1)} + r_{N-s}^{(1)} = n$. In particular, $r_1^{(1)} = 0$. We have the following generalization of Proposition 4.2 of [**DTT2**] to now include non-arithmetic lattices.

PROPOSITION 4. *Suppose that $\mu_1 + \mu_{n+2} > 1$ and that Γ_μ is a lattice in $PU(1, n)$. Then there is an abelian variety A_μ with CM, whose isogeny class depends only on μ, such that $x \in \mathcal{E}_\mu$ if and only if $T(N, \mu, x) \widehat{=} T_0 \times A_\mu^n$.*

PROOF. Suppose that $x \in \mathcal{E}_\mu$. We first treat the case where Γ_μ is arithmetic. By [**Mo2**], Proposition 5.4, the group Γ_μ is arithmetic if and only if $r_s = 0$ or $n+1$ for all $1 < s < N - 1$ coprime to N. By the ball $(n + 3)$-tuple condition, we have $r_1 = 1$, $r_{N-1} = n$. By the discussion preceding the statement of the proposition, we have $r_1 = r_1^{(0)} = 1$ and $r_1^{(1)} r_{N-1}^{(1)} = 0$. It follows that $r_s^{(1)} r_{N-s}^{(1)} = 0$ for all $1 \le s \le N - 1$ coprime to N and therefore that the abelian variety C is isogenous to A_μ^n, where A_μ has CM and is, up to isogeny, independent of x and determined solely by μ (see [**Shi**], [**DTT2**]). Suppose now that Γ_μ is a non-arithmetic lattice in $PU(1, n)$. We first discuss the case $n = 2$, so that μ is a ball quintuple of rational numbers with $\mu_1 + \mu_4 > 1$. By [**CoWo2**], Lemmas 1 and 2, p. 676, with μ_0, μ_2 replaced by μ_1, μ_4, we know that for all μ satisfying the ΣINT condition of Mostow [**Mo1**], we have $\langle s\mu_1 \rangle + \langle s\mu_4 \rangle > 1$ for all $1 \le s \le N - 1$ coprime to N with $r_s = 1$, which implies for these s that $r_s^{(0)} = 1$ and $r_s^{(1)} = 0$. (Note that the references [**CoWo1**] and [**CoWo2**] again feature my maiden name Cohen.) Of course, one can also check this directly for the finite list of all Γ_μ satisfying ΣINT given in [**Mo2**]. The non-arithmetic lattices Γ_μ in $PU(1, n)$ not satisfying ΣINT are listed in [**Mo2**], §5.1, and were also studied by Sauter [**Sa**]. For $n = 2$, there are four such μ which give rise to non-arithmetic groups. One checks directly for these four μ that we always have $\mu_1 + \mu_4 < 1$ for all permutations of the indices of the μ, except in the case $\mu = (\frac{4}{18}, \frac{11}{18}, \frac{5}{18}, \frac{5}{18}, \frac{11}{18})$. For this μ one sees directly that $\langle s\mu_1 \rangle + \langle s\mu_4 \rangle > 1$ for all $1 \le s \le N - 1$ coprime to N with $r_s = 1$. Note that, when $n = 2$ and $r_s r_{N-s} \ne 0$, we must have $r_s = 1$ or 2 and $r_{N-s} = 3 - r_s$. When $n = 2$, for all μ with Γ_μ non-arithmetic, it therefore follows that if $s \in (\mathbb{Z}/N\mathbb{Z})^*$ has $r_s = 1$, then $r_s^{(0)} = 1$ and therefore $r_s^{(1)} = 0$, $r_{N-s}^{(1)} = 2$. Therefore, as in the arithmetic case with $n = 2$, the abelian variety C must always be isogenous to the square A_μ^2 of an abelian variety with CM whose isogeny class is independent of $x \in \mathcal{E}_\mu$ and depends only on μ. There is only one remaining non-arithmetic lattice when $n \ge 3$, namely the sextuple $\mu = (\frac{7}{12}, \frac{5}{12}, \frac{3}{12}, \frac{3}{12}, \frac{3}{12}, \frac{3}{12})$. In this case we have $\mu_i + \mu_j \leq 1$ for all $i \neq j$, which we have excluded.

Conversely, for a lattice Γ_μ in $\mathrm{PU}(1, n)$ with $\mu_1 + \mu_{n+2} > 1$, let A_μ be the abelian variety with CM whose isogeny class depends only on μ and is determined as above. Suppose that $T(N, \mu, x) \hat{=} T_0 \times A_\mu^n$. Therefore $x \in \mathcal{Q}_n \cap \overline{\mathbb{Q}}^n$, and $T(N, \mu, x)$ has CM. As $r_1 = r_1^{(0)} = 1$, the eigendifferentials $\omega(\mu; x)$ and $\omega(\mu; 0)$ generate, on $T(N, \mu, x)$ and T_0, respectively, the 1-dimensional eigenspaces for the action of K on the differentials of the first kind via the identity Galois embedding. Furthermore, as $T(N, \mu, x)$ has CM, the non-zero periods of $\omega(\mu; x)$ are all proportional to each other over $\overline{\mathbb{Q}}^*$ [**Shi**]. Similarly, the non-zero periods of $\omega(\mu; 0)$ are all proportional to each other over $\overline{\mathbb{Q}}^*$. In each case, the 1-dimensional $\overline{\mathbb{Q}}$-vector space generated by these periods is an isogeny invariant. Therefore

$$\int_\gamma \omega(\mu; 0) \sim \int_\gamma \omega(\mu; x),$$

and $F_\mu(x) \in \overline{\mathbb{Q}}^*$, so that $x \in \mathcal{E}_\mu$.

We now complete the proof of Theorem 3. As already remarked, the abelian varieties $T(N, \mu, x)$, $x \in \mathcal{Q}_n$, have generalized CM by K of so-called type $\Phi_\mu = \sum_{s \in (\mathbb{Z}/N\mathbb{Z})^*} r_s \sigma_s$, which encodes the representation of K on the holomorphic 1-forms of the $T(N, \mu, x)$. The data (K, Φ_μ) determine a complex symmetric domain

$$\mathcal{H}(K, \Phi_\mu) = \prod_{s \in (\mathbb{Z}/N\mathbb{Z})^*/\{\pm 1\}} \mathcal{H}_s,$$

where $\mathcal{H}_s$ is a point if $r_s r_{N-s} = 0$ and, otherwise,

$$\mathcal{H}_s = \{z \in M_{r_s, r_{N-s}}(\mathbb{C}) : 1 - z^t \bar{z} \text{ positive definite}\}.$$

As we saw during the course of the proof of Proposition 4, when Γ_μ is an arithmetic lattice, we have $r_s r_{N-s} = 0$, when $s \neq 1, N-1$, $(s, N) = 1$, and $(r_1, r_{N-1}) = (1, n)$, so that $\mathcal{H}(K, \Phi_\mu) = \mathbb{B}_n$. Those non-arithmetic lattices not excluded by the extra condition $\mu_1 + \mu_{n+2} > 1$ all occur for $n = 1, 2$, so that $r_s = 0, 1, 2$ and $r_{N-s} = 2 - r_s$ when $n = 1$, and $r_s = 0, 1, 2, 3$ and $r_{N-s} = 3 - r_s$ when $n = 2$. Therefore, in both the arithmetic and non-arithmetic cases, we have $\mathcal{H}(K, \Phi_\mu) = \mathbb{B}_n^m$, with $m > 1$ when Γ_μ is non-arithmetic. The abelian varieties $T(N, \mu, x)$, $x \in \mathcal{Q}_n$, are principally polarized and we can assume they have lattices isomorphic to $M = \mathbb{Z}[\zeta]^{(n+1)}$. The data (K, Φ_μ, M) determine a Shimura variety S which is the quotient $\Gamma' \backslash \mathcal{H}(K, \Phi_\mu)$ of $\mathcal{H}(K, \Phi_\mu)$ by an arithmetic group Γ'.

When Γ_μ is arithmetic, it is of finite index in Γ' (see, for example, the discussion in [**CoWo2**], §3, which generalizes easily to our case). Therefore there is a Zariski dense subset of $x \in \mathcal{Q}_n$ with $T(N, \mu, x)$ isogenous to $T_0 \times A_\mu^n$.

In the non-arithmetic case, the group Γ_μ is of infinite index in Γ', and by [**CoWo1**], [**CoWo2**], [**DTT2**] there is an embedding of $\mathcal{Q}_n$ into S as a quasi-projective subvariety whose Zariski closure Z is neither a Shimura subvariety of S nor a component of the Hecke image of a Shimura subvariety of S. By the methods and results of [**KY**], [**UY**] on the André–Oort conjecture, the points of Z corresponding to abelian varieties in the same isogeny class as a fixed abelian variety with CM are not Zariski dense in Z.

Combining these facts with the result of Proposition 4 gives Theorem 3.

Notice that the hypergeometric function F_μ depends on the *ordered* $(n+3)$-tuple μ, whereas the monodromy group Γ_μ depends only on the *unordered* $(n+3)$-tuple μ. In other words, we could have stated Theorem 3 as follows: *Suppose $\mu_i + \mu_j > 1$ for some $i \neq j$ and that Γ_μ acts discontinuously on $\mathbb{B}_n$. Let ν be any permutation of μ such that $\nu_1 + \nu_{n+2} > 1$. Then the exceptional set $\mathcal{E}_\nu$ of F_ν is Zariski dense in $\mathcal{Q}_n$ if and only if Γ_μ is an arithmetic lattice in $PU(1,n)$.* When $n = 1$, we have a 4-tuple $\mu_0, \mu_1, \mu_2, \mu_3$ with sum equal 2. Except in the case when all the μ_i equal $1/2$ (so that Γ_μ is the triangle group $\Delta(\infty, \infty, \infty)$, which is arithmetic, and $\mathcal{E}_\mu$ is empty) we always have $\mu_i + \mu_j > 1$ for some $i \neq j$.

The list of ball $(n + 3)$-tuples, $n \geq 2$, such that Γ_μ acts discontinuously on $\mathbb{B}_n$ is given in [**Mo2**]. In light of our results, we have three possibilities for these μ:

Case (I): If $\mu_i + \mu_j \leq 1$ for all $i \neq j$, then the exceptional set $\mathcal{E}_\nu$ of F_ν is empty for any permutation ν of μ.

Case (II): If $\mu_i + \mu_j > 1$ for some $i \neq j$, and Γ_μ is an arithmetic lattice in $PU(1,n)$, then $\mathcal{E}_\nu$ is Zariski dense in $\mathcal{Q}_n$ for all permutations ν of μ with $\nu_1 + \nu_{n+2} > 1$. However $\mathcal{E}_\nu$ is empty for all permutations ν of μ with $\nu_1 + \nu_{n+2} \leq 1$.

Case (III): If $\mu_i + \mu_j > 1$ for some $i \neq j$, and Γ_μ is a non-arithmetic lattice in $PU(1,n)$, then $\mathcal{E}_\nu$ is not Zariski dense in $\mathcal{Q}_n$ for any permutation ν of μ. Moreover $\mathcal{E}_\nu$ is empty for all permutations ν of μ with $\nu_1 + \nu_{n+2} \leq 1$.

As remarked previously, case (I) follows from [**Wu2**], Theorem 5, its statement remaining true even in the non-discontinuous case. Case (II) is already contained in [**DTT2**] and was revisited here in our proof of Theorem 3. The new advance of Part II of the present paper is the unconditional treatment of case (III) (in [**DTT2**] we needed a conjecture of Pink that is still open). Of course, combining case (I) and case (III) we have: *if Γ_μ is a non-arithmetic lattice in $PU(1,n)$ then $\mathcal{E}_\nu$ is not Zariski dense in $\mathcal{Q}_n$ for any permutation ν of μ.*

We show below how the list, given in [**Mo2**], of ball $(n + 3)$-tuples μ with corresponding monodromy group Γ_μ a lattice in $PU(1,n)$ is distributed between these three cases, in particular showing that case (III) is non-empty. In case (I), we write (NA) when Γ_μ is not arithmetic, the rest being arithmetic. Instead of writing μ, we write $(n; N; N\mu_0, \dots, N\mu_{n+2})$, where, as before, N is the least common denominator of the μ_i and $n = n + 3 - 3$ corresponds to the number of variables in the associated hypergeometric functions. Notice that the non-trivial cases, namely (II) and (III), all occur in dimension at most 4. The non-arithmetic entry (2;18;3,3,3,12,15) in case (III) below is incorrect in [**Mo2**], where it is listed as (2;18;3,3,3,13,10), for which the sum of the μ_i does not equal 2, and it is also falsely asserted there that the associated Γ_μ is arithmetic.

Case (I): (9;6;1,1,1,1,1,1,1,1,1,1,1,1), (8;6;1,1,1,1,1,1,1,1,1,1,2), (7;6;1,1,1,1,1,1,1,1,1,3),(7;6;1,1,1,1,1,1,1,1,2,2), (6;6;1,1,1,1,1,1,1,1,1,4), (6;6;1,1,1,1,1,1,1,2,3), (6;6;1,1,1,1,1,1,1,2,2,2), (5;4;1,1,1,1,1,1,1,1), (5;6;1,1,1,1,1,1,1,5), (5;6;1,1,1,1,1,1,2,4), (5;6;1,1,1,1,1,2,2,3), (5;6;1,1,1,1,1,3,3), (5;6;1,1,1,2,2,2,2), (4;4;1,1,1,1,1,1,2), (4;6;1,1,1,1,2,2,4), (4;6;1,1,1,1,2,3,3), (4;6;1,1,1,2,2,2,3), (4;6;1,1,2,2,2,2,2), (4;10;2,3,3,3,3,3,3), (4;12;2,2,2,2,2,2,7,7), (3;12;1,3,5,5,5,5), (3;3;1,1,1,1,1,1), (3;4;1,1,1,1,1,3), (3;4;1,1,1,1,2,2),

(3;6;1,1,1,3,3,3), (3;6;1,1,2,2,2,4), (3;6;1,1,2,2,3,3), (3;6;1,2,2,2,2,3),
(3;8;1,3,3,3,3,3), (3;10;2,3,3,3,3,6), (3;10;3,3,3,3,3,5), (3;12;3,3,3,3,5,7)(NA),
(3;12;3,3,3,5,5,5), (2;21;4,8,10,10,10)(NA), (2;24;5,10,11,11,11)(NA),
(2;30,7,13,13,13,14)(NA), (2;3,1,1,1,1,2), (2;4;1,1,2,2,2), (2;5;2,2,2,2,2),
(2;6;1,2,3,3,3), (2;3;2,2,2,3,3), (2;8;3,3,3,3,4), (2;9;2,4,4,4,4), (2;12;3,5,5,5,6),
(2;12;4,4,4,5,7)(NA), (2;12;4,4,5,5,6)(NA), (2;12;4,5,5,5,5), (2;14;5,5,5,5,8),
(2;15;4,5,5,5,8)(NA); (2;18;5,7,7,7,10), (2;18;7,7,7,7,8)(NA), (2;20;6,6,9,9,10)(NA),
(2;24;7,9,9,9,14)(NA), (2;42;13,15,15,15,26)(NA).

Case (II): (4;6;1,1,1,1,1,2,5), (4;6;1,1,1,1,1,3,4), (3;6;1,1,1,1,3,5),
(3;6;1,1,1,1,4,4), (3;6;1,1,1,2,2,5), (3;6;1,1,1,2,3,4), (3;12;2,2,2,2,7,9),
(3;12;2,2,2,4,7,7), (2;12;1,3,5,5,10), (2;10;1,1,4,7,7), (2;12;1,2,7,7,7),
(2;14;3,3,4,9,9), (2;15;2,4,8,8,8), (2;4;1,1,1,2,3), (2;4;1,1,1,4,5),
(2;6;1,1,2,3,5), (2;5;1,1,2,4,4), (2;6;1,1,3,3,4), (2;6;1,2,2,2,5),
(2;6;1,2,2,3,4), (2;8;1,3,3,3,6), (2;8;2,2,2,5,5), (2;10;1,4,4,4,7), (2;10,2,3,3,3,9),
(2;10;2,3,3,6,6), (2;10;3,3,3,3,8), (2;10,3,3,3,5,6), (2;12;1,5,5,5,8), (2;12;2,2,2,7,11),
(2;12;2,2,2,9,9), (2;12;2,2,4,7,9), (2;12;2,2,6,7,7), (2;12;2,4,4,7,7), (2;12;3,3,3,5,10),
(2;12;3,3,5,5,8), (2;14;2,5,5,5,11), (2;18;1,8,8,8,11), (2;18;2,7,7,10,10),
(2;20;6,6,6,9,13), (2;30;5,5,5,19,26), (2;30;9,9,9,11,22).

Case (III): (2;18;4,5,5,11,11), (2;12;3,3,3,7,8), (2;12;3,3,5,6,7), (2;18;3,3,3,12,15),
(2;18;2,7,7,7,13), (2;20;5,5,5,11,14), (2;24;4,4,4,17,19), (2;30,5,5,5,22,23),
(2;42;7,7,7,29,34).

Acknowledgments. The author was supported in part by NSF Grant DMS-0800311.

References

[An]   Y. André, *G-functions and geometry*, Aspects of Mathematics, E13, Vieweg, Braun-schweig, 1989.

[Ap]   R. Apéry, *L'irrationalité de $\zeta(2)$ et $\zeta(3)$*, Astérisque **61** (1979), 11–13.

[Bak]   A. Baker, *Transcendental number theory*, Cambridge University Press, 1975, reissued with additional material in 1979 and again in the Camb. Math. Library series in 1990.

[Be]   D. Bertrand, *Endomorphismes de groupes algébriques; applications arithmétiques*, Prog. Math. **31** (1983), 1–45.

[BuTu]   E.B. Burger, R. Tubbs, *Making transcendence transparent. An intuitive approach to classical transcendental number theory*, Springer, 2004.

[Co1]   P.B. Cohen, *Humbert surfaces and transcendence properties of automorphic functions*, Rocky Mountain J. Math. **26** (1996), 987–1001.

[CoWo1] P. Cohen, J. Wolfart, *Modular embeddings for some nonarithmetic Fuchsian groups*, Acta Arithmetica **LVI** (1990), 93–110.

[CoWo2] P.B. Cohen, J. Wolfart, *Fonctions hypergéométriques en plusieurs variables et espaces de modules de variétés abéliennes*, Ann. Scient. Éc. Norm. Sup., 4e série **26** (1993), 665–690.

[CoWu]   P.B. Cohen, G. Wüstholz, *Application of the André–Oort conjecture to some questions in transcendence*, Panorama in number theory, A view from Baker's garden (ed. by G. Wüstholz), Cambridge University Press, Cambridge, 2002, pp. 89–106.

[DeMo]   P. Deligne, G.D. Mostow, *Commensurabilities among lattices in $PU(1, n)$*, Ann. Math. Stud. **132**, Princeton University Press, Princeton, New Jersey, USA, 1993.

[De]   G. Derome, *Transcendance des valeurs de fonctions automorphes de Siegel*, J. Number Theory **85** (2000), 18–34.

[DTT1]   P.-A. Desrousseaux, M.D. Tretkoff, P. Tretkoff, *Transcendence of values at algebraic points for certain higher order hypergeometric functions*, IMRN **61** (2005), 3835–3854.

[DTT2]   P.-A. Desrousseaux, M.D. Tretkoff, P. Tretkoff, *Zariski-density of exceptional sets for hypergeometric functions*, Forum Mathematicum **20** (2008), 187–199.

[EdYa]   S. Edixhoven, A. Yafaev, *Subvarieties of Shimura varieties*, Ann. Math. **157** (2003), 621–645.

[Her]   Ch. Hermite, *Sur la fonction exponentielle*, Comptes Rendus de l'Acad. des Sci. (Paris) **77** (1873), 18–24, 74–79, 226–233, 285–293.

[KY]   B. Klingler, A. Yafaev, *The André–Oort conjecture*, preprint.

[Lin]   F. Lindemann, *Über die Zahl π*, Math. Annalen **20** (1882), 213–225.

[Lio1]   J. Liouville, *Sur des classes très-étendues de quantités dont la valeur n'est ni algébriques, ni même reductible à des irrationelles algébriques*, Comptes Rendus de l'Acad. des Sci. (Paris) **18** (1844), 883–885, 910–911.

[Mo1]   G.D. Mostow, *Generalized Picard lattices arising from half-integral conditions*, Publ. Math. IHES **63** (1986), 91–106.

[Mo2]   G.D. Mostow, *On discontinuous action of monodromy groups on the complex n-ball*, J. Amer. Math. Soc. **1** (1988), 555–586.

[Oo]   F. Oort, *Canonical liftings and dense sets of CM points*, Arithmetic geometry (Cortona, 1994) Sympos. Math. XXXVII, Cambridge University Press, Cambridge, 1997, pp. 228–234.

[Pi]   E. Picard, *Sur une extension aux fonctions de deux variables du problème relatif aux fonctions hypergéométriques*, Ann. E.N.S. **10** (1881), 305–321.

[Sa]   J.K. Sauter, Jr., *Isomorphisms among monodromy groups and applications to lattices in PU(1,2)*, Pacific J. Math. **146** (1990), 331–384.

[Sch]   Th. Schneider, *Arithmetische Untersuchungen elliptischer Integrale*, Math. Annalen **113** (1937), 1–13.

[Schw]   H.A. Schwarz, *Über diejenigen Fälle, in welchen die Gaußische hypergeometrische Reihe eine algebraische Funktion ihres vierten Elements darstellt*, J. reine angew. Math. **75** (1873), 1183–1205.

[ShWo]   H. Shiga, J. Wolfart, *Criteria for complex multiplication and transcendence properties of automorphic functions*, J. Reine Angew. Math. **463** (1995), 1–25.

[Shi]   G. Shimura, *On analytic families of polarized abelian varieties and automorphic functions*, Ann. Math. **78** (1963), 149–192.

[Si1]   C.-L. Siegel, *Über einige Anwendungen diophantischer Approximationen*, Abh. Pr. Akad. Wiss. **1** (1929).

[Si2]   C.-L. Siegel, *Über die Perioden elliptischer Funktionen*, J. Reine Angew. Math. **167** (1932), 62–69.

[Ta]   K. Takeuchi, *Arithmetic triangle groups*, J. Math. Soc. Japan **29** (1977), 91–106.

[Te]   T. Terada, *Probléme de Riemann et fonctions automorphes provenant des fonctions hypergéométriques de plusieurs variables*, J. Math. Kyoto Univ. **13** (1973), 557–578.

[UY]   E. Ullmo, A. Yafaev, *Galois orbits and equidistribution: towards the André–Oort conjecture*, preprint.

[Wo]   J. Wolfart, *Werte hypergeometrische Funktionen*, Invent. Math. **92** (1988), 187–216.

[Wu1]   G. Wüstholz, *Zum Periodenproblem*, Invent. Math. **78** (1984), 381–391.

[Wu2]   G. Wüstholz, *Algebraic groups, Hodge theory and transcendence*, Proceedings of the International Congress of Mathematics 1, Berkeley, California, USA, 1986, pp. 476–483.

[Wu3]   G. Wüstholz, *Algebraische Punkte auf analytischen Untergruppen algebraischer Gruppen*, Ann. Math. **129** (1989), 501–517.

[Zu]   V.V. Zudilin, *One of the numbers ζ(5), ζ(7), ζ(9), ζ(11) is irrational*, Russ. Math. Surveys **56** (2001), 774–776.

DEPARTMENT OF MATHEMATICS, MAILSTOP 3368, TEXAS A&M, COLLEGE STATION, TX 77842-3368, USA.

E-mail address: ptretkoff@math.tamu.edu

The Hopf Algebraic Structure of Perturbative Quantum Gauge Theories

Walter D. van Suijlekom

ABSTRACT. This paper gives an overview of renormalization of quantum field theories in terms of Hopf algebras of Feynman graphs. Particular emphasis is on gauge theories, for which a gauge group symmetry implies certain identities between Feynman graphs. These identities generate Hopf ideals in the Hopf algebra, thus reflecting compatibility of gauge symmetries with renormalization. We study the origin of these Hopf ideals through the definition of a coaction of the Hopf algebra on a Gerstenhaber algebra generated by the fields and coupling constants.

1. Introduction

Quantum gauge theories are most successfully described perturbatively, expanding around the free quantum field theory. As a matter of fact, at present its non-perturbative formulation seems to be far beyond reach so it is the only thing we have. On the one hand, many rigorous results can be obtained [**BBH94, BBH95**] using cohomological arguments within the context of the BRST-formalism [**BRS74, BRS75, BRS76**] and independently [**Tyu75**]. On the other hand, renormalization of perturbative quantum field theories has been carefully structured using Hopf algebras [**Kre98, CK99, CK00**]. The presence of a gauge symmetry induces a rich additional structure on these Hopf algebras, as has been explored in [**Kre05, KY06, BKUY08, BKUY09**] and in the author's own work [**Sui07, Sui07c, Sui08**]. All of this work is based on the algebraic transparency of BPHZ-renormalization, with the Hopf algebra reflecting the recursive nature of this procedure.

In this paper, which is based on lectures delivered at Johns Hopkins University in spring 2009, we try to explain what renormalization is and how it works for gauge theories. Hopf algebras nicely structure this procedure, allowing a more mathematically inclined approach than taken in the usual textbooks in physics.

After the introduction of Feynman graphs and their group structure (encoded dually by a commutative Hopf algebra), we formulate (à la Connes-Kreimer) renormalization as a Birkhoff decomposition. Then in Section 3 we introduce gauge theories and point to the problems that arise when going to their quantum versions. This introduces cohomological techniques, entering into Gerstenhaber and BV-algebras.

In Section 4 we describe the Hopf algebra of Feynman graphs for gauge theories and discuss its algebraic structure. In particular, we identify certain Hopf subalgebras generated by so-called Green's functions, and prove the existence of Hopf ideals. The latter implement quadratic relations between the Green's functions, known in the physics literature as Slavnov–Taylor identities for the couplings. They reflect (at the quantum level) the presence of a gauge symmetry at the classical level. The fact that they can be implemented as relations in the Hopf algebras is the statement that renormalization is compatible with the Slavnov–Taylor identities.

This is then nicely connected to the aforementioned Gerstenhaber algebra via a coaction of the Hopf algebra on it. As an application of this construction, we find that the gauge symmetries (as in the Slavnov–Taylor identities) are compatible with the Hopf algebraic structure and thus with renormalization.

2. Renormalization of Perturbative Quantum Field Theories

We start by giving some background from physics and try to explain the origin of Feynman graphs in the perturbative approach to quantum field theory.

We understand *probability amplitudes for physical processes as formal expansions in Feynman amplitudes*, thereby avoiding the use of path integrals. For example, in so-called ϕ^3-scalar field theory, one considers expansions such as

$$G^{\prec} = \quad \prec \quad + \quad \prec\!\!\!\prec \quad + \quad \prec\!\!\!\Diamond \quad + \cdots .$$

Thus, one adds all graphs that have three external lines, with the restriction that all vertices in the graphs have valence 3 only. More generally, one allows for any number of external lines (with the valence of the vertices remaining 3 for ϕ^3-theory). We shall call these expansions **Green's functions**. Of course, this names originates from the theory of partial differential equations and the zeroth order terms in the above expansions are in fact Green's functions in the usual sense.

From these expansions, physicists can actually derive numbers, giving the probability amplitudes mentioned above. The rules of this game are known as the Feynman rules; we briefly list them for our example of ϕ^3-scalar field theory.

Assigning momentum k to each edge of a graph, we have:

$$\underline{\qquad\qquad} = \frac{1}{k^2 + m^2}$$

$$\prec = g\delta(k_1 + k_2 + k_3),$$

where g is the so-called **coupling constant**, a physical parameter measuring the strength of the interaction. In addition to the above assignments, one integrates the above internal momenta k (for each internal edge) over $\mathbb{R}^4$. Finally, one divides by the **symmetry factor**; it is by definition the order of the automorphism group of the graph with fixed external edges.

EXAMPLE. Consider the following graph:

According to the Feynman rules, in 4 dimensions the amplitude for this graph is

$$U(\Gamma) = \frac{g^2}{2}\frac{1}{p^2+m^2}\left(\int d^4k\frac{1}{(p+k)^2+m^2}\frac{1}{k^2+m^2}\right)\frac{1}{p^2+m^2}.$$

As a bookkeeping device for these Feynman rules physicists work with the **action functional**, in this case a functional of the field ϕ:

$$S[\phi] = \int d^4x\left[\phi(-\Delta+m^2)\phi+g\phi^3\right],$$

where Δ is the Laplacian on $\mathbb{R}^4$. The quadratic part corresponds to the edges in the graph; indeed, the above $\frac{1}{k^2+m^2}$ is the Fourier transform of the Green's function for $-\Delta+m^2$, as alluded to before. The remaining cubic part represents the interaction: the coefficient is the value associated to the vertex $\prec$.

2.1. Renormalization

The alert reader may have noted that the above improper integral is actually not well defined. This is the typical situation—it happening for most graphs—and such integrals are the source of the famous divergences in perturbative quantum field theory. This apparent failure can be resolved, leading eventually to spectacularly accurate predictions in physics.

The theory that proposes a solution to these divergences is called *renormalization*. This process consists of two steps: *regularization* and *subtraction*. Let us give two examples of a regularization prescription.

The first we consider is a *momentum cut-off*. This means that we perform the integral above up to a real parameter Λ. More precisely, we make the replacement

$$\int d^4k \rightsquigarrow \int_{|k|\leq\Lambda} d^4k.$$

Let us consider the type of integrations we would like to perform, in its simplest form. If the integrand is $(k^2+m^2)^{-2}$ then

$$\int_{|k|\leq\Lambda} d^4k\frac{1}{(k^2+m^2)^2} \sim \log\Lambda.$$

This explain the divergent integrals we encountered above as $\Lambda\to\infty$, but we now have control of the divergence. Although the momentum cut-off regularization is simple and physically natural, it is not the best regularization prescription for gauge theories since it breaks the gauge invariance. Nevertheless, it is the starting point for the powerful Wilsonian approach to renormalization, which has been studied in the Hopf algebraic setup as well in [**KM08**].

Another regularization prescription is *dimensional regularization*. Instead of integrating in 4 dimensions, one integrates in $4-z$ dimensions, with z a complex number. Of course, this only makes sense after prescribing some rules for such an integration. The key rule is the following:

$$\int d^D k \ e^{-\pi \lambda k^2} = \lambda^{-D/2} \qquad (D \in \mathbb{C}).$$

This formula clearly holds for D a positive integer, where it is just the Gaussian integral. However, if we *demand* it to hold for any complex D, it turns out to provide a very convenient regularization prescription. So, let us consider once more integration over $(k^2 + m^2)^{-2}$, but now in $4 - z$ dimensions. We write using so-called Schwinger parameters, or, equivalently the Laplace transform

$$\frac{1}{k^2 + m^2} = \int_{s>0} ds \ e^{-sk^2 + m^2}.$$

Then, using the above equation we find that

$$\int d^{4-z} k \frac{1}{(k^2 + m^2)^2} = \int_{s>0} ds \int_{t>0} dt \int d^{4-z} k \ e^{-(s+t)(k^2 + m^2)}$$

$$= \pi^{2-z/2} \int_{s>0} ds \int_{t>0} dt \ (s+t)^{-2+z/2} e^{-(s+t)m^2},$$

where we assume that we can interchange the integrals. If we now change variables to $s = \frac{1}{2}(1+x)y$ and $t = \frac{1}{2}(1-x)y$ we obtain

$$\pi^{2-z/2} \int_0^1 dx \int_{y>0} dy y^{-1+z/2} e^{-ym^2} = \pi^{2-z/2} m^{-z} \Gamma(z/2),$$

with Γ the complex gamma function. It has a pole at $z = 0$, reflecting the divergence before regularization. Again, this gives us control over the divergence. In general, for a graph Γ we arrive at regularized Feynman amplitudes $U(\Gamma)(z)$ depending on the complex parameter z.

The second step in the process of renormalization is *subtraction*. For dimensional regularization, we let T be the projection onto the pole part of a Laurent series in z,

$$T \left[\sum_{n=-\infty}^{\infty} a_n z^n \right] = \sum_{n<0} a_n z^n.$$

More generally, we have a projection on the divergent part in the regularizing parameter. This is the origin of the study of Rota-Baxter algebras in the setting of quantum field theories [**EG07**]. We will however restrict ourselves to dimensional regularization, which is a regularization well suited for gauge theories. For the above graph Γ, we define the *renormalized amplitude* $R(\Gamma)$ by simply subtracting the divergent part, that is, $R(\Gamma) = U(\Gamma) - T[U(\Gamma)]$. Clearly, the result is finite for $z \to 0$. More generally, a graph Γ might have subgraphs $\gamma \subset \Gamma$ which lead to subdivergences in $U(\Gamma)$. The so-called BPHZ-procedure (after its inventors Bogoliubov, Parasiuk, Hepp and Zimmermann) provides a way to deal with those subdivergences in a recursive manner. It gives for the renormalized amplitude:

$$R(\Gamma) = U(\Gamma) + C(\Gamma) + \sum_{\gamma \subset \Gamma} C(\gamma) U(\Gamma/\gamma),$$

where C is the so-called counterterm defined recursively by

$$C(\Gamma) = -T\left[U(\Gamma) + \sum_{\gamma \subset \Gamma} C(\gamma) U(\Gamma/\gamma)\right].$$

The two sums here are over all subgraphs in a certain class; we will make this more precise in the next section.

2.2. Group of Feynman Graphs

It turns out that the above BPHZ-procedure can be nicely captured by the mathematical structure of a Hopf algebra. Still working in the case of scalar ϕ^3-theory, we restrict to 1PI graphs. These are graphs that are not trees and cannot be disconnected by cutting a single internal edge. For example, all graphs in this paper are one-particle irreducible, except the following which is one-particle reducible:

Connes and Kreimer then defined the following Hopf algebra.

DEFINITION. *The Hopf algebra H of Feynman graphs is the free commutative $\mathbb{Q}$-algebra generated by all 1PI Feynman graphs, with counit $\epsilon(\Gamma) = 0$ unless $\Gamma = \emptyset$, in which case $\epsilon(\emptyset) = 1$, coproduct,*

$$\Delta(\Gamma) = \Gamma \otimes 1 + 1 \otimes \Gamma + \sum_{\gamma \subsetneq \Gamma} \gamma \otimes \Gamma/\gamma,$$

where the sum is over disjoint unions of subgraphs with residue in R. The antipode is given recursively by

$$S(\Gamma) = -\Gamma - \sum_{\gamma \subsetneq \Gamma} S(\gamma)\Gamma/\gamma.$$

Two examples of this coproduct are:

$$\Delta\left(\text{—⊕—}\right) = \text{—⊕—} \otimes 1 + 1 \otimes \text{—⊕—} + 2\ \text{—◁} \otimes \text{—O—},$$

$$\Delta\left(\text{—⊕—}\right) = \text{—⊕—} \otimes 1 + 1 \otimes \text{—⊕—} + 2\ \text{—◁} \otimes \text{◁—}$$

$$+ 2\ \text{—⊕—} \otimes \text{—◁} + \text{◁—} \otimes \text{—O—}.$$

The above Hopf algebra is an example of a connected graded Hopf algebra, i.e. $H = \oplus_{n\in\mathbb{N}} H^n$, $H^0 = \mathbb{Q}$ and

$$H^k H^l \subset H^{k+l}; \qquad \Delta(H^n) = \sum_{k=0}^{n} H^k \otimes H^{n-k}.$$

Indeed, the Hopf algebra of Feynman graphs is graded by the **loop number** $L(\Gamma)$ of a graph Γ; then H^0 consists of rational multiples of the empty graph, which is the unit in H, so that $H^0 = \mathbb{Q}1$.

Now, consider the collection $G = \mathrm{Hom}_\mathbb{Q}(H, K)$ of multiplicative linear maps from H to an arbitrary field K (containing $\mathbb{Q}$). It is well known that G is a group; indeed, we have the following product, inverse and unit in the group $G(K)$:

$$\phi * \psi(X) = \langle \phi \otimes \psi, \Delta(X) \rangle$$
$$\phi^{-1}(X) = \phi(S(X))$$
$$e(X) = \epsilon(X).$$

A physical example of such a group element arises when K is the field of convergent Laurent series in z. The map U we have defined above is naturally an algebra map. It assigns to each graph Γ its Feynman amplitude as dictated by the Feynman rules, a Laurent series in the regularization parameter z. Naturally, the amplitude for a product of graphs is the product of the amplitudes and hence U is an algebra map. Another, even more interesting example is R, the renormalized Feynman amplitude. We will see soon how they are related in the group $\mathrm{Hom}_\mathbb{Q}(H, K)$.

2.3. Renormalization as a Birkhoff Decomposition

We now demonstrate how to obtain the renormalized amplitude and the counterterm for a graph as a Birkhoff decomposition in the group of characters of H. Let us first recall the definition of a Birkhoff decomposition.

We let $l : C \to G$ be a loop with values in an arbitrary complex Lie group G, defined on a smooth simple curve $C \subset \P_1(\mathbb{C})$. Let $C_\pm$ be the two complements of C in $\P_1(\mathbb{C})$, with $\infty \in C_-$. A *Birkhoff decomposition* of l is a factorization of the form

$$l(z) = l_-(z)^{-1} l_+(z); \qquad (z \in C),$$

where $l_\pm$ are (boundary values of) two holomorphic maps on $C_\pm$, respectively, with values in G. This decomposition gives a natural way to extract finite values from a divergent expression. Indeed, although $l(z)$ might not holomorphically extend to C_+, $l_+(z)$ is clearly finite as $z \to 0$.

We now look at the group $G(K) = \mathrm{Hom}_\mathbb{Q}(H, K)$ of K-valued characters of a connected graded commutative Hopf algebra H, where K is the field of convergent Laurent series in z.[1] The product, inverse and unit in the group $G(K)$ are as in the previous section. We claim that a map $\phi \in G(K)$ is in one-to-one correspondence with loops l on an infinitesimal circle around $z = 0$ and values in $G(\mathbb{Q}) = \mathrm{Hom}_\mathbb{Q}(H, \mathbb{Q})$. Indeed, the correspondence is given by

$$\phi(X)(z) = l(z)(X),$$

and to give a Birkhoff decomposition for l is thus equivalent to giving a factorization $\phi = \phi_-^{-1} * \phi_+$ in $G(K)$. It turns out that for graded connected commutative Hopf algebras such a factorization exists.

---

[1] In the language of algebraic geometry, there is an affine group scheme G represented by H in the category of commutative algebras. In other words, $G = \mathrm{Hom}_\mathbb{Q}(H, \cdot)$ and $G(K)$ are the K-points of the group scheme.

THEOREM. (Connes–Kreimer [**CK99**]) *Let H be a graded connected commutative Hopf algebra. The Birkhoff decomposition of $l : C \to G$ (given by an algebra map $\phi : H \to K$) exists and is given dually by*

$$\phi_-(X) = \epsilon(X) - T\left[m(\phi_- \otimes \phi)(1 \otimes (1 - \epsilon)\Delta(X))\right]$$

and $\phi_+ = \phi_- * \phi$.

The graded connected property of H ensures that the recursive definition of ϕ_- actually makes sense. In the case of the Hopf algebra of Feynman graphs defined above, the factorization takes the following form:

$$\phi_-(\Gamma) = -T\left[\phi(\Gamma) + \sum_{\gamma \subsetneq \Gamma} \phi_-(\gamma)\phi(\Gamma/\gamma)\right]$$

$$\phi_+(\Gamma) = \phi(\Gamma) + \phi_-(\Gamma) + \sum_{\gamma \subsetneq \Gamma} \phi_-(\gamma)\phi(\Gamma/\gamma).$$

The key point is now that the Feynman rules actually define an algebra map $U : H \to K$ by assigning to each graph Γ the regularized Feynman rules $U(\Gamma)$, which are Laurent series in z. When compared with equations for the BPHZ-renormalized value one concludes that the algebra maps U_+ and U_- in the Birkhoff factorization of U are precisely the renormalized amplitude R and the counterterm C, respectively. Summarizing, we can write BPHZ-renormalization as the Birkhoff decomposition $U = C^{-1} * R$ of the map $U : H \to K$ dictated by the Feynman rules.

Although the above construction gives a very nice geometrical description of the process of renormalization, it is a bit unphysical in that the Hopf algebra relies on individual graphs. Rather, as mentioned before, in physics the probability amplitudes are computed from the full expansion of Green's functions. Individual graphs do not correspond to physical processes and therefore a natural question to pose is how the Hopf algebra structure behaves at the level of the Green's functions. We will come back to this point later.

3. Perturbative Quantum Gauge Theories

One of the basic principles in the dictionary between the (elementary particle) physicists' and mathematicians' terminology is that "particles are representations of a Lie group."

This Lie group will be assumed to be reductive. In physics, this is typically $SU(N)$ for some N. For instance, in the case of quantum chromodynamics, it is SU(3) and the *quark* is a $\mathbb{C}^3$-valued function $\psi = (\psi_i)$ on a manifold M. This "fiber" $\mathbb{C}^3$ at each point of spacetime is the defining representation of SU(3). Thus, there is an action on ψ of an SU(3)-valued function on M; let us write this function as U, so that $U(x) \in$ SU(3). In physics, the three components of ψ correspond to the so-called color of the quark, typically indicated by red, green and blue. The *gluon*, on the other hand, is described by a $\mathfrak{su}(3)$-valued one-form on M, that is, a section of $\Lambda^1(\mathfrak{su}(3)) \equiv \Lambda^1 \otimes (M \times \mathfrak{su}(3))$.

Returning to the general case of a reductive Lie group G, a **gauge field** A is a $\mathfrak{g}$-valued one-form on M. We have in components

$$A = A_\mu dx^\mu = A_\mu^a dx^\mu T^a,$$

where the $\{T^a\}$ form a basis for $\mathfrak{g}$. The structure constants $\{f_c^{ab}\}$ of $\mathfrak{g}$ are defined by $[T^a, T^b] = f_c^{ab} T^c$ and the normalization is such that $\text{tr}\,(T^a T^b) = \delta^{ab}$. It is useful to think of A as a connection one-form (albeit on the trivial bundle $M \times G$). The group G acts on the second component $\mathfrak{g}$ in the adjoint representation. This is pointwise on M, leading to an action of $U = U(x) \in G$ on A by

$$A_\mu \mapsto g^{-1} U^{-1} \partial_\mu U + U^{-1} A_\mu U.$$

This is called a *gauge transformation*. The constant g is the so-called **coupling constant**.

As in mathematics, also in physics one is after *invariants*, in this case, one looks for functions—or, rather, functionals—of the quark and gluon fields that are invariant under a local (i.e., x-dependent) action of G. We are interested in the **Yang–Mills action functional**:

$$S_{YM}(A) = -\langle F(A), F(A) \rangle,$$

with $F \equiv F(A) := dA + \frac{1}{2} g[A, A]$ the curvature of A; it is a $\mathfrak{g}$-valued two-form on M. The inner product $\langle \cdot, \cdot \rangle$ is the combination of the inner product of differential forms and the Killing form on $\mathfrak{g}$.

3.1. Ghost Fields and the BRST-formalism

In a path integral quantization of the field theory defined by the above action, one faces the following problem. Since gauge transformations are supposed to act as symmetries on the theory, the gauge degrees of freedom are irrelevant to the final physical outcome. Thus, in one way or another, one has to quotient by the group of gauge transformations. However, gauge transformations are G-valued functions on M, yielding an infinite dimensional group. In order to deal with this infinite redundancy, Faddeev and Popov used the following trick. They introduced so-called *ghost fields*, denoted by ω and $\bar{\omega}$. These are $\mathfrak{g}[-1]$- and $\mathfrak{g}[1]$-valued functions on M, respectively. The shift $[-1]$ and $[+1]$ is to denote that ω and $\bar{\omega}$ have *ghost degree* 1 and -1, respectively. Consequently, they have *Grassmann degree* 1 and -1, respectively. In components, we write

$$\omega = \omega^a T^a; \qquad \bar{\omega} = \bar{\omega}^a T^a.$$

Finally, an auxiliary field h—also known as the Nakanishi–Lantrup field—is introduced; it is a $\mathfrak{g}$-valued function (in ghost degree 0) and we write $h = h^a T^a$.

The dynamics of the ghost fields and their interactions with the gauge field are described by the rather difficult additional term:

$$S_{gh}(A, \omega, \bar{\omega}, h) = -\langle A, dh \rangle + \langle d\bar{\omega}, d\omega \rangle + \frac{1}{2} \xi \langle h, h \rangle + g \langle d\bar{\omega}, [A, \omega] \rangle,$$

where $\xi \in \mathbb{R}$ is the *gauge parameter*.

The essential point about the ghost fields is that in a path integral formulation of quantum gauge field theories, their introduction miraculously takes care of the *fixing of the gauge*, i.e. picking a point in the orbit in the space of fields under the action of the group of gauge transformations. The ghost fields are the ingredients in the BRST-formulation that was developed later by Becchi, Rouet, Stora and independently by Tyutin in [**BRS74, BRS75, BRS76, Tyu75**]. Let us briefly describe this formalism.

Because the gauge has been fixed by adding the term S_{gh}, the combination $S + S_{gh}$ is no longer invariant under the gauge transformations. This is of course precisely the point. Nevertheless, $S + S_{gh}$ possesses another symmetry, which is the BRST-symmetry. It acts on function(al)s in the fields as a ghost degree 1 derivation s, which is defined on the generators by

$$sA = d\omega + g[A,\omega], \qquad s\omega = \frac{1}{2}g[\omega,\omega], \qquad s\bar{\omega} = h, \qquad sh = 0.$$

Indeed, one can check (e.g., see Section 15.7 in [**Wei96**] for details) that $s(S+S_{gh}) = 0$.

The fields generate an algebra, the algebra of local forms $\mathrm{Loc}(\Phi)$. With respect to the above degrees, it decomposes as before into $\mathrm{Loc}^{(p,q)}(\Phi)$ with p the form degree and q the Grassmann degree. The total degree is then $p + q$ and $\mathrm{Loc}(\Phi)$ becomes a graded Lie algebra by setting

$$[X,Y] = XY - (-1)^{\deg(X)\deg(Y)}YX,$$

with the grading given by this total degree. Note that the present graded Lie bracket is of degree 0 with respect to the total degree, that is, $\deg([X,Y]) = \deg(X) + \deg(Y)$. It satisfies graded skew-symmetry, the graded Leibniz identity and the graded Jacobi identity:

$$[X,Y] = -(-1)^{\deg(X)\deg(Y)}[Y,X]$$
$$[XY,Z] = X[Y,Z] + (-1)^{\deg(Y)\deg(Z)}[X,Z]Y$$
$$(-1)^{\deg(X)\deg(Z)}[[X,Y],Z] + (\text{cyclic permutation}) = 0.$$

LEMMA. *The BRST-differential, together with the above bracket, gives* $\mathrm{Loc}(\Phi)$ *the structure of a graded differential Lie algebra.*

Moreover, the BRST-differential s and the exterior derivative d form a double complex, that is, $d \circ s + s \circ d = 0$ and

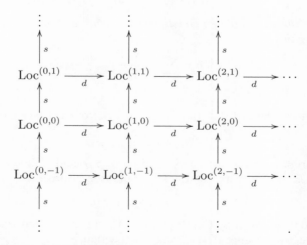

This double complex has quite interesting structure by itself, and was the subject of study in [**BBH94, BBH95**]. These contained further applications in renormalization and the description of anomalies.

3.2. The BV-formalism

A natural mathematical setting for the above BRST-formalism is deformation theory [**Sta97**]. Recall that a Gerstenhaber algebra [**Ger63**] is a graded commutative algebra with a Lie bracket of degree 1 satisfying the graded Leibniz property:

$$(x, yz) = (x, y)z + (-1)^{(|x|+1)|y|}y(x, z).$$

Batalin and Vilkovisky encountered this structure in their study of quantum gauge theories [**BV81, BV83, BV84**]. In fact, they invented what is now called a *BV-algebra* (see, for instance, [**Sta97**]): a Gerstenhaber algebra with an additional operator $\widetilde{\Delta}$ that satisfies:

$$(x, y) = \widetilde{\Delta}(xy) - \widetilde{\Delta}(x)y + (-1)^{|x|}x\widetilde{\Delta}(y).$$

In the BV-formalism (applied to Yang–Mills gauge theories) one introduces additional fields (called **anti-fields**) for each of the above fields $A, \omega, \bar{\omega}$ and h. These are denoted by $A^{\ddagger}, \omega^{\ddagger}, \bar{\omega}^{\ddagger}$ and $h^{\ddagger}$, respectively. The shift in ghost degree is $\deg(\phi^{\ddagger}) = -\deg\phi - 1$. Then we define a **BV-action** by setting

$$S_{BV} = S_{YM} + S_{gf} + \langle sA, A^{\ddagger}\rangle + \langle s\omega, \omega^{\ddagger}\rangle + \langle s\bar{\omega}, \bar{\omega}^{\ddagger}\rangle.$$

After this doubling of the fields, we introduce a bracket on the algebra $\mathfrak{F}([\Phi])$ of functionals of the fields and anti-fields by setting heuristically

$$(\phi_i^{\ddagger}(x), \phi_j(y)) = \delta_{ij}\delta(x-y), \quad (\phi_i^{\ddagger}(x), \phi_j^{\ddagger}(y)) = 0, \quad (\phi_i(x), \phi_j(y)) = 0.$$

The index i combines the type of field (i.e., $A, \omega, \bar{\omega}$ or h) with its differential form and Lie algebra indices. In terms of left and right functional derivatives[2] we have the following result.

PROPOSITION. *The bracket* $(\cdot, \cdot)$, *defined by*

$$(F_1, F_2) = \sum_i \int_M \left[\frac{\delta_R F_1}{\delta\phi_i(x)} \frac{\delta_L F_2}{\delta\phi_i^{\ddagger}(x)} - \frac{\delta_R F_1}{\delta\phi_i^{\ddagger}(x)} \frac{\delta_L F_2}{\delta\phi_i(x)} \right] d\mu(x),$$

gives $\mathfrak{F}([\Phi])$ *the structure of a Gerstenhaber algebra with respect to the ghost degree. Moreover, with*

$$\widetilde{\Delta}(F) = \sum_{i=1}^{N} \frac{\delta_R}{\delta K_{\phi_i}^a(x)} \frac{\delta_L}{\delta\phi_i^a(x)}(F),$$

it becomes a BV-algebra.

It is readily checked that the BRST-operator of the previous section is written in terms of the bracket as

$$s(F) = (S_{BV}, F).$$

BRST-invariance of the action $S = S_{YM} + S_{gf}$ and nilpotence of the BRST-differential is then captured by the single **master equation**

$$(S_{BV}, S_{BV}) = 0,$$

supposed to be satisfied by the BV-action. The upshot is that the master equation replaces gauge invariance of the Yang–Mills action.

---

[2]Recall that the left and right functional derivatives are defined for test functions ψ_ϕ of the same ghost degree as the field ϕ by $\frac{d}{dt}F[\phi+t\psi_\phi] = \int_M \frac{\delta_L F}{\delta\phi^a(x)}\psi_\phi^a(x)d\mu(x) = \int_M \psi_\phi^a(x)\frac{\delta_R F}{\delta\phi^a(x)}d\mu(x).$

4. Renormalization of Perturbative Quantum Gauge Theories Using Hopf Algebras

We now return to the Hopf algebraic description of renormalization. In the case of gauge theories, Feynman graphs are not just built from a single type of vertex or edge. In fact, from the above discussion, it is natural to let S_{BV} dictate the types of vertices and edges that constitute our Feynman graphs. This leads to two sets of types of vertices R_V and edges R_E. The gauge field A is denoted by a wiggly line, the ghost fields $\omega, \bar{\omega}$ by a dotted line and the anti-fields all appear as dashed lines:

$$R_V = \left\{ \text{} \right\}, \qquad R_E = \left\{ \text{} \right\}.$$

Note that the dashed edges do not appear in R_E, i.e. the source terms do not propagate since they do not appear quadratically in the BV-action. Consequently, they will not appear as internal edges of a Feynman graph. Although these explicit sets R_V and R_E motivate our construction, we stress that for the discussion that follows it is not necessary to specify them. The relevant structure is encoded in the action, dictating the vertices and edges in the sets R_V and R_E, respectively.

A **Feynman graph** is a graph built from the types of vertices present in R_V and the types of edges present in R_E. Naturally, we demand edges to be connected to vertices in a compatible way, respecting the type of vertex and edge. As opposed to the usual definition in graph theory, Feynman graphs have no external vertices. However, they do have *external lines* which come from vertices in Γ for which some of the attached lines remain vacant (i.e., with no edge in R_E attached). Implicit in the construction is the fact that source terms only arise as external lines since they are not in R_E, justifying the name "source term."

The **residue** res(Γ) is defined as the vertex or edge that the graph corresponds to after collapsing all its internal points. For example, we have:

$$\text{res}\left(\text{} \right) = \text{} \qquad \text{and} \qquad \text{res}\left(\text{} \right) = \text{}.$$

The definition of the Connes–Kreimer Hopf algebra easily extends to this case.

DEFINITION. [*Connes–Kreimer* [**CK99**]] *The Hopf algebra of Feynman graphs is the free commutative algebra H over $\mathbb{Q}$ generated by all 1PI Feynman graphs with residue in $R = R_V \cup R_E$, with counit $\epsilon(\Gamma) = 0$ unless $\Gamma = \emptyset$, in which case $\epsilon(\emptyset) = 1$, coproduct,*

$$\Delta(\Gamma) = \Gamma \otimes 1 + 1 \otimes \Gamma + \sum_{\gamma \subsetneq \Gamma} \gamma \otimes \Gamma/\gamma,$$

where the sum is over disjoint unions of 1PI subgraphs with residue in R. The quotient Γ/γ is defined to be the graph Γ with the connected components of the subgraph contracted to the corresponding vertex/edge.

Again, the above Hopf algebra is a connected graded Hopf algebra: it is graded by the *loop number* $L(\Gamma)$ of a graph Γ. Let us denote by q_l the projection of H onto graphs of loop number l.

In addition, there is another grading on this Hopf algebra. It is given by the number of vertices and it already appeared in [**CK99**]. However, since we consider vertices and edges of different types (wiggly, dotted, straight, etc.), we extend to a multigrading as follows. As in [**Sui07**], we denote by $m_{\Gamma,r}$ the number of vertices/internal edges of type r appearing in Γ, for $r \in R$. Moreover, let $n_{\gamma,r}$ be the number of connected components of γ with residue r. For each $v \in R_V$ we define a degree d_v by setting

$$d_v(\Gamma) = m_{\Gamma,v} - n_{\Gamma,v}.$$

The multidegree indexed by R_V is compatible with the Hopf algebra structure, as follows easily from the following relation:

$$m_{\Gamma/\gamma,v} = m_{\Gamma,v} - m_{\gamma,v} + n_{\gamma,v},$$

and the fact that $m_{\Gamma\Gamma',v} = m_{\Gamma,v} + m_{\Gamma',v}$, and $n_{\Gamma\Gamma',v} = n_{\Gamma,v} + n_{\Gamma',v}$. This gives a decomposition

$$H = \bigoplus_{(n_1,\dots,n_k) \in \mathbb{Z}^k} H^{n_1,\dots,n_k},$$

where $k = |R_V|$. We denote by $p_{\vec{n}} \equiv p_{n_1,\dots,n_k}$ the projection onto $H^{n_1,\dots,n_k}$. Note that also $H^{0,\dots,0} = \mathbb{C}1$.

There is the following relation between the grading by loop number and the multigrading by number of vertices: $\sum_{v \in R_V} (N(v) - 2)d_v = 2L$, where $N(v)$ is the valence of the vertex v.

4.1. Hopf Subalgebra of Green's Functions

We now address the question of what the Hopf algebra structure is on Green's functions. We define *1PI Green's functions* by

$$G^e = 1 - \sum_{\text{res}(\Gamma)=e} \frac{\Gamma}{\text{Sym}(\Gamma)}, \qquad G^v = 1 + \sum_{\text{res}(\Gamma)=v} \frac{\Gamma}{\text{Sym}(\Gamma)},$$

with $e \in R_E, v \in R_V$. We also introduce the following elements in H:

$$X_v = \left(\frac{G^v}{\prod_e (G^e)^{N_e(v)/2}} \right)^{1/(N(v)-2)},$$

with $N_e(v)$ the number of edges of type e attached to v and $N(v)$ the total number of edges attached to v.

We state the following result, referring to [**Sui08**] for the proof.

THEOREM.

1. The elements $p_{\vec{n}}(G^v)$ and $p_{\vec{n}}(G^e)$ ($\vec{n} \in \mathbb{Z}^k, v \in R_V, e \in R_E$) generate a Hopf subalgebra $H_R \subset H$ with dual group

$$\text{Hom}_{\mathbb{Q}}(H_R, \mathbb{C}) \subset (\mathbb{C}[[x_1,\dots,x_k]]^\times)^N \rtimes \overline{\text{Diff}}(\mathbb{C}^k),$$

with $\overline{\text{Diff}}(\mathbb{C}^k)$ the group of formal diffeomorphisms in k variables.

2. The ideal $J = \langle X_v - X_{v'} \rangle$ in H_R ($v, v' \in R_V$) is a Hopf ideal and

$$\text{Hom}_{\mathbb{Q}}(H_R/J, \mathbb{C}) \subset (\mathbb{C}[[x]]^\times)^N \rtimes \overline{\text{Diff}}(\mathbb{C}).$$

Writing explicitly the elements X_v:

$$X_{\text{---}} = \frac{G^{\text{---}}}{G^{\text{---}} \sqrt{G^{\text{---}}}}, \qquad X_{\prec} = \frac{G^{\prec}}{(G^{\text{---}})^{3/2}}, \qquad X_{\times} = \frac{\sqrt{G^{\times}}}{G^{\text{---}}},$$

the quotienting by the above Hopf ideal J is nothing but implementing the **Slavnov–Taylor identities** for the couplings on the level of the Hopf algebra:

$$G^{\prec}G^{\frown} = G^{\times}\, G^{\frown}\,; \qquad G^{\prec}G^{\prec} = G^{\times}G^{\frown}.$$

4.2. The Comodule BV-algebra of Coupling Constants and Fields

Since the coupling constants measure the strength of the interactions, we label them by the elements $v \in R_V$ and write accordingly λ_v. We consider the algebra A_R generated by local functionals in the fields and formal power series (over $\mathbb{C}$) in the coupling constants λ_v. In other words, we define $A_R := \mathbb{C}[[\lambda_{v_1}, \dots, \lambda_{v_k}]] \otimes_{\mathbb{C}} \mathfrak{F}([\Phi])$, where $k = |R_V|$. The BV-algebra structure on $\mathfrak{F}([\Phi])$ defined in the previous section induces a natural BV-algebra structure on A_R; we denote the bracket on it by $(\cdot, \cdot)$ as well.

THEOREM. ([**Sui08**]) *The algebra A_R is a comodule BV-algebra for the Hopf algebra H_R. The coaction $\rho : A_R \to A_R \otimes H_R$ is given on the generators by*

$$\rho : \lambda_v \longmapsto \sum_{n_1 \cdots n_k} \lambda_v \lambda_{v_1}^{n_1} \cdots \lambda_{v_k}^{n_k} \otimes p_{n_1 \cdots n_k}(X_v^{N(v)-2})$$

$$\rho : \phi \longmapsto \sum_{n_1 \cdots n_k} \phi \, \lambda_{v_1}^{n_1} \cdots \lambda_{v_k}^{n_k} \otimes p_{n_1 \cdots n_k}(G^e)$$

for fields ϕ and corresponding edge e.

As a consequence, the group $G_R := (\mathbb{C}[[x_1, \dots, x_k]]^{\times})^N \rtimes \overline{\mathrm{Diff}}(\mathbb{C}^k)$ acts on A_R. Consider then the following element:

$$S = -\langle dA, dA \rangle - \lambda_{A^3}\langle dA, [A, A] \rangle - \frac{1}{4}\lambda_{A^4}\langle [A, A], [A, A] \rangle$$

$$- \langle A, dh \rangle + \langle d\bar{\omega}, d\omega \rangle + \frac{1}{2}\xi\langle h, h \rangle + \lambda_{\bar{\omega}A\omega}\langle d\bar{\omega}, [A, \omega] \rangle - \langle d\omega, A^{\ddagger} \rangle$$

$$- \lambda_{A\omega A^{\ddagger}}\langle [A, \omega], A^{\ddagger} \rangle - \langle h, \bar{\omega}^{\ddagger} \rangle - \frac{1}{2}\lambda_{\omega^2\omega^{\ddagger}}\langle [\omega, \omega], \omega^{\ddagger} \rangle.$$

In contrast with the usual formula for the Yang–Mills action, we have inserted the different coupling constants λ_v for each of the interaction monomials in the action. We will now show that validity of the master equation $(S, S) = 0$ implies that all these coupling constants are expressed in terms of one single coupling.

In fact, consider the ideal $I = \langle (S, S) \rangle$. One can show [**Sui08**] that I implements the relations

$$\lambda_{A^4} = g^2 \text{ and } \lambda_{\bar{\omega}A\omega} = \lambda_{A\omega A^{\ddagger}} = \lambda_{\omega^2\omega^{\ddagger}} = g.$$

Since G_R acts on A_R, we can consider the subgroup

$$G_R^I = \{g \in G_R : gI \subseteq I\} \subset G_R.$$

THEOREM. ([**Sui08**]) $G^I \simeq \mathrm{Hom}_{\mathbb{C}}(H_R/J, K) \simeq (\mathbb{C}[[x]]^{\times})^N \rtimes \overline{\mathrm{Diff}}(\mathbb{C})$.

In the quotient A_R/I the above functional S satisfies the master equation and is in fact precisely the BV-action S_{BV} considered before. The above result can then be rephrased by saying that the master equation for S implies the Slavnov–Taylor identities for the couplings in the quantum theory (as described by the Feynman graphs in H_R). Moreover, if we assume that the regularization procedure is compatible with gauge invariance, such as dimensional regularization (see also [**Pro07**]), then the map $U : H_R \to K$ defined by the (regularized) Feynman rules vanishes on the ideal J because of the Slavnov–Taylor identities. Hence, it factors through an algebra map from $\widetilde{H}_R/J$ to the field K. Since $\widetilde{H}_R$ is still a commutative connected Hopf algebra, there is a Birkhoff decomposition $U = C^{-1} * R$ as before *with C and R algebra maps from $\widetilde{H}_R$ to K*. This is the crucial point, because it implies that both C and R vanish automatically on J. In other words, both the counterterms and the renormalized amplitudes satisfy the Slavnov–Taylor identities.

References

[BKUY08] G. van Baalen, D. Kreimer, D. Uminsky, and K. Yeats *The QED beta-function from global solutions to Dyson-Schwinger equations,* Ann. Phys. **324** (2009), 205–219.

[BKUY09] G. van Baalen, D. Kreimer, D. Uminsky, and K. Yeats *The QCD beta-function from global solutions to Dyson-Schwinger equations,* Ann. Phys. **325** (2010), 300–324..

[BBH94] G. Barnich, F. Brandt, and M. Henneaux *Local BRST cohomology in the antifield formalism. I. General theorems,* Commun. Math. Phys. **174** (1995), 57–92.

[BBH95] G. Barnich, F. Brandt, and M. Henneaux *Local BRST cohomology in the antifield formalism. II. Application to Yang-Mills theory,* Commun. Math. Phys. **174** (1995), 93–116.

[BV81] I. A. Batalin and G. A. Vilkovisky *Gauge algebra and quantization,* Phys. Lett. **B102** (1981), 27–31.

[BV83] I. A. Batalin and G. A. Vilkovisky *Feynman rules for reducible gauge theories,* Phys. Lett. **B120** (1983), 166–170.

[BV84] I. A. Batalin and G. A. Vilkovisky *Quantization of gauge theories with linearly dependent generators,* Phys. Rev. **D28** (1983), 2567–2582.

[BRS74] C. Becchi, A. Rouet, and R. Stora *The abelian Higgs-Kibble model. Unitarity of the S operator,* Phys. Lett. **B52** (1974), 344.

[BRS75] C. Becchi, A. Rouet, and R. Stora *Renormalization of the abelian Higgs-Kibble model,* Commun. Math. Phys. **42** (1975), 127–162.

[BRS76] C. Becchi, A. Rouet, and R. Stora *Renormalization of gauge theories,* Annals Phys. **98** (1976), 287–321.

[CK99] A. Connes and D. Kreimer *Renormalization in quantum field theory and the Riemann-Hilbert problem. I: The Hopf algebra structure of graphs and the main theorem,* Commun. Math. Phys. **210** (2000), 249–273.

[CK00] A. Connes and D. Kreimer *Renormalization in quantum field theory and the Riemann-Hilbert problem. II: The beta-function, diffeomorphisms and the renormalization group,* Commun. Math. Phys. **216** (2001), 215–241.

[EG07] K. Ebrahimi-Fard and L. Guo *Rota-Baxter algebras in renormalization of perturbative quantum field theory,* Fields Inst. Commun. **50** (2007), 47–105.

[Ger63] M. Gerstenhaber *The cohomology structure of an associative ring,* Ann. Math. **78** (1963), 267–288.

[KM08] T. Krajewski and P. Martinetti *Wilsonian renormalization, differential equations and Hopf algebras,* arXiv:0806.4309.

[Kre98] D. Kreimer *On the Hopf algebra structure of perturbative quantum field theories,* Adv. Theor. Math. Phys. **2** (1998), 303–334.

[Kre05] D. Kreimer *Anatomy of a gauge theory,* Ann. Phys. **321** (2006), 2757–2781.

[KY06] D. Kreimer and K. Yeats *An étude in non-linear Dyson-Schwinger equations,* Nuclear Phys. B Proc. Suppl. **160** (2006), 116–121.

[Pro07] D. V. Prokhorenko *Renormalization of gauge theories and the Hopf algebra of diagrams,* arXiv:0705.3906 [hep-th].

[Sta97] J. Stasheff *Deformation theory and the Batalin-Vilkovisky master equation,* In Deformation theory and symplectic geometry (Ascona, 1996), volume 20 of Math. Phys. Stud., Kluwer Academic, Dordrecht, 1997, pp. 271–284.

[Sui07] W. D. van Suijlekom *Renormalization of gauge fields: A Hopf algebra approach* Commun. Math. Phys. **276** (2007), 773–798.

[Sui07c]   W. D. van Suijlekom, *Renormalization of gauge fields using Hopf algebras*, In J. T. B. Fauser and E. Zeidler, editors, Quantum Field Theory, Birkhäuser Verlag, Basel, 2008.

[Sui08]   W. D. van Suijlekom *The structure of renormalization Hopf algebras for gauge theories. I: Representing Feynman graphs on BV-algebras*, Commun. Math. Phys. **290** (2009), 291–319.

[Tyu75]   I. V. Tyutin *Gauge invariance in field theory and statistical physics in operator formalism*, LEBEDEV-75-39.

[Wei96]   S. Weinberg *The quantum theory of fields*. Vol. II. Cambridge University Press, 1996.

INSTITUTE FOR MATHEMATICS, ASTROPHYSICS AND PARTICLE PHYSICS, FACULTY OF SCIENCE, RADBOUD UNIVERSITY NIJMEGEN, HEYENDAALSEWEG 135, 6525 ED NIJMEGEN, THE NETHERLANDS.

E-mail address: waltervs@math.ru.nl